EVOLUTION OF INSECT PESTS

Patterns of Variation

EVOLUTION OF INSECT PESTS

Patterns of Variation

Edited by

KE CHUNG KIM

BRUCE A. McPHERON

JOHN WILEY & SONS, INC.

New York • Chichester • Brisbane • Toronto • Singapore

This text is printed on acid-free paper.

Library of Congress Cataloging in Publication Data:

Evolution of insect pests : patterns of variation / edited by Ke Chung Kim and Bruce A. McPheron.
p. cm. — (Environmental science and technology)
Includes index.
ISBN 0-471-60077-6
1. Insect pests—Evolution. 2. Insect pests—Variation. 3. Insect pests—Control. I. Kim, Ke Chung. II. McPheron, Bruce A. III. Series.
SB931.E86 1993
632′.7—dc20 92-18178

Printed in the United States of America

10 9 8 7 6 5 4 3 2 1

CONTRIBUTORS LIST

MARK C. ANDERSEN, Department of Ecology and Evolutionary Biology, School of Biological Sciences, University of California at Irvine, Irvine, CA 92717

DAVID A. ANDOW, Department of Entomology, 219 Hodson Hall, University of Minnesota, St. Paul, MN 55108

J. R. BRADLEY, JR., Department of Entomology, North Carolina State University, Raleigh, NC 27695-7630

M. W. BROWN, Appalachian Fruit Research Station, USDA/ARS, Rt. 2, Box 45, Kearneysville, WV 25430

B. A. CROFT, Department of Entomology, Oregon State University, Corvallis, OR 97331

ROBERT F. DENNO, Department of Entomology, University of Maryland, College Park, MD 20742

J. DUNLEY, Department of Entomology, Oregon State University, Corvallis, OR 97331

FRED L. GOULD, Department of Entomology, North Carolina State University, Raleigh, NC 27695-7630

LAWRENCE M. HANKS, Department of Entomology, University of California at Riverside, Riverside, CA 92521

TING-HUAN HSIAO, Department of Biology UMC 53, Utah State University, Logan, UT 84322

KENNETH Y. KANESHIRO, Hawaiian Evolution Biology Program, Pacific Biomedical Research Center, University of Hawaii at Manoa, 3050 Maile Way, Gilmore 310, Honolulu, HI 96822

PETER M. KAREIVA, Department of Zoology, NJ-15, University of Washington, Seattle, WA 98195

GEORGE G. KENNEDY, Department of Entomology, North Carolina State University, Raleigh, NC 27695-7630

KE CHUNG KIM, Department of Entomology, The Pennsylvania State University, University Park, PA 16802

JAMES L. KRYSAN, Yakima Ag Research Lab, USDA/ARS, 3706 W. Nob Hill Blvd., Yakima, WA 98902

SIMON A. LEVIN, Section of Ecology and Systematics, Ecosystems Research Center, Cornell University, Ithaca, NY 14853

DAVID C. MARGOLIES, Department of Entomology, Kansas State University, Manhattan, KS 66506

BRUCE A. MCPHERON, Department of Entomology, The Pennsylvania State University, University Park, PA 16802

A. OKUBO, Marine Science Research Center, State University of New York, Stony Brook, NY 11794

DOROTHY P. PASHLEY, Department of Entomology, Louisiana State University, Baton Rouge, LA 70803

DAVID PIMENTEL, Department of Entomology, Cornell University, Ithaca, NY 14853

M. D. RAUSHER, Department of Zoology, Duke University, Durham, NC 27706

MARGARET E. SAKS, Division of Biology 147-75, California Institute of Technology, Pasadena, CA 91125

WILLIAM B. SHOWERS, JR., Corn Insects Research Lab, USDA/ARS, Iowa State University, Ames, Iowa 50011

WILLIAM W. M. STEINER, Biological Control of Insects Research Lab, USDA/ARS, P.O. Box 7629, University of Missouri Research Park, Columbia, MO 65205

WALTER J. TABACHNICK, Arthropod-borne Animal Diseases Research Lab, USDA/ARS, P.O. Box 3965, University Station, Laramie, WY 82071

CATHERINE A. TAUBER, Department of Entomology, Comstock Hall, Cornell University, Ithaca, NY 14853

MAURICE J. TAUBER, Department of Entomology, Comstock Hall, Cornell University, Ithaca, NY 14853

CONTENTS

PREFACE

Genetic variation is the central issue of evolutionary biology, and it poses a basic problem for systematics. Issues of population variation bring forth both theoretical and practical challenges to ecological, evolutionary, and taxonomic perspectives on species. In recent years, an increased awareness of infraspecific variation, a broadly inclusive term for heterogeneity within a species, in insect pests has added an economic and practical significance to these issues. Page after page in many entomological and ecological journals have documented new case studies and the problem of infraspecific variation. Different conspecific populations of insect pests differ in so many biological traits that one wonders if they are, in fact, different organisms.

We have long been fascinated by the problem of infraspecific variation. Since Kim worked with biological differences of European corn borer biotypes in Waseca, Minnesota, in 1964, he has been convinced that the issue of infraspecific variation should be confronted directly if rational, long-term pest management strategies were to be adopted and sound autecological research was to prevail. When he served on the Program Committee for the 1986 National Conference of the Entomological Society of America, Kim had an opportunity to help organize a series of symposia related to infraspecific variation under the conference theme, "Habitat: Biology of Habitat-related Variation," with the following specific topics:

1. Patterns of Habitat-Related Variation in Insects
2. Evolution and Ecology of Immigrant Species
3. Implications of Pest Management
4. Foreign Pest Introductions: Invasion of New Habitats

This conference provided an excellent opportunity for an in-depth look at various aspects of infraspecific variation. Late in the following year, the concept behind this volume was developed to bring together information on infraspecific variation in insect pests. Most of the contributors to this volume are those who participated in the 1986 ESA National Conference symposia. This volume, *Evolution of Insect Pests: Patterns of Variation*, is the culmination of the subsequent writing efforts by these scientists.

This book will focus on the evolution of insect pests by evaluating existing theories, documenting case studies of diverse pest species, and presenting new thoughts on the problem of infraspecific variation and implications for pest management strategies. It consists of five sections. Section I deals with general overviews of genetic variation in insect pests. Contributions deal with variation in genetic markers, as well as variation in ecologically important traits. The second group of chapters extend this discussion of variation in ecologically meaningful traits. Section II addresses features of the habitat that are particularly germane to insects in agricultural situations. Section III addresses ecological perspectives of infraspecific variation in insect pests, focusing on species extending their range, either geographically or ecologically. The process and mechanism of invasions by insects are also discussed. Section IV consists of a set of eight case studies involving diverse pest species. This section provides a flavor of the many characters in which genetic variation has been documented and a sense of the variety in how scientists approach empirical studies of infraspecific variation. The final section, Section V, provides overviews for integrated pest management (IPM) strategies and future perspectives in entomology and biosystematics.

The chapters were reviewed by two to three reviewers, including other contributing authors. We thank the reviewers for their many excellent suggestions. It would have been difficult to complete this book without their prompt response and cooperation. We thank Ms. Thelma Brodzina, Ms. Ellen Rockey, and Ms. Jackie Wolfe of the Department of Entomology, Pennsylvania State University, for typing assistance. Our thanks also go to Mary M. Conway, formerly Life Sciences Editor, and the production staff of John Wiley & Sons, Inc., for their excellent cooperation in all aspects of preparing this book.

KE CHUNG KIM

BRUCE A. MCPHERON

University Park, Pennsylvania

SECTION I

GENETIC VARIATION AND EVOLUTION

1

INSECT PESTS AND EVOLUTION

Ke Chung Kim

The Pennsylvania State University

INTRODUCTION

Insect pests have been with us for a long time. As early as 10,000 years ago, farmers in primitive agricultural societies in areas of the Near and Middle East already had to reckon with insect pest problems to increase harvest (Jones, 1973). Tremendous efforts and expenses have been put into containment of insect pests throughout human history, and some significant successes were made in this century (Klassen, 1989). For the last 50 years, although such alternate approaches as biological control have been promoted, control and management strategies for insect pests have relied heavily on chemical pesticides. Despite control efforts directed against many serious insect pests, no single insect pest species has been completely eradicated. Insect pest problems are still here and will remain with us indefinitely. Insect pest problems have become increasingly more complicated and difficult to deal with because of the burdens of immigrant species (Kogan, 1982; Horsfall, 1983; CAST, 1987), pesticide resistance (NRC, 1986), economic factors (Southwood and Norton, 1973), and environmental degradation (NRC, 1986; Green et al., 1990). These may become further aggravated by the global climate changes.

Insect pests are a dynamic group of organisms with a long evolutionary history. They continue to evolve in response to changing environmental conditions, particularly of anthropocentric origin. However, the pest concept is a human construct: the problem of insect pests is defined by a value judgement and cultural criteria (Southwood and Norton, 1973). Insect pests counter new strategies as they are introduced. The fate of DDT was a case in point. DDT, which was discovered in 1939 and hailed as the "ultimate" weapon against insect pests in the 1940s, was banned by the 1960s because of the environmental and public health problems that it caused. By then, many insect pests had already developed a resistance to DDT and many other pesticides. Subsequent pest management technologies, such as integrated pest management (IPM), still do not consider evolutionary dynamics of pest species; rather, they rely on contemporary ecology and anthropocentric criteria. Rational solutions to the problem of insect pests must be based on knowledge of genetics and evolutionary strategy as much as on biology and ecology of pest species (Stenseth, 1987).

Insects as Pests

Insects, the most abundant group of organisms, are widely distributed throughout the world. They inhabit diverse habitats and serve many important functional roles in ecosystems. Their small size and high mobility enable long-distance dispersal and invasion of new territories and habitats (Wilson and Graham, 1983; Ehrlich, 1986). Insects are able to resist substantial environmental changes and yet, under certain conditions, are highly adaptable to changing environmental conditions (Parsons, 1983, 1987, 1991; Hoffmann and Parsons, 1991).

Insect pest fauna consists of both endemic and exotic species. Endemic species may become pests when a new crop species or variety is planted in an area (Elton, 1958), whereas immigrant species may become pests after they have successfully colonized the new territory (Wilson and Graham, 1983; Pimentel, 1986). Indigenous insects, either endemic or immigrant species, that have been established for a long time in the territory but not are directly associated with human enterprise or are of minimal economic importance can become major pests following environmental perturbance (Gould, 1991; Hoffman and Parsons, 1991).

The insect pest fuuna has become more homogeneous throughout the world by introductions, accidental or intentional. Highly mobile insects persistently move about between the nations through international traffic and commerce (Sailer, 1978, 1983; Kim, 1983), and continue to invade new territories and habitats, particularly those modified by human disturbance (Wilson and Graham, 1983; Margolies, Chapter 6; Pimentel, Chapter 8; Andow et al., Chapter 11; Rauscher, Chapter 13 in this book). For example, the North American insect fauna contains approximately 2,000 immigrant species (Kim and Wheeler, 1991), which constitute about 40% of the North American agricultural insect pests (Pimentel, 1986). Many of these introduced species are the most serious agriculture pests (Sailer, 1983; Lattin and Oman, 1983).

Of the major arthropod pest species, approximately 64% have a worldwide distribution (Pimentel, 1986). For example, most animal pests were introduced with the importation of livestock; cattle lice and hog lice are now distributed throughout the world. Many phytophagous and vector species have a worldwide or transoceanic distribution, such as the European corn borer, *Ostrinia nubilalis*

(Hübner) (Showers, Chapter 14 in this book), and yellow fever mosquito, *Aedes aegypti* (Linnaeus) (Tabachnick, Chapter 21 in this book). They show an extensive geographic variation (Gould and Johnston, 1972; Parsons, 1983; Brussard, 1984; Futuyma and Peterson, 1985).

Insect Pests Evolving

Through a long evolutionary process, insects have developed the ability, dynamic genetic norms of reactions, to adapt to changing environmental conditions (Hedrick et al., 1976; Hoffmann and Parsons, 1991; Parsons, 1991). As described in this book, a wide range of genetic diversity is shown in major insect pests; for example, European corn borer, by Showers et al. (1975) and Shower (Chapter 14 in this book); Colorado potato beetle, *Leptinotarsa decemlineata* (Say), by Hsiao (1985); alfalfa weevil, *Hypera postica* (Gyllenhal) by Hsiao (Chapter 15 in this book); Mexican bean beetle, *Epilachna varivestis* Mulsant,by Saks (Chapter 16 in this book); yellow fever mosquito, *Ades aegypti* (Linnaeus) by Tabachnick (Chapter 21 in this book); Greenbug, *Schizaphis graminum* (Rondani), by Beregovoy and Starks (1986) and Inayatullah et al. (1987); and pear psylla, *Cacopsylla pyricola* Forester, by Unruh (1990).

Insect pests are evolving rapidly with much higher genetic variability under stressed enviromental conditions (Parsons, 1983, 1987, 1988; Croft and Dunley, Chapter 7 in this book; Krysan, Chapter 18 in this book), created by anthropogenic factors such as pesticide application, habitat degradation, and environmental pollutions.

In this chapter, the theme of the book, patterns of genetic variation and evolution of insect pests, is defined, and the biological nature of insect pests is discussed as background for the synthesis of succeeding chapters.

THE NATURE OF INSECT PESTS

The Concept of Pest

The *Random House Dictionary* defines *pest* as a destructive or troublesome organism; any species of insects is a potential pest. Insects become pests when they compete with humans for the common resources or their activity interferes with human interests. In other words, the state of being a pest is a human construct, and insect pest problems are defined by anthropocentric criteria.

The degree of insect pest problems is defined by specific economic criteria and cultural value systems, including the nuisance level. For example, although it may not be a big problem to the indigenous people of the Ethiopian highland, the human body louse, *Pediculus humanus* Linnaeus, is a serious pest and is important to Americans and Europeans as a potential disease vector. Evaluation of the impact or damage of a pest species by the measure of economic threshold may also differ by the type of crop species and local value criteria.

Origin of Insect Pests

Insect pests have a diverse origin, which involves endemics, immigrants, and mutants. Indigenous insects consist of endemic species and those of foreign origin that are well established in a region. An indigenous species that feeds on native plants may become a serious pest when a preferred crop species is introduced to the agroecosystem. For example, the Colorado potato beetle (CPB), which is native to the semiarid region of eastern and southern Colorado and fed on the sandbur (*Solanum rostratum*), colonized potato (*Solanum tuberosum*) when this plant was first introduced into the southwestern United States, perhaps 135 years ago (Riley, 1877; Tower, 1906). Since then, CPB has extended its range throughout the United States (Hsiao, 1985) and has become the most serious pest of potato. Similarly, sugarcane has a local set of pest species unique to each region in which it is grown, because indigenous insect species have become associated with the crop (Strong et al., 1977).

Insect pest and host plant, both of exotic origin, may become associated in a new territory (Rausher, Chapter 13; Saks, Chapter 16 in this book). For example, corn (*Zea mays*) was originally introduced into the United States from tropical America, whereas the European corn borer came from central and southern Europe. The European corn borer became one of the most serious pests of corn in north-central North America (Hudon et al., 1989). Similarly, Hessian fly, *Mayetiola destructor* (Say), introduced into North America from Europe around 1779, became one of the most destructive pests of wheat, which originated in the Middle East (Allan et al., 1959).

Immigrant pest species may come from either the original species distribution or its secondary range. For example, the Mediterrenean fruit fly, *Ceratitis capitata* (Wiedmann), found sporadically in southern California, may come from Sub-Saharan Africa, the original distribution, or from Latin America or the Mediterranean, secondary distribution areas.

A major category of insect pests that are directly related to anthropogenic environmental perturbances are indigenous species that were not considered economically important or were associated with cultivated crops and have become serious pests under stressed or perturbed environmental conditions (Kennedy, Chapter 22 in this book). For example, the sweetpotato whitefly, *Bemisia tabaci* (Gennadius), which has been in California since the 1920s but was not considered a pest of field-grown crops (Byrne et al., 1990), has become a destructive pest of many field crops since 1981 because of intensive pesticide application.

INSECT PEST FAUNA

Most important pests are of exotic origin, but insect pests constitute a small part of the larger insect fauna of a region (Sailer, 1983; Lattin and Oman, 1983). The North American insect fauna contains approximately 2,000 immigrant species, which is equivalent to about 2% of the total diversity (Kim and Wheeler, 1991). Of the 148 major insect pest species, about 40% are introduced species and are among the most serious pests in North American agriculture (Pimentel, 1986).

North American Fauna

The known diversity of North American insects consists of approximately 91,800 species, which represent perhaps 50% of the extant fauna (Kosztarab and Schaefer, 1990); Coleoptera is the largest order, with more than 23,600 species (Arnett, 1990), followed by Diptera, 19,562 species (Thompson, 1990); Hymenoptera, 17,429 species (Masner, 1990); and Lepidoptera, 11,233 species (Covell, 1990).

The North American fauna has changed since Columbus's 1492 discovery of the Americas. The magnitude of human effects on the natural environment of North America are comparable to those of the Pleistocene glaciations (Howden, 1969). One of the most important effects is the introduction of numerous insects into the New World. With continued immigration and commerce from around the world, the entry of exotic species from other continents has been constant, despite all measures taken to prevent their entry (Herrick, 1929; Kim, 1983). These invaders have caused major losses to agricultural production and forests (Pimentel et al., 1980; Pimentel, 1986; Klassen, 1989) and also have had a detrimental effect on environmental quality, human health, and aesthetics (Reynolds et al., 1982).

During colonial times, many European insects managed to become established in North America, such as the Hessian fly, *Mayetiola destructor* (Say), Angoumois grain moth, *Sitotroga cerealella* (Olivier), and codling moth, *Laspeyresia pomonella* (Linnaeus). Pierce (1917) listed 103 important foreign insects introduced into the United States, although about four times that number had already become established (Sailer, 1983). In the 1920s, 81 species of the 182 (44.3%) most economically important insect plant pests were of definite foreign origin (Smith, 1929). Recent assessment of the North American fauna for nonindigenous insects recorded 1,115 (McGregor, 1973), 1,385 (Sailer, 1978), and 1,683 species (Sailer, 1983). The most recent study revealed over 2,000 exotic species (Kim and Wheeler, 1991).

Distribution of Nonindigenous Insects

The geographical distribution of our nonindigenous insects is not even. The Hawaiian Islands, for example, having no history of a continental connection, are characterized by an immigrant biota (Elton, 1958; Carlquist, 1981). In continental North America, three principal areas are considered historically to be centers of introduction: the Canadian Maritime Provinces, the seaports of the Northeastern states, and the Pacific Northwest. For example, the carabid fauna of Newfoundland now contains 14.5% European species (Larson and Langor, 1982) and its leafhopper fauna has 17.2% introduced species (Hamilton and Langor, 1987). California and Florida are probably characterized by a relatively high proportion of immigrant taxa (Frankie et al., 1982; Dowell and Gill, 1989; Ewel, 1986), and the Great Lakes region also seems vulnerable to invasion by exotic species (Davis, 1962).

Habitats and disturbed sites made by humans, such as railroad and highway rights-of-way, contain an unusually high percentage of immigrants (Elton, 1958). Eurasian weeds in many disturbed areas, now well established among the native biota, are fed on by various Old World insects. For example, noxious weeds like Canada thistle and spotted knapweed are attacked by accidentally introduced generalist and specialist herbivores and also by a suite of biocontrol species deliberately introduced (Kelleher and Hulme, 1981; Wheeler, 1985, 1988; Wheeler and Whitehead, 1985). A similar pattern is also found in agroecosystems (Pimentel and Wheeler, 1973). Because about 99% of the cultivated area in North America is planted with introduced crops, this agroecosytem is usually vulnerable to immigrant pests (CAST, 1987). However, few immigrant species may be found in natural plant communities other than in areas modified by humans (Elton, 1958; McGregor, 1973), and annual field crops appear relatively less vulnerable to successful colonization by immigrant pests than perennial plants (Kogan, 1982).

GENETIC VARIATION

Genetic variation provides the basis for evolutionary change, and the genetic structure of populations is shaped by natural selection through the interaction of genetic forces and changing environments in space and time. Thus, the primary force in maintaining and structuring the genetic diversity in natural populations is environmental heterogeneity (Nevo, 1988). In a constantly changing environment, molecules, chromosomes, and morphological traits evolve at very different rates (Beaverstock and Adams, 1987), and evolutionary rates are not related to the population size and the mutation rate (Gillespie, 1985; Nevo, 1988). Molecular evolution is closest to being constant in time (Beaverstock and Adams, 1987) and protein evolution usually follows a specific model of the evolutionary process (Gillespie, 1985; Nevo, 1988), whereas chromosomal and morphological evolution vary greatly with time (Beaverstock and Adams, 1987).

Genetic differentiation between populations largely depends on the interacting balance between gene flow (homogenizing force) and genetic drift and natural selection (diversifying mechanism) (Futuyma and Peterson, 1985). Although natural selection directly acts on phenotypes, it is a major factor causing genetic differentiation at the protein and DNA levels within and between populations (Gillespie, 1985).

Genetic Diversity

Genetic diversity in insects in highly polymorphic (Selander, 1976) and genetic variation is expressed in numerous biological (e.g., Denno and Dingle, 1980; Tauber et al., 1986; Taylor and Karban, 1986; Hudon et al., 1989) and genetic traits (e.g., Lewontin, 1974; Nevo et al., 1984), and genetic diversity in natural populations is well documented for diverse groups of organisms by biochemical analyses of allozymic (Lewontin, 1974, 1985; Powell, 1975; Nevo, 1983a, 1983b; Steiner, Chapter 2 in this book) and DNA diversity (Nevo, 1988) and morphometric analyses (Daly, 1985). Diverse biological traits involved include the following: morphological, sexual behavior, geotactic behavior, developmental rate, fecundity, dispersal ability, seasonal adaptation such as diapause, feeding preference, host plants, resistance to pesticides and other toxins, and chromosomal such as crossover and inversion rates, and genic traits like allozyme polymorphism and DNA sequence.

The following examples illustrate a diversity of genetic variation in insect and acarine pests.

Electrophoretic Studies on Allozymes

ACARI

Ixodidae: *Boophilus microplus* (Canestrini) (Sattler et al., 1986)

COLEOPTERA

Chrysomelidae: *Leptinotarsa decemlineata* (Say) (Jacobson and Hsiao, 1983; Zehnder et al., 1992)

Curculionidae: *Hypera postica* (Gyllenhal) (Hsiao and Stutz, 1985; Hsiao, Chapter 15 in this book)

DIPTERA

Culicidae: *Aedes triseriatus* (Say) (Matthews and Craig, 1980)

Aedes aegypti (Linnaeus) (Wallis et al., 1984; Tabachnick and Wallis, 1985; Tabachnick, Chapter 21 in this book)

Tephritidae: *Rhagoletis pomonella* (Walsh) (McPheron et al., 1988)

HOMOPTERA

Aphididae: *Macrosiphum rosae* (Linnaeus) (Rhomberg et al., 1985)

Myzus persicae (Sulzer) (May and Holbrook, 1978; Wool et al., 1978; Tomiuk and Wohrmann, 1983; Brooks and Loxdale, 1987)

Rhopalosiphum padi Linnaeus (Martinez et al., 1992)

Schizaphis graminum (Rondani) (Beregovoy and Starks, 1986)

Psyllidae: *Cacopsylla pyricola* (Foerster) (Unruh, 1990)

LEPIDOPTERA

Gelechiidae: *Pectinophora gossypiella* (Saunders) (Bartlett, 1981)

Geometridae: *Alsophila pometaria* (Harris) (Mitter et al., 1979)
Noctuidae: *Heliothis zea* (Boddie) (Sell et al., 1975)
Spodoptera (den Boer, 1978; Pashly et al., 1985)

Chromosomal Studies

COLEOPTERA
Curculionidae: *Hypera postica* (Hsiao and Hsiao, 1984)
DIPTERA
Culicidae: *Anopheles* (Tadei, 1982)
Drosophilidae: *Drosophila* (Parsons, 1988)

Phenotypic Studies

COLEOPTERA
Chrysomelidae: *Diabrotica* (crop rotation; Krysan et al., 1986; Krysan, Chapter 18 in this book)
DIPTERA
Calliphoridae: *Calliphora* (diapause; Vinogradova, 1986)
Culicidae: *Wyeomyia smithii* (Coquillett) (development rate; Istock, 1980)
Tephritidae: *Rhagoletis pomonella* (Walsh) (seasonal asynchrony; Smith, 1988)
HOMOPTERA
Aleyrodidae: *Bemisia tabaci* (Gennadius) (morphometry; Bethke et al., 1991)
Aphididae: *Schizaphis* (three biotypes; Inayatullah et al., 1987)
LEPIDOPTERA
Pyralidae: *Ostrinia nubilalis* (Hübner) (morphometry, diapause, development, seasonal biology; Hudon et al., 1989; Showers, Chapter 14 in this book)

Genetic variation is expressed in specific traits by different species, depending upon the forces that direct natural selection in the changing environment. The greenbug, *S. graminum,* corn leaf aphid, *Rhopalosiphum maidis* (Fitch), and Hessian fly, *M. destructor,* the three major cereal pests, show genetic variation in different biological traits. Three biotypes of the greenbug differ in their ability to live on and damage varieties of wheat, barley, and sorghum, whereas the corn leaf aphid, another sorghum pest, has four or more biotypes that differ in fecundity or survival capacity on different host plants. Hessian fly biotypes interact differently with varieties of wheat (Gallun et al., 1975).

Genetic variation between populations at molecular levels differ by species. Local populations may not have differentiated, and yet a significant allele frequency between different populations does exist. For example, sexual forms of *Alsophila pometaria* Harris (fall cankerworm) show no allozyme differences, while parthogenetic forms show a variety of genetic differences in emergence and hatching time among local populations (Mitter et al., 1979). Populations of *Pectinophora gossypiella* (pink bollworm) are highly polymorphic, with a high heterozygosity, and yet are not genetically isolated (Bartlett, 1981). *Rhagoletis pomonella* (apple maggot) does not show genetic isolation between host-associated fly populations (hawthorn and apples), although a significant allele frequency exists between fly populations (McPheron et al., 1988), and temporal asynchrony and seasonal genetic divergence may restrict gene flow between apple and hawthorn host races (Smith, 1988).

Populations of the sweetpotato whitefly, *Bemisia tabaci,* on cotton have a larger body length for 1–4 instars, a longer and wider pupal case, and lower fecundity, while nymphal instars of the poinsettia populations are smaller with a high fecundity (Bethke et al., 1991). It is likely that these differences are genetically fixed.

Geographic Variation

Geographic variation, which represents a wide range of variation, both ecological and genetic, is found in almost all species of insects. Every species has a different degree of variation by location and time. There is a considerable body of literature on geographic variation (e.g., Gould and Johnston, 1972; Powell and Taylor, 1979; Slatkin, 1985; Futuyma and Peterson, 1985; Nevo, 1988; Steiner, Chapter 2 in this book). Geographic variation between central and marginal populations is well documented in terms of gene and chromosomal polymorphisms in *Drosophila* (Parsons, 1983, 1991).

Geographic variation is developed by local evolutionary processes that involve interactions of genetic mechanisms and environmental processes (see Chapters 4–13 in this book). Local populations evolve as those forces interact within the range of genetic variability ("norm of reactions") of the species. Genetic forces such as mutation and genetic drift interact with natural selection favoring adaptation to local environmental conditions, leading a genetic differentiation of local populations, whereas the force of gene flow opposes such differentiation (Slatkin, 1985, 1987). A significant genetic bottleneck likely occurs when an expanding population colonizes a new locality, and the short-term effects of a genetic bottleneck are a modest reduction of overall heterozygosity and a loss of rare alleles (Nei et al., 1975). In this process, environmental diversity leads to genetic variation with little genetic load (Parsons, 1983). In other words, the degree of genetic variation in local populations is a balance of the intensity of these forces, and it is expressed in biological characteristics of local populations that affect fitness, such as morphology, physiology, behavior, life history traits, and gene frequency (Barker and Thomas, 1987). The taxonomic status of local populations, such as Rassenkreis, Artenkreis, biotype, ecotype, and variety, is the manifestation of such evolutionary processes (Endler, 1977).

Among widespread species, the genetic variability levels differ by local populations. In *Drosophila,* central populations in the range of species distribution are contiguous at high density and show high levels of genetic and phenotypic variation, whereas populations on the margin usually are small and isolated with monomorphic chromosomes but no reduction in allozyme heterozygosity (Brussard, 1984). Heterozygosity is correlated positively with environmental heterogeneity and habitat unpredictability (Nevo, 1976), and high genic heterozygosity is found among marginal populations (Lewontin, 1974). Although there is no reduction in electrophoretic polymorphisms, most marginal populations show a reduction of chromosomal polymorphisms and also a reduction of lethal and semilethal alleles (Parsons, 1983). Yet, the pattern of DNA polymorphism is similar to that of protein diversity (Nevo, 1988).

The pattern of geographic variation greatly varies and is not necessarily similar between closely related species. The fall armyworm, *Spodoptera frugiperda* (J. E. Smith), shows a significant heterogeneity in allozymes among different populations that are related to host associations (Pashley et al., 1985; Pashley, Chapter 17 in this book), whereas the related African armyworm, *Spodoptera exempta* (Walker), shows practically no heterogenity in 17 population samples (den Boer, 1978). Natural populations of tree-hole mosquito, *Aedes triseriatus* (Say), showed a high genetic variation with average heterozygosity of 21.5% (Matthews and Craig, 1980), and, similarly, another mosquito, *Anopheles darlingi* Root, was highly polymorphic in inversion frequency, with a high heterozygosity (Tadei, 1982). However, many other species are monomorphic or have relatively low polymorphism in space and time: *Heliothis zea* (Boddie) (Sell et al., 1975); *Boophilus microplus* (Canestrini) (Sattler et al., 1986); *Schizaphis graminum* (Rondani) (Beregovoy and Starks, 1986); *Myzus persicae* (Sulzer) (May and Holbrook, 1978; Wool et al., 1978; Tomiuk and Wohrmann, 1983; Brooks and Loxdale, 1987); *Macrosiphum rosae* (Linnaeus) (Rhomberg et al., 1985).

Geographic variation appears in both phenotypic and genotypic systems in most species; phenotypic characters generally include structure/function and behavior. For example, the European corn borer, *O. nubilalis* (Hübner), whose three ecotypes are defined by diapause-based responses (Showers et al., 1975; Showers, 1981; Showers, Chapter 14 in this book), differ among geographic populations by host plant preference (Hudon et al., 1989), development and seasonal biology (diapause) (Beck, 1989), and morphology (Kim et al., 1967; Chiang et al., 1970). These variations, however, are not necessarily expressed similarly in both phenotypic and genotypic systems. In *Schizaphis graminum,* three biotypes (B, C, E) show distinct morphometric differences (Inayatullah et al., 1987) but not enzyme patterns (Beregovoy and Starks, 1986). In the Colorado potato beetle, *Leptinotarsa decemlineata* (Say), which is widely distributed in North America and continental Europe and feeds both on potatoes and alternate solanaceous plants, different populations show considerable difference in ecological adaptation and host plant relationships. However, of 12 different populations studied, all populations were alike in genetic diversity, except the southern Mexican populations, which are very different from all others, with moderately high genetic variability (Jacobson and Hsiao, 1983; Hsiao, 1985). Similarly, geographic variation among the alfalfa weevil (*Hypera postica*) populations were not very extensive in allozyme patterns, karyotype, chiasmata frequency, and Giemsa C-banding pattern (Hsiao and Hsiao, 1984; Hsiao and Stutz, 1985; Hsiao, Chapter 15 in this book).

Phenotypic variation involves both genetic and environmental components that closely interact. These factors may compete to produce a selected level of phenotypic variance (Bull, 1987), as the direction and degree of responses to environmental factors are genetically variable (Gupta and Lewontin, 1982). A genotype may express an individual character differently in response to different environments (evolutionary plasticity) (Bradshaw, 1985), whereas at another level a genotype may produce more than one alternative phenotype, such as structure/function, physiological state, and behavior, in response to environmental conditions (phenotypic plasticity) (West-Eberhard, 1989). Phenotypic plasticity is one of the most common types of variability influenced by local environmental conditions and subjected to natural selection and evolutionary change (Williams, 1966). Thus, phenotypic variation is selectively maintained in a population by one of the two components; selection may favor maintaining only the environmental component, only the genetic component, or may be indifferent to the composition of the variance (Bull, 1987).

EVOLUTION OF INSECT PESTS

Insect pests are by definition associated with anthropogenic environments; they no longer live in natural ecosystems. Natural ecosystems are rapidly transformed into human ecosystems, the environments strongly impacted by human enterprise, in which insect pests must adapt to the changing environmental conditions for survival. Insect pests have a great innate capacity to adapt to environmental stress (Hedrick et al., 1976; Hoffmam and Parsons, 1991; Parsons, 1991).

Stressed Environments

Transformation of nature into human ecosystems is continually accelerated by rapid increase in the global human population, urbanization, and economic development. The effects of global changes are environmental stresses to insects as well as to humans. In a stressed environment, anthropogenic input, such as pesticide application, habitat destruction, acid rain, and climatic changes, is the *stressor*, while *stress*, which is defined as a detrimental or disorganizing influence, is the *output* or *response*. Odum (1985) identified 18 trends expected in stressed ecosystems, of which 5 components likely occur at the community level: (1) r-strategists increase proportionally; (2) organisms get smaller; (3) life spans decrease; (4) energy flow at high trophic levels reduces and food chains shorten; and (5) if original diversity is high, species diversity declines and dominance increases (Woodwell, 1983).

At the ecosystem level, the trends include (1) community respiration increases; (2) nutrient turnover increases with more nutrient loss (Mooney and Godron, 1983); (3) resource use becomes less efficient; (4) parasitism and other negative interactions increase, but positive interactions, such as mutualism, decrease (Axelrod and Hamilton, 1981); and (5) functional properties, such as community metabolism, are more robust than structural properties, such as species composition (Odum, 1985). As we, the human species, must adapt to the human ecosystem, insect pests are forced to cope with anthropogenic stresses.

Differentiation of Populations

Local populations have adapted uniquely to their habitats, and their differentiation is shaped by the interaction between gene flow and genetic drift and/or mutations and natural selection (Nevo et al., 1984; Futuyma and Peterson, 1985; Parsons, 1991). Genetic divergence can come about by changes in environment or genetic system (Kirkpatrick, 1982). For example, as landscape or environmental condition changes, the equilibrium between genetic and environmental systems sustained by the population under an initial environment becomes unstable and deterministic selection pressure will force the population to adapt to the new environment (Templeton, 1980).

The differentiation of populations involves both genetic and environmental systems. However, the process is ultimately genetic, even though natural selection acts on phenotypes (Barker and Thomas, 1987). Populations of a species can be differentiated genetically by one of the following events: (1) colonization of a new habitat (habitat, hosts) (Bartlett and Richardson, 1986; Holt, 1987; Rice and Salt, 1988); (2) colonization of a new territory or region (founder effect) (Barton and Charlesworth, 1984; Carson and Templeton, 1984; Bartlett and Richardson, 1986); (3) landscape changes (bottleneck effect) (Nei et al., 1975; Hedrick, 1986); (4) genetic changes by stochastic events, such as genetic drift and mutation (Templeton, 1980; Bush, 1981) or natural selection (directional and disruptive) (Rice, 1987; Barker and Thomas, 1987).

Barton and Charlesworth (1984) summarized potential mechanisms that drive genetic divergence for major speciation models. They include (1) drift with background homozygosity (peripatric model) (Mayr, 1954), relaxation of selection (founder-flush model) (Carson, 1968, 1975), and deviation from Hardy-Weinberg Law (genetic transilience) (Templeton, 1980); (2) accumulation of mutations and changing selection (allopatric speciation) (Mayr, 1963); (3) spatial variation in selection (parapatric model) (Endler, 1977); (4) drift and meiotic drive (stasipatric model) (White, 1968, 1978); (5) drift and fluctuations in the adaptive landscape (shifting-balance model) (Wright, 1932); (6) disruptive selection (sympatric speciation model) (Maynard-Smith, 1966). However, the level of differentiation differs for individual local populations because of varying interactive forces of genetic and environmental systems, and the differentiation of populations does not necessarily lead to speciation. As a result, these processes will produce diverse infraspecific taxa that have been classified as biotype, ecotype, formenkeris, rassenkreis, race, or subspecies, based on the degree of genetic and phenotypic divergence.

The differentiation of populations by colonization involves niche shift, invasion of disturbed or newly created habitats by humans or immigration and founder effect (Parsons, 1983). Successful colonizers are usually generalists with the following attributes: (1) widely distributed and abundant in the original range of distribution, (2) polyphagous, (3) having a short generation time, and (4) possessing a wide range of genetic and phenotypic variability, which allows them to function in a wide range of physical conditions (Ehrlich, 1986). Local populations may be differentiated by habitat specialization (Rice, 1987; Rice and Salt, 1988) or host shift (Bush, 1981; McPheron et al., 1988; and Kaneshiro, Chapter 4; Tauber and Tauber, Chapter 5; Margolies, Chapter 6; Inmentel, Chapter 8; Rausher, Chapter 13 in this book).

Populations of a species may simply be subdivided by establishment of a new population from founder individuals (Carson and Templeton, 1984). After the successful founder event, the populations go through genetic changes (recombination) and produce descendent populations that are well adapted to the local environment. In this process, the genetic system interacts with the environment, and this causes gradual genetic changes over many generations (Carson and Templeton, 1984). At this time, a genetic bottleneck may occur, resulting in modest reduction of overall heterozygosity and a loss of rare alleles (Nei et al., 1975). At a locality, the populations may be drastically reduced in size by landscape changes, such as flooding, drought, and other natural factors, or by environmental changes in the habitat. Here, the surviving individuals essentially become a random sample of the original population in which a genetic bottleneck likely occurs (Nei et al., 1975).

Genetic changes may occur in populations by stochastic events in the genetic system, such as genetic drift or mutation (Lewontin, 1974; Templeton, 1980; Bush, 1981), and, similarly, genetic divergence may develop by natural selection (directional and disruptive), such as the constant application of chemical pesticides (Rice, 1987; Barker and Thomas, 1987; Shelton and Wyman, 1991; Croft and Dunley, Chapter 7 in this book). For example, the sweetpotato whitefly (*Bemisia tabaci*), which has been in California since the 1920s, has become highly resistant to most chemical pesticides widely used in agriculture. Thus, its pest status in field crops has increased since 1981 (Byrne et al., 1990). In the last three years, this species has been a serious threat to California agriculture, and it is now perhaps the most damaging pest of field crops in that state.

Evolution in the Stressed Environment

The level of stress that insects must face in stressful environments is much greater than in natural ecosystems. In stressed environments, insect pests must continually adapt to severe environmental conditions for survival. As stress and variability are correlated, genotypic and phenotypic variabilities under severe environmental stress are usually much higher than in stable environments, particularly for those traits that are important in determining survival (Parsons, 1987, 1991).

Under stressed environments, additive genetic variability for recombination is higher than in constant environments, such as laboratory conditions, and substantial recombination can be induced under the combination of karyotype and environments (Parsons, 1988). For example, in *Drosophila*, the genetic system of recombination rapidly reorganized and recombination increased in the chromosomes 2 and 3 when subjected to selection for resistance to DDT (Laurie-Ahlberg and Merrell, 1979; Flexon and Rodell, 1982), for variable temperature stress (Zhuchenko et al., 1985), and for nutritional and behavioral stress (Parsons, 1988). Chromosomal mutation rates also are known to vary considerably over time and space within species (Bush, 1981).

Insect pests have rapidly evolved and are evolving at an unprecedented rate as they are subjected to stressed environments. Although no direct experimental evidence exists as yet, insect pests are forced to evolve rapidly to counter anthropogenic stresses, such as pesticides, temperature, and habitat degradation. Despite all the efforts invested for controlling insect pests for this century, no single species of insect pest has ever been completely eradicated

Some insect pest species are resistant to most pesticides ever used (NRC, 1986); for example, *Tetranychus urtical* Koch, *Psylla pyricola* Foerster, phytoseiid mites, and many tortricoid moths (Croft and Dunley, Chapter 7 in this book). The populations of pest species that were introduced and established in the United States for a long time are no longer the same organisms as their ancestral populations. For example, European corn borers today are different from the populations that existed in 1930, even at the same locality, and are certainly different from those in Iowa and Minnesota in many biological traits (Hudon et al., 1989). This genetic differentiation has developed in less than 50 generations. Today's Hessian fly is not the same species as that first brought into the United States during colonial times (Gallun et al., 1975). The poinsettia population of the sweetpotato whitefly no longer behaves as did its ancestral population of many years ago (Bethke et al., 1991).

CONCLUSION AND SUMMARY

Insect pests are species of insects closely associated with anthropogenic activities. They are groups of dynamic organisms with a powerful genetic system and evolutionary strategy that have developed through a long evolutionary process. Insect pests are highly polymorphic and have broad genetic diversity, expressed in a multitude of biological traits.

Geographic variation is found in almost all species of insect pests, and populations of every species have different degrees of variation in space and time. The pattern of geographic variation greatly varies among species, and geographic variation appears in both phenotyic and genetic characters.

Human ecosystems are stressed environments for insect pests. In stressed environments insect pests are forced to evolve rapidly in response to anthropogenic stress, such as pesticides and environmental degradation. Insect pests have evolved at an unprecedented rate and are also rapidly evolving as they are subjected to stressed environmental conditions. Local populations of many insect pests have been differentiated, and some are recognized as genetically distinct infraspecific taxa.

The concept of pest is a human construct, and the problem of insect pests is defined by specific economic criteria and a cultural value system. However, pest management strategy should not be based on anthropocentric criteria, because insect pests are dynamic organisms with the genetic capacity to adapt rapidly to changing environments.

Conventional pest control practices, based primarily on shortsighted economics and a heavy use of diverse chemical pesticides, resulted in rapid adaptations of insect pests to diverse local environmental conditions and the anthropocentric control measures, such as DDT and crop rotations. Enormous efforts to contain insect pests have not succeeded in permanently eradicating a single species of insect pest. Genetic variation is important in determining the pest status and evolutionary response of various insect species (Kennedy, Chapter 22 in this book). Thus, sound pest management strategies must be based on the knowledge of genetic mechanisms and evolutionary strategies of the pest species.

REFERENCES

Allan, E. R., E. G. Heyne, E. T. Jones, and C. O. Johnson. 1959. Genetic analysis of ten sources of Hessian fly resistance, their interrelationships and association with leaf rust reaction in wheat. *Bull. Kans. Agric. Expt. Sta.* 104.

Arnett, R. H., Jr. 1990. Present and future of systematics of the Coleoptera in North America. Pages 165–173 in Kosztarab, M., and C. W. Schaefer (eds.), *Systematics of the North American Insects and Arachnids: Status and Needs*. Va. Agric. Exp. Sta. Infor. Ser. 90–1. Blacksburg, Va.: Va. Polytech. Inst. St. Univ.

Axelrod, R., and W. D. Hamilton. 1981. The evolution of cooperation. *Science* 211: 1390–1396.

Barker, J. S. F., and R. H. Thomas. 1987. A quantitative genetic perspective on adaptive evolution. Pages 3–23 in Loeschcke, V. (ed.), *Genetic Constraints on Adapative Evolution*. Berlin: Springer-Verlag.

Bartlett, A. 1981. Isozyme polymorphisms in populations of the pink bollworm. *Ann. Entomol. Soc. Am.* 74:9–13.

Bartlett, D. C. H., and B. J. Richardson. 1986. Genetic attributes of invading species. Pages 21–33 in Groves, R. H., and J. J. Burdon (eds.), *Ecology of Biological Invasions*. Cambridge: Cambridge Univ. Press.

Barton, N. H., and B. Charlesworth. 1984. Genetic revolutions, founder effects and speciation. *Annu. Rev. Ecol. Syst.* 15: 133–164.

Beaverstock, P. R., and M. Adams. 1987. Comparative rates of molecular, chromosomal and morphological evolution in some Australian vertebrates. Pages 175–188 in Campbell, K. S. W., and M. F. Day (eds.), *Rates of Evolution*. London: Allen & Unwin.

Beck, S. D. 1989. Developmental and seasonal biology of *Ostrinia nubilalis*. Pages 45–82 in Russell, G. E. (ed.), *Biology and Population Dynamics of Invertebrate Crop Pests*. Andover, Hampshire: Intercept, Ltd.

Beregovoy, V. H., and K. J. Starks. 1986. Enzyme pattern in biotypes of the greenbug, *Schizaphis graminum* (Rondani) (Homoptera: Aphididae). *J. Kan. Entomol. Soc.* 59(3):517–523.

Bethke, J. A., T. D. Paine, and G. S. Nuessly. 1991. Comparative biology, morphometrics, and development of two populations of *Bemisia tabaci* (Homoptera: Aleyrodidae) on cotton and poinsettia. *Ann. Entomol. Soc. Am.* 84(4):407–411.

Bradshaw, A. D. 1985. Evolutionary significance of phenotypic plasticity in plants. *Adv. Genet.* 13:115–155.

Brooks, C. P., and H. D. Loxdale. 1987. Survey of enzyme variation in British populations of *Myzus persicae* (Sulzer) (Hemiptera: Aphididae) on crops and weed hosts. *Bull. Entomol. Res.* 77:83–89.

Brussard, P. F. 1984. Geographical patterns and environmental gradients: The central and marginal model in *Drosophila* revisited. *Annu. Rev. Ecol. Syst.* 15:25–64.

Bull, J. J. 1987. Evolution of phenotypic variance. *Evolution* 41 (2):303–315.

Bush, G. L. 1981. Stasipatric speciation and rapid evolution in animals. Pages 201–218 in Atchley, W. R., and D. S. Woodruff (eds.), *Evolution and Speciation in Honor of M. J. D. White*. Cambridge: Cambridge Univ. Press.

Byrne, D. N., T. S. Bellows, Jr., and M. P. Parrella. 1990. Whiteflies in agricultural systems. Pages 227–261 in Gerling, D. (ed.), *Whiteflies: Their Bionomics, Pest Status and Management*. Andover, Hamphshire: Intercept.

Carlquist, S. 1981. Chance dispersal. *Am. Sci.* 69:509–515.

Carson, H. L. 1968. The population flush and its genetic consequences. Pages 123–137 in Lewontin, R. C. (ed.), *Population Biology and Evolution*, Syracuse, N.Y.: Syracuse Univ. Press.

———. 1975. The genetics of speciation at the diploid level. *Am. Nat.* 109:73–92.

Carson, H. L., and A. R. Templeton. 1984. Genetic revolutions in relation to speciation phenomena; the founding of new populations. *Annu. Rev. Ecol. Syst.* 15:97–131.

Chiang, H. C., K. C. Kim and B. W. Brown. 1970. Morphometric variability related to ecological conditions of three biotypes of *Ostrinia nubilalis* in North Central United States. *Ann. Entomol. Soc. Am.* 63:1013–1016.

Council for Agricultural Science and Technology (CAST). 1987. *Pests of Plants and Animals: Their Introduction and Spread*. Report 112. 40 pp.

Covell, C. V., Jr. 1990. The status of our knowledge of the North American Lepidoptera. Pages 211–230 in Kosztarab, M., and C. W. Schaefer (eds.), *Systematics of the North American Insects and Arachnids: Status and Needs*. Va. Agric. Exp. Sta. Infor. Ser. 90–1. Blacksburg, Va.: Va. Polytech. Inst. St. Univ.

Daly, H. V. 1985. Insect morphometrics. *Annu. Rev. Entomol.* 30:415–438.

Davis, L. G. 1962. The need for pest detection follow-ups. *Bull. Entomol. Soc. Am.* 8:59–62.

den Boer, M. H. 1978. Isoenzyme and migration in the African armyworm, *Spodoptera exempta* (Lepidoptera, Noctuidae). *J. Zool.*, London 185:539–553.

Denno, R. F., and H. Dingle (eds.). 1980. *Insect Life History Patterns.* New York: Springer-Verlag.

Dowell, R. V., and R. Gill. 1989. Exotic invertebrates and their effects on California. *Pan-Pac. Entomol.* 65:132–145.

Ehrlich, P. R. 1986. Which animal will invade? Pages 79–95 in Mooney, H. A., and J. A. Drake (eds.), *Ecology of Biological Invasions of North America and Hawaii.* New York: Springer-Verlag.

Elton, C. S. 1958. *The Ecology of Invasions by Animals and Plants.* London: Methuen.

Endler, J. A. 1977. *Geographic Variation, Speciation, and Clines.* Princeton: Princeton Univ. Press.

Ewel, J. J. 1986. Invasibility: Lessons from South Florida. Pages 214–230 in Mooney, H. A., and J. A. Drake (eds.), *Ecology of Biological Invasions of North America and Hawaii.* New York: Springer-Verlag.

Flexon, P. B., and C. F. Rodell. 1982. Genetic recombination and directional selection of DDT resistance in *Drosophila melanogaster. Nature* (London) 298:672–674.

Frankie, G., W. R. Gill, C. S. Kochler, D. Dilly, J. O. Washburne, and P. Hamman. 1982. Some considerations for the eradication and management of introduced insect pest in urban environments. Pages 237–255 in Battenfield, S. L. (ed.), *Proceedings, Symposium Imported Fire Ant,* Atlanta, Ga., June 7–10, 1982. USEPA & USDA, APHIS, Washington, D. C.

Futuyma, D. J., and S. C. Peterson. 1985. Genetic variation in the use of resources by insects. *Annu. Rev. Entomol.* 30:217–238.

Gallun, R. L., K. J. Starks, and W. D. Guthrie. 1975. Plant resistance to insects attacking cereals. *Annu. Rev. Entomol.* 20:337–357.

Gillespie, J. H. 1985. The interaction of genetic drift and mutation with selection in a fluctuating environment. *Theor. Popul. Biol.* 27:222–237.

Gould, F. 1991. The evolutionary potential of crop pests. *Amer. Sci.* 79:496–507.

Gould, S. J., and R. F. Johnston. 1972. Geographic variation. *Annu. Rev. Ecol. Syst.* 3:457–498.

Green, N. G., H. M. LeBaron, and L. W. K. Moberg. 1990. *Managing Resistance to Agrochemicals: From Fundamental Research to Practical Strategies.* Washington, D. C.: American Chemical Society.

Gupta, A. P., and R. C. Lewontin. 1982. A study of reaction norms in natural populations of *Drosophila pseudoobscura. Evolution* 36:934–948.

Hamilton, K. G. A., and D. W. Langor. 1987. Leafhopper fauna of Newfoundland and Cape Breton Islands (Rhynchota: Homoptera: Cicadellidae). *Can. Entomol.* 119:663–695.

Hedrick, P. H. 1986. Genetic polymorphism in heterogeneous environments: A decade later. *Annu. Rev. Ecol. Syst.* 17:535–566.

Hedrick, P. W., M. E. Ginevan, and E. P. Ewing. 1976. Genetic polymorphism in heterogeneous environments. *Annu. Rev. Ecol. Syst.* 7:1–32.

Herrick, G. W. 1929. The procession of foreign insect pests. *Sci. Mon.* 29:269–274.

Hoffmann, A. A., and P. A. Parsons. 1991. *Evolutionary Genetics and Environmental Stress.* Oxford: Oxford Univ. Press.

Holt, R. D. 1987. Population dynamics and evolutionary process: The manifold roles of habitat selection. *Evol. Ecol.* 1:331–347.

Horsfall, J. G. 1983. Impact of introduced pests on man. Pages 1–13 in Wilson, C. L., and C. L. Graham (eds.), *Exotic Plant Pests and North American Agriculture.* New York: Academic Press.

Howden, H. F. 1969. Effects of man on the Ontario insect fauna. *Proc. Entomol. Soc. Ont.* 99:51–53.

Hsiao, T. H. 1985. Ecophysiological and genetic aspects of geographic variations of the Colorado potato beetle. Pages 63–77 in Ferro, D. N., and R. H. Voss (eds.). *Proceedings, Symposium on Colorado Potato Beetle,* XVIIth International Congress of Entomology. Amherst, Mass.: Mass. Agric. Exp. Sta. Res. Bull. No. 704.

Hsiao, C., and T. H. Hsiao. 1984. Cytogenetic studies of alfalfa weevil (*Hypera postica*) strains (Coleoptera: Curculionidae). *Can. J. Genet. Cytol.* 26:348–353.

Hsiao, T. H., and J. M. Stutz. 1985. Discrimination of alfalfa weevil strains by allozyme analysis. *Entomol. Exp. Appl.* 37:113–121.

Hudon, M., E. J. LeRoux, and D. G. Harcourt. 1989. Seventy years of European corn borer (*Ostrinia nubilalis*) research in North America. Pages 1–44 in Russell, G. E. (ed.). *Biology and Population Dynamics of Invertebrate Crop Pests.* Andover, Hamphshire: Intercept.

Inayatullah, C., W. S. Fargo, and J. A. Webster. 1987. Use of multivariate models in differentiating greenbug (Homoptera: Aphididae) biotypes and morphs. *Environ. Entomol.* 16:839–846.

Istock, C. A. 1980. Natural selection and life history variation: Theory plus lessons from a mosquito. Pages 113–127 in Denno, R. F., and H. Dingle (eds.), *Insect Life History Patterns.* New York: Springer-Verlag.

Jacobson, J. W., and T. H. Hsiao. 1983. Isozyme variation between geographic populations of the Colorado potato beetle, *Leptinotarsa decemlineata* (Coleoptera: Chrysomelidae). *Ann. Entomol. Soc. Am.* 76:162–166.

Jones, D. P. 1973. Agricultural entomology. *Annu. Rev. Entomol.* 307–332.

Kelleher, J. S., and M. A. Hulme. 1981. *Biological Control Programs Against Insects and Weeds in Canada 1969–1980.* Slough, England: Commonwealth Agricultural Bureaux, 410 pp.

Kim, K. C. 1983. How to detect and combat exotic pests. Pages 261–319 in Wilson, C. L., and C. L. Graham (eds.), *Exotic Plant Pests and North American Agriculture.* New York: Academic Press.

Kim, K. C., H. C. Chiang, and B. W. Brown, Jr. 1967. Morphometric differences among four biotypes of *Ostrinia nubilalis. Ann. Entomol. Soc. Am.* 60:796–801.

Kim, K. C., and B. A. McPheron (eds.). 1993. *Evolution of Insect Pests: The Patterns of Variation.* New York: John Wiley & Sons.

Kim, K. C., and A. G. Wheeler, Jr. 1991. Pathways and Consequences of the Introduction of Non-Indigenous Insects and Arachnids in the United States. Report to the U. S. Congress, Office of Technology Assessment, Washington, D. C. (unpublished).

Kirkpatrick, M. 1982. Quantum evolution and punctuated equilibria in continuous genetic characters. *Am. Nat.* 119:833–848.

Klassen, W. 1989. Eradication of introduced arthropod pests: Theory and historical practice. *Misc. Publ. Entomol. Soc. Am.* 73:1–29.

Kogan, M. 1982. Impact and management of introduced pests in agriculture. Pages 226–235 in Battenfield, S. S. (ed.), *Proceedings of the Symposium on Imported Fire Ant,* Atlanta, Ga., June 7–10, 1982. USEPA & USDA, APHIS, Washington, D. C.

Komai, T., N. Chino, and Y. Hosino. 1950. Contributions to the evolutionary genetics of the lady beetle, *Harmonia I. Genetics* 36:382–390.

Kosztarab, M., and C. W. Schaefer (eds.). 1990. *Systematics of the North American Insects and Arachnids: Status and Needs.* Va. Agric. Exp. Sta. Infor. Ser. 90–1. Blacksburg, Va.: Va. Polytech. Inst. St. Univ.

Krysan, J. L., D. E. Foster, T. F. Branson, K. R. Ostlie, and W. S. Cranshaw. 1986. Two years before the hatch: Rootworm adapt to crop rotation. *Bull. Entomol. Soc. Am.* 32:250–253.

Larson, D. J., and D. W. Langor. 1982. The carabid beetles of insular Newfoundland (Coleoptera: Carabidae: Cicindellidae)—30 years after Lindroth. *Can. Entomol.* 114:591–597.

Lattin, J. D., and P. Oman. 1983. Where are the exotic insect threats? Pages 93–137 in Wilson, C. L., and C. L. Graham (eds.), *Exotic Plant Pests and North American Agriculture.* New York: Academic Press.

Laurie-Ahlberg, C. C., and D. J. Merrell. 1979. Aldehyde oxidase allozyme, inversion and DDT resistance in some laboratory populations of *Drosophila melanogaster. Evolution* 33(1):342–349.

Lewontin, R. C. 1974. *The Genetic Basis of Evolutionary Changes.* New York: Columbia Univ. Press.

———. 1985. Population genetics. *Annu. Rev. Genet.* 19:81–102.

McGregor, R. C. 1973. The Emigrant Pests: A Report to Dr. Francis J. Mulhern, Administrator, Animal and Plant Health Inspection Service, U. S. Dept. Agr. APHIS, Hyattsville, Md. 167 pp. (unpublished).

McPheron, B. A., D. C. Smith, and S. H. Berlacher. 1988. Genetic differences between host races of *Rhagolitis pomonella. Nature* 336:64–66.

Martinez, D., A. Moya, A. Latorre, and A. Fereres. 1992. Mitochondrial DNA variation in *Rhopalosiphum padi* (Homoptera: Aphididae) population from four Spanish localities. *Ann. Entomol. Soc. Am.* 85(2):241–246.

Masner, L. 1990. Status report on taxonomy of Hymenoptera in North America. Pages 231–240 in Kosztarab, M., and C. W. Schaefer (eds.), *Systematics of the North American Insects and Arachnids: Status and Needs.* Va. Agric. Exp. Sta. Infor. Ser. 90–1. Blacksburg, Va.: Va. Polytech. Inst. St. Univ.

Matthews, T. C., and G. B. Craig, Jr. 1980. Genetic heterozygosity in natural populations of tree-hole mosquitoes, *Aedes triseriatus. Ann. Entomol. Soc. Am.* 73:739–743.

May, B., and F. R. Holbrook. 1978. Absence of genetic variability in the green peach aphid, *Myzus persicae* (Hemiptera: Aphididae). *Ann. Entomol. Soc. Am.* 71:809–812.

Maynard-Smith, J. 1966. Sympatric speciation. *Am. Nat.* 100:637–650.

Mayr, E. 1954. Change of genetic environment and evolution. Pages 156–180 in Huxley, J. S., A. C. Hardy, and E. B. Ford (eds.), *Evolution as a Process.* London: Allen & Unwin.

———. 1963. *Animal Species and Evolution.* Cambridge, Mass.: Harvard Univ. Press.

Mitter, C., D. J. Futuyma, J. C. Schneider, and J. D. Hare. 1979. Genetic variation and host plant relations in a parthenogenetic moth. *Evolution* 33(3):777–790.

Mooney, H. A., and M. Godron (eds.). 1983. *Disturbance and Ecosystems*. Berlin: Springer-Verlag.

National Research Council (NRC). 1986. *Pesticide Resistance: Strategies and Tactics for Management. Committee on Strategies for the Management of Pesticide Resistant Pest Populations*. Washington, D. C.: National Acad. Press.

Nei, M., T. Maruyama, and R. Chakraborty. 1975. The bottleneck effect and genetic variability in population. *Evolution* 29: 1–10.

Nevo, E. 1976. Adaptive strategies of genetic systems in constant and varying environments. Pages 141–158 in Karlin, S., and E. Nevo (eds.), *Population Genetics and Ecology*. New York: Academic Press.

———. 1983a. Population genetics and ecology: The interface. Pages 287–321 in Bendall, D. S. (eds.). *Evolution from Molecules to Men*. Cambridge: Cambridge Univ. Press.

———. 1983b. Adaptive significance of protein variation. Pages 239–282 in Oxford, G. S., and D. Rollinson (eds.), *Protein Polymorphism: Adaptive and Taxonomic Significance*. New York: Academic Press.

———. 1988. Genetic diversity in nature: Patterns and theory. *Evol. Biol.* 23:217–246.

Nevo, E., A. Beiles, and R. Ben-Shlomo. 1984. The evolutionary significance of genetic diversity: Ecological, demographic and life-history correlates. Pages 13–213 in Mani, G. S. (ed), *Evolutionary Dyanamics of Genetic Diversity*. Berlin: Springer-Verlag.

Odum, E. P. 1985. Trends expected in stressed ecosystems. *Bioscience* 35(7):419–422.

Parsons, P. A. 1983. *The Evolutionary Biology of Colonizing Species*. Cambridge: Cambridge Univ. Press.

———. 1987. Evolutionary rates under environmental stress. *Evol. Biol.* 21:311–347.

———. 1988. Evolutionary rates: Effects of stress upon recombination. *Biol. J. Linn. Soc.* 35:49–68.

———. 1991. Evolutionary rates: Stress and species boundaries. *Annu. Rev. Ecol. Syst.* 22:1–18.

Pashley, D. P., S. J. Johnson, and A. N. Sparks. 1985. Genetic population structure of migratory moths: The fall armyworm (Lepidoptera: Noctuidae). *Ann. Entomol. Soc. Am.* 78:756–762.

Pierce, W. D. 1917. *A Manual of Dangerous Insects Likely to Be Introduced in the United States Through Importations*. Washington, D. C.: Government Printing Office. 256 pp.

Pimentel, D. 1986. Biological invasions of plants and animals in agriculture and forestry. Pages 149–162 in Mooney, H. A., and J. A. Drake (eds.), *Ecology of Biological Invasions of North America and Hawaii*. New York: Springer-Verlag.

Pimentel, D., and A. G. Wheeler, Jr. 1973. Species and diversity of arthropods in the alfalfa community. *Environ. Entomol.* 2:659–668.

Pimentel, D., D. Andow, R. Dyson-Hudson, D. Gallahan, S. Jacobson, M. Irish, S. Kroop, A. Moss, I. Schreiner, M. Shepard, T. Thompson, and B. Vinzant. 1980. Environmental and social costs of pesticides: A preliminary assessment. *Oikos* 34:127–140.

Powell, J. R. 1975. Protein variation in natural populations of animals. *Evol. Biol.* 8:79–119.

Powell, J. R., and C. E. Taylor. 1979. Genetic variation in ecologically diverse environments. *Amer. Sci.* 67:590–596.

Reynolds, H. T., R. Clark, G. Frankie, F. E. Gilstrap, P. J. Hamman, M. Kogan, E. H. Smith, and D. E. Wiedhaas. 1982. Management of established vs. introduced pests. Pages 82–90 in Battenfield, S. L. (ed.). *Proceedings of the Symposium on Imported Fire Ant,* Atlanta, Ga., June 7–10, 1982. USEPA & USDA/APHIS, Washington, D.C.

Rhomberg, L. R., S. Joseph, and R. S. Singh. 1985. Seasonal variation and clonal selection in cyclically parthenogenetic rose aphid (*Macrosiphum rosae*). *Can. J. Genet. Cytol.* 27:224–232.

Rice, W. R. 1987. Speciation via habitat specialization. *Evol. Ecol.* 1:301–314.

Rice, W. R., and G. W. Salt. 1988. Speciation via disruptive selection on habitat preference: Experimental evidence. *Am. Nat.* 131:911–917.

Riley, C. V. 1877. *The Colorado Beetle.* London: George Routledge & Sons.

Sailer, R. I. 1978. Our immigrant insect fauna. *Bull. Entomol. Soc. Am.* 24:3–11.

———. 1983. History of insect introductions. Pages 15–38 in Wilson, C. L., and C. L. Graham (eds.), *Exotic Plant Pests and North American Agriculture.* New York: Academic Press.

Sattler, P. W., L. R. Hilburn, R. B. Davey, J. E. George, and J. B. Rajas Avalos. 1986. Genetic similarity and variability between natural populations and laboratory colonies of North American *Boophilus* (Acari: Ixodidae). *J. Parasitol.* 72(1):95–100.

Selander, R. K. 1976. Genic variation in natural populations. Pages 21–44 in Ayala, F. J. (ed.), *Molecular Evolution.* Sunderland, Mass.: Sinauer Associates.

Sell, D. K., G. S. Whitt, and W. H. Luckmann. 1975. Esterase polymorphism in the corn earworm, *Heliothis zea* (Boddie): A survey of temporal and spatial allelic variation in natural populations. *Biochem. Genet.* 13(11/12):885–898.

Shelton, A. M., and J. A. Wyman. 1991. Insecticide resistance of diamondback moth (Lepidoptera: Plutellidae) in North America. *Proc. 2nd Internat. Diamondback Moth Workshops.* Taiwan.

Showers, W. B. 1981. Geographical variation of the diapause response in the European corn borer. Pages 97–111 in Denno, R. F., and H. Dingle (eds.), *Insect Life History Patterns: Habitat and Geographic Variation.* New York: Springer-Verlag.

Showers, W. B., H. C. Chiang, A. J. Keaster, R. E. Hill, G. L. Reed, A. N. Sparks, and G. J. Musick. 1975. Ecotypes of the European corn borer in North America. *Environ. Entomol.* 4:753–760.

Slatkin, M. 1985. Gene flow in natural populations. *Annu. Rev. Ecol. Syst.* 16:393–430.

———. 1987. Gene flow and the geographic structure of a natural population. *Science* 236:787–792.

Smith, D. C. 1988. Heritable divergence of *Rhagolitis pomonella* host races by seasonal asynchrony. *Nature* 336:66–67.

Smith, H. S. 1929. On some phases of preventive entomology. *Sci. Mon.* 29:177–184.

Southwood, T. R. E., and G. A. Norton. 1973. Economic aspects of pest management strategies and decisions. Pages 164–184 in Geier P. W., C. R. Clark, D. F. Anderson, and H. A. Nix (eds.), *Insects: Studies in Pest Management.* Ecol. Soc. Austr., Canberra-City.

Stenseth, N. C. 1987. Evolutionary processes and insect outbreaks. Pages 533–563 in Barbosa, P., and J. C. Schultz, *Insect Outbreaks*. New York: Academic Press.

Strong, D. R., E. P. McCoy, and J. R. Rey. 1977. Time and the number of herbivore species. The pests of sugarcane. *Ecology* 58: 167–175.

Tabachnick, W. G., and G. P. Wallis. 1985. Genetic structure of *Aedes aegypti* populations in the Caribbean: Ecological considerations. Pages 371–381 in Lounibos, L. P., J. R. Rey, and J. F. Frank (eds.), *Ecology on Mosquitoes: Proceedings of a Workshop*. College Park, Md.: Entomol. Soc. Am. Publ.

Tadei, W. P. 1982. Biologia de anofelinos amazonicos. VII. Estudo da variacao de frequenmcias das inversoes cromossomicas de *Anopheles darlingi* Root (Diptera, Culicidae). *Acta Amazonica* 12(4):759–785.

Tauber, M. J., C. A. Tauber, and S. Masaki. 1986. *Seasonal Adaptation of Insects*. New York: Oxford Univ. Press.

Taylor, H., and R. Karban (eds.) 1986. *The Evolution of Insect Life Cycles*. New York: Springer-Verlag.

Templeton, A. R. 1980. The theory of speciation via the founder principle. *Genetics* 94:1011–1038.

———. 1981. Mechanisms of speciation—A population genetic approach. *Annu. Rev. Ecol. Syst.* 12:23–48.

Thompson, F. C. 1990. Biosystematic information: Dipterists ride the third wave. Pages 179–201 in Kosztarab, M., and C. W. Schaefer (eds.), *Systematics of the North American Insects and Arachnids: Status and Needs*. Va. Agric. Exp. Sta. Infor. Ser. 90–1. Blacksburg, Va.: Va. Polytech. Inst. St. Univ.

Tomiuk, J., and K. Wohrmann. 1983. Enzyme polymorphism and taxonomy of aphid species. *Z. Zool. Syst. & Evolutionsforsch*. 21:266–274.

Tower, W. L. 1906. An investigation of evolution in chrysomelid beetles of the genus *Leptinotarsa*. Washington, D. C.: Carnegie Institute.

Unruh, T. R. 1990. Genetic structure among 18 West Coast pear psylla populations: Implications for the evolution of resistance. *Amer. Entomologist* Spring 1990: 37–48.

Vinogradova, E. B. 1986. Geographical variation and ecological control by diapause in flies. Pages 35–47 in Taylor F., and R. Karban (eds.), *The Evolution of Insect Life Cycles*. New York: Springer-Verlag.

Wallis, G. P., W. J. Tabachnick, and J. R. Powell. 1984. Genetic heterogeneity among Caribbean populations of the yellow fever mosquito, *Ades aegypti*. *Am. J. Trop. Med. Hyg.* 33:492–498.

West-Eberhard, M. J. 1989. Phenotypic plasticity and the origins of diversity. *Annu. Rev. Evol. Syst.* 20:249–278.

Wheeler, A. G., Jr. 1985. Seasonal history, host plants, and nymphal descriptions of *Orthocephalus coriaceus*, a plant bug pest of herb garden composites (Hemiptera: Miridae). *Proc. Entomol. Soc. Wash.* 87:85–93.

———, Jr. 1988. An exotic honeysuckle aphid established in Pennsylvania. *Regul. Hortic.* 14(2):21–22.

Wheeler, A. G., Jr., and D. R. Whitehead. 1985. *Larinus planus* (F.) in North America (Coleoptera: Curculionidae: Cleoninae) and comments on biological control of Canada thistle. *Proc. Entomol. Soc. Wash.* 87:751–759.

White, M. J. D. 1968. Models of speciation. *Science* 158: 1065–1070.

———. 1978. *Modes of Speciation.* San Francisco: Freeman.

Williams, G. C. 1966. *Adaptations and Natural Selections.* Princeton: Princeton Univ. Press.

Wilson, C. L., and C. L. Graham (eds.). 1983. *Exotic Plant Pests and North American Agriculture.* New York: Academic Press.

Woodwell, G. M. 1983. The blue planet: Of wholes and parts and man. Pages 2–10 in Mooney, H. A., and M. Godron (eds.), *Disturbance and Ecosystems.* Berlin: Springer-Verlag.

Wool, D., S. W. Bunting, and H. F. Van Emden. 1978. Electrophoretic study of genetic variation in British *Myzus persicae* (Sulz.) (Hemiptra: Aphidiae). *Biochem. Genet.* 16:987–1006.

Wright, S. 1932. The role of mutation, inbreeding, crossbreeding and selection in evolution. *Proc. 6th Int. Congr. Genet.* 1:356–366.

Zehnder, G. W., L. Sandall, A. M. Tisler, and T. O. Powers. 1992. Mitochondrial DNA diversity among 17 geographic populations of *Leptinotarsa decemlineata* (Coleoptera: Chrysomelidae). *Ann. Entomol. Soc. Am.* 85(2):234–240.

Zhuchenko, A. A., A. B. Koral, and L. P. Kovtyukh. 1985. Changes for crossing over frequency in *Drosophila* during selection for resistance to temperature fluctuation. *Genetica* 67:73–78.

2

GENETIC DIFFERENTIATION AND GEOGRAPHIC DISTRIBUTION IN INSECTS: PATTERNS AND INFERENCES

William W. M. Steiner

U.S. Department of Agriculture, Agricultural Research Service, Biological Control of Insects Laboratory

> . . . the genes in question, insignificant as are their visible effects, have an important influence on the physiology of the organism, modifying profoundly the individual as a unit upon which selection operates.
>
> —E. B. Ford, 1965

INTRODUCTION

Since the rediscovery of Mendel's primogenial ideas, genetic variation has been central to understanding heredity. Without genetic variation (e.g., chromosomal aberrations, protein polymorphisms, DNA variants, and transposons), we simply would not have the understanding of how genes are structured, and how they function and evolve.

However, understanding the role of the high levels of genetic variation observed in nature is not so easy. On the one hand, we are restricted by typological thinking so common to the scientific approach. Occam's Razor, which bids us to consider the simplest answer as the most likely, can explain genetic variation in terms of the Neutral Gene Theory (Lecam et al., 1972). The Neutral Gene Theory suggests that most genetic variation has no selective value (is neutral) and is primarily maintained by genetic drift and chance events. This would explain the paradox placed upon classical Darwinian theory, which predicts that survival of the fittest should result in optimized, nonvarying genotypes, seemingly resulting in a lack of genetic polymorphism. In classical theory, the paradigm is that selection guides mutation in creating, editing, and fitting life forms into workable relationships while conserving the whole. On the other hand, we simply do not have strong insight into how protein polymorphisms differ in their responses to either internal or external environments. The experimental evidence at hand often flies in the face of expectations predicated by the Neutral Gene Theory.

If, in fact, the Neutralist theory is wrong, then it must be replaced with workable and testable hypotheses that are robust enough to account for many varied and simultaneous observations drawn from within and across species, yet that still satisfy neo-Darwinian tenets of evolution (see Riddiford and Penny, 1984). Here, I examine the experimental evidence from studies of genetic variation in insects, which suggests that protein polymorphism is adaptive in nature and therefore involved in microevolutionary processes. The evidence as presented suggests the type of studies that should be undertaken and holds within it implications for evolutionary processes in general, and for future attempts at genetic engineering.

PROTEIN POLYMORPHISM AS SINE QUA NON

Studying the genetic basis of adaptation requires genetic polymorphism. Interest must center on any form of genetic variation that alters an organism's genetic fitness, ultimately changing survival probabilities. There are many types of genetic variants resulting from mutation processes that affect fitness. These fall into seven categories and are not necessarily mutually excludable: (1) lethals and semilethals; (2) physiological, including those affecting circadian rhythms and diapause; (3) behavioral, including those affecting mate choice, oviposition site, host selection, and so on; (4) morphological, including mutations affecting structure, numbers of bodily organs, segments, hairs, color, and so on; (5) chromosomal, including translocations, inversions, deletions and those affecting number; (6) molecular, such as DNA polymorphisms; and (7) biochemical, including protein (allozyme) polymorphisms. Protein polymorphisms may underly the first four types. Protein variation is the first choice for a number of reasons, as follows.

Of the seven categories, the first, lethals and semilethals, can be dismissed. These make up a special category of gene that is clearly recognized as lowering population fitness but whose study is limited to the genus *Drosophila*. These may occur in up to 48% frequency (Band and Ives, 1968) and are correlated with different climatic and environmental changes. It is very difficult, in the studies to date, to determine what their nature is, what products they produce, how they act, and even how many are contributing to a particular lethal load. Likewise, the second and third categories, behavioral and physiological mutants, need not be considered here, even though their effects may be far-reaching. These have mostly a polygenic basis and are probably epistatic in effect. Their study at present is restricted mostly to *Drosophila* under laboratory conditions. Morphological mutants, the fourth category, can be dismissed, since population studies indicate that most occur at less than 4% frequency and may be under strong directional or normalizing selection (see Sokolova and Golubovskii, 1979). It should be noted, however, that morphological variance and enzyme heterozygosity are correlated in *Drosophila* (Lizana and Prado, 1986) and in the monarch butterfly (Eanes, 1978). The distribution of morphologi-

Table 2.1. Enzymes and Proteins Commonly Detected by Electrophoretic and Histochemical Techniques. (All have been genetically mapped in *Drosophila* to confirm their genetic bases with the exception of those marked with an asterisk.)

Enzyme or Protein	Acronym	I.U.B. Number[a]
Acetylcholinesterase	ACHE	3. 1. 1. 7
Acid phosphatase	ACPH	3. 1. 3. 2
Aconitase*	ACON	4. 2. 1. 3
Adenylate kinase*	ADK	2. 7. 4. 3
Alcohol dehydrogenase	ADH	1. 1. 1. 1
Aldehyde oxidase	AO	1. 2. 1. 3
Aldolase	ALD	4. 1. 2.13
Alkaline phosphatase	APH	3. 1. 3. 1
Arginine kinase	AK	2. 7. 3. 3
Carbonic anhydrase*	CA	4. 2. 1. 1
Catalase	CAT	1.11. 1. 6
Diaphorase*	DIA	1. 6. 2. 2
Dipeptidase	DIP	3. 4.11
Esterase	EST	3. 1. 1. 1
α-Fucosidase	α-FUC	3. 2. 1.51
Fumarase	FUM	4. 2. 1. 2
β-Galactosidase	β-GAL	3. 2. 1.23
α-Glucosidase*	α-GLU	3. 2. 1.20
Glucose oxidase	GO	1. 1. 3. 4
Glucose-6-phosphate dehydrogenase	G-6-PDH	1. 1. 1.49
β-Glucuronidase	β-GUS	3. 2. 1.31
Glutamate dehydrogenase*	GLUD	1. 4. 1. 3
Glutamate oxaloacetate transaminase	GOT	2. 6. 1. 1
Glutamate pyruvate transaminase	GPT	2. 6. 1. 2
α-Glycerophosphate dehydrogenase	α-GPDH	1. 1. 1. 8
α-Glycerophosphate oxidase	α-GPO	1. 1.99. 5
Glyceraldehyde-3-phosphate dehydrogenase	G-3-PDH	1. 2. 1.12
Hexokinase	HK	2. 7. 1. 1
Hydroxy acid dehydrogenase	HAD	1. 1. 1.45
Isocitrate dehydrogenase	IDH	1. 1. 1.42
Lactate dehydrogenase*	LDH	1. 1. 1.27
Leucine aminopeptidase	LAP	3. 4.11. 1
Larval protein	LPT	—
Malate dehydrogenase	MDH	1. 1. 1. 3
Malic enzyme	ME	1. 1. 1.40
Mannose phosphate isomerase*	MPI	5. 3. 1. 8
α-Mannosidase*	α-MAN	3. 2. 1.24
Octanol dehydrogenase	ODH	1. 1. 1.73
Phenol oxidase	IPO	1.10. 3. 1
Phosphoglucoisomerase	PGI	5. 3. 1. 9
Phosphoglucomutase	PGM	2. 7. 5. 1
3-Phosphoglycerate kinase	PGK	2. 7. 2. 3
Sorbital dehydrogenase*	SDH	1. 1. 1.14

(Continued)

Table 2.1. *(Continued)*

Enzyme or Protein	Acronym	I.U.B. Number[a]
6-Phosphogluconate dehydrogenase	6-PGDH	1. 1. 1.43
Tetrazolium oxidase	TO	1.15. 1. 1
Trehalase	TRE	3. 2. 1.28
Xanthine dehydrogenase	XDH	1. 2. 3. 2

[a]International Union of Biochemistry.

cal variants can be of evolutionary interest, as witnessed by Willey (1987) for wing color polymorphism in the North American grasshopper, *Arphia pseudonietana* (Thomas).

Chromosome aberrations suffer from two drawbacks with respect to studying the basis of population adaptation. The amount of work spent in preparing chromosome slides limits the numbers that can be studied quickly. In addition, the aberration usually encompasses a great many genes, for many of which the function and epistatic relationships remain unknown. They may possibly encompass transposons and genes involved in hybrid dysgenesis. Thus, although chromosome aberrations may show unique geographic patterns and unique internal genetic contents on comparisons of the same chromosome (Prakash and Levitan, 1973), the basis of the observed patterns is especially difficult to discern. This is true even where linkage disequilibrium between enzyme polymorphisms and chromosome inversions are found (Watanabe and Watanabe, 1977; Voelker et al., 1978; Knibb, 1983). Interestingly, when chromosomes have been "extracted" from diverse populations of a species and tested against each other for viability and other fitness parameters (see Takano et al., 1987; Hiraizumi and Crow, 1960; Dobzhansky and Levene, 1955), they appear to differ, suggesting that locale-dependent selection has acted on the gene pool to "fine tune" its genetic nature for the extant conditions. Molecular genetic polymorphisms include DNA restriction fragment length polymorphisms. These can be readily studied in a wide variety of organisms. Restriction fragment length polymorphisms are especially valuable for determination of genetic relatedness among organisms (Avise et al., 1979; DeSalle et al., 1987) and for studying speciation processes (Solignac and Monnerot, 1986; Gale and Crampton, 1987). We must dismiss restriction fragment length polymorphisms, because the enzymes used to restrict or "cut" DNA into different length fragments do so across the entire genome and may cover many genes of different and mostly unknown function.

We are left, then, with biochemical genetic variation. To study the genetic basis of any adaptive process requires identification of the genes of interest and an understanding of their function. Proteins offer precisely these advantages. Proteins also have biochemical attributes that are easy to measure and are important to physiological functioning. These features enable, in certain cases, identification of selective advantages and can be compared to geographic correlates for further analysis. I will limit my discussion, then, to protein variation in arthropods and examine how they may provide evidence indicating the existence of an adaptive process. The most studied allozymes and the abbreviations used throughout this chapter are listed in Table 2.1.

PROTEIN VARIATION IN NATURAL POPULATIONS OF INSECTS

Wagner and Selander (1974) pointed out the various uses to which isozymes have been put in the study of insect organization, development, gene duplication, genetic regulation, protein synthesis, and evolution. However, since the summary of protein variation in *Drosophila* by Johnson (1974), many more insects of a wide variety of orders have been examined (Nevo, 1983, 1985; Powell, 1975a). Insects and invertebrates were originally thought to have levels of protein variation a third higher than mammals (Johnson, 1975). This has been questioned by Nevo (1978), who found that, by removing *Drosophila* from the data, insect species heterozygosity was not significantly higher than most vertebrate classes. Simon and Archie (1985) have dramatically shown that choice of loci may be a factor as well.

Nevertheless, the belief that invertebrate variation is higher than other organisms has led to interesting speculation. Selander and Kaufman (1973) suggested that larger animals perceive their environment to be fine-grained (predictable) and thus need less variation for adaptive purposes; put simply, their size "dampens internal metabolic losses due to environmental perturbation" (Wooten and Smith, 1985). The idea has been criticized by the latter authors and by Nevo et al. (1984). Wooten and Smith (1985) summarized data available for 133 species of mammals and found a significant negative correlation of size with heterozygosity across all species. However, when species were broken into smaller, more distinct groupings (such as the Geomyidae, with 18 species, and the Heteromyidae, with 17 species) and analyzed for size and heterozygosity relationships, nonsignificant values resulted.

It may be that certain invertebrates do tend to have higher levels of protein variation, but this in turn may depend on the taxon under investigation; thus, evolution within a taxon may be somehow constrained in its potential by limitations due to the nature of the genome itself (Zouros, 1975). This might explain why *Drosophila* has apparently high levels of protein polymorphism. It would also explain why, breeding structure notwithstanding, the eusocial Hymenoptera generally tend to have low levels of genetic variation (Berkelhamer, 1983; Reeve et al., 1985). Insect samples to date (Table 2.2) suggest that the most variable may include not only *Drosophila* but also Heteroptera, Coleoptera, Lepidoptera, and Diptera.

TYPES OF GENE-ENVIRONMENT RELATIONSHIPS

Some aspects of gene-environment relationships have been previously described by Ayala (1976) and Powell (1975b). Somero and Hochachka (1971) have laid the foundation for studying biochemical adaptation from the perspective of varying environments, and Gillespie and Langley (1974) have provided a general model to account for enzyme variation in natural populations. Early workers in this area were quick to recognize that enzymes involved in energy production differed in the degree of variation realized when compared with enzymes that were not found in metabolic pathways (Kojima et al., 1970; Gillespie and Kojima, 1978; Chianchi et al., 1980). More recently, Moon (1988) has identified the adaptive responses of gluconeogene-

Table 2.2 Genetic Variation in Different Insects and Arthropods Versus Selected Vertebrates

Taxa	Number of Loci Examined	Number of Species Examined	Total Number Populations Examined	H Range	Reference
Ephemerellida	20–26	5	40	0.04–0.11	Sweeney et al., 1987
Eurylophella	19	14	46	0.03–0.21	Funk et al., 1988
Orthoptera	16–26	25	—	0.02–0.17	Compiled from Graur, 1985
Homoptera	15–23	6	—	0–0.17	Compiled from Graur, 1985
Heteroptera	15–23	18	—	0–0.26	Compiled from Graur, 1985
Coleoptera	15–25	5	—	0.13–0.31	Compiled from Graur, 1985
	17	4	32	0.11–0.30	Liebherr, 1986a, 1986b
					Bentz and Stock, 1986
Lepidoptera	15–31	34	—	0–0.39	Compiled from Graur, 1985
Diptera	15–26	54	40	0.02–0.24	Compiled from Graur, 1985
					Steiner et al., 1982
Drosophila	20–68	31	—	0.01–0.24	Compiled from Graur, 1985
Hymenoptera					
Solitary	15–29	14	—	0.02–0.07	Compiled from Graur, 1985
	8–29	11	21	0–0.08	Berkelhamer, 1983; data corrected as suggested by Reeve et al., 1985 and for species overlap with Graur, 1985

Presocial	19	9	18	0.02–0.08	Wood and Guttman, 1987
Eusocial	15–28	20	—	0–0.07	Compiled from Graur, 1985; includes primitively eusocial
					Kukuk and May, 1985
					Ross et al., 1987
	8–39	27	57	0–0.10	Berkelhamer, 1983; data corrected as suggested by Reeve et al., 1985 and for species overlap with Graur, 1985
Insecta	10–51	122	439	0–0.23	Nevo et al., 1984; excludes *Homo sapiens* for which a manifold difference exists in numbers of loci sampled.
Mammalia	10–53	184	504	0.002–0.08	Nevo et al., 1984
Aves	15–40	46	152	0–0.15	Nevo et al., 1984
Echinodermata	13–30	17	22	0.02–0.27	Nevo et al., 1984
Crustacea	10–39	126	334	0–0.47	Nevo et al., 1984

sis when glucose levels are ". . . challenged by changes in diet . . . activity level, and by environmental disturbances." Five such patterns are identified, including ". . . quantitative changes in gluconeogenic enzyme activities and their subcellular distribution, alterations in tissue demand for glucose, existence of in situ skeletal muscle lactate cycling, quantity and type of gluconeogenic precursors, and regulation of gluconeogenisis." These studies are important in suggesting that energy requirements per se somehow play a role in maintaining genetic variability.

Several types of gene-environment relationships have been found and examined. These can be broken down into the following categories: (1) gene frequency or heterozygosity correlated with latitude or longitude, climate or environmental variables, host specificity, niche width, or resource index; (2) biochemical attribute of a variable protein correlated with metabolic function or dysfunctional state; (3) gene frequency correlated with some population or life cycle (intraspecific) attribute usually associated with genetic fitness; and (4) genetic type correlated with metabolic function.

Clinal Variation

Clinal variation is the most common type of relationship observed and is the case when allele frequencies or heterozygosity varies continuously in a directional manner over some geographic distance, usually latitudinal. This definition is more specific than the original definition given by Huxley (1938). When observed genetic variation is multiple-character in nature, it may be more accurate to call it "intergradational," after the definition of Hubbell (1956).

The theoretical work on gene flow, which lies at the base of observations on clinal variation, has been summarized by Slatkin (1985, 1987), who has given consideration to methods as well as concept. He aptly points out that gene flow opposes population differentiation on the one hand, while spreading new genetic combinations through a population on the other. Other than directly observing migrating individuals, he recognizes two indirect methods of estimating gene flow, which are dependent on obtaining gene frequencies, easily done using electrophoretic techniques. The first is Slatkin's own "frequency of rare alleles" method, which assumes that rare alleles in a single population are a function of *Nm*, the average rate of immigration.

The second method is Wright's (1951) F_{ST} statistic

$$F_{ST} = \sigma^2 p \,/\, p(1-p)$$

which estimates the standardized variance (σ^2_{ST}) in allele frequencies where p is one frequency for a two-allele case among local populations. The statistic is an estimator of population size as well as of *Nm*. It assumes the existence of subdivided populations that are equally distant from each other. Both methods appear to estimate the same magnitude of effects and are robust estimators when population structure, selection, and mutation are considered.

Clines, thought to be indicative of gene flow, were among the first descriptions of protein gene-environment relationships to be found in insects (*Drosophila*: O'Brien and MacIntyre, 1969; Prakash et al., 1969) about the same time that it was demonstrated in humans by Bajatzadeh and Walter (1968) and Harris (1969) for allele frequencies at hemoglobin and G-6-PDH. The literature is now replete with such descriptions, most observed using *Drosophila* (Table 2.3). Virtually every study done utilizing geographically disparate populations has found one or more allozyme loci showing this relationship (Table 2.3). The recent survey of Singh and Rhomberg (1987) is exemplary; of 114 loci analyzed, these workers found that 25 of 61 polymorphic allozyme systems in *D. melanogaster* varied with a clinal nature. Eighteen loci showed clinal variation consistent across several continents.

The value of the F_{ST} calculations become apparent in the above study. Singh and Rhomberg (1987) found that 66% of all loci fell with a bell-shaped distribution of F_{ST} values at 8–10%, suggesting extensive gene flow between populations. Further, a high value was indicative of a clinally varying locus.

Still, it is especially difficult to demonstrate this type of relationship as meaningful to adaptation, because there are factors that might give rise to a genetic cline independent of direct gene-environment interaction. For example, the original idea behind the concept of a cline was that a population at one extreme of a range might be fielding migrants to other points of the range and that any genetic effect was due simply to movement of genetically different individuals [this concept was the basis of studies aimed at determining the origin of the migratory corn aphid *Rhopalosiphum maidis* arriving in Illinois in the spring (Steiner et al., 1985; Voegtlin et al., 1987) and by Pashley (1985) for migration of three noctuid moths]. Or, the allelic gene in question might not be directly involved in any interaction but might be "hitchhiking," being dragged along simply because it is closely linked to, and on the same chromosome with, an unidentified gene or chromosome aberration that is interacting with some environmental variable. In this regard, it is encouraging to note that studies in *Drosophila* support the idea that allozymes closely linked or even trapped within chromosome inversions do not necessarily mimic changes in inversion frequencies with latitude (Knibb, 1983; Mukai and Voelker, 1977; Zouros and Krimbas, 1973; Singh and Rhomberg, 1987). In fact, these allozymes often show linkage disequilibrium to each other with respect to gametic frequencies, indicating that nonrandom genetic associations exist between genes, possibly as coadapted gene complexes (Zouros and Johnson, 1976; Alahiotis and Pelecanos, 1978; Mukai et al., 1974; Zouros and Krimbas, 1973).

There are theoretical cases in which a cline may be the result of selection, however, indicated by strong differences between F_{ST} values for different loci. Assume, for example, that two populations exist, separated on a North–South axis by several hundred miles. Samples in between show a gradual genetic change with a frequency of allele *"n"* at some locus *"L"*, going from 0.90 in the North to 0.10 in the South, where it is replaced by allele *"s"* [as shown by Johnson (1976) for α-GPDH in *Colias meadii*, we might even find such a cline over short distances and associated with elevation). For simplification, let us further assume that the locus is not linked to any chromosomal mutation.

At least four explanations can be seen to fit the observations. Going from South to

Table 2.3. Allozyme Variation Demonstrating Clines in Gene Frequencies in Different Species of Insects, Mostly *Drosophila*

Locus	Species	Continent	Reference
AAT[a]	*De. ponderosae*	NA	Stock and Guenther, 1979
AMY	*Po. badius*	NA	Tomaszewski et al., 1973
ACPH	*D. melanogaster*	NA, E, ASA	Oakeshott et al., 1983a
ADH	*D. melanogaster*	NA	Pipkin et al., 1973
			Vigue and Johnson, 1973
		AF	David, 1982
		NA, E, ASA	Oakeshott et al., 1982
APH	*D. obscura*	E	Lakovaara and Saura, 1971
	D. pavani	NA	Kojima et al., 1972
CAT	*Pl. tenuicollis*	NA	Liebherr, 1986b
EST	*D. melanogaster*[b]	NA	Johnson and Schaffer, 1973
		NA, E, ASA	Anderson and Oakeshott, 1984
			Oakeshott et al., 1981
	D. simulans	Greece	Triantaphyllidis et al., 1982
	D. paulistorum	SA	Richmond, 1972
	D. ananassae	SPI	Johnson, 1971
	D. pavani	NA	Kojima et al., 1972
	D. obscura	E	Lakovaara and Saura, 1971
GAM	*Pl. tenuicollis*	NA	Liebherr, 1986
α-GPDH	*D. simulans*	Greece	Triantaphyllidis et al., 1982
G-6-PDH	*D. melanogaster*	NA, E, ASA	Oakeshott et al., 1983b
G-3-PDH	*D. melanogaster*	NA, ASA	Oakeshott et al., 1984
		E	Oakeshott et al., 1982
LAP	*D. subobscura*	E, AF	Pinsker and Sperlich, 1979
MDH	*Pl. tenuicollis*	NA	Liebherr, 1986b
	D. subobscura	E, AF	Pinsker and Sperlich, 1979
MPI	*Pl. tenuicollis*	NA	Liebherr, 1986b
ODH	*D. melanogaster*	NA	Johnson and Schaffer, 1973
		NA, E, ASA	Oakeshott et al., 1983a
PGK	*Pl. tenuicollis*	NA	Liebherr, 1986b
PGM	*D. pavani*	NA	Kojima et al., 1972
6-PGDH	*D. melanogaster*	AN, E, ASA	Oakeshott et al., 1983b
	D. subobscura	E, AF	Pinsker and Sperlich, 1979
DIP	*Pl. tenuicollis*	AN	Liebherr, 1986b
TPI	*D. melanogaster*	ASA	Oakeshott et al., 1984

Abbreviations: NA = North America, AF = Africa, E = Europe, ASA = Australasian Bioregion, A = Australia only, SA = South America, D. = *Drosophila*, *Pl.* = *Platynus*, *De.* = *Dendroctonus*, *Po.* = *Pogonomyrmex*.

[a]Aspartate aminotransferase.

[b]See Girard et al. (1977), who failed to find clinal-related variation in seven European populations.

North, case 1 might involve unequal migratory ability, where one genotype (say, "*nn*") is better able to make the trip due to a metabolic system selected for long-distance travel. There need be no actual loss of individuals from the extended gene pool except perhaps at the margins. Case 2 would exist when selection occurred en route and/or at the point of arrival, affecting one genotype (assume "*ss*") more strongly than the other (assume "*nn*"). In this case, we expect some environmental or biotic factor to be the root cause, its increase being negatively correlated with the decrease and possible demise of genotype "*ss*." Case 3 might be a specialized case in which we see a lack of intermediate populations (no apparent migration) and that isolation has allowed genetic differences to accumulate differentially within the northern versus southern gene pool. Here, some genetic drift, as well as selection, may be driving the changes. Finally, in case 4, the northern population may actually be a sibling (cryptic) species, reproductively isolated from the southern population but allowing sterile hybrids to be formed and distributed along the cline. In this case, selection is assumed a priori as a historical event.

Besides the complication of determining which is the founding population, in each case, specific tests can be devised to determine which is applicable to the observations. In case 1, the approach takes a biochemical nature and is exemplified by the excellent studies of Watt and his colleagues for PGI in *Colias* butterflies (Watt, 1977, 1983, 1985; Watt et al., 1983; Kingsolver and Watt, 1983). They noticed that significant heterozygote excesses occurred that increased with season at PGI in different species of *Colias* (Watt 1977). The excesses appeared for the most part to be independent of any Wahlund effect (population substructuring). These workers studied the biochemical nature of each allele at this locus. Of the four major alleles found in *C. eurytheme*, electrophoretically faster migrating allozymes were more heat resistant. In addition, within the 10 genotypes resulting from these four alleles, substrate binding and coupling effectiveness were a function of genotype and temperature. Studies were then made to test the assumptions and to fit predictive mathematical models (Watt, 1983; Kingsolver and Watt, 1983), suggesting that heterozygotes were better fliers in nature under a wide variety of conditions. Additional findings were that "cool" genotypic combinations would begin flying earlier in the day; kinetic differences in PGI genotypes could be reflected as heterozygote advantage for flight initiation, breadth of flight time, and/or overall flight density through the day; and, under high temperature stress, the usual survivorship advantage of kinetically favored genotypes is reversed, with the most thermally stable genotypes showing better survivorship. Recently, studies of the different PGI genotypes have been extended to include their role in mating success (Watt et al., 1986).

Case 2 is determined by showing that fitness can be lowered with respect to an environmental or biotic factor. One should see bimodal tendencies in the F_{ST} values as well. Cases 1, 3, and 4 can be differentiated from case 2 and each other because of differences in disequilibrium gene-frequency patterns as specific genotypes become lost due to selection. Thus, in case 1, we might expect genotypes to seek their most adaptive conditions but to remain in the extant population; in this case, one must be careful to design representative sampling techniques in order to determine gene frequencies. In cases 3 and 4, heterozygote deficiencies should be found, reflected to some degree in the calculation of F_{ST} values, which would be strongly skewed toward the high side. We can differentiate between the two quite easily if hybrids of the northern *x* southern populations have reduced fitness and fertility.

In summary, the demonstration of a cline in gene or phenotype frequencies may indicate that selection is present in cases in which bimodal or skewed tendencies occur in relative F_{ST} values and/or genotypes are not in Hardy-Weinberg equilibrium. These values do not tell us per se what the gene-enzyme relationship is but may give us a clue as to what varying environmental parameter over the species distribution may be important for further study. It may be possible to use protein genes to trace the movement and migration of insects of economic importance.

Gene-Environment Correlations

Two recent reviews sufficiently cover a portion of this topic for insects, mostly *Drosophila* (Hedrick, 1986; McDonald, 1983). A third review by Nevo et al. (1983) covers marine invertebrates. Here, I shall summarize some of the data that support the idea that allozyme polymorphisms in insects are adaptive in nature. In the process, it will become evident that genetic or reproductive fitness must be a considered component in any approach that attempts to identify a gene-environment relationship.

Here, most of the studies have used *Drosophila* as the experimental organism. These species are superb to study for the traditional reasons of short generation time, ease of colony maintenance, and sheer quantity of background information. The effects of environmental perturbation on population density and fitness can be easily studied (Dominguez and Albornoz, 1987; Verdonck, 1987; Siddiqui and Barlow, 1972) and can be easily extended to strains differing only in allozymic content.

Powell (1971) was the first to use this approach in cage populations, varying media and temperature while simultaneously monitoring an array of polymorphic enzyme systems. His demonstration of correlated responses raised the question of whether the changes were due to individual gene responses to environmental vicissitudes or were a response based on epistatic relationships, whole organism in nature (i.e., whether the gene or the organism was the unit of selection). The question is still unanswered today, and it has sparked much study of individual genes. These latter studies primarily take the form of characterizing the alleles at a locus for their binding affinities, speed of reaction, pH optima, temperature sensitivity, and so on. In turn, these characteristics may then be related to natural variation in environmental components deemed important to the locus in question (i.e., alcohol levels in fermenting fruit for the ADH locus of *D. melanogaster* larvae). For the most part, these studies have been one- or two-dimensional; that is, they have left out biologically meaningful components, usually fitness related, that might give insight into how gene frequencies and population density are manipulated by selection. Thus, temperature effects on biochemical parameters might be studied, but survival rates of different genotypes are not.

The more relevant studies are summarized in Table 2.4, which is not meant to be comprehensive. These clearly fall into several categories. First attempts at studying

Table 2.4. Relationships Reported Between Specific Allozyme Polymorphisms and Abiotic, Biotic, and Fitness Parameters for Insects (All correlations are significant.)

Species[a]	Correlation	Reference
D. melanogaster	ADH allozyme Km values differ; fast electrophoretic alleles are more cold tolerant.	Alahiotis, 1982
	Study of ADH allozymes in cage and natural populations show fast allele increases in frequency in fall populations.	Gilbert and Richmond, 1982a
	ADH allozymes in isofemales correlate with media type.	Oakeshott et al., 1983b
	ADH genotypes have relative fitness differences when grown on different alcohols.	Morgan, 1975
	ADH genotypes differ in metabolism rates of ethanol and a long-chain alcohol.	Heinstra et al., 1987
	Preference for ethanol environment associated with ADH polymorphism.	Cavener, 1979
	ADH tracks environmental diversity in six replicated population cages.	Mercot, 1985
	Ethanol selection results in better ability to utilize ethanol; ADH fast genotype has higher relative fitness.	Dorado and Barbancho, 1984
	α-GPDH allozymes differ in temperature dependence, Km, and reaction rate constancy.	Miller et al., 1975
	α-GPDH genotypes associated with seasonal change.	Berger, 1971
	TPI genotypes associated with total monthly rainfall; α-GPDH associated with mean monthly temperature.	Nielsen et al., 1985
D. melanogaster	AMY allozymes differ in Vmax and Km, suggesting that temperature-dependent selective advantages may exist.	Hoorn and Scharloo, 1978
	AMY allozyme with highest Km has higher selection coefficient, with increasing starch levels of larval food.	DeJong and Scharloo, 1976
	EST-6 active genotypes copulate sooner for shorter periods and	Gilbert and Richmond, 1982b

(Continued)

Table 2.4. *(Continued)*

Species[a]	Correlation	Reference
	produce more progeny at lower temperatures when compared with EST-6 null genotypes.	
	EST-6 catalyzes synthesis of a sex pheromone in females after being passed in the ejaculate by the male.	Mane et al., 1983
	PGM alleles differ in kinetic and stability parameters.	Fucci et al., 1979
	At PGM, various metabolites have different effects on activities of two allozymes.	Pontecorvo et al., 1986
	PGM and PGI have a strong reciprocal relationship in heterozygosity values in many *Drosophila*, other insects, and other organisms.	Gauldie, 1984
	Selection experiments show that PGM converges to the same set of equilibrium frequencies in six replicates; relative fitness differences exist between genotypes.	Carfagna et al., 1980
	G-6-PDH responds to high sucrose selection, and 6-PGDH responds to ethanol selection.	Cavener, 1984
	G-6-PDH genotypes differ in relative fitness, with heterozygote fitness increasing with population density.	Giesel, 1976
	ACHE associated with malathion resistance.	Morton and Singh, 1982
	ACHE and EST-6 have parallel changes associated with rise in resistance during selection with organophosphorus.	Salam and Pinsker, 1981
D. pavani	EST-6 genotype frequencies and G-6-PD genotype frequencies associated with seasonal change.	Kojima et al., 1972
D. mimica	IDH, PGM, and EST-2 all show variation correlated to environmental fluctuations and rainfall over 20-month period.	Steiner, 1979a
	In samples from natural populations of *D. mimica*, variation at PGM, ACPH, APH, and EST is correlated to survival time at 0% relative humidity; desiccation resistance is highly heritable (midparent, and	Steiner, 1974

Table 2.4. ***(Continued)***

Species[a]	Correlation	Reference
	daughters on mothers estimates range from 53 to 84); in closed test systems that prevent escape, the more heterozygous the individual fly is across seven polymorphic loci when tested for resistance, the faster the fly desiccates; desiccation resistance increases during the dry season, and polymorphism at PGM and ACPH also track dry season changes; these results are consistent across two isolated populations and also occur in the following, sympatric species.	
D. engyochracea	PGM variation correlated to rainfall; populations sympatric with *D. mimica* above and sampled over 18-month period overlapping with this species.	Steiner, 1979b
	Similar results for desiccation resistance obtained as in *D. mimica*.	Steiner, 1974
D. pachea	Variation at four allozymes correlate with non-saponifiable product in cactus host plant.	Rockwood-Sluss et al., 1973
D. persimilis	PGM and ME-2 variation associated with seasonal change.	Dobzhansky and Ayala, 1973
D. pseudoobscura	PGM and ME-2 variation associated with seasonal change.	Dobzhansky and Ayala, 1973
	α-GPDH, ADH, and ACPH genotypes differ in viability under competition conditions.	Seradilla and Ayala, 1983
	MDH and PGM genotypes differ in egg-to-adult survival under differences in temperature and competitive conditions.	Snyder and Ayala, 1979
	EST, ODH, and MDH genotypes differ in female fecundity, male mating capacity, egg-to-adult survival, and rate of development under competitive conditions.	Marinkovic and Ayala, 1975a
	PGM and ME-2 genotypes differ for seven fitness components such that no single genotype is best; average	Marinkovic and Ayala, 1975b

(Continued)

Table 2.4. *(Continued)*

Species[a]	Correlation	Reference
	egg-to-adult survival and rate of development are better when different genotypes are reared together.	
D. buzzati	A naturally isolated population polymorphic for EST, PYR, and ADH was repeatedly perturbed by addition of flies to alter the equilibrium gene frequencies; population returned to a preexisting equilibrium state.	Barker and East, 1980
H. zea	EST allozymes correlated with principle component made up of latitude, longitude, temperature, and precipitation.	Sell at al., 1975
A. citricola	Fixed genetic differences occur that are host–plant dependent.	Steiner et al., 1985
My. persicae	EST allozymes correlated with plant host type.	Baker, 1979
	EST allozymes correlated with insecticide usage.	Brooks and Loxdale, 1987
M. euphorbiae	Fixed genetic differences occur that are host–plant dependent.	Steiner et al., 1985
M. rosae	EST allozymes correlate with seasonal change in nine populations.	Rhomberg et al., 1985
Colias butterflies	PGI phenotypes differ in fitness-related properties for flight capacity, survivorship, and mating success.	Watt et al., 1985
	Older male PGI phenotypes with above advantages mate more often with older females.	Watt et al., 1986
Coch. hominivorax	Reduction in α-GPDH activity reduces fitness by lowering flight ability.	Bush et al., 1976
De. ponderosae	LAP, PGI, EST, AAT, and PeP show host-dependent genetic differentiation.	Stock and Amman, 1980
Po. barbatus	EST correlated with weather patterns.	Johnson et al., 1969

[a] *D.* = *Drosophila*, *M.* = *Macrosiphum*, *My.* = *Myzus*, *H.* = *Heliothis*, *A.* = *Aphis*, *Coch.* = *Cochliomyia*, *De.* = *Dendroctonus*, *Po.* = *Pogonomyrmex*.

genetic polymorphism in the ecological context demonstrated correlations of genetic polymorphisms to specific environmental parameters. For example, Rockwood-Sluss et al. (1973) found that, although the gene frequencies across the range of distribution for *D. pachea* (Patterson and Wheeler) in the Southwest United States were remarkably homogenous, patterns of allele frequencies were significantly correlated to nonsaponifiable components of the host plant breeding site but not to alkaloids, fatty acids, or climatic variation (with the possible exception of an esterase with temperature differences). This suggests that the homogeneity is due to a nutrient source that does not vary across the range.

Other studies that relate genetic variation to resource-related environmental components are listed in Table 2.4. A good example concerns the widely distributed genetic polymorphism at the ADH locus of *D. melanogaster*, involving many laboratories and a variety of approaches. These laboratories found that different genotypes display (1) effects due to differential metabolic rates, depending on substrate type (Heinstra et al., 1987; Singh, 1976; Morgan, 1975; Oakeshott, 1976); (2) effects due to pH change and temperature (Chambers, 1984; Sampsell and Sims, 1982; Sangiorgi et al., 1981; Oakeshott, 1976); feeding (Starmer and Barker, 1986), and ethanol tolerance (McKenzie and McKechnie, 1978; Cavener, 1979; David et al., 1986); (3) effects due to ethanol stress (Dorado and Barbancho, 1984; McElfresh and McDonald, 1983); and (4) effects due to availability of the oxido-reductant pool for metabolism (McElfresh and McDonald, 1983). However, the overall picture relating resource components of the environment to genetic variation is complicated by effects due to initial genetic composition of the gene pool (Sangiorgi et al., 1981), genetic background (Barnes and Birley, 1978; McKechnie and Geer, 1986; Alonso-Moraga et al., 1985), and behavioral effects due to differential responses in mating (Kojima and Tobari, 1969).

Physiological Function and Fitness

Most interesting are those studies that suggest that fitness differences can be found between allelic types (Table 2.4). Sometimes, these can be related to biotic or abiotic components in the environment. Seager and Anderson's (1987) paper on viabilities of different amylase genotypes on different substrates at different temperatures is a good example. The series of studies on fitness components of eight polymorphic allozyme loci in *D. pseudoobscura* (Frolova) demonstrates that, at virtually any locus one examines in a representative insect, some fitness parameter affecting reproduction can be found (Table 2.4; Dobzhansky and Ayala, 1973; Seradilla and Ayala, 1983; Snyder and Ayala, 1979; Marinkovic and Ayala, 1975a, 1975b). The parameters studied included viability, rate of development, fecundity, male mating capacity, egg-to-adult survival, and egg hatchability. Comparisons were made between genotypes at individual loci under varying conditions of competition and temperature.

The role of enzyme polymorphism in metabolic processes is difficult to define. It can be studied, because many of the enzymes are well characterized with respect to their biochemical profiles. The significance of many isoenzyme polymorphisms lies in energy production, central to an organism's function and survival. Recent studies shed some light on the problem. Clark and Keith (1988) have demonstrated that significant differences can occur in enzyme activities and amounts of stored metabolic products (such as triglycerides and carbohydrates) between *Drosophila* lines. They found significant genetic correlations for a number of enzyme pairs and raised interesting questions concerning how these relate to fitness, metabolic flux within biochemical pathways, and physiology. Other studies in Table 2.4 suggest that the biochemical characteristics associated with differing genotypes or alleles do have biochemical and metabolic ramifications.

No beneficial or pest insect has been genetically studied in the same depth as *Drosophila*. Table 2.4 and the above discussion demonstrate the type of work that lies ahead for researchers who are interested in insects of agricultural importance. The proper approach, after first demonstrating the genetic basis for the polymorphism, is to begin studying how it relates to physiology, survival, reproduction, behavior, internal metabolite pools, stress, environment, and so on. In some cases, educated guesses can be made about what the important biotic or abiotic parameter might be. For example, for ADH, this is temperature and alcohol (Table 2.4). For AMY, it is temperature and starch type (Seager and Anderson, 1987; Hoorn and Scharloo, 1978; DeJong and Scharloo, 1976).

The relationship to fitness is an important one. It is here that the applied biologist may find a use for potential differences underlying polymorphic allozyme systems. In an era in which biological control of insects is gaining credibility against the known harmful effects of most insecticides, it becomes critical to understand the physiological basis underlying fitness responses. Insight can be gained concerning biochemical manipulation of the insect genome. The following section shows how critical such an approach might be. In it, I suggest how we should design our experiments to understand genotype-environment interactions lying at the base of ecological genetics and to ground future syntheses solidly in biochemistry.

THE BIOCHEMICAL NATURE OF ADAPTATION

There are two attributes of life that are of primary significance. First, life forms organize and produce energy. Second, as part of an ecological hierarchy, these forms consume energy. It is these attributes that make many of the enzyme polymorphisms that we see especially important, for many of these enzymes are embedded in metabolic pathways that mobilize energy production. Relative positions of some of these key enzymes are depicted in a simplified set of pathways in Figure 2.1 (after White et al., 1968; Passmore and Robson, 1969; and Harper, 1969). Complete details on structure, function, mode of action, interactions, and so forth, for the enzymes of interest are common scientific knowledge (White et al., 1968; Harper, 1969). Details concerning insect biochemistry, such as oxidative phosphorylation (Shu-sen et al., 1973), pentose shunt-lipid synthesis relationships (Geer et al., 1979), role of dietary ethanol in lipid synthesis (Geer et al., 1985), and genetic control of carbohydrate and lipid metabolism (Lucchesi et al., 1979), are available. A host of studies now exist that examine relationships between energy production and genes. For example, Clark and Keith (1988) examined the importance of genetic background on stored triacylglycerol and carbohydrate. Teague et al. (1986) studied the effect of developmental mutants on lipid production. Laurie-Ahlberg et al. (1985) examined the influence of genetic variation on power output, while Cavener and Clegg (1981)

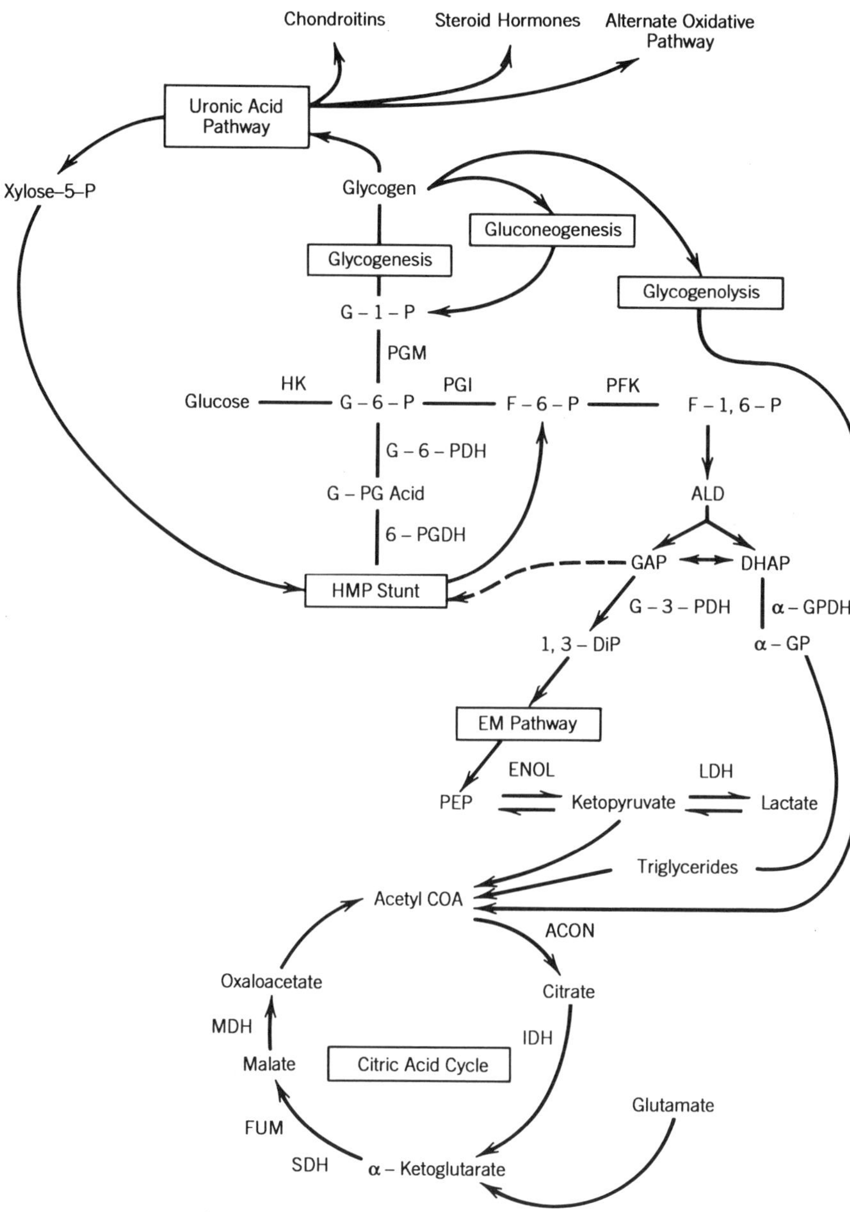

Figure 2.1. Metabolic flow chart demonstrating approximate energy pathway relationships, and their enzymes which demonstrate genetic polymorphism.

found evidence of interactive relationships between enzyme genotypes and physiological function. Hadley's (1985) review of the adaptive significance of lipids offers serious insight as well. Bioenergetics, the study of energy flow through living systems, must become a central part of evolutionary study if we are to understand adaptive processes properly (Watt, 1985).

If enzymes of metabolic pathways have some evolutionary significance with respect to adaptation, and if energy considerations lie at the heart of the matter, then various taxa may be expected to have genetic variations that might reflect different requirements, depending on the ecological niche they occupy. This is because, for some organisms, glycogen storage may be an important strategy to meet environmental vicissitudes, while, for another group, lipids might be more important. Still another group may need immediate energy at all times (e.g., compare the different predatory behaviors of the ladybug beetle with a constantly flying dragonfly or the migratory monarch butterfly with the overwintering bagworm).

Table 2.5 comprises a brief literature survey that attempts to compare taxa for energetically important polymorphic enzymes. Although it is not comprehensive, it appears that some patterns are apparent. For example, α-GPDH is highly polymorphic in *Culex* but not in the two other main mosquito groups or in the remaining Diptera. PGM is highly variable in all insects listed, with the exception of the Homoptera (eusocial wasps are too few to quantify). IDH is not polymorphic in the presocial Hymenoptera, or *Eurylophella* (Ephemeroptera).

There are at least two explanations for these kinds of patterns. On the one hand, evolutionary constraints at the level of an insect order may limit which enzymes will vary in a particular taxon (e.g., 6-PGDH for *Eurylophella*, as compared with Coleoptera). It is difficult, however, to envision what these constraints may be, unless they deal with positioning and site recognition of biochemicals that increase mutation rates or those that affect crossover rates. Here, the alternative explanation is favored; differences like the 6-PGDH example may reflect different biochemical strategies in meeting environmental and biotic requirements.

In Table 2.6, the data are examined in another way; the question I investigate is whether different metabolic pathways have differences in overall genetic variation that might provide insight into the relationship. It appears that the Emden-Meyerhof (EM) pathway enzymes have lower genetic variations than the citric acid cycle enzymes or those of the hexose-monophosphate (HMP) shunt. Questions like these are of interest because of the observations (e.g., Band, 1975) demonstrating the increase in variation of glucose-metabolizing enzymes that occur in natural populations as one progresses away from the equator. The preceding relationships may be significant if the increase in heterozygosity involves enzymes of importance to lipid storage rather than ATP generation. Identifying metabolic pathway components takes us one step closer to describing in detail the nature of the adaptive response we may be observing.

How does one relate biochemistry, genetic variation, physiology, and environmental variation? How does behavior affect these interactions? One approach is that of Watt and his colleagues (Watt, 1977, 1983; Watt et al., 1983, Watt et al., 1986)

Table 2.5. Energy-related Proteins and Their Relative Average Expected Heterozygosity Values for Selected Insect Orders and Families (Number of species analyzed is in parentheses. The literature review was not comprehensive. See text for details.)

	Taxa						
	Culicidae			Other		Hymenoptera	
Locus	Anopheles	Culex	Aedes	Drosophila	Diptera	Presocial	Eusocial
ALD	0.250 (2)	0 (1)	——	0.050 (9)	0 (2)	0.143 (14)	0 (1)
FUM	——	——	0.021 (6)	0.040 (7)	0.087 (2)	0 (14)	0.020 (1)
G-6-PDH	0.020 (2)	0 (1)	0 (1)	0.110 (8)	——	——	0 (1)
G-3-PDH	0.032 (1)	0.102 (1)	——	0.045 (4)	0 (2)	0 (14)	0 (1)
GOT	0.298 (6)	0.408 (1)	0.252 (3)	0.043 (13)	0.231 (5)	0.921 (14)	——
α-GPDH	0.005 (12)	0.198 (7)	0 (19)	0.010 (24)	0.079 (5)	——	0.043 (1)
IDH	0.294 (6)	0.407 (3)	0.163 (14)	0.134 (23)	0.181 (6)	0 (14)	0.058 (1)
LDH	0 (6)	0 (6)	0 (1)	——	0.065 (3)	0.376 (14)	0.116 (1)
MDH	0.167 (6)	0.352 (7)	0.233 (20)	0.010 (15)	0.040 (5)	0.139 (14)	0 (1)
PGI	0.064 (13)	0.301 (6)	0.059 (9)	0.250 (5)	0.273 (7)	0.529 (14)	0 (1)
PGM	0.288 (16)	0.441 (3)	0.313 (20)	0.183 (11)	0.249 (4)	0.314 (14)	0.010 (3)
6-PGDH	0.010 (12)	0.365 (7)	0.071 (19)	0.030 (6)	0.391 (3)	——	0 (1)

Locus	Coleoptera	Eurylophella	Homoptera	Lepidoptera
ALD	——	0.643 (15)	0.144 (4)	0.048 (24)
FUM	——	——	——	0.030 (25)
G-6-PDH	——	——	0.222 (2)	0.390 (4)
G-3-PDH	0.412 (6)	0.108 (15)	0.241 (2)	0.022 (24)
GOT	0.776 (1)	0.144 (15)	0.191 (6)	0.334 (27)
α-GPDH	——	0 (15)	0.100 (9)	0.130 (27)
IDH	0.451 (16)	0 (15)	0.193 (6)	0.215 (29)
LDH	——	——	——	0.027 (26)
MDH	0.565 (12)	0.543 (15)	0.033 (14)	0.089 (25)
PGI	0.502 (16)	0.889 (15)	0 (3)	0.326 (29)
PGM	0.717 (6)	0.126 (15)	0.103 (7)	0.422 (29)
6-PGDH	0.677 (6)	0 (15)	0.177 (6)	0.455 (2)

References for this compilation are listed in Appendix I.

Enzyme abbreviations in order (DH = dehydrogenase, I = isomerase): Aldolase (ALD), fumarase (FUM), glucose-6-phosphate DH (G-6-PDH), glyceraldehyde-3-phosphate DH (G-3-GPDH), glutamate oxaloacetate transaminase (GOT), α-glycerophosphate DH (α-GPDH), isocitrate DH (IDH), lactate DH (LDH), malate DH (MDH), phosphoglucose I (PGI), phosphoglucomutase (PGM), and 6-phosphogluconate DH (6-PGDH). Dotted line indicates that no information is available.

Estimates for other Diptera include species of the genera *Glossina, Musca, Rhagoletis, Anastrepha, Stomoxys, Plecia, Tabanus, Cochliomyia, Mayetiola,* and *Haematobia.*

Table 2.6. Energetic Significance of Enzymes of Metabolic Pathways That Are Found to Be Allozymic (Polymorphic) in Some Insect Species (Glucose metabolism by hexokinase lies at the base of all pathways and with fatty acids is estimated to provide 85% of energy intake from all diets).[a]

Metabolic Pathway	Significance	Allozymes	He[b]
Citric acid cycle[c]	1. Generates 24 ATP.	IDH	0.178
	2. Provides malate, citrate, aspartate, α-ketoglutarate, and glutamate for extramitochondrial transfer toward oxalocetate synthesis in gluconeogenesis.	ACON FUM SDH ME, AAT	NA 0.033 NA NA 0.358
	3. Malate shuttle.	MDH	0.197
HMP shunt (Direct oxidative pathway)	1. Generates 36 ATP.	G-6-PDH	0.106
	2. Most of energy used in fatty acid and steroid production.	6-PGDH	0.218
	3. Provides pentoses for nucleotide and nucleic acid syntheses.		
Emden-Meyerhof[c] (Glycolysis or pentose phosphate pathway)	1. Generates 12 ATP.	PGI	0.290
	2. Generates pyruvate for citric acid cycle and other pathways.	ALD G-3-PDH ENOLASE	0.142 0.096 NA
	3. Lactic acid shuttle.	LDH	0.083
	4. α-gpdh shuttle.	α-GPDH	0.071
	5. Phosphoenolpyruvate catabolism.	Py. KINASE	NA
Glycogenesis	1. Provides glycogen as a high-energy storage compound.	PGM	0.288
	2. Provides products for uronic acid pathway important for hormone production.	Not studied.	
Glycogenolysis	Provides fatty acids for glycolysis and the citric acid cycle (glucose, pyruvate, lactate).	Not studied, but α-GPDH is important.	

[a]Summarized from Harper (1969), Passmore and Robson (1969), and White et al. (1968).
[b]Expected heterozygosity (HE), summarized across all species in Table 2.2; NA = not analyzed here.
[c]Dehydrogenations plus phosphorylations at substrate levels account for 42% of the available free energy from combustion of glucose.

discussed previously. Here, beginning with field observations, the story is built gradually, moving from ecological study through genetics to biochemistry. A similar approach has been taken by Powers and Powers (1975) for LDH-B in the fish *Fundulus heteroclitus* (Linnaeus). What is unique about the latter study, however, is the development of an equation

$$q = E^n / E^n + E^n$$

where

q = the equilibrium gene frequency in a population
E = some environmental variable affecting the frequency
n = a system constant

In this research, E related to environmental temperature, to which q was sensitive. The equation is important, because it was used to predict gene frequencies in 16 populations, which fit very well with the observed values (Powers and Powers, 1975). The mathematical correlation between genetic type and an environmental feature is satisfyingly testable in this situation.

Another approach is that of Steiner (1974), in which physiology, genetics, and variable environmental parameters were studied simultaneously over time. In this case, the biochemical perspective was not included. However, the results are informative, for they emphasize the adaptive significance of a physiological process.

The approach in the latter case was to monitor two sympatrically occurring species of Hawaiian *Drosophila*, located on the island of Hawaii as well-defined and naturally isolated populations. These were sampled for 100+ individuals every 6 weeks (about the length of time for a generational turnover) for 20 months (*D. engyochracea* Hardy) and 18 months (*D. mimica* Hardy). The samples were returned to the laboratory, where each individual was tested for its ability to survive at 0% relative humidity and then was frozen for electrophoretic analysis of 20 allozyme systems. Simultaneously, weather data for six climatic parameters were collected on a daily basis. Additional information was available on breeding ecology, relative population densities, and so forth. Heritability of desiccation resistance was estimated and was equivalent to that found by Parsons (1970) for *D. melanogaster* Meigen.

The genetic statistics for these species have been previously published (Steiner, 1979a, b) and have included correlations to a reduced set of the weather variables. The physiological statistics for one species, *D. mimica*, are presented in Table 2.7 (phenotypic contributions at individual enzyme loci to variability in survival) and in Table 2.8 (mean survival, in minutes, of phenotypes contributing significantly to survival) after stepwise multiple regression analyses. In both populations for both species, the ability to resist desiccation tracked pertinent moisture variables in the environment; the most resistant individuals and the increased mean survival rate for each population and species occurred during periods of low humidity and rainfall. Physiologically, the ability to resist desiccation appeared adaptive.

Interestingly, heterozygotes contribute to variation in survival at 0% relative humidity more so than homozygotes (Table 2.7). For example, at PGM-1, about 1% of the variance in survival time can be accounted for by all six phenotypes, and, of these, individuals heterozygous for a fast electrophoretic allele and the most common allele (phenotype PGM-1^{34}) account for almost 75% of this variance. Not all loci contribute significantly, but, across the seven polymorphic loci in this study,

Table 2.7. Phenotypic Contributions at Individual Enzyme Loci to the Variation in Survival Time at 0% Relative Humidity for *D. mimica* (from Steiner, 1974)

Locus	Number	Number of Alleles	Source	R^2	Phenotype Frequency	Correlation
PGM-1	809	3	All 6 phenotypes	0.007		
			Phenotype 34*	0.005	0.074	0.073
HK-3	315	3	All 6 phenotypes	0.018		
			Phenotype 33	0.008	0.010	−0.096
IDH-1	758	2	All 3 phenotypes	0.010		
			Phenotype 45		0.198	−0.034
ACPH-1	1049	3	All 6 phenotypes	0.009		
			Phenotype 35**	0.005	0.016	0.071
APH-3	386	6	All 21 phenotypes	0.043		
			Phenotype 56**	0.018	0.075	−0.135
EST-2	756	6	13 of 21 phenotypes	0.035		
			Phenotype 24**	0.014	0.024	−0.117
LAP-1	491	4	All 10 phenotypes	0.011		
			Phenotype 24	0.005	0.012	0.068

*P < 0.05, partial regression coefficient.
**P < 0.01, partial regression coefficient.
R^2 Coefficient of determination.

13.3% of the variance in survival time is accounted for. This seems very high when compared with the thousands of genes an organism carries, but it may be expected if energy metabolism is central to survival, as I postulate. The question of how these phenotypes relate to desiccation resistance in the functional sense remains to be answered, however.

The discovery that heterozygosity at energy-metabolizing enzymes might be important prompted examination of individuals heterozygous at one, two, or three glucose-metabolizing loci. It was found that the more heterozygous the individual was, the less likely was the individual to survive (Table 2.9). This discovery was at odds with the finding that heterozygote frequency increased during times of low environ-

Table 2.8. Mean Survival (Minutes) of Significant Phenotypes Contributing to Survival at 0% Relative Humidity in *D. mimica* (from Steiner, 1974)

Locus	Phenotype	Difference from Homozygote Mean	Mean Survival	Significance
PGM-1	34	40.85 ± 20.46	389.11	P < 0.05
IDH-1	45	−13.01 ± 11.91	354.44	P < 0.01
ACPH-1	35	84.74 ± 37.63	452.88	P < 0.01
APH-3	56	82.85 ± 30.92	262.81	P < 0.01
EST-2	24	−124.88 ± 36.31	251.39	P < 0.01

Table 2.9. Linear Least-Squares Estimates of Heterozygote Survival When Compared with the Mean Homozygote Survival for Three Polymorphic Glucose-metabolizing Enzymes at 0% Relative Humidity in *D. mimica* (after Steiner, 1974)

Heterozygote Combination[a]	Mean Survival	Difference from Homozygote Mean	Frequency of Heterozygotes
0,0,0	358.87	0	
1,0,0 (HK-3)	356.71	− 2.16	0.250
0,1,0 (IDH-1)	309.48	− 49.39	0.250
0,0,1 (PGM-1)	309.73	− 49.14	0.103
1,1,0	307.32	− 51.55	
1,0,1	307.57	− 51.30	
0,1,1	260.34	− 98.53	
1,1,1	258.18	− 100.69	

[a]0 = homozygous, 1 = heterozygous.

mental moisture. The explanations offered were twofold. First, under desiccation stress in the laboratory, water generation as a metabolic side product could become important. If heterozygotes were more efficient metabolically, they would desiccate more rapidly. Second, the test conditions prevent a fly from escaping. If heterozygotes were more active in trying to seek an escape route (the tests were done in closed containers), again they would require more energy than homozygotes and would survive for shorter periods under desiccation.

The significance of these approaches is the combination of techniques utilized. The resulting information can be integrated in biologically more interesting ways. In certain instances, the results allow prediction of genotypic occurrence in the natural setting; certain PGI genotypes in the *Colias* will occur at certain periods and altitudes; certain heterozygous combinations in *D. engyochracea* will occur during dry periods in Hawaii; certain LDH genotypes in *Fundulus* will occur under certain temperature regimes in the ocean. They do suffer certain drawbacks; for example, no actual fitness estimates are made in these studies. Any experimental design to examine genetics of an adaptive process should take this into account, for it provides a quantifiable and, therefore, comparative basis to the resulting data.

IMPLICATIONS OF GENETIC POLYMORPHISM FOR INSECT PEST MANAGEMENT

My purpose in describing and summarizing the genetic results of a wide range of genetic studies on a wide range of insects, including *Drosophila*, was to present evidence that genetic polymorphism of enzymes, especially those involved in metabolism, are adaptive. Certainly, strong arguments now exist against the selective neutrality of the more common variants. We are now beginning to ask the right questions and to use better, integrated approaches that should yield definable results concerning the nature of genetic adaptation.

From the applied biologist's viewpoint, establishing interactive relationships between gene, gene product, physiology, and environmental fluctuation has important implications, especially given today's bioengineering techniques, which include the potential for engineering protein molecules. We are approaching the ability to tailor metabolic pathways to fit particular environments. Inherent in this are insect replacement schemes to reduce the range and activity of pest insects or to improve the range and activity of beneficial insects.

Given these potentials, it is important to continue describing the nature of genetic polymorphism in both types of insects and to look for taxa-wide genetic patterns, as well as functional relationships. Given the tools, we can bring better understanding to microevolution as a process; we can also develop potentially new ways to handle the insect problems we are sure to face in the future as resistance to insecticides continues to develop.

Acknowledgments

I would like to thank Dr. K. C. Kim for asking me to contribute to this volume and for his patience during the preparation of this manuscript. Drs. C. Ignoffo, M. Greenstone, I. McDonald, E. Spiess, and M. R. Lee provided many useful comments, but the interpretations herein remain my own. Dr. Robert Willey provided several timely reprints for the review process. Portions of this paper were presented at the 1986 Entomological Society of America National Conference, Reno, NV.

Mention of a proprietory product in this paper does not constitute a recommendation for use by the USDA.

REFERENCES

Alahiotis, S., and M. Pelecanos. 1978. Induction of gene pool differentiation in *Drosophila melanogaster. Can. J. Genet. Cytol.* 20:265–273.

Alahiotis, S. N. 1982. Adaptation of *Drosophila* enzymes to temperature. IV. Natural selection at the alcohol dehydrogenase locus. *Genetica* 59:81–87.

Alonso-Moraga, A., A. Munoz-Serrano, and A. Rodero. 1985. Variation of allozyme frequencies in Spanish field and cellar populations of *D. melanogaster. Genet. Sel. Evol.* 17:435–442.

Anderson, P. R., and J. G. Oakeshott. 1984. Parallel geographic patterns of allozyme variation in two sibling *Drosophila* species. *Nature* 308:729–730.

Avise, J. C., R. A. Lansman, and R. O. Shade. 1979. The use of restriction endonucleases to measure mitochondrial DNA sequence relatedness in natural populations. I. Population structure and evolution in the genus *Peromyscus. Genetics* 92:279–295.

Ayala, F. J. (ed.). 1976. *Molecular Evolution*. Sunderland, Mass.: Sinauer Associates.

Bajatzadeh, M., and H. Walter. 1968. Serumprotein polymorphism in Iran. *Humangenetik* 6:40–54.

Baker, J. P. 1979. Electrophoretic studies on populations of *Myzus persicae* in Scotland from October to December 1976. *Ann. Appl. Biol.* 91:159–164.

Band, H. T. 1975. A survey of isozyme polymorphism in a *Drosophila melanogaster* natural population. *Genetics* 80:761–771.

Band, H. T., and P. T. Ives. 1968. Genetic structure of populations. IV. Summer environmental variables and lethal and semilethal frequencies in a natural population of *D. melanogaster*. *Evolution* 22:633–641.

Barker, J. S. F., and P. D. East. 1980. Evidence for selection following perturbation of allozyme frequencies in a natural population of *Drosophila*. *Nature* 184:166–168.

Barnes, B. W., and A. J. Birley. 1978. Genetic variation enzyme activity in a population of *Drosophila melanogaster*. IV. Analysis of alcohol dehydrogenase activity in chromosome substitution lines. *Heredity* 40:51–57.

Bentz, B. J., and M. W. Stock. 1986. Phenetic and phylogenetic relationships among ten species of *Dendroctonus* bark beetles (Coleoptera: Scotylidae). *Ann. Entomol Soc. Am.* 79:527–534.

Berger, E. M. 1971. A temporal survey of allelic variation in natural and laboratory populations of *Drosophila melanogaster. Genetics* 67:121–130.

Berkelhamer, R. C. 1983. Intraspecific genetic variation and haplodiploidy, eusociality, and polygyny in the Hymenoptera. *Evolution* 37:540–545.

Brooks, C. P., and H. D. Loxdale. 1987. Survey of enzyme variation in British populations of *Myzus persicae* (Sulzer) (Hemiptera: Aphididae) on crops and weed hosts. *Bull. Entomol. Res.* 77:83–89.

Bush, G. L., R. W. Neck, and G. B. Kitto. 1976. Screwworm eradication: Inadvertent selection for noncompetitive ecotypes during mass rearing. *Science* 193:491–493.

Carfagna, M., L. Fucci, L. Gaudio, G. Pontecorvo, and R. Rubino. 1980. Adaptive value of PGM polymorphism in laboratory populations of *Drosophila melanogaster. Genet. Res.* 36:265–276.

Cavener, D. 1979. Preference for ethanol in *Drosophila melanogaster* associated with the alcohol dehydrogenase polymorphism. *Behav. Genet.* 9:359–365.

———. 1984. Response of the *G6pd* and *6Pdg* polymorphisms in *Drosophila melanogaster* to dietary selection. *Genetica* 63:81–83.

Cavener, D., and M. T. Clegg. 1981. Evidence for biochemical and physiological differences between enzyme genotypes of *Drosophila melanogaster. Proc. Natl. Acad. Sci. USA* 78:4444–4447.

Chambers, G. K. 1984. The purification and biochemical properties of alcohol dehydrogenase-"fast (Chateau Douglas)" from *Drosophila melanogaster. Genet.* 22:529–549.

Chianchi, R., S. Maini, and L. Bullini. 1980. Genetic distance between pheromone strains of the European corn borer, *Ostrinia nubilalis*: Different contribution of variable substrate, regulatory, and non-regulatory enzymes. *Heredity* 45:383–388.

Clark, A. G., and L. E. Keith. 1988. Variation among extracted lines of *Drosophila melanogaster* in triacylglycerol and carbohydrate storage. *Genetics* 119:595–607.

David, J. 1982. Latitudinal variability of *Drosophila melanogaster*: Allozyme frequency divergence between European and Afrotropical populations. *Biochem. Genet.* 20:747–761.

David, J.R., H. Mercot, P. Capy, S. F. McEvey, and J. Van Herrewege. 1986. Alcohol tolerance and *Adh* gene frequencies in European and African populations of *Drosophila melanogaster. Genet. Sel. Evol.* 18:405–416.

DeJong, G., and W. Scharloo. 1976. Environmental determination of selective significance or neutrality of amylase variants in *Drosophila melanogaster. Genetics* 84:77–94.

DeSalle, R., A. Templeton, I. Mori, S. Pletscher, and J. S. Johnston. 1987. Temporal and spatial heterogeneity of mtDNA polymorphism in natural populations of *Drosophila mercatorum. Genetics* 116:215–223.

Dobzhansky, T., and H. Levene. 1955. Genetics of natural populations. XXIV. Developmental hemeostasis in natural populations of *Drosophila pseudoobscura. Genetics* 40:797–808.

Dobzhansky, T., and F. J. Ayala. 1973. Temporal frequency changes of enzyme and chromosomal polymorphisms in natural populations of *Drosophila pseudoobscura. Genetics* 40:797–808.

Dominguez, A., and J. Albornoz. 1987. Environmental-dependent heterosis in *Drosophila. Proc. Natl. Acad. Sci. USA* 70:680–683.

Dorado, G., and M. Barbancho. 1984. Differential responses in *Drosophila melanogaster* to environmental ethanol: Modification of fitness components at the *ADH* locus. *Heredity* 53:309–320.

Eanes, W. F. 1978. Morphological variance and enzyme heterozygosity in the monarch butterfly. *Nature* 276:263–264.

Ford, E. B. 1965. *Genetic Polymorphism.* All Soul's Studies. London: Faber and Faber.

Fucci, L., L. Gaudio, R. Rao, A. Spano, and M. Carfagna. 1979. Properties of two common electrophoretic variants of phosphoglucomutase in *Drosophila melanogaster. Biochem. Genet.* 17:825–836.

Funk, D. H., B. W. Seeney, and R. L. Vannote. 1988. Electrophoretic study of Eastern North American *Eurylophella* (Ephemeroptera: Ephemerellidae) with the discovery of morphologically cryptic species. *Ann. Entomol. Soc. Am.* 81:174–186.

Gale, K. R., and J. M. Crampton. 1987. A DNA probe to distinguish the species of *Anopheles quadriannulatus* from other species of the *Anopheles gambiae* complex. *Trans. R. Soc. Trop. Med. Hyg.* 81:842–846.

Gauldie, R. W. 1984. A reciprocal relationship between heterozygosities of the phosphoglucomutase and glucose phosphate isomerase loci. *Genetica* 63:93–103.

Geer, B. W., D. L. Lindel, and D. M. Lindel. 1979. Relationship of the oxidative pentose shunt pathway to lipid synthesis in *Drosophila melanogaster. Biochem. Genet.* 17:881–895.

Geer, B. W., M. L. Langevin, and S. W. McKechnie. 1985. Dietary ethanol and lipid synthesis in *Drosophila melanogaster. Biochem. Genet.* 23:607–622.

Giesel, J. T. 1976. Biology of a duplicate gene system with glucose 6-phosphate dehydrogenase activity in *Drosophila melanogaster*: Genetic analysis and differences in fitness components and reaction to environmental parameters among *Zw* genotypes. *Biochem. Genet.* 14:823–833.

Gilbert, D. G., and R. C. Richmond. 1982a. Studies of esterase 6 in *Drosophila melanogaster.* XII. Evidence for temperature selection of *Est 6* and *Adh* alleles. *Genetic* 58:109–119.

———. 1982b. Esterase 6 in *Drosophila melanogaster*: Reproductive function of active and null males at low temperature. *Proc. Natl. Acad. Sci. USA* 79:2962–2966.

Gillespie, J. H., and C. H. Langley. 1974. A General model to account for enzyme variation in natural populations. *Genetics* 76:837–848.

Gillespie, J. H., and K. Kojima. 1978. The degree of polymorphisms in enzymes involved in energy production compared to that in nonspecific enzymes in two *Drosophila ananassae* populations. *Proc. Natl. Acad. Sci. USA* 61:582–585.

Girard, P., L. Palabost, and C. Petite. 1977. Enzymatic variation at seven loci in nine natural populations of *Drosophila melanogaster. Biochem. Genet.* 15:589–599.

Graur, D. 1985. Gene diversity in Hymenoptera. *Evolution* 39:190–199.

Hadley, N. F. 1985. *The Adaptive Role of Lipids in Biological Systems*. New York: John Wiley and Sons.

Harper, H. A. 1969. *A Review of Physiological Chemistry*, 12th ed. Palo Alto, Calif.: Lange Med. Publ.

Harris, H. 1969. Enzyme and protein polymorphism in human populations. *Br. Med. Bull.* 25:5–13.

Hedrick, P. W. 1986. Genetic polymorphism in heterogeneous environments: A decade later. *Annu. Rev. Ecol. Syst.* 17:535–566.

Heinstra, P. W. H., W. Scharloo, and G. E. W. Thorig. 1987. Physiological significance of the alcohol dehydrogenase polymorphism in larvae of *Drosophila. Genetics* 117:75–84.

Hiraizumi, Y., and J. F. Crow. 1960. Heterozygous effects of viability, fertility, rate of development and longevity of *Drosophila* chromosomes that are lethal when homozygous. *Genetics* 45:1071–1083.

Hoorn, A. J. W., and W. Scharloo. 1978. The functional significance of amylase polymorphism. I. Properties of two amylase variants. *Genetica* 49:173–180.

Hubbell, T. H. 1956. Some aspects of geographic variation in insects. *Annu. Rev. Entomol.* 1:71–88.

Huxley, J. S. 1938. Clines; an auxilliary method in taxonomy. *Nature* 142:219–220.

Johnson, F. M. 1971. Isozyme polymorphisms in *Drosophila ananassae*: Genetic diversity among islands in the South Pacific. *Genetics* 68:77–95.

Johnson, F. M., H. E. Schaffer. 1973. Isozyme variability in species of the genus *Drosophila*. VII. Genotype-environment relationships in populations of *D. melanogaster* from the eastern United States. *Biochem. Genet.* 10:149–163.

Johnson, F. M., and H. E. Schaeffer, J. E. Gillaspy, and E. S. Rockwood. 1969. Isozyme genotype-environment relationships in natural populations of the harvester ant, *Pogonomyrmex barbatus* from Texas. *Biochem. Genet.* 3:429–437.

Johnson, G. B. 1974. Enzyme polymorphism and metabolism. *Science* 184:28–37.

———. 1975. Enzyme polymorphism and adaptation. *Stadler Genet. Symp.* 7:91–116.

Johnson, G. B. 1976. Polymorphism and predictability at the α-Glycerophosphate dehydrogenase locus in *Colias* butterflies: Gradients in allele frequency within single populations. *Biochem. Genet.* 14:403–426.

Kingsolver, J. G., and W. B. Watt. 1983. Thermoregulatory strategies in *Colias* butterflies: Thermal stress and the limits to adaptation in temporally varying environments. *Am. Nat.* 121:32–55.

Knibb, W. R. 1983. Chromosome inversion polymorphisms in *Drosophila melanogaster.* III. Gametic disequilibrium and the contributions of inversion clines to the *Adh* and *Gpdh* clines in Australia. *Genetica* 61:139–146.

Kojima, K., J. Gillespie, and Y. N. Tobari. 1970. Profile of *Drosophila* species' enzymes assayed by electrophoresis. I. Number of alleles, heterozygosity and linkage disequilibrium in glucose-metabolizing systems and some other enzymes. *Biochem. Genet.* 4:627–637.

Kojima, K., P. Smouse, S. Y. Yang, P. S. Nair, and D. Brncic. 1972. Isozyme frequency patterns in *Drosophila pavani* associated with geographical and seasonable variables. *Genetics* 72:721–731.

Kojima, K., and Y. N. Tobari. 1969. The pattern of viability changes associated with genotype frequency at the alcohol dehydrogenase locus in a population of *Drosophila melanogaster. Genetics* 61:201–209.

Kukuk, P., and B. May. 1985. A re-examination of genetic variability in *Dialictus zephyrus* (Hymenoptera: Halictidae). *Evolution* 39:226–228.

Lakovaara, S., and A. Saura. 1971. Genetic variation in natural populations of *Drosophila obscura. Genetics* 69:377–384.

Laurie-Ahlberg, C. C., P. T. Barnes, J. W. Curtsinger, T. H. Emigh, B. Karlin, R. Morris, R. A. Norman, and A. N. Wilton. 1985. Genetic variability of flight metabolism in *Drosophila melanogaster*. II. Relationship between power output and enzyme activity levels. *Genetics* 111:845–868.

LeCam, L. M., J. Neyman, and E. L. Scott. 1972. Proc. 6th Berkeley Symposium on Mathematical Statistics and Probability. Vol. V. Darwinian, Neo-Darwinian and Non-Darwinian Evolution. Berkeley: Univ. California Press.

Liebherr, J. K. 1986a. Cladistic analysis of North American Platynini and revision of the *Agonum extensicolle* species group (Coleoptera: Carabidae). *Univ. Calif. Publ. Entomol.* 106:198.

———. 1986b. Comparison of genetic variation in two carabid beetles (Coleoptera) of differing vagility. *Ann. Entomol. Soc. Am.* 79:424–433.

Lizana, G. B., and J. A. S. Prado. 1986. Enzymatic heterozygosity and morphological variance in synthetic populations of *Drosophila melanogaster. Genet. Sel. Evol.* 18:417–426.

Lucchesi, J. C., M. B. Hughes, and B. W. Geer. 1979. Genetic control of pentose phosphate pathway enzymes in *Drosophila. Curr. Top. Cell Regul.* 15:143–154.

Mane, S. D., L. Tompkins, and R. C. Richmond. 1983. Male esterase 6 catalyzes the synthesis of a sex pheromone in *Drosophila melanogaster* females. *Science* 222:419–421.

Marinkovic, D., and F. J. Ayala. 1975a. Fitness of allozyme variants in *Drosophila pseudoobscura*. I. Selection at the *Pgm-1* and *Me-2* loci. *Genetics* 79:85–95.

———. 1975b. Fitness of allozyme variants in *Drosophila pseudoobscura*. II. Selection at the *Est-5, Odh* and *Mdh-2* loci. *Genet. Res.* 24:137–149.

McDonald, J. F. 1983. The molecular basis of adaptation: A critical review of relevant ideas and observations. *Annu. Rev. Ecol. Syst.* 14:77–102.

McElfresh, K. C., and J. F. McDonald. 1983. The effect of alcohol stress on nicotinamide adenine dinucleotide (NAD+) levels in *Drosophila. Biochem. Genet.* 21:365–374.

McKechnie, S. W., and B. W. Geer. 1986. sn-Glycerol-3-phosphate oxidase and alcohol tolerance in *Drosophila melanogaster. Biochem. Genet.* 24:859–868.

McKenzie, J. A., and S. W. McKechnie. 1978. Ethanol tolerance and the *ADH* polymorphism in a natural population of *Drosophila melanogaster. Nature* 272:75–76.

Mercot, H. 1985. Allozyme frequency changes in two inverse sequences of environments in *Drosophila melanogaster. Genet. Sel. Evol.* 17:59–72.

Miller, S., R. W. Pearcy, and E. Berger. 1975. Polymorphism at the alpha-glycerophosphate dehydrogenase locus in *Drosophila melanogaster.* I. Properties of adult allozymes. *Biochem. Genet.* 13:175–188.

Miyashita, N., and C. C. Laurie-Ahlberg. 1986. Developmental variation in effects of the second and third chromosomes on the activities of the glucose-6-phosphate and 6-phosphogluconate dehydrogenases of *Drosophila melanogaster. Biochem. Genet.* 24:447–467.

Moon, T. W. 1988. Adaptation, constraint, and the function of the gluconeogenic pathway. *Can. J. Zool.* 66:1059–1068.

Morgan, P. 1975. Selection acting directly on an enzyme polymorphism. *Heredity* 34:124–127.

Morton, R. A., and R. S. Singh. 1982. The association between malathion resistance and acetylcholinesterase in *Drosophila melanogaster. Biochem. Genet.* 20:179–198.

Mukai, T. and R. Voelker. 1977. The genetic structure of natural populations of *Drosophila melanogaster*. XIII. Further studies on linkage disequilibrium. *Genetics* 86:175–185.

Mukai, T., T. K. Watanabe, and O. Yamaguchi. 1974. The genetic structure of natural populations of *Drosophila melanogaster*. XII. Linkage disequilibrium in a large local population. *Genetics* 77:771–793.

Nevo, E. 1978. Genetic variation in natural populations: Patterns and theory. *Theor. Pop. Biol.* 13:121–177.

———. 1983. Adaptive significance of protein variation. Pages 239–282 in Oxford, G. S., and D. Rollinson (eds.), *Protein Polymorphism: Adaptive and Taxonomic Significance*. New York: Academic Press.

———. 1985. Ecological and populational correlates of allozyme polymorphism in mammals. *Acta Zool. Fenn.* 170:25–29.

Nevo, E., A. Beiles, and R. B. Shlomo. 1984. The evolutionary significance of genetic diversity: Ecological, demographic, and life history correlates. Pages 13–213 in Mani, G. S. (ed.), *Evolutionary Dynamics of Genetic Diversity, Lecture Notes in Biomathematics*, 53. Berlin: Springer Verlag.

Nevo, E., B. Lavie, and R. B. Shlomo. 1983. Selection of allelic isozyme polymorphisms in marine organisms: Pattern, theory, and application. Pages 69–92 in Ratazzi, M. C., J. G. Scandalies, and G. S. Whitt (eds.), *Isozymes: Current Topics in Biological and Medical Research*, Vol. 10: *Genetics and Evolution*. New York: Alan R. Liss.

Nielsen, K. M., A. A. Hoffman, and S. W. McKechnie. 1985. Population genetics of the metabolically related *Adh, Gpdh* and *Tpi* polymorphisms in *Drosophila melanogaster*. II. Temporal and spatial variation in orchard populations. *Genet. Sel. Evol.* 17:41–58.

Oakeshott, J. G. 1976. Biochemical differences between alcohol dehydrogenases of *Drosophila melanogaster. Aust. J. Biol. Sci.* 29:365–373.

Oakeshott, J. G., G. K. Chambers, J. B. Gibson, and D. A. Willcocks. 1981. Latitudinal relationships of esterase-6 and phosphoglucomutase gene frequencies in *Drosophila melanogaster. Heredity* 47:385–396.

Oakeshott, J. G., G. K. Chambers, J. B. Gibson, W. F. Eanes, and D. A. Willcocks. 1983c. Geographic variation in *G-6-pd* and *6-pgd* allele frequencies in *Drosophila melanogaster. Heredity* 50:67–72.

Oakeshott, J. G., J. B. Gibson, P. R. Anderson, W. R. Knibb, D. G. Anderson, and G. K. Chambers. 1982. Alcohol dehydrogenases and glycerol-3-phosphate dehydrogenase clines in *Drosophila melanogaster* on different continents. *Evolution* 36:86–96.

Oakeshott, J. G., J. B. Gibson, D. A. Willcocks, and G. Chambers. 1983a. Latitudinal variation in octanol dehydrogenase and acid phosphatase allele frequencies in *Drosophila melanogaster. Theor. Appl. Genet.* 65:191–196.

Oakeshott, J. G., S. R. Wilson, and J. B. Gibson. 1983b. An attempt to measure selection coefficients affecting the alcohol dehydrogenase polymorphism in *Drosophila melanogaster* populations maintained on ethanol media. *Genetica* 61:151–159.

Oakeshott, J. G., S. W. McKechnie, and G. K. Chambers. 1984. Population genetics of the metabolically related *Adh, Gpdh* and *Tpi* polymorphisms in *Drosophila melanogaster*. I. Geographic variation in *Gpdh* and *Tpi* allele frequencies in different countries. *Genetica* 63:21–29.

O'Brien, S. J., and R. J. MacIntyre. 1969. An analysis of gene-enzyme variability in natural populations of *Drosophila melanogaster* and *D. simulans*. *Am. Nat.* 103:97–113.

Parsons, P. A. 1970. Genetic heterogeneity in natural populations of *Drosophila melanogaster* for ability to withstand desiccation. *Theor. Appl. Genet.* 40:261–266.

Pashley, D. P. 1985. The use of population genetics in migration studies: A comparison of three noctuid species. Pages 305–324 in MacKenzie, D. R., C. S. Barfield, G. C. Kennedy, R. D. Berger, and D. J. Taranto (eds.), *The Movement and Dispersal of Agriculturally Important Biotic Agents*. Baton Rouge, La.: Claitor's Publ.

Passmore, R., and J. S. Robson (eds.). 1969. *Companion to Medical Studies*, Vol. 1. Philadelphia, Pa.: F. A. Davis Co.

Pinsker, W., and D. Sperlich. 1979. Allozyme variation in natural populations of *Drosophila subobscura* along a North-South gradient. *Genetica* 50:207–219.

Pipkin, S. B., C. Rhodes, and N. Williams. 1973. Influence of temperature on *Drosophila* alcohol dehydrogenase polymorphism. *J. Hered.* 64:181–185.

Pontecorvo, G., M. Carfagna, L. Fucci, and L. Gaudio. 1986. Effects of various metabolites on two phosphoglucomutase allozyme activities from *Drosophila melanogaster. Biochem. Genet.* 24:397–403.

Powell, J. R. 1971. Genetic polymorphism in varied environments. *Science* 174:1035–1036.

———. 1975a. Protein variation in natural populations of animals. *Evol. Biol.* 8:79–119.

———. 1975b. Isozymes and non-Darwinian evolution: A reevaluation. Pages 9–26 in Market, C. L. (ed.), *Isozymes IV. Genetics and Evolution*. New York: Academic Press.

Powers, D. A., and D. Powers. 1975. Predicting gene frequencies in natural populations: A testable hypothesis. Pages 63–84 in Market, C. L. (ed.), *Isozymes IV. Genetics and Evolution*. New York: Academic Press.

Prakash, S., and M. Levitan. 1973. Associations of alleles of esterase-1 locus with gene arrangements of the left arm of the second chromosome in *Drosophila robusta*. *Genetics* 75:371–379.

Prakash, S., R. C. Lewontin, and J. L. Hubby. 1969. A molecular approach to the study of genic heterozygosity in natural populations. IV. Patterns of genic variation in central, marginal and isolated populations of *Drosophila pseudoobscura*. *Genetics* 61:841–858.

Reeve, H. K., J. S. Reeve, and D. W. Pfennig. 1985. Eusociality and genetic variability: A re-evaluation. *Evolution* 39:200–201.

Rhomberg, L. R., S. Joseph, and R. S. Singh. 1985. Seasonal variation and clonal selection in cyclically parthenogenetic rose aphids *(Macrosiphum rosae)*. *Can. J. Genet. Cytol.* 27:224–232.

Richmond, M. W., and J. D. Guenther. 1979. Enzyme variability in the *Drosophila willistoni* group. III. Amounts of variability in the superspecies *D. paulistorum*. *Genetics* 70:87–112.

Richmond, R. C. 1972. Enzyme variability in the *Drosophila willistoni* group III. Amounts of variability in the superspecies, *D. paulistorum*. *Genetics* 70:87–112.

Riddiford, A., and D. Penny. 1984. The scientific status of modern evolutionary theory. Pages 1–38 in Pollard, J. W. (ed.), *Evolutionary Theory: Paths into the Future*. New York: John Wiley & Sons.

Rockwood-Sluss, E. S., J. S. Johnston, and W. B. Heed. 1973. Allozyme genotype-environment relationships. I. Variation in natural populations of *Drosophila pachea*. *Genetics* 73:135–146.

Ross, K. G., E. L. Vargo, and D. J. C. Fletcher. 1987. Comparative biochemical genetics of three fire ant species in North America, with special reference to the two social forms of *Solenopsis invicta* (Hymenoptera: Formicidae). *Evolution* 41:979–990.

Salam, A. Z. El-Abidin, and W. Pinsker. 1981. Effects of selection for resistance to organophosphorus insectides on two esterase loci in *Drosophila melanogaster. Genetica* 55:11–14.

Sampsell, B., and S. Sims. 1982. Effect of *adh* genotype and heat stress on alcohol tolerance in *Drosophila melanogaster. Nature* 296:853–855.

Sangiorgi, S., E. Pieragostini, L. Prosperi, P. Leggieri, and S. Cavicchi. 1981. Mimicry of isozyme adaptive advantage by gene association. III. *Adh* genotype frequency variations in *Drosophila* cage populations reared at different temperatures. *Genetica* 56:229–234.

Seager, R. D., and W. W. Anderson. 1987. Viabilities of amylase genotypes in *Drosophila pseudoobscura* on starch and maltose food. *Evolution* 41:199–208.

Selander, R. K., and D. W. Kaufman. 1973. Genic variability and strategies of adaptation in animals. *Proc. Natl. Acad. Sci. USA* 70:1875–1877.

Sell, D. K., G. S. Whitt, and W. H. Luckman. 1975. Esterase polymorphism in the corn earworm, *Heliothis zea* (Boddie): A survey of temporal and spatial allelic variation in natural populations. *Biochem. Genet.* 13:885–898.

Serradilla, J. M., and F. J. Ayala. 1983. Effects of allozyme variation on fitness components in *Drosophila melanogaster. Genetica* 62:139–146.

Shu-sen, L., T. Hui-lo, H. Mu-yu, W. Fu-yun, and H. Chien-ying. 1973. Oxidative phosphorylation and ultrastructure of flight muscle mitochondria in the housefly *Musca domestica* Vinca Macq. *Acta Sinica* 16:1–14.

Siddiqui, W. H., and C. A. Barlow. 1972. Population growth of *Drosophila melanogaster* (Diptera: Drosophilidae) at constant and alternating temperatures. *Ann. Entomol. Soc. Am.* 65:993–1001.

Simon, C., and J. Archie. 1985. An empirical demonstration of the lability of heterozygosity estimates. *Evolution* 39:463–467.

Singh, R. S. 1976. Substrate-specific enzyme variation in natural populations of *Drosophila pseudoobscura*. *Genetics* 82:507–526.

Singh, R. S., and L. R. Rhomberg. 1987. A comprehensive study of genic variation in natural populations of *Drosophila melanogaster.* II. Estimates of heterozygosity and patterns of geographical differentiation. *Genetics* 117:255–271.

Slatkin, M. 1985. Gene flow in natural populations. *Annu. Rev. Ecol. Syst.* 16:393–430.

———. 1987. Gene flow and the geographic structure of natural populations. *Science* 236:787–792.

Synder, T. P., and F. J. Ayala. 1979. Temperature and density effects on fitness at the MDH-2 and PGM-1 loci of *Drosophila pseudoobscura*. *Genetica* 51:59–67.

Sokolova, K. B., and M. D. Golubovskii. 1979. Manifestation, interaction and population distribution of alleles at the "giant larvae" locus in *Drosophila melanogaster. Genetika* 15:233–243.

Solignac, M., and M. Monnerot. 1986. Race formation, speciation, and introgression within *Drosophila simulans, D. muritiana*, and *D. sechellia* inferred from mitochondrial DNA analysis. *Evolution* 40:531–539.

Somero, G. N., and P. W. Hochachka. 1971. Biochemical adaptation to the environment. *Am. Zool.* 11:159–167.

Starmer, W. T., and J. S. F. Barker. 1986. Ecological genetics of the *Adh-1* locus of *Drosophila bazzati. Biol. J. Linn. Soc.* 28:373–385.

Stebbins, G. L., and F. J. Ayala. 1981. Is a new evolutionary synthesis necessary? *Science* 213:967–971.

Steiner, W. W. M. 1974. Enzyme Polymorphism and Desiccation Resistance in Two species of Hawaiian *Drosophila*. Ph.D. Dissertation, Dept. of Genetics, Univ. of Hawaii. 301 pp.

———. 1979a. Genetic variation in Hawaiian *Drosophila*. VI. Seasonally dependent gene changes in *Drosophila mimica. Evolution* 33:543–562.

———. 1979b. Genetic variation in Hawaiian *Drosophila*. VIII. Heterozygosity and genic changes in isolated populations of *D. engyochracea. Biochem. Genet.* 17:645–664.

Steiner, W. W. M., S. Narang, J. B. Kitzmiller, and D. L. Swofford. 1982. Genetic diversity and evolution in the *Neotropical Anopheles* (Subgenus Nyssorhynchus). Pages 523–550 in Steiner, W. W. M., W. J. Tabachnick, K. S. Rai, and S. Narang (eds.), *Recent Developments in the Genetics of Insect Dieases Vectors*. Champaign, Ill.: Stipes Publ. Co.

Steiner, W. W. M., D. J. Voegtlin, and M. E. Irwin. 1985. Genetic differentiation and its bearing on migration in North American populations of the corn leaf aphid, *Rhopalosiphum maidis* (Fitch) (Homoptera: Aphididae). *Ann. Entomol. Soc. Am.* 78:518–525.

Steiner, W. W. M., D. J. Voegtlin, M. E. Irwin, and G. Kampmeier. 1985. Electrophoretic comparison of aphid species: Detecting differences based on taxonomic status and host plant. *Comp. Biochem. Physiol.* 81B:295–299.

Stock, M. W., and G. D. Amman. 1980. Genetic differentiation among mountain pine beetle populations from lodgepole pine and ponderosa pine in Northeast Utah. *Ann. Entomol. Soc. Am.* 73:472–478.

Stock, M. W., and J. D. Guenther. 1979. Isozyme variation among mountain pine beetle (*Dendroctonus ponderosae*) populations in the Pacific Northwest. *Environ. Entomol.* 8:889–893.

Sweeney, B. W., D. H. Funk, and R. L. Vannote. 1987. Genetic variation in stream mayflies (Insecta: Ephemeroptera) populations of eastern North America. *Ann. Entomol. Soc. Am.* 80:600–612.

Takano, T., S. Kusakabe, and T. Mukai. 1987. The genetic structure of natural populations of *Drosophila melanogaster.* XX. Comparison of genotype-environment interaction in viability between a northern and a southern population. *Genetics* 117:245–254.

Teague, M. B., A. G. Clark, and W. W. Doane. 1986. Developmental analysis of lipids in wild-type and adipose-60 *Drosophila melanogaster. J. Exp. Zool.* 240:95–104.

Tomaszewski, E., H. E. Schaffer, and F. M. Johnson. 1973. Isozyme genotype-environment associations in natural populations of the harvestor ant, *Pogonomyrmex badius*. *Genetics* 75:405–421.

Triantaphyllidis, C. D., Z. G. Scouras, J. N. Panourgias, and G. C. Ioannidis. 1982. Allozyme variation in Greek wild populations of *Drosophila melanogaster* and *D. simulans* along a North-South gradient. *Genetica* 58:129–136.

Vanderplank, F. L. 1947. Experiments in the hybridization of tsetse-flies (Glossina: Diptera) and the possibility of a new method of control. *Trans. R. Enomol. Soc. Lond.* 98:1–18.

Verdonck, M. V. 1987. Adaptation to environmental heterogeneity in populations of *Drosophila melanogaster. Genet. Res.* 49:1–10.

Vigue, C. L., and F. M. Johnson. 1973. Isozyme variability in species of the genus *Drosophila*. IV. Frequency-property-environment relationships of allelic alcohol dehydrogenases in *D. melanogaster. Biochem. Genet.* 9:213–227.

Voegtlin, D. J., W. W. M. Steiner, and M. E. Irwin. 1987. Searching for the source of the annual spring migrants of Rhopalosiphum maiis (Homoptera: Aphididae) in North America. Pages 000–000 in Holman, J. (ed.), *Proc. Int. Symp. on Popul. Structure, Genetics and Taxonomy of Aphids*. Yugoslavia: Smolenice, Chez.

Voelker, R. A., C. C. Cockerham, F. M. Johnson, H. E. Schaeffer, T. Mukai, and L. E. Mettler. 1978. Inversions fail to account for allozyme clines. *Genetics* 88:515–527.

Wagner, R. P., and R. K. Selander. 1974. Isozymes in insects and their significance. *Annu. Rev. Entomol.* 19:117–138.

Watanabe, T. K., and T. Watanabe. 1977. Enzyme and chromosome polymorphisms in Japanese natural populations of *Drosophila melanogaster. Genetics* 85:319–329.

Watt, W. B. 1977. Adaptation at specific loci. I. Natural selection on phosphoglucose isomerase of *Colias* butterflies: Biochemical and population aspects. *Genetics* 87:177–194.

———. 1983. Adaptation at specific loci. II. Demographic and biochemical elements in the maintenance of the *Colias* PGI polymorphism. *Genetics* 103:691–724.

———. 1985. Bioenergetics and evolutionary genetics: Opportunities for new synthesis. *Am. Nat.* 125:118–143.

Watt, W. B., P. A. Carter, and S. M. Blower. 1985. Adaptation at specific loci. IV. Differential mating success among glycolytic allozyme genotypes of *Colias* butterflies. *Genetics* 109:157–167.

Watt, W. B., P. A. Carter, and K. Donohue. 1986. Females' choice of "good genotypes" as mates is promoted by an insect mating system. *Science* 233:1187–1189.

Watt, W. B., R. C. Cassin, and M. S. Swan. 1983. Adaptation at specific loci. III. Field behavior and survivorship differences among *Colias* PGI genotypes are predictable from in vitro biochemistry. *Genetics* 103:725–739.

White, A., P. Handler, and E. L. Smith. 1968. *Principles of Biochemistry.* New York: McGraw-Hill.

Willey, R. B. 1987. Patterns of gene flow disruption in the evolution of the genus *Arphia* (Acrididae: Oedipodinae). Pages 592–600 in *Evolutionary Biology of Orthopteroid Insects*. Chichester, England: Ellis Horwood Ltd.

Wood, P. E., and S. I. Guttman. 1987. Genetic variation in *Neodiprion* (Hymenoptera: Symphyta: Dipronidae) sawflies and a comment on low levels of genetic diversity within the Hymenoptera. *Ann. Entomol. Soc. Am.* 80:590–599.

Wooten, M. C., and M. H. Smith. 1985. Large mammals are genetically less variable? *Evolution* 39:210–212.

Wright, S. 1951. The genetical structure of populations. *Ann. Eugen.* 15:323–354.

Zouros, E. 1975. Electrophoretic variation in allelozymes related to function or structure? *Nature* 254:446–448.

Zouros, E., and W. Johnson. 1976. Linkage disequilibrium between functionally related enzyme loci of *Drosophila mojavensis*. *Can. J. Genet. Cytol.* 18:245–254.

Zouros, E., and C. B. Krimbas. 1973. Evidence for linkage disequilibrium maintained by selection in two natural populations of *Drosophila subobscura*. *Genetics* 73:659–674.

APPENDIX REFERENCES FOR TABLE 2.5

Abou-Youssef, A. Y. 1978. Genetic studies on isozyme polymorphism in *Drosophila*. I. Survey of *D. Virilis* natural populations for amylase isozymes. *Egypt. J. Genet. Cytol.* 7:56–68.

Bentz, B. J., and M. W. Stock. 1986. Phenetic and phylogenetic relationships among ten species of *Dendroctonus* bark beetles (Coleoptera: Scotylidae). *Ann. Entomol. Soc. Am.* 79:527–534.

Berlocher, S. H. 1984. Genetic changes coinciding with the colonization of California by the walnut husk fly, *Rhagoletis completa*. *Evolution* 38:906–918.

Black, IV, W. C. and E. S. Krafsur. 1986. Population biology and genetics of winter house fly (Diptera: Muscidae) populations. *Ann. Entomol. Soc. Am.* 79:636–644.

Bush, G., and R. W. Neck. 1976. Ecological genetics of the screwworm fly, *Cochliomyia hominivorax* (Diptera: Calliphoridae) and its bearing on the quality control of mass-reared insects. *Environ. Entomol.* 5:821–826.

Chambers, S. M. 1977. Genetic characteristics of a colonizing episode in the "Lovebug," *Plecia nearctica*. *Ann. Entomol. Soc. Am.* 70:537–540.

Chianchi, R., A. Sabatini, M. Coluzzi, and L. Bullini. 1980. Divergenza genetica tradue specie gemelle del genere *Anopheles: An. claviger e An. petragnanii* (Diptera: Culicidae). *Atti. XII Congr. Naz. Ital. Entomol. Roma* 2:261–263.

Daly, J. C., and P. Gregg. 1985. Genetic variation in *Heliothis* in Australia: Species identification and gene flow in the two pest species *H. armigera* (Hübner) and *H. punctigera* Wallengren (Lepidoptera: Noctuidae). *Bull. Entomol. Res.* 75:169–184.

Dubach, J. M., D. B. Richman, and R. B. Turner. 1988. Genetic and morphological variation among geographical populations of the range caterpillar, *Hemileuca oliviae*. (Lepidoptera: Saturniidae). *Ann. Entomol. Soc. Am.* 81:132–137.

Eldridge, B. F., L. E. Munstermann, and G. B. Craig. Enzyme variation in some mosquito species related to *Aedes (Ochlerotatus) stimulans* (Walker) (Diptera: Culicidae). In manuscript.

Foey, D. H. 1986. Allozyme variation in laboratory-reared and field-collected potato tuberworm moth (Lepidoptera: Gelechiidae). *Ann. Entomol. Soc. Am.* 79:80–83.

Funk, D. H., B. W. Sweeney, and R. L. Vannote. 1988. Electrophoretic study of Eastern North American *Eurylophella* (Ephemeroptera: Ephemerellidae) with the discovery of morphologically cryptic species. *Ann. Entomol. Soc. Am.* 81:174–186.

Johnson, G. B. 1974. Enzyme polymorphism and metabolism. *Science* 184:28–37.

Johnson, W. E., H. L. Carson, K. Y. Kaneshiro, W. W. M. Steiner, and M. M. Cooper. 1975.

Genetic variation in Hawaiian Drosophila. II. Allozyme differentiation in the *D. planitibia* subgroup. Pages 563–584 in Market, C. L. (ed.), "Third Int. Conf. on Isozymes," Vol. IV. New York: Academic Press.

Jones, C. J., R. S. Patterson, and D. E. Milne. Electrophoretic comparisons of isozymes from selected populations of *Stomoxys calcitrans* (Diptera: Muscidae). In manuscript.

Kuenzi, F. M., and H. C. Coppel. 1986. Isozymes of the sawflies *Neodiprion* and *Diprion similis*: Diagnostic characters and genetic distance. *Biochem. Syst. Evol.* 14:423–429.

Kukuk, P., and B. May. 1985. A re-examination of genetic variability in *Dialictus zephyrus* (Hymenoptera: Halictidae). *Evolution* 39:226–228.

Liebherr, J. K. 1986a. Cladistic analysis of North American Platynini and revision of the *Agonum extensicolle* species group (Coleoptera: Carabidae). *Univ. Calif. Publ. Entomol.* 106:198.

———. 1986b. Comparison of genetic variation in two carabid beetles (Coleoptera) of differing vagility. *Ann. Entomol. Soc. Am.* 79:424–433.

Loxdale, H. D., I. J. Tarr, C. P. Weber, C. P. Brookes, and P. G. N. Digby. 1985. Electrophoretic study of enzymes from cereal aphid populations. III. Spatial and temporal genetic variation of populations of *Sitobion avenae* (F.) (Hemiptera: Aphididae). *Bull. Entomol. Res.* 75:121–141.

Malavasi, A., and J. S. Morgante. 1983. Population genetics of *Anestrepha fraterculus* (Diptera, Tephritidae) in different hosts: Genetic differentiation and heterozygosity. *Genetica* 60:207–211.

Matthews, T. C., and L. E. Munstermann. 1983. Genetic diversity and differentiation in northern populations of the tree-hole mosquito, *Aedes hendersoni* (Diptera: Culicidae). *Ann. Entomol. Soc. Am.* 76:1005–1010.

May, B., and F. R. Holbrook. 1978. Absence of genetic variability in the green peach aphid, *Myzus persicae* (Hemiptera: Aphididae). *Ann. Entomol. Soc. Am.* 71:809–812.

McDonald, I. C., and O. A. Johnson. 1976. Isozyme variability studies of translocation homozygotes in the house fly, *Musca domestica* L. In Packer, J. S. and S. E. Kunz (eds.). Proc. XV Int. Congr. Entomol., Washington, D.C., pp. 140–145.

McDonald, P. T., L. R. Hilburn, and S. E. Kunz. 1987. Genetic similarities among natural populations of the horn fly, *Haematobia irritans* (Diptera: Muscidae). *Ann. Entomol. Soc. Am.* 80:288–292.

Miles, S. J. 1978. Protein variation in the *Anopheles gambiae* group of species. *Bull. Entomol. Res.* 68:85–96.

Miles, S. J., and H. E. Paterson. 1979. Protein variation and systematics in the *Culex pipiens* group of species. *Mosq. Syst.* 11:187–202.

Moroni, A., O. Ravera, and A. Anelli. 1981. *Atti Primo Congr. Naz. Soc. Ital. Ecol.* pp. 305–316.

Narang, S. K., J. A. Seawright, N. L. Willis, and M. F. Aurez. 1987. Genetic variability and mapping of peptidase-4 in *Anopheles alibmanus*. *J. Hered.* 78:278–279.

Pinsker, W., and D. Sperlich. 1979. Allozyme variation in natural populations of *Drosophila subobscura* along a North-South gradient. *Genetica* 50:207–219.

Powell, J. R., M. Rico, and M. Andjelkovic. 1980. Population genetics of *Drosophila* amylase. III. Interspecific variation. *Evolution* 34:209–213.

Rajendram, G. F. 1988. Esterase variation in *Neophotettix nigropictus* (Stal) population from Australia. Proc. 15th Auchennoryncha Meeting, Turin, Italy.

Richardson, R. H., and F. M. Johnson. 1967. Isozyme variability in species of the genus *Drosophila*. II. A multiple allelic isozyme system in *Drosophila busckii*: A stable polymorphic system. *Biochem. Genet.* 1:73–79.

Saura, A., O. Halkka, and J. Lokki. 1973. Enzyme gene heterozygosity in small island populations of *Philaenus spumarius* (L.) (Homoptera). *Genetica* 44:459–473.

Schultz, J. H., P. G. Meier, and H. D. Newson. 1986. Evolutionary relationships among the salt marsh *Aedes* (Diptera: Culicidae). *Mosq. Syst.* 18:145–195.

Singh, S. M., and T. K. Cunningham. 1981. Morphological and genetic differentiation in aphids (Aphididae). *Can. Entomol.* 113:539–549.

Sluss, H., M. Graham, and M. DuBois. 1978. Allozyme differences between *Heliothis virescens* and *H. Zea*. *Ann. Entomol. Soc. Am.* 71:191–195.

Sluss, T. P., E. S. Sluss, and H. M. Graham. 1979. Allozyme variation in natural populations of *Heliothis virescens*. *Ann. Entomol. Soc. Am.* 72:317–322.

Snyder, T. P., and M. C. Linton. 1984. Population structure in black flies: Allozymic and morphological estimates for *Prosimulium mixtum* and *P. fuscum* (Diptera: Simuliidae). *Evolution* 38:942–956.

Sofield, R. K., N. E. Buroker, E. J. Hansens, and R. C. Vrijenhoek. 1984. Genetic diversity within and between sibling species of salt marsh horseflies (Diptera: Tabanidae). *Ann. Entomol. Soc. Am.* 77:663–668.

Steiner, W. W. M. 1979a. Genetic variation in Hawaiian *Drosophila*. VI. Seasonally dependent gene changes in *Drosophilia mimica*. *Evolution* 33:543–562.

———. 1979b. Genetic variation in Hawaiian *Drosophila*. VIII. Heterozygosity and genic changes in isolated populations of *D. engyochracea*. *Biochem. Genet.* 17:654–664.

Steiner, W. W. M., and D. J. Joslyn. Genetic variation in *Anopheles crucians*. Unpublished manuscript.

Steiner, W. W. M., J. B. Kitzmiller, and D. L. Osterbur. 1980. Gene differentiation in chromosome races of *Anopheles nuneztovari* (Gabaldon). *Mosq. Syst.* 12:306–319.

———. 1981. On the genetic identity and evolution of the malaria vectors *Anopheles aquasalis* Curry and *Anopheles emilianus* Komp. Pages 75–90 in Pal, R., J. B. Kitzmiller, and T. Kanada (eds.), *Genetics and Cytogenetics of Dieases Vectors*. Tokyo: Kodansha Press, and New York: Elsevier Biomedical Press.

Steiner, W. W. M., D. J. Voegtlin, and M. E. Irwin. 1985. Genetic differentiation and its bearing on migration in North American populations of the corn leaf aphid *Rhopalosiphum maidis* (Fitch) (Homoptera: Aphididae). *Ann. Entomol. Soc. Am.* 78:518–525.

Steiner, W. W. M., D. J. Voegtlin, M. E. Irwin, and G. Kampmeier. 1985. Electrophoretic comparison of aphid species: Detecting differences based on taxonomic status and host plant. *Comp. Biochem. Physiol.* 81B:295–299.

Sweeney, B. W., D. H. Funk, and R. L. Vannote. 1987. Genetic variation in stream mayflies (Insecta: Ephemeroptera) populations of eastern North America. *Ann. Entomol. Soc. Am.* 80:600–612.

Szmczak, L. J., L. R. Hilburn, and K. S. Rai. 1986. Genetic differentiation in the *Aedes atropalpus* complex. I. Isozyme variability and genetic distance between *Ae. atropalpus* and *Ae. epactius J. Genet.* 65:93–204.

Tabachnick, W. J., and J. R. Powell. 1979. A worldwide survey of genetic variation in the yellow fever mosquito, *Aedes aegypti*. *Genet. Res.* 34:215–229.

———. 1983. Genetic analysis of *Culex pipiens* in the Central Valley of California. *Ann. Entomol. Soc. Am.* 76:715–720.

Tomiuk, J., and K. Wohrmann. 1984. Genotypic variability in natural populations of *Macrosiphum rosae* (L.) in Europe. *Biol. Zentralbl.* 103:113–122.

Urbanelli, S., A. Sabatini, and L. Bullini. 1980. Divergenza e variabilita genetica in *Culex pipiens e Culex torrentium* (Diptera: Culicidae). *Atti. XII Congr. Naz. Ital. Enotmol. Roma* II:281–283.

Wellso, S. G., D. J. Howard, S. L. Adams, and J. Arnold. 1988. Electrophoretic monomorphism in six biotypes and two populations of the Hessian fly (Diptera: Cecidomyiidae). *Ann. Entomol. Soc. Am.* 81:50–53.

Woods, P. E., and S. I. Guttman. 1987. Genetic variation in *Neodiprion* (Hymenoptera: Symphyta: Dipronidae) sawflies and a comment on low levels of genetic diversity within the Hymenoptera: *Ann. Entomol. Soc. Am.* 80:590–599.

3

THE SPATIAL SCALE OF GENETIC VARIATION IN INSECT POPULATIONS

Fred Gould

North Carolina State University

INTRODUCTION

In the past 20 years, entomologists have moved a long way from the typological view of species. Instead of treating variation in the morphological and ecological characteristics of insect species as troublesome "noise," entomologists are using this variation as a helpful tool in their quest to understand the ecology and evolution of these organisms.

In some cases, one must sample individuals from locations that are hundreds of kilometers apart before detecting quantitative variation in a genetically controlled trait. In other cases, genetically controlled traits of insects have been found to vary quantitatively within areas that are smaller than 5 square kilometers. In this chapter, I will review some cases in which the spatial pattern of genetic variation in an ecologically relevant trait has been explored. I will discuss some of the factors that may influence these patterns, as well as the implications of different patterns of variation for the way in which we study populations of insects.

THE BASIC THEORETICAL FRAMEWORK

Before I discuss empirical studies, it is useful to briefly examine some theoretical expectations regarding patterns of genetic variation. Mutation and recombination of genetic material are constantly generating variation within a species. However, the actual level of variation that we find within a species and the spatial pattern of this variation depend on how this initial variation is affected by natural selection and historical events (i.e., genetic drift).

Most genetic mutants are thought to be less fit than common genotypes and should usually be removed from a population by natural selection (Dobzhansky, 1970). Even if a mutant is equal in fitness to common genotypes, it is likely to be lost from the population due to random mortality. If one of the 1,000 eggs laid by a female is a genetic mutant and, on average, only 2 of these 1,000 eggs escape mortality factors before they reproduce, there is only 1 chance in 500 that the mutant will not be lost from the population in the first generation. The actual probability of a mutation being lost from a population is as follows:

$$1 - \frac{1}{2N}$$

where N is the population size.

If the mutant is more fit than other individuals, its chance of surviving is somewhat greater, but it is still likely to be lost from the population. For example, let us consider two mutants, one with half the fitness of average individuals in a population (mutant H) and the other with twice the fitness of average individuals in a population (mutant T). Each is the offspring of a single female that lays 1,000 eggs. Mutant H has a (0.5) × (1/500) or 1 in a 1,000 chance of surviving to reproduce, whereas mutant T has a (2.0) × (1/500) or 1 in 250 chance of surviving to reproduce in the first generation.

In large populations of millions of insects, a given type of mutation at a locus is likely to recur, so over time, the chance of one mutation like mutant T getting past the first generation is much higher than 1 in 250. Given its high fitness, mutant T should attain a high frequency or become fixed in large populations. Mutations like mutant H are also likely to recur many times in a large population, but, even if one of these mutations survives through one generation, its final fate is lost from the population because of its low fitness. A mutant whose fitness is equal to that of the average individual in the population may enter the population by chance but is likely to remain at a low frequency or be lost from the population, again, by chance.

The situation can be quite different for small populations. In these populations, genetic drift in allelic frequency, or what could be called "historical accidents," could be much more important than small differences in the fitness of mutants. For an extreme example, consider a single female aphid that established a new colony. Of 1,000 offspring, 2 survive. If by chance 1 of those 2 was a mutant with lower than average fitness, its frequency in the population would be 0.50. By the next generation, chance events could lead to fixation of this mutant in the population. A mutant with approximately average fitness that entered a population by chance would have an even greater chance of becoming the predominant genotype in some small populations.

What does this all mean in terms of spatial variation? Let us assume a neutral mutation (i.e., one with average fitness for the population) in two sympatric insect species. One species has large populations every 10 kilometers (km) along a transect, and the other has small populations every 10km. If we sampled individuals of both species over a 100-km transect, it is likely that we would find more variation among populations in the species with small population size (Fig. 3.1). (Within-population variation would, however, be greater in the large populations.) I have treated this problem in a qualitative fashion to present the general concept. Many detailed quantitative treatments of this problem are available elsewhere (Hedrick, 1983; Hartl, 1980; Slatkin, 1985).

If, instead of assuming a totally neutral mutation, we assume that the fitness of individuals with the mutation varies over space, a different picture emerges. Consider two species over the 100-km transect where there is a linear cline in the fitness of individuals with the mutation that has become established in an area. Both species have one identifiable population every 10 km. Species A has large populations, with moderate movement between them. Species B has small populations, with low rates of movement between populations. Let fitness of these individuals with the mutation be 10% less than the average population fitness at one end of the transect, 10% more

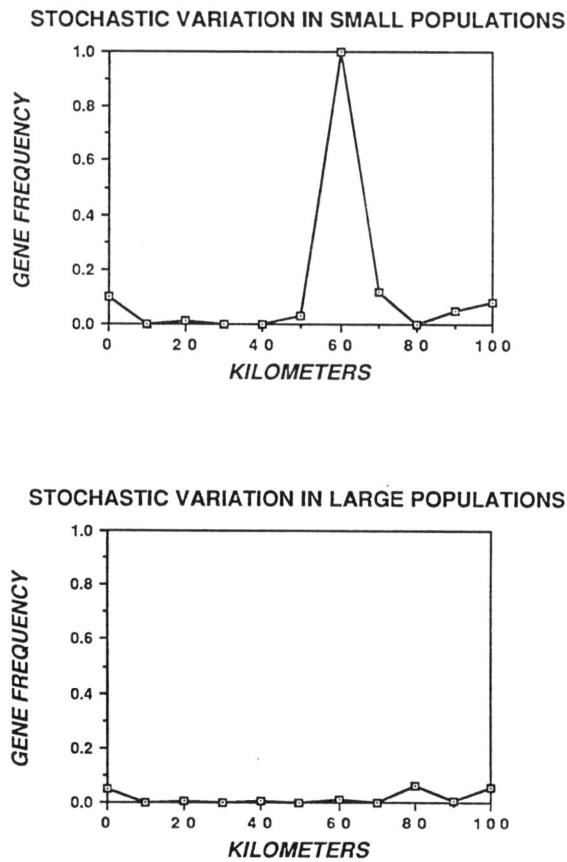

Figure 3.1. Hypothetical gene (allelic) frequencies for a neutral mutation at a single locus across a 100-km transect in a population with small (top) and large (bottom) effective population size.

than the average fitness at the other end, and equal to the population average at the 50-km point. In the species with large population size and moderate rates of migration among populations, one would expect a reasonably straight sloping line to describe the frequency of the trait along the transect (Fig. 3.2b). Even though individuals with the mutation had low fitness at one end of the transect, it would remain in the population due to migration. In the species with small population size and a lower migration rate, one would expect a much more jagged line to describe gene frequencies along the transect (Fig. 3.2a). We would also expect to find a number of samples along the transect to produce all individuals with or without the mutation. This fixation or loss of the mutant could be due to selection, genetic drift, or a combination of the two within the small populations.

The assumption in Figure 3.2 is that there is a smooth, linear cline in the fitness of the mutant. If, instead, there is unpredictable, fine-scale variation in the mutant's

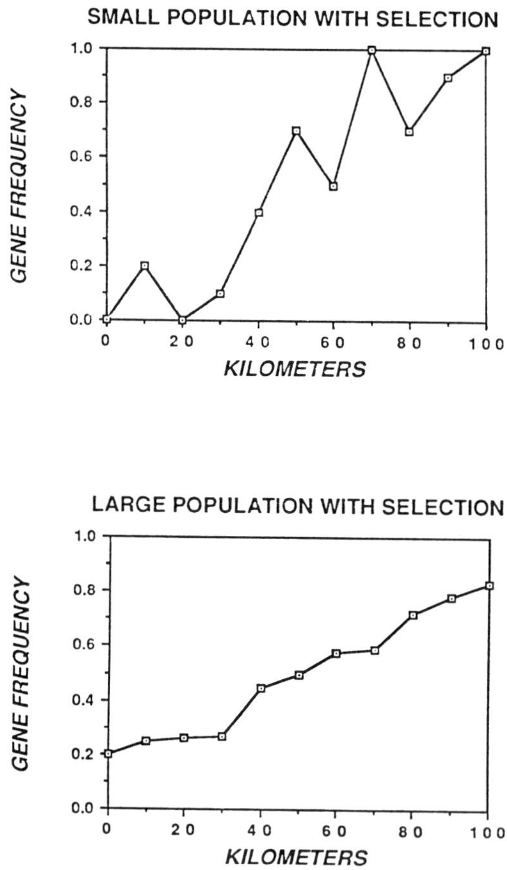

Figure 3.2. Hypothetical gene frequencies for a trait controlled by a single locus across a 100-km transect. Fitness of insects homozygous for the trait is lower than average at the origin of the transect and is higher than average at the end of the transect.

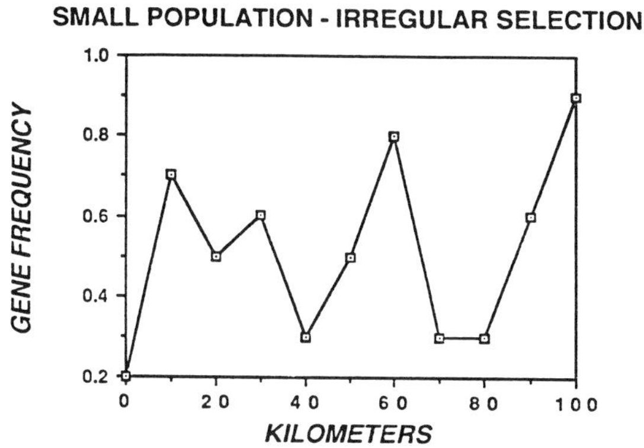

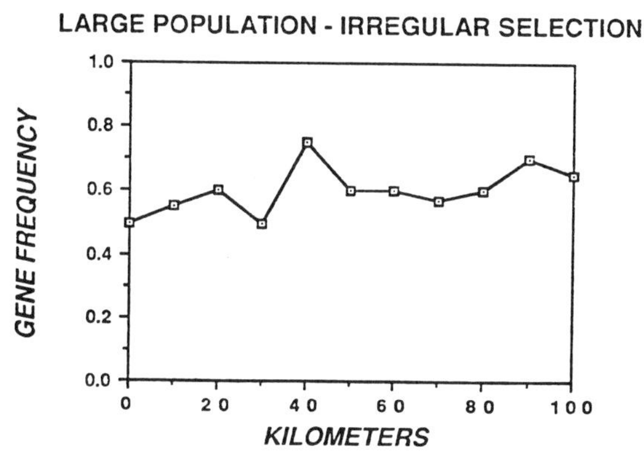

Figure 3.3. Hypothetical gene frequencies for a trait controlled by a single locus across a 100-km transect when fitness conferred by the trait varies over space in a random fashion. Top: small effective population size with low migration rate. Bottom: large effective population size with relatively high migration rate.

fitness along the transect, the frequency of the mutant in the population would not have an overall slope (Fig. 3.3b). Yet, even in this case, the variation among populations in the frequency of the mutant is expected to be greater in the species with small populations because of lower gene flow and drift (Fig. 3.3a).

I have restricted this discussion to traits controlled by single genes. Variation in many ecologically relevant traits is governed by alleles at multiple loci (polygenic traits). Although, in a general qualitative fashion, population structure affects variance in polygenic traits similarly to the way it affects monogenic traits, the reader should refer to more detailed treatments in order to appreciate certain important quantitative differences (Keightley and Hill, 1987; Madalena and Hill, 1972; Madalena and Robertson, 1975; Wright, 1951).

EMPIRICAL OBSERVATIONS

In the preceding section, I described some of the expectations from theory. In this section, I will describe patterns of phenotypic variation that have been observed in a number of selected cases. In presenting these cases, I have assumed that they are a representative sample and that evidence presented for a genetic basis to phenotypic variation is reasonable, if not rigorous. I will divide the cases into those that have looked at variation on a large geographical scale (i.e., more than 50 km), those that have examined fine-scale variation (sometimes less than 5 square km), and those that combine analysis at both scales. The literature review for this chapter was conducted in 1986–87.

Large-Scale Variation

Diapause Many insects use photoperiod as a cue for initiating diapause. The optimal photoperiod for initiating diapause often varies in a predictable manner along a latitudinal gradient. A number of detailed studies of mite and insect species have shown that populations vary genetically in a manner that reflects predictions based on climate and photoperiod (Tauber et al., 1986). For example, Bradshaw and his colleagues studied diapause initiation in a number of mosquito species (Bradshaw, 1976; Jordan and Bradshaw, 1978; Holzapfel and Bradshaw, 1981). In these species, decreasing photoperiod is involved in diapause induction. In all cases, diapause induction required a shorter photoperiod at lower latitudes. In *Aedes triseriatus* (Say), critical photoperiod increased by one hour for every 2.06 degree increase in latitude (Holzapfel and Bradshaw, 1981). The photoperiod required to induce diapause also increased as elevation increased (one hour per 549 meters). Bradshaw (1976) found a similar pattern for *Wyeomyia smithii* (Coquillett) and showed a high correlation ($r = 0.96$) between a population's critical photoperiod and the number of freeze-free days in its habitat. Trimble and Smith (1979) did not find differences in the critical photoperiod of the mosquito *Toxorhynchites rutilus septentrionalis* (Dyar and Knab) when raised at 19°C, but, at 27°C, a population from Delaware had more of a tendency to enter diapause than a population from Louisiana.

Voltinism In the higher latitudes, most insect species are not capable of completing two generations in a season. Instead of using environmental cues for initiating diapause, many northern populations are genetically programmed to enter diapause once per generation. For species in more moderate climates, diapausing every generation is generally less adaptive than diapausing only when environmental cues indicate that conditions will become poor. Tauber et al. (1986) offer a good review of this general area.

One of the most detailed series of studies on latitudinal clines in voltinism and diapause is found in the work of Masaki (1972, 1978). Masaki studied the life history of cricket populations over a North-to-South transect in Japan. In the species *Pteronemobius fascipes*, geographic variation was found in photoperiodic responses, controlling induction of diapause and development of nymphs, and in the determination of wing form. The most northern populations were univoltine. Under conditions of long-day length, these populations have an extended period of nymphal development that results in production of large adults. In southern areas, long-day length results in an opposite effect, a relatively shorter period of nymphal growth and small adults. This allows completion of the first generation in time to produce a midsummer generation. At intermediate latitudes, Masaki (1972) found populations composed of both univoltine and bivoltine crickets. Working with a second cricket species, *Pteronemobius taprobanensis*, Masaki (1978) showed similar patterns of latitude-related changes in life history characteristics.

Response to Temperature There seems to be a great deal of geographic variation within insect species for many ecologically relevant traits. An exception to this pattern involves the growth response of individuals to temperature. Although a number of studies of poikilotherms have documented geographic variation in the effects of temperature on development (e.g., Moore, 1949; Birch et al., 1963; Lonsdale and Levinton, 1985; Baldwin and Dingle, 1986), most studies have found little variation in this trait (Tauber and Tauber, 1987). In the work of Baldwin and Dingle (1986) and Lonsdale and Levinton (1985), it was found that populations from cold climates grew faster at low temperatures than populations from areas with warm climates; the converse was also found. The work of Tauber and Tauber (1987) found a more common result. There was a 17% difference in the thermal requirements (for complete development) of five populations of *Chrysopa oculata* Say from Manitoba, Canada, to Coahuila, Mexico. Even this small difference could be attributed to variation in adaptation of the strains in laboratory conditions (the population requiring the most thermal units had an average laboratory mortality rate of 42% across all temperatures, while the population requiring the least thermal units had an average mortality rate of 12%).

Ritland and Scriber (1985) analyzed the thermal response of tiger swallowtail larvae collected from 27 degrees north latitude in Florida to 50 degrees north latitude in Manitoba, Canada. These larvae, which are considered to come from distinct subspecies, had essentially the same developmental response to temperatures ranging from 16°C to 28°C.

Host Utilization Large-scale variation in host utilization has been recorded for a number of parasites and phytophagous insects (Mitter and Futuyma, 1983; Gould, 1983; Diehl and Bush, 1984; Futuyma and Peterson, 1985). For example, Hsiao (1978) documented host use differences among geographic strains of the Colorado potato beetle collected from Arizona, Utah, and Texas. The host utilization patterns of these strains in part reflected the hosts that were available to them. More recent work by Hare and Kennedy (1986) found a similar pattern of variation in this beetle's ability to utilize horsenettle and related the differences to seasonal availability of potato, which is an alternate host (Kennedy, Chapter 22 in this book).

Scriber (1983) documented variation in host use by swallowtail butterflies (*Papilio* spp.) in the United States and Canada. This variation also reflected changes in host availability. For example, *Papilio glaucus glaucus* L. and its northern subspecies, *P. glaucus canadensis*, differ in their host utilization abilities. *P. glaucus canadensis* cannot survive on tulip trees but can survive on quaking aspen. The reverse is true of larvae of *P. glaucus glaucus*. These patterns of larval qualities reflect the availability of the two host plants in the geographic areas where these subspecies occur.

It is comforting to find patterns of host use that follow changes in host availability. Unfortunately, not all patterns of host adaptation are easy to interpret. Detailed studies of host use patterns of the cabbage aphid, *Brevicoryne brassicae* (L.), and the brown planthopper, *Nilaparvata lugens* Stål, reveal complex patterns of variation in host utilization ability. In their study of the cabbage aphid, Dunn and Kempton (1972) tested the ability of seven geographic aphid clones to utilize eight genetically distinct Brussels sprout clones. Each of the aphid clones exhibited a unique pattern of host utilization ability. Reanalysis of these data (Gould, 1983) revealed no relationship between the geographic proximity of clones and the similarity of their host use pattern. Indeed, two aphid clones from the same district in England were far more dissimilar in host use than two other aphid clones collected from different areas of the country. The factors responsible for this pattern are unknown.

Claridge et al. (1982) presented data on the ability of six strains of *N. lugens* to grow on nine types of rice. The planthopper strains that were collected from geographic locations in Sri Lanka varied in their host use patterns in a manner unrelated to location. In this study, differences between the types of rice from which strains were collected explained some of the variation among clones.

Small-Scale Variation

In most of the studies referred to earlier, individuals representing a location were derived from one sample, and the locations were generally over 50-km apart. This information gives us a general understanding of the maximum range of variation in a trait, but we cannot determine from these studies whether this range of variation can be found at a more localized scale.

Diapause Most published studies of diapause variation have examined large-scale variation. The dearth of published studies on local variation in diapause could be due to the fact that such investigations usually come up with negative results or simply that few of these studies have been conducted. One interesting exception to this tendency is the work by Cranham (1973) on intensity of diapause of winter eggs of the spider mite, *Panonychus ulmi* (Koch). Cranham collected winter eggs from six orchards in Kent, England. The eggs were kept in an open shelter and hatch time was recorded. The dates at which 50% of the eggs from the six orchards hatched were May 24, May 28, and May 29, and June 1, June 10, and June 14, a range of three weeks. Cranham concluded that any predictions of hatch for control purposes would have to be on an orchard-by-orchard basis. Finch and Collier (1983) came to a similar conclusion for cabbage root fly over winter emergence, but they worked on a somewhat larger spatial scale.

Istock (1981) examined genetic variation in the mosquito *Wyeomyia smithii* (Coquillet) on a local scale. He found significant genetic variation for diapause traits at a very localized level, but the extent of variation was not as great as the variation found by Bradshaw (1976) on a larger geographic scale.

Host Utilization In the past 10 years, there has been an increase in interest regarding genetic variation in host use within populations. Much of this work has been reviewed in detail (Gould, 1983; Mitter and Futuyma, 1983; Diehl and Bush, 1984; Futuyma and Peterson, 1985). Two approaches have been taken in these studies. One approach involves selection for a change in host use (e.g., Gould, 1979; Wasserman and Futuyma, 1981); the other involves assessment of differences in host use by progeny from different females that come from one population (e.g., Tabashnik et al., 1981; Moran, 1981; Stanton and Cook, 1984; Via, 1984; Rausher, 1984; Weber, 1986). The general result from these studies is that genetic variation for host use is likely to be found within herbivore populations.

On a slightly larger scale, a number of studies compared host use by strains collected within a small area. For example, Service (1984) found that aphid clones, *Uroleucon rudbeckiae* (Fitch), collected from four locations in the small town of Chapel Hill, North Carolina, varied genetically in their ability to utilize four genetically distinct clones of *Rudbeckia laciniata* L. Similarly, Gould (1978) and Tavormina (1982) (and personal communication) found genetic variation in host use ability within areas smaller than 5 square kilometers.

Comparative Studies of Small- and Large-Scale Variation

The studies just mentioned examine large- or small-scale geographic variation. Given the amount of variation found in some of the small-scale studies, it becomes difficult to determine if the variation found in some large-scale studies is due to small-scale noise. For example, if one sample is taken from each of 10 large geographic areas and the range of growth on a host plant is found to be four-fold, how do we know that this same four-fold variation could not have been found within a square kilometer?

Host Utilization The first study I could find that addressed this question quantitatively was conducted by Bournoville (1981). Bournoville assessed variation in the pea aphid's ability to use *Medicago sativa* L. at three levels: between aphid populations, between individuals in a single population (one field), and within individuals (by assessing asexually produced offspring of single females). By using an analysis of variance design, Bournoville was able to assess the significance of the variance contributed at the within- and among-population level. Almost all of the variation found in this study could be attributed to the within-population component of variance.

This result may seem somewhat surprising. However, a number of less quantitative studies of host use support this finding. For example, Pathak and Painter (1959) examined the ability of corn leaf aphids from many areas of Kansas to grow on sorghum, barley, and wheat. Their graphical presentation of the results indicate that local variation was as great as large-scale variation. Similarly, Gould (1978) found that populations of the mite *Tetranychus urticae* (Koch) collected from fields in Suffolk Co., New York, were as different in their ability to survive on resistant cucumber as populations from the East and West Coasts of the United States. Jaenike and Grimaldi (1983) found that strains of the fungus and fruit-feeding insect, *Drosophila tripunctata* Loew, collected from one area in the Smokey Mountains were almost as variable in host use as strains from wide areas of the East Coast. More detailed studies on this system (Jaenike, in review) confirm this observation.

Weber (1986) assessed the reproductive output of *Myzus persicae* (Sulzer) clones reared on sugar beet and potato. Clones were collected from sugar beet and potato fields in growing regions that specialized in potato or sugar beet production or that produced both crops. Weber found significant differences in reproductive rates that could be explained by the type of crop grown in the field from which the clone originated. Clones from large regions that produced only sugar beets or potato were no more adapted to their respective hosts than clones from the mixed production area that were taken from the middle of fields that were larger than 5 hectares in size. He concluded that "the spatial unit of differentiation was below an average field size of 3–5 hectares." Interestingly, there was temporal variation in the extent of differentiation, even in specialized cropping regions. This indicated that aphids adapted to the crop planted in a field over the season.

Insecticide Resistance Populations of agricultural insect pests have long been known to vary in their tolerance of synthetic pesticides, but, until recently, little effort was directed at analyzing the geographic pattern of variation with regard to this characteristic. Follett et al. (1985) assessed insecticide resistance of pear psylla populations among farms within production areas and among production areas that differed in intensity of insecticide use. The average resistance of populations from different production areas did vary with pesticide use patterns. Overall, the variance in resistance within production areas was less than the variance among areas, although statistical significance was not established.

Rosenheim and Hoy (1986) conducted a similar study on pesticide resistance in the parasitoid, *Aphytis melinus* DeBach. Although they did not use an analysis of variance approach to separate variation into within and among county components, their multiple-regression approach showed a contribution at both the among-farm and among-county levels.

In a statistically rigorous analysis of patterns of insecticide resistance in the diamondback moth, Tabashnik et al. (1987) found significant variation among populations within the islands of Hawaii. They did not, however, find a significant contribution to variation in resistance at the among-island level. Of 11 possible comparisons of neighboring diamondback moth populations within 10 km of each other, Tabashnik et al. (1987) found 5 that differed significantly in their LC_{50}'s. The authors attributed this localized pattern of variation to local differences in past and present intensity of insecticide treatments.

Vector Competence An important ecological characteristic of mosquitoes involves their ability to act as vectors of human and animal diseases. Tabachnick et al. (1985) reviewed the literature on geographic variation in vector competence and conducted a worldwide survey of the ability of *Aedes aegypti* (L.) to become orally infected with yellow-fever virus. Results of the survey of 28 populations indicated that oral infection of *A. aegypti* strains ranged from 7% in one Senegal, Africa, strain to 57% in a strain from Kenya, Africa. Although populations from some geographic areas were generally homogeneous in percent infection (e.g., the Pacific region ranged from 21% to 30%), other areas were quite variable (southeastern United States, 12% to 41%, small sample sizes; Kenya, 29% to 57%, large sample sizes).

In a companion study, Wallis et al. (1985) selected strains of *A. aegypti* for high and low infection rate. The parent strain initially had a 15% infection rate. The selected lines ranged from 11% to 29%; thus, one strain contained genetic variants that were more dissimilar than field strains from the Pacific region. (It is important to realize that the laboratory-selected variation may have carried negative pleiotropic effects in terms of fitness in the field.)

Response to Temperature Recently, Lamb et al. (in press) examined the temperature-mediated development of the aphid *Acyrthosiphon pisum* (Harris) from latitudes of 39 degrees to 53 degrees North. They tested samples from three clones from each of five locations at temperatures ranging from 9.8°C to 27.9°C. They found no biologically significant relationship between latitude and response to temperature. They did, however, find that clones from a single location varied in their responses to temperature. They concluded from their work that findings of previous studies demonstrating geographic differences in response of this aphid to temperature may have resulted from inadequate sampling of the aphid population from each local area.

GENETIC VERSUS ENVIRONMENTAL DETERMINANTS OF SPATIAL VARIATION

In the section, "Empirical Observations," I made the assumption that the reported phenotypic variation reported among and within populations has a genetic basis. Indeed, it is difficult to prove that such phenotypic variation is the result of genetic differences. While I will not attempt to dissect the evidence for genetic differences presented in each study, it is worth noting some of the limitations involved in these studies.

The simplest approach used to establish differences between strains involves rearing them in the laboratory for one or more generations and noting differences in the traits of interest. If, for example, we were measuring growth rate on a certain plant, genes directly coding for growth rate on that plant would be the likely cause of the phenotypic differences. It is possible, however, that one strain may be more adapted to laboratory rearing than another, which may cause more rapid growth. Maternal and paternal effects could also cause phenotypic differences among strains (e.g., Greenblatt and Barbosa, 1980; Gould, 1988). If parents of one strain were taken from a physically stressful environment, offspring vigor could be lowered. In many cases, vigor should return after a number of generations in the laboratory, unless laboratory conditions were also stressful for the species and the differences in the initial vigor of the strains became exaggerated.

In many cases, laboratory strains are found to be infected with nonlethal pathogens, such as microsporidians, that persist in a colony and reduce vigor (Gaugler and Brooks, 1975; Wilson, 1980; Siegel et al., 1986). Since some pathogens can be sexually transmitted by males, even half-sibling, quantitative genetic designs are not perfect. Such half-sibling designs are, however, more robust than parent offspring regressions that are commonly used to infer a genetic basis for a phenotypic difference. In such designs, direct maternal effects are not separated from genetic effects.

Before we assumed that a phenotypic difference was due to genetic causes, it would seem reasonable to do simple F_1 crosses to determine if the male origin affects the trait in its offspring. In parthenogenic species, this is not possible, although Lynch (1985), working with copepods, presents a possible alternative for such cases. In all cases, it is advisable to check the disease status of strains by conferring with an appropriate invertebrate pathologist. Such precautions could save wasted effort in the long run.

SPATIAL SCALE OF VARIATION AND THE STUDY OF INSECT ECOLOGY

From the brief review of case studies presented in the section, "Empirical Observations," it is obvious that ecologically important traits often vary over space. Sometimes the space is smaller than 5 km; at other times, it is on the scale of hundreds of kilometers. It is often difficult to judge whether spatial variation is due strictly to genetic differences, to maternal effects, to persistent environmental effects (including nonlethal diseases), or to combinations of these factors. Although understanding the cause of these differences could be very useful, it is important to deal with how the variation itself will affect ecological phenomena and the way we study them.

In cases in which we have no capability of predicting the pattern of variation in a biologically important trait [e.g., egg hatch of overwintering mites (Cranham, 1973)], developing general population dynamics models for a species would seem foolhardy. In other cases in which the magnitude of variation is great but predictable [e.g., diapause induction in some mosquitoes (Bradshaw, 1976)], models could "correct" for this variation.

Given the common occurrence of spatial variation in ecological traits, how should a population ecologist begin to study an insect species? It takes years to develop a data set on the ecological parameters of a single population. If one attempts to collect the same data on even three or four populations, the burden of data collection could become enormous. There are no simple solutions to this problem, but careful planning of experimental design could be useful.

In initiating the study of a species population biology, it may be useful to get baseline data on one population. Once these baseline data have been used to develop a population dynamics model, a sensitivity analysis (for a given trait) could be used to determine how big a difference between populations would be necessary to cause an important change in population dynamics. Knowing how large the difference would have to be, experiments on other populations could be designed to be only powerful enough (Zar, 1984) to detect such differences.

Another approach for laboratory study would be to use two replicates for each experiment. Insects for one replicate would come from a strain originating from one small, localized sample of one insect population. The second replicate would use insects from a strain composed of individuals originating from many geographic areas and environments. If either the mean or the variance for the two replicates differed significantly, spatial heterogeneity in the characteristic would be implicated. Such a finding could be followed up by more rigorous experiments.

In many cases, entomologists do not need a detailed population dynamics model. For example, they may only need to know about pesticide resistance levels or diapause initiation conditions. In these situations where variation for the important traits is commonly present, it may be wise to study a number of populations from the start.

SPATIAL SCALE OF VARIATION AND RESPONSES OF POPULATIONS TO SELECTION

Most models of natural population responses to selection assume infinitely large populations with random mating. Given these assumptions, the frequencies of alleles coding for an evolving trait are considered homogeneous within a population at a given time. Even models that assume subdivided natural populations only consider two evolving subpopulations, and each subpopulation is treated as being very large (i.e., has the population genetic characteristics of a population with infinite size) (Comins, 1977; Gould, 1986). Animal breeders and quantitative geneticists have examined the effects of population subdivision on response to selection but have not included genotype-specific population dynamics or low rates of gene flow in their models (Robertson, 1952; Madalena and Hill, 1972; Bryant et al., 1986a, 1986b; Wade and McCauley, 1984).

The spatial variation described in some of the previous studies indicates that some insect populations are not genetically homogeneous. Furthermore, studies on insect movement and mating (McCauley et al., 1981; McCauley, 1987) have estimated small effective population sizes in some cases. As described qualitatively in the first section of this chapter, the smaller the population size is, the greater is the expected spatial variation. On a quantitative level for diploid populations, we have

$$\sigma^2_{qt} = (q_o)(1 - q_o) \left[1 - (1 \frac{1}{2N_e})^t \right]$$

where σ^2_{qt} is the variance in gene frequency q among populations due to t generations of drift. N_e is effective population size (Hartl, 1980). The smaller N_e is, the larger the variance in q is likely to be among populations. Mitter and Schneider (1987) have pointed out that, for effectively haplodiploid populations, this variation can be one-third larger at any standardized population size. Schneider (unpublished data) has examined how genetic variation among subdivided Hessian fly populations could contribute to the rapid rate at which they adapt to resistant wheat cultivars. Assuming that Hessian fly alleles for adaptation are recessive, the expected increase in allelic frequency due to selection is very slow when the baseline allelic frequency is below 0.05. Schneider (unpublished data) points out that, if a Hessian fly population is composed of many small subpopulations with low rates of migration between them, stochastic factors should cause some subpopulations to have allelic frequencies higher than 0.1 before they are subjected to selection. These subpopulations could respond rapidly to selection.

Paulson and Gould (unpublished data), building on Schneider's idea, have used a simulation model to show how the rate of migration between subdivided insect populations could contribute to the rate at which such recessive traits evolve. As migration increases, the high allelic frequencies in outlier subpopulations are swamped out and the rate of adaptation slows down dramatically. Figure 3.4 shows the results of "pseudostochastic" computer simulations in which the rates of migration varied between subpopulations with and without high initial allelic frequencies. We assumed a population that had an average frequency of adaptive alleles at locus 1 and 2 of 0.006. The population was divided into 25 subpopulations, with migration of mated females between them. If all subpopulations have an initial gene frequency of 0.006, it takes over 300 generations of selection for adaptive alleles to become fixed. If, instead of homogenous subpopulations, we assumed that 1 subpopulation had an allelic frequency of 0.15 and the remaining 24 had a frequency of 0.00 (average allelic frequency = 0.006), the rate of fixation of the adaptive alleles would depend on migration rates between subpopulations. In the figures presented, the migration rate of 8.5% per generation led to rapid fixation, whereas a migration rate of 8.7% led to very slow adaptation, because migration equalized gene frequencies before they had time to increase in the single high-frequency subpopulation. If, instead of the high frequency subpopulation having an allelic frequency of 0.15, the frequency was 0.16, then migration of 11% or less would still allow rapid fixation. If the initial frequency was 0.14, a migration rate of 6.4% or more would lead to a very slow rate of fixation.

The effect of subdivision on rates of population response to selection could be important when a large component of the genetic variance in a trait is due to dominance variation or epistatic interaction of two or more loci. When traits are, for the

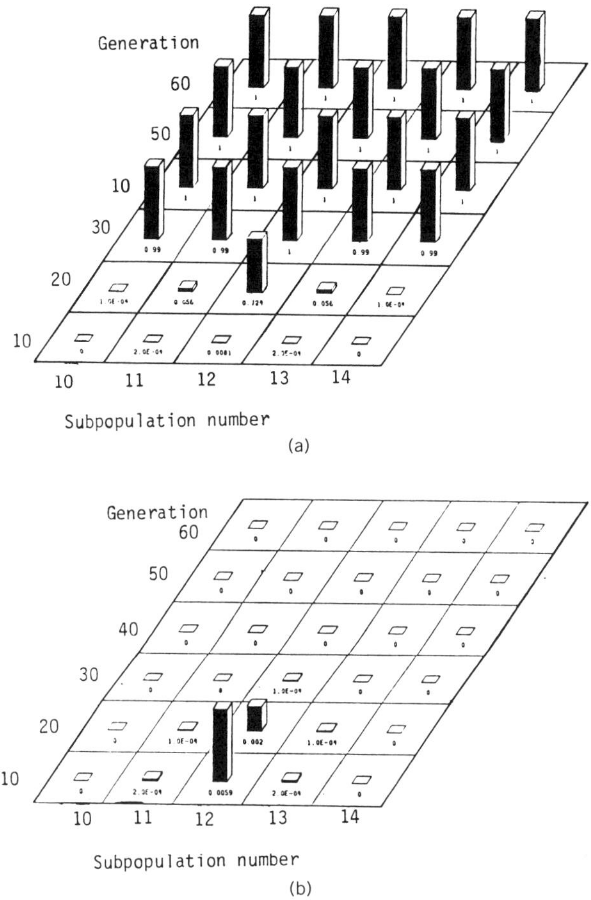

Figure 3.4. Results of computer simulations depicting the effects of population heterogeneity and migration on the rate of which a population responds to selection. Initial average gene frequency at locus 1 and 2 is 0.006. Fitness of the genotype, homozygous for adaptive alleles at both loci, is 1.0. Other genotypes have fitness of 0.04 as in Gould (1986). Each figure depicts the frequency of the adaptive alleles in 5 of the 25 subpopulations over time. In A and B, all subpopulations except the central one have an initial frequency of 0.00. The central subpopulation has an initial frequency of 0.15. In A, the migration rate to adjacent subpopulations is 8.5%. In B, the migration rate is 8.7% (From Paulson and Gould, unpublished data).

most part, inherited in an additive fashion, subdivision of the population is much less important to the overall rate at which the population adapts to a stress.

Even if the genetic variance for a trait were additive, subdivision of a population could influence equilibrium gene frequencies, as depicted in Figs. 3.2 and 3.3. Thus, an understanding of population structure might help us understand the patterns of genetic variation that we find in such traits as diapause initiation and termination where the average selective regime is not undergoing obvious and dramatic changes.

CONCLUDING REMARKS

Studies that have surveyed ecologically relevant traits in insect species have often found spatial variation in these traits. In studying an insect species, it is no longer realistic to assume that a species or even a population is genetically homogeneous for ecologically relevant traits.

Once we give up the assumption of homogeneity, we are faced with the question of how to gain a good understanding of a species population ecology. Population biologists who are wrestling with the lack of homogeneity of an insect species environment must also deal with genetic heterogeneity in the insect and the interactions between a variable environment and a varying insect. This is not a simple challenge, but it is a challenge that we will need to meet if we are to understand the ecology of insects more accurately.

Acknowledgments

Helpful comments on earlier versions of this manuscript by J. R. Bradley, P. Follett, G. G. Kennedy, T. MacKay, A. Sheck, K. Suiter, and anonymous reviewers were much appreciated. This work was supported in part by grant no. USDA-CRG 85CRCR-1-1687.

REFERENCES

Baldwin, J. D., and H. Dingle. 1986. Geographic variation in the effects of temperature on life-history traits in the large milkweed bug *Oncopeltus fasciatus*. *Oecologia* 69:64–71.

Birch, L. C., T. Dobzhansky, P. O. Elliott, and R. C. Lewontin. 1963. Relative fitness of geographic races of *Drosophila serrata*. *Evolution* 17:72–83.

Bournoville, R. 1981. Variability of the net reproductive rate of clones of the pea aphid on lucerne. Pages 129–132 in *Breeding Resistance Insects Mites*, Bull. SROP WPRS Bull., Rep. 2nd Meet., 1980.

Bradshaw, W. E. 1976. Geography of photoperiodic response in diapausing mosquito. *Nature* 262:384–385.

Bryant, E. H., S. A. McCommas, and L. M. Combs. 1986a. The effect of an experimental bottleneck upon quantitative genetic variation in the housefly. *Genetics* 114:1191–1211.

———. 1986b. Morphometric differentiation among experimental lines of the housefly in relation to a bottleneck. *Genetics* 114:1213–1223.

Claridge, M. F., J. Den Hollander, and I. Furet. 1982. Adaptations of brown planthopper *Nilaparvata lugens* populations to rice varieties in Sri Lanka. *Entomol. Exp.* Appl. 32:222–226.

Comins, H. N. 1977. The management of pesticide resistance. *J. Theor. Biol.* 65:399–420.

Cranham, J. E. 1973. Variation in the intensity of diapause in winter eggs of fruit tree red spider mite, *Panonychus ulmi*. *Ann. Appl. Biol.* 75:173–182.

Diehl, S. R., and G. L. Bush. 1984. An evolutionary and applied perspective of insect biotypes. *Annu. Rev. Entomol.* 29:471–504.

Dobzhansky, T. 1970. *Genetics of the Evolutionary Process*. New York: Columbia Univ. Press.

Dunn, J. A., and D. P. H. Kempton. 1972. Resistance to attack by *Brevicoryne brassicae* among plants of Brussels sprouts. *Ann. Appl. Biol.* 72:1–11.

Finch, S., and R. H. Collier. 1983. Emergence of flies from overwintering populations of cabbage root fly pupae. *Ecol. Entomol.* 8:29–36.

Follett, P. A., B. A. Croft, and P. H. Westigard. 1985. Regional resistance to insecticides in *Psylla pyricola* from pear orchards in Oregon. *Can. Entomol.* 117:565–573.

Futuyma, D. J., and S. C. Peterson. 1985. Genetic variation in the use of resources by insects. *Annu. Rev. Entomol.* 30:217–238.

Gaugler, R. R., and W. M. Brooks. 1975. Sublethal effects of infection by *Nosema heliothidis* in the corn earworm, *Heliothis zea*. *J. Invert. Pathol.* 26:57–63.

Gould, F. 1978. Predicting the future resistance of crop varieties to pest populations: A case study of mites and cucumbers. *Environ. Entomol.* 7:622–626.

———. 1979. Rapid host range evolution in a population of the phytophagous mite, *Tetranychus urticae* Koch. *Evolution* 33:791–802.

———. 1983. Genetics of plant-herbivore systems: Interactions between applied and basic study. Pages 599–653 in *Denno, R., and B. McClure (eds.), Variable Plants and Herbivores in Natural and Managed Systems*. New York: Academic Press.

———. 1986. Simulation models for predicting durability of insect-resistant germ plasm: Hessian fly (Diptera: Cecidomyiidae)-resistant winter wheat. *Environ. Entomol.* 15:11–23.

———. 1988. Stress specificity of maternal effects in *Heliothis virescens* (Lepidoptera: Noctuidae) larvae. *Mem. Can. Entomol. Soc.* 146: 191–197.

Greenblatt, J. A., and P. Barbosa. 1980. Interpopulation quality in gypsy moths with implications for success of two pupal parasitoids: *Brachymeria intermedia* (Nees) and *Coccygomimus turionellae* (L.). *Ecol. Entomol*. 5:31–38.

Hare, J. D., and G. G. Kennedy. 1986. Genetic variation in plant-insect associations: Survival of *Leptinotarsa decemlineata* populations on *Solanum carolinense*. *Evolution* 40:1031–1043.

Hartl, D. L. 1980. *Principles of Population Genetics*. Sunderland, Mass.: Sinauer Assoc.

Hedrick, P. W. 1983. *Genetics of Populations*. Boston: Jones & Bartlett Publ.

Holzapfel, C. M., and W. E. Bradshaw. 1981. Geography of larval dormancy in the tree-hole mosquito, *Aedes triseriatus* (Say). *Can. J. Zool.* 59:1014–1021.

Hsiao, T. H. 1978. Host plant adaptations among geographic populations of the Colorado potato beetle. *Entomol. Exp. Appl.* 24:437–447.

Istock, C. A. 1981. Natural selection and life history variation: Theory plus lessons from a mosquito. Pages 113–127 in *Denno, R. F., and H. Dingle (eds.), Insect Life History Patterns*. New York: Springer Verlag.

Jaenike, J., and D. Grimaldi. 1983. Genetic variation for host preference within and among populations of *Drosophila tripunctata*. *Evolution* 37:1023–1033.

Jordan, R. G., and W. E. Bradshaw. 1978. Geographic variation in the photoperiodic response of the western tree-hole mosquito, *Aedes sierrensis*. *Ann. Entomol. Soc. Am.* 71:487–490.

Keightley, P. D., and W. G. Hill. 1987. Directional selection and variance in finite populations. *Genetics* 117:573–582.

Lamb, R.J., P.A. MacKay, and G.H. Gerber. In press. Are development and growth of pea aphids, *Acyrthosiphon pisum*, in North America adapted to local temperatures? *Oecologia*.

Lonsdale, D. J., and J. S. Levinton. 1985. Latitudinal differentiation in copepod growth: An adaptation to temperature. *Ecology* 66:1397–1407.

Lynch, M. 1985. Spontaneous mutations for life-history characters in an obligate parthenogen. *Evolution* 39:804–818.

Madalena, F. E., and W. G. Hill. 1972. Population structure in artificial selection programmes: Simulation studies. *Genet. Res.* 20:75–99.

Madalena, F. E., and A. Robertson. 1975. Population structure in artificial selection: Studies with *Drosophila melanogaster*. *Genet. Res.* 24:113–126.

Masaki, S. 1972. Climatic adaptation and photoperiod response in the band-legged ground cricket. *Evolution* 26:587–600.

———. 1978. Climatic adaptation and species status in the lawn ground cricket. II. Body size. *Oecologia* 35:343–356.

McCauley, D. E. 1987. Population genetic consequences of local colonization: Evidence from the milkweed beetle *Tetraopes tetraophthalmus*. *Fla. Entomol.* 70:21–30.

McCauley, D. E., J. R. Ott, A. Stine, and S. McGrath. 1981. Limited dispersal and its effect on population structure in the milkweed beetle *Tetraopes tetraophthalmus*. *Oecologia* 51:145–150.

Mitter, C., and J. Futuyma. 1983. An evolutionary-genetic view of host plant utilization by insects. Pages 427–459 in *Denno, R. F., and M. S. McClure (eds.), Herbivores and Their Variable Hosts in Natural and Managed Systems*. New York: Academic Press.

Mitter, C., and J. C. Schneider. 1987. Genetic change and insect outbreaks. Pages 505–532 in *Barbosa, P., and J. Schultz (eds.), Insect Outbreaks*. New York: Academic Press.

Moore, J. A. 1949. Geographic variation of adaptive characters in *Rana pipiens* Schreber. *Evolution* 3:1–24.

Moran, N. 1981. Intraspecific variability in herbivore performance and host quality: A field study of *Uroleucon caligatum* (Homoptera: Aphididae) and its *Solidago* hosts (Asteraceae). *Ecol. Entomol.* 6:301–306.

Pathak, M. D., and R. H. Painter. 1959. Geographical distribution of the four bio types of corn leaf aphid, *Rhopalosiphum maidis* (Fitch), in Kansas. *Trans. Kans. Acad. Sci.* 62:1–8.

Rausher, M. D. 1984. Trade-offs in performance on different hosts: evidence from within- and between-site variation in the beetle *Deloyala guttata*. *Evolution* 38:582–595.

Ritland, D. B., and J. M. Scriber. 1985. Larval developmental rates of three putative subspecies of tiger swallowtail butterflies, *Papilio glaucus*, and their hybrids in relation to temperature. *Oecologia* 65:185–193.

Robertson, A. 1952. The effect of inbreeding on the variation due to recessive genes. *Genetics* 37:189–207.

Rosenheim, J. A., and M. A. Hoy. 1986. Intraspecific variation in levels of pesticide resistance in field populations of a parasitoid, *Aphytis melinus* (Hymenoptera: Aphelinidae): The role of past selection pressures. *J. Econ. Entomol.* 79:1161–1173.

Scriber, J. M. 1983. The evolution of feeding specialization, physiological efficiency and host races in selected Papilionidae and Saturniidae. Pages 373–412 in *Denno, R. F. and M. S. McClure (eds.), Impact of Variable Host Quality on Herbivorous Insects*. New York: Academic Press.

Service, P. 1984. Genotypic interactions in an aphid-host plant relationship: *Uroleucon rudbeckiae* and *Rudbeckia laciniata*. *Oecologia* 61:271–276.

Siegel, J. P., J. V. Maddox, and W. G. Ruesink. 1986. Lethal and sublethal effects of *Nosema pyrausta* on the European corn borer (*Ostrinia nubilalis*) in central Illinois. *J. Invert. Pathol.* 48:167–173.

Slatkin, M. 1985. Gene flow in natural populations. *Annu. Rev. Ecol. Syst.* 16:393–430.

Stanton, M. L., and R. E. Cook. 1984. Sources of intraspecific variation in the host plant-seeking behavior of *Colias* butterflies. *Oecologia* 60:365–370.

Tabachnick, W. J., G. P. Wallis, T. H. G. Aitken, B. R. Miller, G. D. Amato, L. Lorenz, J. R. Powell, and B. J. Beaty. 1985. Oral infection of *Aedes aegypti* with yellow fever virus: Geographic variation and genetic considerations. *Am. J. Trop. Med. Hyg.* 34:1219–1224.

Tabashnik, B. E., N. L. Cushing, and M. W. Johnson. 1987. Diamondback moth (Lepidoptera: Plutellidae) resistance to insecticides in Hawaii: Intra-island variation and cross-resistance. *J. Econ. Entomol.* 80:1091–1099.

Tabashnik, B. E., H. Wheelock, J. D. Rainbolt, and W. B. Watt. 1981. Individual variation in oviposition preference in the butterfly, *Colias eurytheme*. *Oecologia* 50:225–230.

Tauber, C. A., and M. J. Tauber. 1987. Thermal requirements for developmemt in *Chrysopa oculata*: A geographically stable trait. *Ecology* 68:1479–1487.

Tauber, M. J., C. A. Tauber, and S. Masaki. 1986. *Seasonal Adaptations of Insects*. New York: Oxford Univ. Press.

Tavormina, S. J. 1982. Sympatric genetic divergence in the leaf-mining insect *Liriomyza brassicae* (Diptera: Agromyzidae). *Evolution* 36:523–534.

———. 1990. Personal communication.

Trimble, R. M., and S. M. Smith. 1979. Geographic variation in the effects of temperature and photoperiod on dormancy induction, development time, and predation in the tree-hole mosquito, *Toxorhynchites rutilus septentrionalis* (Diptera: Culicidae). *Can. J. Zool.* 57:1612–1618.

Via, S. 1984. The quantitative genetics of polyphagy in an insect herbivore. II. Genetic correlations in larval performance within and among host plants. *Evolution* 38:896–905.

Wade, M. J., and D. E. McCauley. 1984. Group selection: The interaction of local deme size and migration in the differentiation of small populations. *Evolution* 38:1047–1058.

Wallis, G. P., T. H. G. Aitken, B. J. Beaty, L. Lorenz, G. D. Amato, and W. J. Tabachnik. 1985. Selection for susceptibility and refractoriness of *Aedes aegypti* to oral infection with yellow fever virus. *Am. J. Trop. Med. Hyg.* 34:1225–1231.

Wasserman, S. S., and D. J. Futuyma. 1981. Evolution of host plant utilization in laboratory populations of the southern cowpea weevil, *Callosobruchus maculatus* Fabricius (Coleoptera: Bruchidae). *Evolution* 35:605–617.

Weber, G. 1986. Ecological genetics of host plant exploitation in the green peach aphid, *Myzus persicae*. *Entomol. Exp. Appl.* 40:161–168.

Wilson, G. G. 1980. III. Note. Effects of *Nosema fumiferanae* (Microsporidia) on rearing stock of spruce budworm, *Choristoneura fumiferana* (Lepidoptera: Tortricidae). *Proc. Entomol. Soc. Ont.* 111:115–116.

Wright, S. 1951. The genetical structure of populations. *Ann. Eugenics* 15:323–354.

Zar, Z. H. 1984. *Biostatistical Analysis*. Englewood Cliffs, N.J.: Prentice-Hall.

SECTION II

THE NATURE OF GENETIC VARIATION

4

HABITAT-RELATED VARIATION AND EVOLUTION BY SEXUAL SELECTION

Kenneth Y. Kaneshiro

University of Hawaii

INTRODUCTION

Insect pests cause appalling economic losses for humankind, yet surprisingly little attention has been given to investigating the origin of pest populations. In a recent paper, Mayr (1982) makes the following statement: "Speciation . . . now appears as the key problem of evolution. It is remarkable how many problems of evolution cannot be fully understood until speciation is understood. . . ." Thus, research on the mechanisms of speciation in insect pest species is an important first step in understanding the basic biology of pest populations.

It is a well-accepted theorem that *adaptation* is the central theme of the evolutionary process and that natural selection results from the interaction between heritable phenotypic variation within a population and the extrinsic environment in which it lives. In discussing the importance of variation in the development of his theory of natural selection, Darwin (1859) makes the following observations:

> It may metaphorically be said that natural selection is daily and hourly scrutinising, throughout the world, the slightest variations; rejecting those that are bad, preserving and adding up all that are good; silently and insensibly working, whenever and wherever opportunity offers, at the improvement of each organic being in relation to its organic and inorganic conditions of life. We see nothing of these slow changes in progress, until the hand of time has marked the lapse of ages, and then so imperfect is our view into long-past geological ages, that we see only that the forms of life are now different from what they formerly were.

There is little doubt that the selective forces exerted by the external environment (Darwin's "organic and inorganic conditions of life") of a population play an important role in regulating its effective size and that the plasticity of the population in adjusting to rapid changes in its environment influences its direction of evolution. On the other hand, based on recent studies of mating systems in some endemic species of Hawaiian *Drosophila*, I suggest that adjustments to the intrinsic "sexual environment" of the population play an even more significant role in rapid changes in the population than previously believed; that is, although it is a well-accepted notion that natural selection is a dominant force in maintaining the optimum phenotype for survival in its environment, I suggest that the forces of sexual selection may be playing a more important role in the initial stages of species formation, as well as maintaining the optimum phenotype for that environment.

The tremendous diversity we find in the endemic fauna of Hawaiian *Drosophila* may provide some insights into understanding the role of sexual selection versus natural selection in molding the evolutionary history of this group of insects. While this group is of little economic or medical importance to humankind, more than 20 years of basic research on the evolutionary biology of Hawaiian drosophilids have made significant contributions to the understanding of the dynamics of evolutionary phenomena (see reviews in Carson et al., 1970; Carson and Kaneshiro, 1976; Hardy, 1971; Kaneshiro, 1983; Williamson, 1981; Zimmerman, 1970). Williamson (1981) writes of the supremacy of the Hawaiian Drosophilidae for understanding island biology:

> Of all the groups of organisms, plants or animals, that can be studied on islands, the Hawaiian Drosophilidae are supreme. . . . Any other study of evolution must surely seem inadequate after the study of the Hawaiian Drosophilidae. In no other group is so much known, are there so many species, and can the ecological relationships be clarified so well. They also show the advantage of studying evolutionary phenomena on islands where one group dominates the evolutionary scene in a limited theatre. There is still an immense amount of work to be done on the group, but the work carried out so far clearly established their supremacy.

What is even more significant about the results of basic research, such as on the Hawaiian drosophilids, is that this kind of information can now be applied to other organisms, such as insect pest populations, and can provide a solid foundation upon which more mission-oriented research programs can be developed. In this chapter, I will present an intuitive model of the dynamics of the sexual selection system in the Hawaiian *Drosophila* and suggest how such a system may be responsible for maintaining a normal distribution of mating types in the population. I will then develop a more mechanistic model of how differential selection for opposite ends of the mating type distribution in the two sexes and the pleiotropic effects of the sexual selection system on other parts of the genome may play an important role in maintaining levels of genetic variability in the population.

THE HAWAIIAN DROSOPHILIDAE

To date, 484 species in the family Drosophilidae have been described as being endemic (approximately 25 others are exotic) to the Hawaiian Islands (Hardy and Kaneshiro, 1981), and at least 250–300 more undescribed species are already in the collection at the University of Hawaii, awaiting taxonomic treatment. New species are still being collected as previously unsampled localities are studied. It is estimated that there may be as many as 800 or perhaps even 1,000 species present in these islands. Yet, there are several lines of evidence that the entire endemic fauna may have arisen from a single ancestral form that arrived in the Hawaiian Archipelago several million years ago. Considering the small landmass and the geological newness of these islands, the endemic drosophilid fauna constitute one of the most striking examples of adaptive radiation (Carson et al., 1970; Carson and Kaneshiro, 1976; Hardy, 1971, 1974; Kaneshiro, 1983; Williamson, 1981; Zimmerman, 1958, 1970).

The orographic rainfall pattern in these islands presents a unique situation in which extreme climatic conditions can be found within short distances. The prevailing trade winds that blow across the Pacific Ocean 80–95% of the time (Blumenstock and Price, 1972) carry warm, moist air to the islands, most of which precipitates on the windward sides of the island chain. Thus, we find wet, rain forest conditions on the windward side of an island and dry, desertlike conditions on the leeward side only a few kilometers away. The wide diversity of habitats presented by the topographic and climatic features of these islands and the large number of drosophilid species support Raven and Johnson's (1986) statements that ". . . in general, rapid evolution takes place in areas where the habitats are varied and located near one another. The pace of evolution is accelerated further by rapid climatic change."

Indeed, the Hawaiian drosophilids have developed forms that have adapted to the wide diversity of habitats present in these islands. Furthermore, these flies have been opportunistic in exploiting a wide diversity of microniches as larval breeding substrates (Heed, 1968, 1971; and Montgomery, 1975). There is a group of species in the genus *Scaptomyza (Exalloscaptomyza)* that inhabit morning glory flowers from sea level to more than 1,200 meters in elevation. There are species in the genus *Drosophila* that utilize fermenting flowers, leaves, fruits, and bark of many of the endemic plants found in deep rain forest jungles, while the larvae of others may be found in the slime flux exudates formed as secondary yeast or bacterial infections on a number of endemic trees in the dry alpine forests on the slopes of Mauna Kea on the island of Hawaii. Still others utilize various fungi and mushrooms that may be found in the Hawaiian ecosystem. A small group of species belonging to the endemic genus *Titanochaeta* is reported to be predaceous in the egg masses of the endemic spider family Thomisidae (Wirth, 1952; Carson, 1971a). There have even been unpublished observations that a species in the genus *Scaptomyza* may have immature stages that are semiaquatic. Although adults of this species were found in large numbers sitting on large boulders in the middle of a mountain stream, it is believed that the immature stages occur below the surface of the water. Dipteran larvae, which are believed to be those of the boulder-sitting *Scaptomyza* species, were discovered in the green algae scraped from the boulder below the surface of the water. Although we have not been able to confirm that these larvae are indeed immature forms of the adult flies on the boulders (this species has presumably adapted to an aquatic life form in a well-aerated situation and the journey back to the laboratory resulted in an oxygen-starved condition in the plastic bags in which the algae with larvae were collected; all of the larvae suffocated), this potentially represents another bizarre adaptive shift by the Hawaiian drosophilids into an unconventional habitat for Drosophilidae.

There is little reason to deny the fact that the Hawaiian Drosophilidae represent an extraordinary example of explosive adaptive radiation. Clearly, the tremendous diversity of habitats and the availability of a wide assortment of microniches for larval breeding have played an important role in the evolution of such a large number of closely related species coinhabiting such a restricted geographic area. However, recent studies of sexual behavior in this remarkable group of insects indicate that response to a changing intrinsic sexual environment is mostly responsible for the initiation of the speciation process (Carson, 1978, 1985, 1987; Carson and Bryant, 1979; Carson and Lande, 1984; Carson and Teramoto, 1984; Kaneshiro, 1976, 1980, 1983, 1987, 1988; Kaneshiro and Giddings, 1987).

SEXUAL SELECTION THEORY

More than a century ago, Darwin (1859) proposed the concept of sexual selection to explain the sexually dimorphic features that he observed in a wide diversity of organisms. In an attempt to differentiate between sexual and natural selection, Darwin (1859) makes the following statements:

> Sexual selection depends on the success of certain individuals over others of the same sex, in relation to the propagation of the species; while natural selection depends on the success of both sexes, at all ages, in relation to the general conditions of life. Sexual selection is a struggle between individuals of one sex, generally the males, for the possession of the other sex. The result is not death to the unsuccessful competitor, but few or no offspring.

Evolutionary biologists, for the most part, paid little attention to the significance of Darwin's theory of sexual selection, and theoreticians such as A. R. Wallace (1889) and J. Huxley (1938) severely criticized Darwin's ideas. Until recently, most evolutionary biologists accepted the notion that natural selection is the most dominant force in the evolutionary history of the species and that sexual selection is less important than natural selection in bringing about evolutionary changes. Even Darwin himself (1871) stated that ". . . sexual selection will also be dominated by natural selection tending towards the general welfare of the species."

Within the last decade, there has been renewed interest in the significance of sexual selection in evolutionary processes. Thornhill and Alcock's (1983) book on mating systems in insects and a series of papers published in three books edited by Campbell (1972), Blum and Blum (1979), and Bateson (1983) present excellent reviews of recent research results on sexual selection theory.

Associated with various models of sexual selection (Fisher, 1930; O'Donald, 1977, 1980; Lande, 1981, 1982; Kirkpatrick, 1982; including that of Darwin's) is the basic assumption that the forces of natural selection act to counterbalance the runaway process of sexual selection. Thus, it is a well-accepted notion that, while female preference for a certain male phenotype results in selection for extreme forms of that character, natural selection exerts its forces to maintain the optimum male phenotype to survive in its environment. In this chapter, I will suggest an alternative explanation for the maintenance of optimum phenotypes in the population, based solely on the dynamics of the sexual selection process without having to rely on natural selection as a stabilizing force.

ASYMMETRICAL SEXUAL ISOLATION AND SPECIATION VIA THE FOUNDER PRINCIPLE

In a recent series of papers, Kaneshiro (1976, 1980, 1983), Ahearn (1980), Arita and Kaneshiro (1979), Watanabe and Kawanishi (1979), Ohta (1978), Wasserman and Koepfer (1980), Markow (1981), and Heth and Nevo (1981) reported that a frequent outcome of mate preference experiments between a pair of closely related species is asymmetrical sexual isolation; that is, females of one of the populations may accept the courtship displays of males from the second population, resulting in successful matings, but, in the reciprocal combination, females from the second population are less likely to mate with males from the first population. While mating asymmetries in mate preference experiments had been observed in earlier studies (e.g., Dobzhansky and Streisinger, 1944; Bastock, 1956; Reed and Reed, 1950; Rendel, 1951), the significance of such asymmetries was not clearly understood. Differential mating propensity and/or sexual maturation were frequently used to explain these kinds of observations. Kaneshiro (1976, 1980, 1983) first suggested that the courtship requirements of females in derived populations are simplified during the early stages of founder events when the population size is reduced. It is hypothesized that a relaxation of sexual selection may accompany founder events, since successful colonization by the new population is likely to be enhanced if the behavioral pattern is simplified. In other words, during periods of small population size, females that are less discriminating will be most likely to mate and leave progeny, whereas highly discriminating females may never encounter males that are capable of satisfying their courtship requirements. Thus, there may be a shift in the distribution of female mating types toward females that are less discriminating than those that are highly discriminating. During the genetic revolution that accompanies founder events (Carson, 1971b; Mayr, 1963), such a shift in the mating system may become fixed in the new population.

BEHAVIORAL STUDIES OF HAWAIIAN *DROSOPHILA*

The tremendous morphological diversity of the Hawaiian *Drosophila*, especially in secondary sexual structures in the males, gave a clue to early investigators of this group that sexual selection may play a significant role in the evolutionary history of this group of flies. Indeed, the complex courtship patterns observed in these flies provide evolutionists an opportunity to test some of the classical concepts of sexual selection theory. Following the descriptive work of Spieth (1966, 1968, 1974) on the courtship displays of many of the species, Ahearn (1980), Ahearn et al. (1974), Arita and Kaneshiro (1979), Kaneshiro (1976, 1980), Kaneshiro and Kurihara (1981), and Ohta (1978) conducted mate preference experiments on various species of Hawaiian *Drosophila* in order to understand the role of sexual selection in the speciation process of these species. In general, the results of these studies supported the hypothesis put forth by Kaneshiro (1976, 1980) that females from ancestral populations were more discriminating in mate choice than were females from derived populations and that males from the derived populations were less successful in mating with females from either population than were males from ancestral populations. These kinds of data suggested to Kaneshiro (1976, 1980) that asymmetrical mating isolation between a group of closely related species or even between populations of the same species could be used to infer direction of evolution within the group.

The ability to infer "direction of evolution" within a group of related species based on asymmetrical mating isolation has been controversial with approximately equal numbers of papers in support of Kaneshiro's hypothesis (Ahearn, 1980; Arita and Kaneshiro, 1979; Dodd and Powell, 1986; Giddings and Templeton, 1983; Ohta, 1978; and Powell, 1978) as there were against (Watanabe and Kawanishi, 1979; Wasserman and Koepfer, 1980; Markow, 1981; Barton and Charlesworth, 1984; Erhman and Wasserman, 1987). The discussions resulting from the controversy have indeed been lively and have led to increased research effort in this area. Three chapters in the 1987 issue of *Evolutionary Biology* are devoted to discussions for and against the Kaneshiro hypothesis (Erhman and Wasserman, 1987; DeSalle and Templeton, 1987; Kaneshiro and Giddings, 1987). Only rigorous studies of mating preference among groups other than the Hawaiian *Drosophila* can provide further insights into the generality of Kaneshiro's ideas. However, the real challenge is not only in determining whether mating asymmetries are reliable as predictors of the direction of evolution but also in understanding why such asymmetries exist and the mechanisms by which they arise.

THE DIFFERENTIAL SELECTION MODEL OF SEXUAL SELECTION

Kaneshiro (1988) proposed a model that might explain the existence of mating asymmetries between populations and how such a mechanism might be responsible for selection on other parts of the genome. It is suggested that there is a range of mating types segregating in both sexes within a single interbreeding population; that is, within a population, there are males that are highly successful in lowering the receptivity threshold of most of the females in the population, but there are also males that are less successful in doing so. Within the same population, there are females that are highly discriminating in mating with most of the males in the population, but there are also females that are less discriminating in mating with males in the population.

Numerous models have been developed to explain the genetic consequences of Darwin's theory. For example, Fisher (1930) presented a model that suggested that female choice and secondary sexual structures in males would coevolve very rapidly within an interbreeding population. This has been referred to as "Fisher's runaway selection," and theoreticians (O'Donald, 1977, 1980; Lande, 1981, 1982; Kirkpatrick, 1982) have applied mathematical treatments to this model in an attempt to confirm that the evolution of a male-limited character could result in a correlated response in the female's preference for that character. Furthermore, there appears to be a high genetic correlation between homologous characters of the two sexes, such as body size, which may be due to pleiotropy (Lande, 1981). If indeed there is a strong genetic correlation between male mating success and female discrimination, the sexual selection process itself may act as a stabilizing force in maintaining a normal distribution of mating types in both sexes in subsequent generations. If we assume that there is a range of mating types segregating in both sexes within a single interbreeding population, it is suggested that sexual selection may be operating in the manner proposed by Rose's (1982) model of antagonistic pleiotropy; that is, there would be strong selection for the genotypes of males that are highly successful in mating with females but selection against those that are not. On the other hand, there would also be strong selection for female genotypes that are less discriminating in mating with males than those that are highly discriminating in mate choice. Such a model would avoid the need to invoke the forces of natural selection in maintaining a balanced mate recognition system in the population with the range of variability segregating in the two sexes.

An appealing feature of the differential selection model is that it characterizes the sexual selection process as a truly dynamic system. During periods of small population size, there might be shifts in the frequency of mating types segregating in the population, because there would be strong selection against females that are highly discriminating in mate choice. Thus, as discussed earlier, immediately following founder events, there would be a rise in the frequency of less discriminating females, since these are the females that are most likely to mate and to contribute genetic material to the next generation. Such shifts in the distribution of female mating types (which, if indeed there is a genetic correlation between female discrimination and male mating success, will result in simplification of male courtship pattern) may become fixed in the founder population during the genetic revolution (Mayr, 1963) or genetic transilience (Templeton, 1980) that accompanies founder events.

PLEIOTROPIC EFFECTS OF THE SEXUAL SELECTION SYSTEM

Secondary sexual characters found in males are believed to be under strong sexual selection, and it is thought that female preference for that character maintains such features in the males. However, mate preference experiments among a group of Hawaiian *Drosophila* species appear to indicate that some of these characters are not by themselves necessarily important in mating success of the males. For example, in Kaneshiro's (1976) study of four species of the planitibia complex of picture-winged *Drosophila*, he showed that a secondary sexual feature in *D. planitibia* and *D. silvestris* was not essential in the mating success of the males. Males of both of these species have velvety-black faces, whereas the females have yellow-brown faces.

Presumably, the black faces of the males play a role in mate recognition during the frontal-display segment of the courtship of both species. However, the results of mate preference experiments with a third closely related species, *D. differens*, which lack the black face character in the males [whose face color is similar to females of *D. planitibia* (Hardy) and *D. silvestris* (Perkins) indicate that females of *D. planitibia* and *D. silvestris* mated even more frequently with males of *D. differens* Hardy and Kaneshiro than they did with males of their own species.

In yet another example, Carson and Bryant (1979) reported the discovery of an incipient morphotype in populations of *D. silvestris* and described the evolution of a secondary sexual character that they observed to be important in the courtship ritual of the males. Males from populations from one side of the island of Hawaii had only two rows of cilia on the dorsal surface of the front tibia, whereas males from the other side of the island had an extra (third) row of cilia on the dorsal surface. Carson and Teramoto (1984) conducted selection experiments for high and low cilia number in the third row of bristles of *D. silvestris* and found that ". . . selection in either direction away from the population mean produces dysgenic lines that cannot be maintained in the laboratory." They further suggested that, since the cilia on the front tibia are used in the final and most crucial stages of the courtship repertoire of the males, sexual selection operated as a stabilizing force in maintaining the optimum number of cilia in the population. However, Kaneshiro and Kurihara (1981) conducted mate preference tests among the different populations of *D. silvestris*, including one comparison between two-row and three-row populations, and found that females from the three-row population actually preferred to mate with males from the two-row population more than they did with males from their own populations.

Both of the preceding examples in which secondary sexual characters did not seem to be crucial criteria for mate choice by females seem to indicate that these features in the males may not by themselves be under strong sexual selection. Rather, it is suggested that these characters are being maintained in a balanced state as a pleiotropic effect of some other feature of the sexual selection process; that is, it may be that the extra row of cilia in *D. silvestris* males and the black faces of *D. planitibia* and *D. silvestris* males are pleiotropic effects of male mating ability but not necessarily characters that are directly acted upon by sexual selection. If, indeed, males with superior mating ability had a higher number of cilia than less successful males and if the hypothesis of differential selection between the sexes is the stabilizing force, then the polymorphism for cilia number observed by Carson and Bryant (1979) could easily be maintained. Furthermore, under this model, it would not be necessary to invoke natural selection to counter the runaway process proposed by Fisher and others.

HABITAT-RELATED VARIATION VIA SEXUAL SELECTION

Given the possibility of differential selection between the two sexes, there can be shifts in the frequency of mating types segregating in the population during small population size as a result of environmental stress conditions. As discussed earlier, during small population size, there would be selection against those females that are highly discriminating in mate choice, since the probability of encountering males that would satisfy their courtship requirements would be much reduced. While there is a relaxation of the normal sexual selection process under these conditions, there is selection for less discriminating females in the population. When the population size returns to a more stable condition (i.e., reaches the carrying capacity for that environment), sexual selection may once again play an active role in regulating effective reproductive population size. Also, if there are indeed pleiotropic effects of the sexual selection process on other parts of the genome, then shifts in gene frequency may be an indirect result of a dynamic sexual selection system. Thus, secondary sexual structures, such as the extra row of tibial cilia in males of certain populations of *D. silvestris* and the black faces of males of *D. silvestris* and *D. planitibia*, may be maintained in pleiotropic balance through sexual selection for males with high mating ability as well as for females with low discrimination.

Some of the results from mate preference experiments in the Hawaiian *Drosophila* suggest that sexual selection itself, rather than reliance on natural selection to counter the runaway process, may be more important in maintaining secondary sexual characters in a balanced polymorphic state. Rather than being under direct selection, it is suggested that secondary sexual characters are being maintained in some form of pleiotropic balance within the mate recognition system of the population. Could parts of the genome other than just secondary sexual characters be affected in the same way? Could it be that the dynamics of the sexual selection process play an even greater role in maintaining levels of genetic variability in the population? Could it be that sexual selection is a powerful stabilizing force in maintaining the kinds of balanced polymorphisms that are at times difficult to explain as a result of natural selection? Could it be that, despite the tremendous habitat diversity associated with the Hawaiian *Drosophila*, the primary mechanism for genetic changes during the initial stages of the speciation process can be found in the dynamics of the sexual selection process?

CONCLUDING REMARKS

Clearly, more empirical data are needed to obtain answers to these questions. We need to conduct experiments to determine whether genetic changes that occur in populations may be due to shifts in the distribution of the mating types in the population. This might be accomplished by comparative studies of the genetic composition of lines selected for different mating types. For example, it should be possible to select for high and low female discrimination and to determine whether there are detectable gene frequency shifts via the various molecular techniques available today. Basic evolutionary studies of insect fauna, such as the Hawaiian Drosophilidae, have been extremely important in elucidating some of the mechanisms by which new species are formed. More important, research that addresses basic biological concepts forms a solid foundation upon which research on more applied areas of biology can be based. A clear understanding of the dynamics of biological phenomena in pest populations of insects would be extremely important in the development of effective control programs and, in fact, should form the initial stages of all research on any pest species. It is remarkable that so little attention has been given to investigating the circumstances surrounding the evolution of the annoying properties in pest organisms, especially in view of the fact that an understanding of how such genetic shifts come about might provide important information for the development of control programs that might minimize the indiscriminant use of poisonous pesticides. It is hoped that grant support through various governmental and private agencies would provide the necessary impetus for entomologists to conduct these kinds of basic research even before considering control programs. Only then will we be able to begin to reverse the flagrant contamination of our fragile ecosystem with massive amounts of chemical pesticides.

Acknowledgments

I would like to thank two anonymous reviewers for their helpful comments and suggestions on an earlier draft of this manuscript. Partial support for some of the research reported here came from NSF grants BSR-8614970 and BSR-8600120 to the author.

REFERENCES

Ahearn, J. N. 1980. Evolution of behavioral reproductive isolation in a laboratory stock of *Drosophila silvestris*. *Experientia* 36:63–64.

Ahearn, J. N., H. L. Carson, T. Dobzhansky, and K. Y. Kaneshiro. 1974. Ethological isolation among three species of the planitibia subgroup of Hawaiian *Drosophila*. *Proc. Natl. Acad. Sci. USA* 71:901–903.

Arita, L. H., and K. Y. Kaneshiro. 1979. Ethological isolation between two stocks of *Drosophila adiastola* Hardy. *Proc. Hawaii. Entomol. Soc.* 13:31–34.

Barton, N. H., and B. Charlesworth. 1984. Genetic revolutions, founder effects, and speciation. *Annu. Rev. Ecol. Syst.* 15:133–164.

Bastock, M. 1956. A gene mutation which changes a behavior pattern. *Evolution* 10:421–439.

Bateson, P. 1983. *Mate Choice*. London: Cambridge Univ. Press.

Blum, M. S., and N. A. Blum. 1979. *Sexual Selection and Reproductive Competition*. New York: Academic Press.

Blumenstock, D. I., and S. Price. 1972. Climates of the states: Hawaii. Pages 155–204 in Kay, E. A. (ed.), *A Natural History of the Hawaiian Islands*. Honolulu, Hawaii: Univ. Hawaii.

Campbell, B. 1972. *Sexual Selection and the Descent of Man 1871–1971*. Chicago: Aldine Publishing Co.

Carson, H. L. 1971a. The ecology of *Drosophila* breeding sites. *Univ. Hawaii, Harold L. Lyon Arboretum Lecture No.* 2:1–27.

———. 1971b. Speciation and the founder principle. *Stadler Genet. Symp.* 3:51–70.

———. 1978. Speciation and sexual selection in Hawaiian *Drosophila*. Pages 93–107 in Brussard, P. F. (ed.), *Ecological Genetics: The Interface*. New York: Springer-Verlag.

———. 1985. Genetic variation in a courtship-related male character in *Drosophila silvestris* from a single Hawaiian locality. *Evolution* 39:678–686.

———. 1987. The contribution of sexual behavior to Darwinian fitness. *Behav. Genet.* 17:597–611.

Carson, H. L., and P. J. Bryant. 1979. Genetic variation in Hawaiian *Drosophila*. VI. Change in a secondary sexual character as evidence of incipient speciation in *Drosophila*. *Proc. Natl. Acad. Sci. USA* 76:1929–1932.

Carson, H. L., and K. Y. Kaneshiro. 1976. *Drosophila* of Hawaii: Systematics and ecological genetics. *Annu. Rev. Ecol. Syst.* 7:311–346.

Carson, H. L., and R. Lande. 1984. Inheritance of a secondary sexual character in *Drosophila silvestris*. *Proc. Natl. Acad. Sci. USA* 81:6904–6907.

Carson, H. L., and L. T. Teramoto. 1984. Artificial selection for a secondary sexual character in males of *Drosophila silvestris* from Hawaii. *Proc. Natl. Acad. Sci. USA* 81:3915–3917.

Carson, H. L., D. E. Hardy, H. T. Spieth, and W. S. Stone. 1970. The evolutionary biology of Hawaiian Drosophilidae. Pages 437–543 in Hecht, M. K., and W. C. Steere (eds.), *Essays in Evolution and Genetics in Honor of Theodosius Dobzhansky*. New York: Appleton-Century-Crofts.

Darwin. C. 1859. *The Origin of Species*. New York: Modern Library.

———. 1871. *The Descent of Man and Selection in Relation to Sex*. New York: Modern Library.

DeSalle, R., and A. R. Templeton. 1987. Comments on "The Significance of Asymmetrical Sexual Isolation." Pages 21–27 in Hecht, M. K., B. Wallace, and G. T. Prance (eds.), *Evolutionary Biology*, Vol. 21. New York: Plenum.

Dobzhansky, T., and G. Streisinger. 1944. Experiments on sexual selection in *Drosophila*. II. Geographic strains of *Drosophila prosaltans*. *Proc. Natl. Acad. Sci. USA* 30:340–345.

Dodd, D. M. B., and J. R. Powell. 1986. Founder-flush speciation: An update of experimental results with *Drosophila*. *Evolution* 39:1388–1392.

Erhman, L., and M. Wasserman. 1987. The significance of asymmetrical sexual isolation. Pages 1–20 in Hecht, M. K., B. Wallace, and G. T. Prance (eds.), *Evolutionary Biology*, Vol. 21. New York: Plenum.

Fisher, R. A. 1930. *The Genetical Theory of Natural Selection*. Oxford: Clarendon Press.

Giddings, L. V., and A. R. Templeton. 1983. Behavioral phylogenies and the direction of evolution. *Science* 220:372–377.

Hardy, D. E. 1971. Evolution of the Hawaiian Drosophilidae (Insecta: Diptera). Symposium of Indian Ocean and Adjacent Seas. *Mar. Biol. Assoc. India, Cochin, India* 228:147–148.

———. 1974. Evolution in the Hawaiian Drosophilidae: Introduction and background information. Pages 71–80 in White, M. J. D. (ed.), *Genetic Mechanisms of Speciation in Insects*. Sydney: Australia and New Zealand Book Co.

Hardy, D. E., and K. Y. Kaneshiro. 1981. The Drosophilidae of Pacific Oceania. Pages 309–342 in Ashburner, M., H. L. Carson, and J. N. Thompson, Jr. (eds.), *The Genetics and Biology of Drosophila*. New York: Academic Press.

Heed, W. B. 1968. Ecology of the Hawaiian Drosophilidae. *Univ. Tex. Publ.* 6818:387–419.

———. 1971. Host specificity and adaptive radiation in Hawaiian Drosophilidae. *Taxon* 20:115–121.

Heth, G., and E. Nevo. 1981. Origin and evolution of ethological isolation in subterranean mole rats. *Evolution* 35:259–274.

Huxley, J. 1938. Darwin's theory of sexual selection and the data subsumed by it in light of recent research. *Am. Nat.* 72:416–433.

Kaneshiro, K. Y. 1976. Ethological isolation and phylogeny in the *planitibia* subgroup of Hawaiian *Drosophila*. *Evolution* 30:740–745.

———. 1980. Sexual behavior and phylogeny of Hawaiian *Drosophila*. Page 114 in *Abstracts, Second Int. Congr. Syst. Evol. Biol.* Vancouver, B.C.: Univ. British Columbia.

———. 1983. Sexual selection and direction of evolution in the biosystematics of Hawaiian Drosophilidae. *Annu. Rev. Entomol.* 28:161–178.

———. 1987. The dynamics of sexual selection and its pleiotropic effects. *Behav. Genet.* 17:559–569.

———. 1989. The dynamics of sexual selection and founder effects in species formation. Pages 279–296 in Giddings, L. V., K. Y. Kaneshiro, and W. W. Anderson (eds.), *Genetics, Speciation and the Founder Principle*. New York: Oxford Univ. Press.

Kaneshiro, K. Y., and L. V. Giddings. 1987. The significance of asymmetrical sexual isolation. Pages 29–44 in Hecht, M. K., B. Wallace, and G. Prance (eds.), *Evolutionary Biology*, Vol. 21. New York: Plenum.

Kaneshiro, K. Y., and J. S. Kurihara. 1981. Sequential differentiation of sexual isolation in populations of *Drosophila silvestris*. *Pac. Sci.* 35:177–183.

Kirkpatrick, M. 1982. Sexual selection and the evolution of female choice. *Evolution* 36: 1–12.

Lande, R. 1981. Models of speciation by sexual selection on polygenic traits. *Proc. Natl. Acad. Sci.* 78:3721–3725.

———. 1982. Rapid origin of sexual isolation and character divergence in a cline. *Evolution* 36:213–223.

Markow, T. A. 1981. Mating preferences are not predictive of the direction of evolution in experimental populations of *Drosophila*. *Science* 213:1405–1407.

Mayr, E. 1963. *Animal Species and Evolution*. Cambridge, Mass.: Harvard Univ. Press.

———. 1982. Processes of speciation of animals. Pages 1–19 in Barigozzi, C. (ed.), *Mechanisms of Speciation*. New York: Alan R. Liss.

Montgomery, S. L. 1975. Comparative breeding site ecology and the adaptive radiation of picture-winged *Drosophila* (Diptera: Drosophilidae) in Hawaii. *Proc. Hawaii. Entomol. Soc.* 22:65–102.

O'Donald, P. 1977. Theoretical aspects of sexual selection. *Theor. Pop. Biol.* 12:298–334.

———. 1980. *Genetic Models of Sexual Selection*. Cambridge, Mass.: Cambridge Univ. Press.

Ohta, A. T. 1978. Ethological isolation and phylogeny in the grimshawi species complex of Hawaiian *Drosophila*. *Evolution* 32:485–492.

Powell, J. R. 1978. The founder-flush speciation theory: An experimental approach. *Evolution* 32:465–474.

Raven, P. H., and G. B. Johnson. 1986. *Biology*. St. Louis, Mo.: Times Mirror/Mosby Coll. Publ. Co.

Reed, S. C., and E. W. Reed. 1950. Natural selection in laboratory populations of *Drosophila*. II. Competition between white-eye gene and its wild type allele. *Evolution* 4:34–42.

Rendel, J. M. 1951. Mating of ebony, vestigial and wild type *Drosophila melanogaster* in light and dark. *Evolution* 5:226–230.

Rose, M. R. 1982. Antagonistic pleiotropy, dominance, and genetic variation. *Heredity* 48:63–78.

Spieth, H. T. 1966. Courtship behavior of endemic Hawaiian *Drosophila. Univ. Tex. Publ.* 6615:245–313.

———. 1968. Evolutionary implications of the mating behavior of the species of *Antopocerus* (Drosophilidae) in Hawaii. *Univ. Texas Publ.* 6818:319–333.

———. 1974. Courtship behavior in *Drosophila. Annu. Rev. Entomol.* 19:385–405.

Templeton, A. R. 1980. The theory of speciation via the founder principle. *Genetics* 94:1011–1038.

Thornhill, R., and J. Alcock. 1983. *The Evolution of Insect Mating Systems*. Cambridge, Mass.: Harvard Univ. Press.

Wallace, A. R. 1889. *Darwinism, 3rd ed.* London: Macmillan and Co.

Wasserman, M., and H. R. Koepfer. 1980. Does asymmetrical mating preference show the direction of evolution? *Evolution* 34:1116–1124.

Watanabe, T. K., and M. Kawanishi. 1979. Mating preference and the direction of evolution in *Drosophila. Science* 205:906–907.

Williamson, M. 1981. *Island Populations*. London: Oxford Univ. Press.

Wirth, W. W. 1952. Two new spider egg predators from the Hawaiian Islands (Diptera: Drosophilidae). *Proc. Hawaii. Entomol. Soc.* 14:415–417.

Zimmerman, E. C. 1958. Three hundred species of *Drosophila* in Hawaii? A challenge to geneticists and evolutionists. *Evolution* 12:557–558.

———. 1970. Adaptive radiation in Hawaii with special references to insects. *Biotropica* 2:32–38.

5

ADAPTATIONS TO TEMPORAL VARIATION IN HABITATS: CATEGORIZING, PREDICTING, AND INFLUENCING THEIR EVOLUTION IN AGROECOSYSTEMS

Maurice J. Tauber and Catherine A. Tauber

Cornell University

INTRODUCTION

The ancient Greek philosopher Heraclitus reasoned that all matter remains in perpetual flux; a person does not step into the same river twice, because fresh waters flow by continuously. By analogy, a biologist does not deal with identical biological events twice because of the never ending changes that organisms, populations, communities, and the environment undergo. Both ancient and modern observers confronted this formidable problem in their attempts to generalize, to predict events, and to create order from the raw rubble of observations derived from dynamic systems.

Darwin revolutionized modern biological thinking when he proposed the concept of evolutionary change—gradual, step-by-step transformation—in organisms over time (Darwin, 1859). He conceived the now generally accepted, yet still refreshingly revolutionary concept: the diverse kinds of organisms have descended through modification of common ancestors via the action of natural selection on intrinsic variation among individuals. Darwin's writings, and those of many evolutionary biologists that followed, emphasized the dynamic nature of natural selection, and they brought forth the idea that evolution is not a series of events solely confined to the past, but a continuing, vigorous process.

Given sufficient time, all habitats on our planet undergo change. Indeed, all organisms are creatures of time; they continually experience transformation. As a result, biologists seek to understand the physiological, behavioral, ecological, and evolutionary processes that contribute to the adaptation and success of organisms in changing environments. Understanding these processes and defining their limits are universal requisites for the advancement of both basic and applied biology; they constitute a *unifying* theme for all biological endeavour.

Temporal Variability in Agroecosystems

When we consider issues in applied biology, we readily recognize that temporal variability characterizes all interactions in agroecosystems—among individuals, populations, and communities. Moreover, the success of a pest or a natural enemy often resides in its ability to adapt to the human-made and natural perturbations in its environment. This applies to pests and natural enemies associated with annual or perennial crops, as well as to exotic and native species invading new territory. Thus, temporal variability (whether within or between seasons, over many years, or through evolutionary time) has become an important consideration in the development of effective pest management strategies and tactics (van den Bosch, 1971; Rabb and Kennedy, 1979; Branson and Krysan, 1981; Kennedy, 1986; Tauber et al., 1986).

Insects have evolved a remarkable array of adaptations to the temporal variability of their habitats. These diverse adaptations, which are described in a number of books (e.g., Lees, 1955; Danilevsky, 1965; Dingle, 1978; Beck, 1980; Saunders, 1982; Brown and Hodek, 1983; Prosser, 1986; Tauber et al., 1986; Danks, 1987) reflect enormous phenotypic, genetic, and evolutionary malleability. And they lead us to inquire whether there are limitations on the evolutionary flexibility of insect life cycles. If so, what factors restrict a species's ability to adapt to new environmental conditions?

Although biologists have only recently addressed the evolutionary limits to adaptation, the problem is vital to both basic and applied biology. In theoretical terms, it stands at the nexus of our understanding of (1) the distribution of populations geographically and among habitats, (2) the evolution and diversification of life histories, and (3) the processes of diversification and speciation. In practical terms, it underlies our ability to understand, predict, and manipulate (1) the geographic spread of pests, (2) the evolution of resistance to chemical, cultural, and other management practices, and (3) the colonization and adaptation of exotic and native natural enemies. In practice, addressing the problem is essential to the development of reliable, long-lasting procedures for biological control and integrated pest management.

Goals

Within this setting, our chapter considers theoretical and practical concepts concerning the limits to the evolutionary flexibility in insect life cycles. The main question we examine (What restricts a species's ability to adapt to environmental change?) is a complex one. Its investigation requires an approach that combines population ecology and population genetics (Istock, 1984). Moreover, it must be explored within the context of the two key elements of evolution: (1) the genetic variability underlying temporal (and spatial) adaptations (e.g., various authors in Loeschcke, 1987) and (2) the selective and other forces that determine the speed and direction of evolutionary change (e. g., Endler, 1986).

In our chapter, we suggest a sequence of four practical steps that field biologists and entomologists can use to approach the problem of exploring the limits of adaptation. In actuality, these steps have as their ultimate endpoint the prediction and manipulation of evolutionary adaptation in insect pests and natural enemies. Their immediate goal is to identify questions, as well as specific biological systems, for future research. The four steps we propose are

1. Define the *types of temporal adaptations* that dominate the species's life history.
2. Elucidate the *variability in the ecophysiological mechanisms* that underlie the dominant temporal adaptations.
3. Establish the *genetic basis for the variability* in the ecophysiological mechanisms.
4. Identify *"trade-offs"* among various temporal adaptations and between temporal adaptations and other life-history traits.

The initial part of our chapter concentrates on the first step: defining the types of temporal adaptations. In a general way, we categorize habitats on the basis of temporal variability and briefly describe the major adaptations that insects have evolved to cope with the various types of environmental changes. Subsequently, we review, in some detail, findings from our ecophysiological and genetic studies with a group of predaceous insects that specifically impinge on the second, third, and fourth steps. Finally, we discuss the implications of our findings to the genetic improvement of natural enemies and to integrated pest management, and we identify areas of research for the future.

HABITATS AND TEMPORAL ADAPTATIONS

Categorization of Habitats and Adaptations

Evidence is strong that some species keep track of small environmental changes very closely; other species do so more generally or on a longer time scale (Southwood, 1977, 1988; Wiens et al., 1986). Therefore, it is useful to begin analyses of organisms in temporally variable habitats by determining the time scale and the major temporal adaptations that dominate the organism's life cycle. Generally, the evolution of adaptation cannot be studied directly through a priori methods, but it is handled indirectly via a posteriori approaches with correlational and comparative techniques (see discussion by Sibly and Calow, 1986). Although such an approach cannot establish that a particular trait is evolutionarily labile, it may provide insight into traits of probable adaptive significance to organisms in temporally varying habitats.

Both direct and indirect methods for examining the evolution of adaptations require a substantial understanding of the phylogenetic relationships of the organisms under study. A sound systematic component is essential to such studies, because it provides the framework for establishing both the direction and the pattern of evolution. (See Tauber et al., 1986, for a brief discussion of the use and misuse of comparative ecophysiological studies.) Indeed, sound systematic analyses form the essential foundation for the vast majority of biologically related endeavors. Ironically, although this idea often finds expression within editorials of leading scientific journals (e.g., Wilson, 1971, 1985), relatively limited substantive support and funding follow at either federal, state, or institutional levels.

In this section, we suggest a practical context within which to categorize temporal adaptations, particularly within the temperate zones. To do so, we ask, "What characteristics are useful in describing and quantifying changes in habitats?" This question has been addressed numerous times and in relation to a variety of organisms (e.g., Grime, 1977; Southwood, 1977, 1981, 1988; Greenslade, 1983; Holm, 1988). Our emphasis differs from previous studies in that it focuses on *mechanisms* for dealing with environmental change, rather than on the effects that environmental changes have on traditional life-history traits (fecundity, developmental time, survival), and it is restricted to insects and mites.

We suggest that environmental changes have four properties: (1) predictability, (2) duration, (3) frequency, and (4) magnitude. Two of these—predictability and duration—are the most frequently measured; ultimately, all four categories require emphasis. Moreover, all four should be evaluated in relative terms, that is, in relation to spatial variation in the habitat and in relation to the life cycles and phylogenetic histories of the organisms involved (Roughgarden, 1979; Gould and Stinner, 1984; Strong et al., 1984; Southwood, 1988).

The first characteristic of temporal variability is predictability. Does the change occur repeatedly, either at regular or irregular intervals? Is the change preceded by "token stimuli" that herald the change? As a rough measure of predictability, we categorize environmental changes as *acyclic* and *cyclic* (see Fig. 5.1). *Cyclic* changes occur on a regular basis and usually within the context of specific environmental conditions that forewarn of their arrival. *Acyclic* changes, such as those associated with sudden (within season) changes in temperature or humidity or with fire or drought, are less predictable.

Both the second and third characteristics of temporal change involve timing: how long does the condition persist (duration), and how often does the change occur (frequency)? Each of these characteristics can be viewed in relation to the life cycle of the insect species (e.g., see Southwood, 1988). For example, typical multivoltine species undergo several generations within an annual cycle. They are developmentally and reproductively active during much of the growing season and thus are particularly susceptible to within-season changes in the environment. In contrast, each generation of univoltine species experiences the complete seasonal cycle; however, during much of the year, individuals may be in diapause and thus relatively uninflu-

TEMPORAL CHANGES IN HABITATS

TIMING	PREDICTABILITY	
	Acyclic	Cyclic
Short-term or Frequent	Sudden change in temperature Deterioration/ loss of food Lack of mates	Diurnal, lunal or tidal cycles in temperature, food, humidity, enemies, competitor
Long-term or Occasional	Drought Forest/prairie fire Agriculture, logging Urbanization Climatic cooling/ warming	Seasonal cycles in temperature, food, humidity, enemies, competitors

Figure 5.1. Categorization of temporal changes in habitats based on predictability and timing

enced by many within-season changes. Therefore, for consistency in dealing with the various environmental changes, we combined aspects of frequency and duration under the category of "timing" (Fig. 5.1). We refer to short term as hours or days; we arbitrarily designate long term as months and years.

Magnitude, the fourth category, indicates the severity of change. How harsh or subtle is the change? Whether in the laboratory or in the field, the magnitude of environmental changes is difficult to quantify, because it must be done through the "eyes" of the organism. For example, does the change in conditions affect survival? Does it influence development or reproduction? Because magnitude is the least well studied of the four categories, we can offer few generalizations, and we deal with it indirectly.

As Southwood (1988) recently emphasized, the pattern of life-history variation is constrained both by the available genetic variability (which is a reflection of phylogenetic history) and by the template of habitat characteristics. Therefore, although multiple pathways may lead to evolutionary adaptation, it should be possible to detect patterns in the way in which organisms adapt to various changes in their environments. Thus, after considering the temporal variability in the habitat itself, we asked, "What adaptations have insects and mites evolved to cope with the various types of changes in their habitats (Fig 5.2)?"

First are the responses to short-term, acyclic changes. If physical conditions change (e.g., if there is a sudden drop or rise in temperature or humidity), insects or mites may become quiescent or seek shelter until temperatures return to the favorable range. If localized depletion of food occurs, insects may move from unfavorable areas and seek more favorable conditions elsewhere. This type of behavior typifies the movements of insect pests between fields during the growing season,

ADAPTATIONS TO TEMPORAL VARIATION

TIMING	PREDICTABILITY	
	Acyclic	Cyclic
Short-term or Frequent	*Adaptations:* Quiescence Aseasonal migration Aseasonal polyphenism	*Adaptations:* Diurnal, lunar or tidal rhythms in physiological and behavioral processes
	Primary control of adaptation: NERVOUS SYSTEM physiological & behavioral adjustments	*Primary control of adaptation:* BIOLOGICAL CLOCK behavioral & physiological adjustments
Long-term or Occasional	*Adaptations:* Prolonged dormancy (diapause or cryptobiosis) Dispersal Genetic polymorphism for dormancy or migration Evolutionary change	*Adaptations:* Diapause-mediated dormancy seasonal migration seasonal polyphenism
	Primary control of adaptation: GENETIC physiological and evolutionary alterations	*Primary control of adaptation:* NEUROHORMONAL diapause

Figure 5.2. Categorization of insect adaptations to temporal variation in habitats

and it involves a very strong component of spatial adaptation (Kennedy and Margolies, 1985; Kareiva, 1986; Margolies, Chapter 6 in this book).

Despite their importance to population biology, many problems pertaining to short-term or localized changes in the environment remain unaddressed and unanswered. Very few studies define or quantify the behavioral and physiological mechanisms whereby insects overcome short-term, unpredictable stress (e.g. Wellington and Trimble, 1984). For example, how long can insects remain quiescent (Hinton, 1960)? How quickly can they adjust to sudden changes (e.g. Lee et al., 1987)? How long can insects survive without food during their developmental or reproductive stages (e.g. Moeur and Istock, 1980; Lounibos et al., 1982)? How far can they travel before finding food (e.g. Fleschner, 1950)? Can they switch to other food or hosts (e.g. Doutt et al., 1976; Jermy et al., 1982; Slansky, 1982; Wiklund, 1982)? What are the effects of movement and periods of quiescence on subsequent development and reproduction (e.g. Keiser and Ashraf, 1987)? In summary, how phenotypically flexible are the developmental and reproductive requirements of the species?

These behavioral, physiological, and genetic questions are of key importance to basic and applied biology. Responses to short-term, localized changes in the environment involve the temporary postponement of development or reproduction, and they may result in significant demographic changes (e.g. Solbreck, 1985). Furthermore, the diverse response patterns that subserve survival during short-term, acyclic perturbations represent complex and highly evolved adaptations, all of which need to be studied further.

Now, let us turn from short-term, acyclic changes to short-term, cyclic changes in environmental factors, such as temperature, humidity, and the occurrence of food, mates, natural enemies, and competitors (Fig. 5.2). Responses to these changes largely involve physiological processes or behavioral patterns having diurnal, tidal, and lunar rhythms. Such rhythms are highly evolved adaptations, the functions and mechanisms of which have been well reviewed (e.g. Beck, 1980; Saunders, 1982). However, the effects of the rhythms on life-history parameters are not well known; also unknown is the extent of their evolutionary flexibility (Lankinen and Lumme, 1984).

The third category—long-term, unpredictable temporal changes in the habitat—encompasses a variety of phenomena; that is, it includes the irregular occurrence of drought, forest fires and prairie fires, logging, agriculture or other disturbances caused by humans, and long-term climatic changes (Fig. 5.1). To cope with such exigencies, insects can become dormant or disperse, but the unpredictability of these changes precludes organisms from relying on environmental cues as harbingers of their occurrence. In many instances, adaptation to long-term unpredictable changes involves genetic polymorphism for diapause-controlled dormancy or migration. During each generation, a portion of the population remains in prolonged diapause, despite conditions favorable for growth and reproduction. Thus, if conditions suddenly deteriorate, the dormant part of the population acts as a replenisher (e.g. Dingle, 1984; Ingrisch, 1986; Tauber et al., 1986). Such temporal adaptations, whether they occur in pests or in natural enemies (Krysan et al., 1986; Krysan, Chapter 18 in this book), have profound implications for integrated pest management.

Other responses to long-term, unpredictable environmental changes include those that occur when species colonize or are introduced into new areas (through natural dispersal and migration or through movement by humans). These changes are reflected in alterations in the species composition of localities (i.e., changes in community structure, succession) and with modification of the genetic composition of populations (e.g., adaptation, speciation). Current research dealing with the adaptations of colonizing species has focused on developing methods for predicting which species can invade successfully, which communities can be invaded successfully, and the influence of the successful invasion on the invaded community. A substantial body of literature addresses these issues (see Baker and Stebbins, 1965; Hidaka,

1977; Parsons, 1983; Howarth, 1985; Diamond and Case, 1986; Groves and Burdon, 1986; Mooney and Drake, 1986; Tauber et al., 1986; Southwood, 1988).

Our final category considers the adaptations of insects and mites to long-term, cyclic, or predictable temporal changes in the habitat (Fig. 5.2). Here, the predominant environmental challenges are seasonally related. Because these changes occur on a regular, predictable basis, insects can prepare for them, and, in most cases, the preparations involve the neurohormonal system. Although biological clocks and genetic variability may be vital components in the preparatory functions, the proximal and most noticeable aspect is the neurohormonally controlled, physiological state of diapause (Saunders, 1984; Tauber et al., 1986).

Diapause, viewed as a syndrome, has many reliable symptoms or expressions, including low metabolic activity, reduced morphogenesis, and increased resistance to environmental extremes. It forms the basis for most known cases of dormancy in insects, and it also mediates seasonal migration and seasonal polyphenism [i.e., environmentally induced alterations in the phenotype, including structure (winged versus wingless forms, long versus short wings), color, physiology, and behavior]. Diapause constitutes one of the most important examples of phenotypic plasticity among insects; its variability can be subject to genetic analysis (see Smith-Gill, 1983; Istock, 1984; Michod, 1984; Tauber and Tauber, 1987; Via, 1987).

Mechanisms Involved

Our preceding analysis indicates that insects can measure and respond to environmental variation over extremely broad ranges of time and magnitude—from abrupt, small changes with immediate effects, to gradual but large, changes with delayed effects. The diversity of temporal variations in habitats is associated with an astonishingly broad range of adaptations to these temporal variations. However, there is a unifying factor within this diversity: at the nexus of all of the adaptations lie physiological processes that serve as vital connecting links between ecology (selective forces) and evolution (change in gene frequency) (Stearns, 1983; Tauber and Tauber, 1986a). Thus, a comprehensive understanding of evolutionary adaptations to temporal variation in habitats requires a precise and thorough knowledge of the mechanisms that exist at the ecological and physiological interface.

With this in mind, let us clarify the use of two ecophysiological terms: dormancy and diapause (Tauber et al., 1986). We use the term *dormancy* in a general, ecological sense, to mean any recurring period in the life cycle during which growth, development, and reproduction are suppressed by either internal or external factors. As commonly used, it does not identify the physiological basis for suppressed development. Thus, the term dormancy includes both quiescence as well as diapause.

In contrast, the term *diapause* describes a physiological state—a stage-specific, neurohormonally mediated, state of low metabolic activity that occurs in response to environmental (token) stimuli that precede unfavorable conditions (Lees, 1955; Beck, 1980; Tauber et al., 1986). The most common token stimulus for regulating diapause is photoperiod. However, temperature, moisture, food, and other factors may also act (singly or in concert) as significant token stimuli.

This view of diapause encompasses seemingly contradictory concepts—a dynamic state and low metabolic activity. Suppressed metabolic activity can be viewed as dynamic for a number of reasons. First, although metabolic activity may be reduced during diapause, it is not eliminated; important processes, termed *diapause development* or *physiogenesis*, proceed. Furthermore, in many cases, growth and development, even moulting, occur during diapause (Tauber et al., 1986).

Second, as diapause is induced, maintained and terminated, insects go through a series of very distinct periods (Fig 5.3). During these periods, changes occur in the insects' responses to various stimuli. For example, photoperiod may induce and maintain diapause for a given period. Subsequently, other stimuli, such as temperature or food (acting as token stimuli), may maintain a second phase of diapause. In nature, diapausing insects pass through these phases in regular, species-specific patterns, and thus, diapause constitutes a dynamic process.

We emphasize that diapause forms the basis—the physiological basis—for a variety of ecological expressions; as such, it plays an integral role in adaptation to both temporal and spatial variability. For example, diapause can play an essential role in the expression of dormancy, seasonal migration, and seasonal polyphenism—the triumvirate of seasonal adaptations. We and others have stressed the close relationship between diapause and migration (Kennedy, 1961; Dingle, 1972, 1980, 1984; see Tauber et al., 1986, for a discussion of the close relationship between diapause, seasonal migration, and seasonal polyphenism). Migration involves changes in behavior and physiology that are commonly associated with diapause. The token stimuli that induce diapause-mediated dormancy often also induce migration. Both result in the postponement of development and reproduction for varying periods of time. And both are highly evolved, correlated adaptations to environmental variability.

Insects escape in both space and time, and they endure and evolve in both space and time (Dingle, 1978; Price, 1980; Denno, 1983, and others in Denno and McClure, 1983; Tauber et al., 1986; Margolies, Chapter 6 in this book). The all-important connection between temporal and spatial adaptations is illustrated by the close physiological and ecological relationship between diapause and migration. From evolutionary and ecological perspectives, these two components of the habitat are intimately related; they should not be artificially separated.

A CASE STUDY

Given the preceding discussion, we now examine the four steps we proposed for exploring the limits to the evolutionary flexibility in insect life cycles. To do this, we will focus on our work with a group of predaceous insects (green lacewings in the *Chrysoperla carnea* species complex) that are natural enemies of many arthropod pests of crops. First, we review the biology and seasonal cycles of the lacewings. Next, we consider the types of temporal adaptations they have evolved, the ecophysiological mechanisms that underlie the adaptations, and, finally, the genetic variation and tradeoffs among the mechanisms. Although our discussion places emphasis on beneficial insects, the type of information and the methods we discuss are directly applicable to pest species.

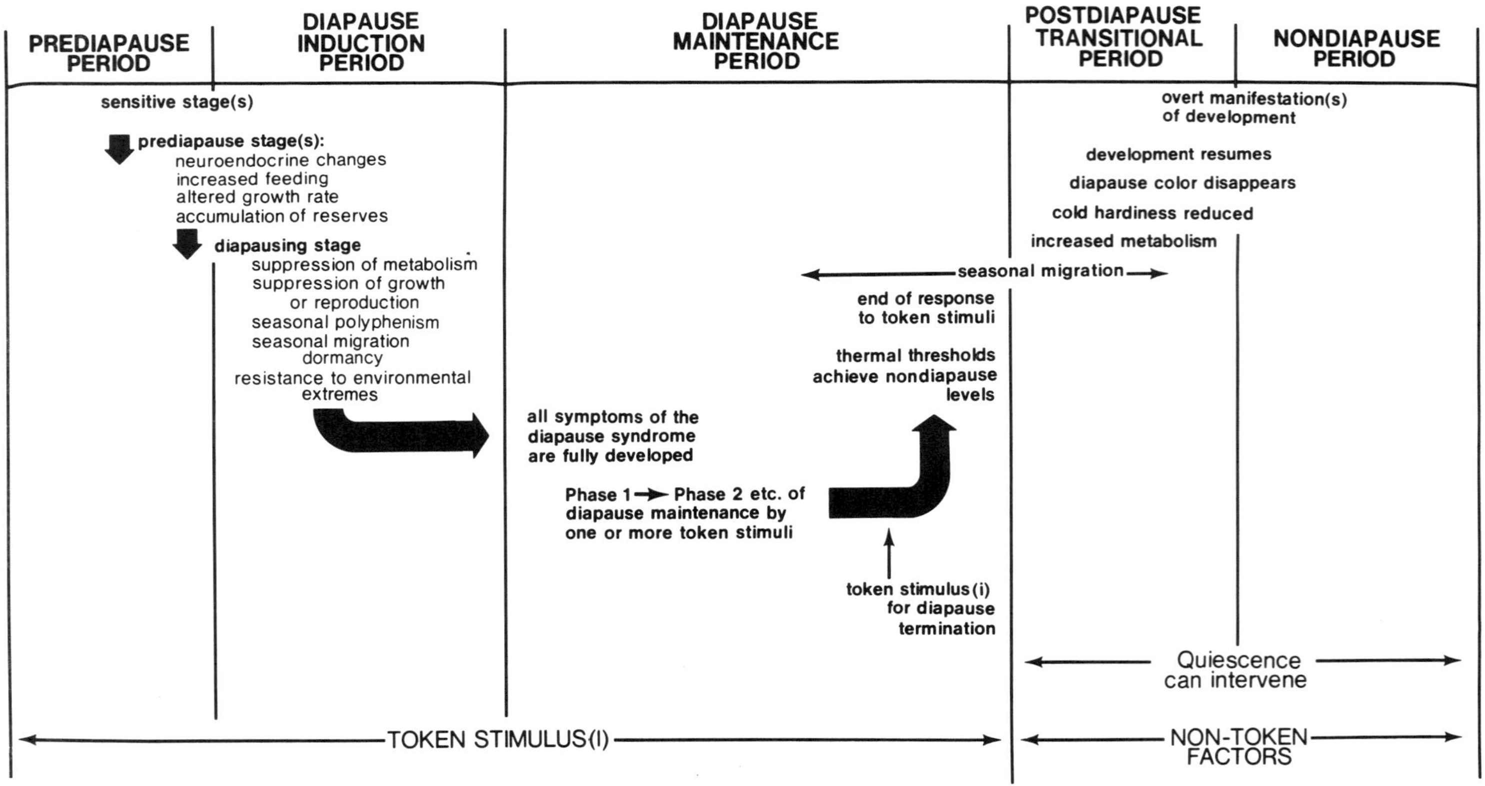

Figure 5.3. The progression of insect seasonal cycles

In contiguous United States and southern Canada, the *Chrysoperla carnea* species complex includes three major biotypes (*sensu* Bush and Diehl, 1982): *Chrysoperla carnea* (*carnea* strain), *Chrysoperla carnea* (*mohave* strain), and *Chrysoperla downsei*. Each of these biotypes is typified by ecological, behavioral, and morphological traits (Tauber and Tauber, 1982, 1986a; Henry, 1985). The degree of reproductive isolation between the biotypes varies geographically (Tauber and Tauber, 1986a, 1989), and thus the taxonomic relationships and species status of the biotypes within the species complex are open to investigation and interpretation (Tauber and Tauber, 1987, 1989).

Ecophysiological Mechanisms

Like other chrysopids, *Chrysoperla* females lay their eggs on silken stalks. The predaceous larvae feed on various soft-bodied arthropods, including many economically important lepidopteran and homopteran pests. With the exception of one biotype (*C. carnea, mohave* strain), *Chrysoperla* adults do not take prey. They ingest honeydew and pollen. Diapause occurs only in the adult stage; overwintering and aestivation (when it occurs) are accomplished by the adults.

The *carnea* strain of *C. carnea*, the common green lacewing currently used in biological control programs in the United States, has a multivoltine life cycle. Overwintered, postdiapause adults undergo reproduction in spring; subsequently, there are several generations during summer. In autumn, adults enter reproductive diapause that is expressed in a variety of ways. First, the insects avert reproductive activity; they neither mate nor oviposit. Second, they migrate from meadows and fields to forest edges, where they overwinter. Third, they change color. During the summer, the lacewings are light green, and, in autumn, when diapause is induced, they assume a reddish-brown hue. Within a seasonally changing environment, the alterations in the lacewings' seasonal color provide a high degree of crypsis (at least to the human eye).

These seasonal changes are largely controlled by photoperiod and temperature. Long day lengths, above a critical photoperiod, result in reproduction and the maintenance of the green summer coloration. Short day lengths induce diapause and facilitate the assumption of autumnal coloration. Temperature influences the timing of reproduction in spring, the number of generations produced per year, and the rate and degree of color change. Both the critical photoperiod for diapause induction and the duration (or intensity) of diapause exhibit considerable, genetically based, geographic variability (Tauber and Tauber, 1972); this variability contributes in a major way to the widespread geographic distribution of the biotype (Fig. 5.4).

C. downesi differs from *C. carnea* (*carnea* strain) in its habitat association, color, and seasonal cycle. First, *C. downesi* occurs primarily in coniferous evergreen trees. Its adults are dark green, a color that makes them cryptic in their specific habitat. Like the evergreen trees they inhabit, these lacewings do not change color in autumn. Second, *C. downesi* is univoltine. The overwintered adults do not break diapause until early spring. Mating and oviposition occur in midspring, larvae develop during late spring, and the resulting adults emerge in early to midsummer. Immediately after emergence, the adults enter diapause, and they remain in diapause

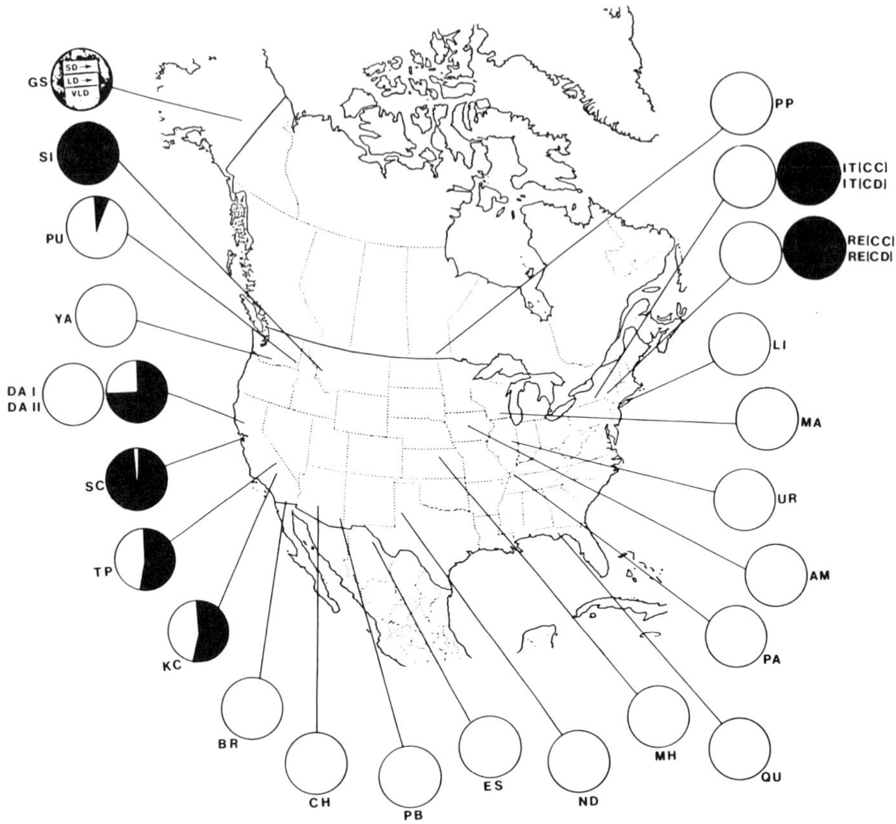

Figure 5.4. Variation in photoperiodic responses among some North American populations of the *Chrysoperla carnea* species complex. Light sections of circles represent the proportion of individuals with long-day reproduction; dark sections represent the proportion with a short-day/long-day requirement for reproduction. Completely light circles designate *C. carnea* (*carnea* strain); completely dark circles indicate *C. downesi*; partially dark circles indicate *C. carnea* (*mohave* strain). The Alaskan (GS) population has a short-day/long-day/very-long-day requirement for reproduction. Davis, California, is represented by two different, apparently sympatric populations (DAI, DAII)

throughout the remainder of the summer, autumn, and winter (Tauber and Tauber, 1976a).

C. downesi's seasonal cycle is controlled primarily by photoperiod. In contrast to *C. carnea* (*carnea* strain), which continues to reproduce when day lengths remain above a certain critical level, *C. downesi* reproduces only if it has experienced short days followed by long days (Tauber and Tauber, 1976b). This is called the short-day/long-day requirement for reproduction, and it ensures that reproduction by *C. downesi* is confined to late spring and early summer.

The final biotype to be considered here, *C. carnea* (*mohave* strain), is restricted to western North America. It possesses a remarkable set of integrated adaptations, but one of the most intriguing is its variability. It is polymorphic for a number of important behavioral and life-history traits (Tauber and Tauber, 1986a). First, it occurs in a variety of habitats: evergreen oak and coniferous forests, fields, and meadows. Associated with this range in habitats is variability in adult coloration. Adults of most populations are light to bright green during spring; pale, waxy green during summer; and reddish brown during autumn. Adults from other populations, especially those from conifers, are darker green during spring and summer; they may remain green or turn reddish brown during autumn.

Polymorphism and phenotypic plasticity in the seasonal reproductive cycle also characterize the *mohave* strain. Like the other two biotypes, the *mohave* strain begins reproduction in the spring and undergoes photoperiodically maintained diapause in autumn. But, unlike the other biotypes, the *mohave* strain has three reproductive pathways during the summer.

First, each population contains some individuals that reproduce under constant long-day conditions; long day lengths are sufficient to stimulate reproduction in these individuals, and reproduction by this part of the population may continue all summer long. This life cycle resembles the multivoltine life cycle of the *carnea* strain.

Other individuals within each *mohave* population have a short-day/long-day requirement for reproduction; thus, summer photoperiods are not sufficient to avert diapause. However, not all of these individuals aestivate. Some respond to the presence of prey by averting diapause and undergoing reproduction. Thus, a portion of each *mohave* population enters aestival diapause or undergoes aestival reproduction, depending on prey abundance. The remainder of the population aestivates even if prey levels are high.

Geographic populations of the *mohave* strain express considerable variability in the response patterns that control their seasonal cycles. Both the responses to photoperiod and to prey are all-or-none responses; that is, after the insect perceives the appropriate stimulus, it responds in either of two distinct ways: by averting or entering diapause (or conversely, by reproducing or not reproducing). Thus, geographic variability is expressed as differences among populations in the incidence of the various response mechanisms.

Only two of the biotypes, the multivoltine *C. carnea* (*carnea* strain) and the univoltine *C. downesi*, inhabit eastern and midwestern North America (Fig. 5.4). In these parts of the continent, the two biotypes exist as separate and pure populations that do not interbreed even where they occur sympatrically.

A much more complex situation occurs in western North America, where a mosaic of populations exists (Fig. 5.4). In this region, the *carnea* strain inhabits agricultural and nonagricultural areas (e.g., DAI, BR, YA, CH, ND on Fig. 5.4). Pure *C. downesi* populations also occur in the West (SI). However, in most nonagricultural, relatively undisturbed areas of the West, populations are made up of the genetically polymorphic *mohave* strain (PU, DAII, SC, TP, KC). Many of these western habitats are characterized by unpredictable periods of drought and low prey availability (see, e.g. Shapiro, 1979). The *mohave* polymorphism for the three reproductive pathways represents an important adaptation that enables it to "hedge its bets" in such variable environments. This "bet-hedging tactic" involves responsiveness both to photoperiod and to prey, and there are "trade-offs" among the component traits that apparently maintain the polymorphism over time.

Where do these biotypes fit into our original consideration of temporal changes in the habitat (Figs. 5.1 and 5.2)? Diapause in *C. carnea* (*carnea* strain) and in *C. downesi* largely serves one vital function: to synchronize reproduction with regularly occurring, cyclic (seasonal) changes in the environment. It is not polymorphic, nor is it known to persist for more than one year; thus it would not provide protection from unpredictable exigencies during the annual period of reproduction or from unfavorable conditions extending more than one year.

In contrast, diapause in the *mohave* strain serves two primary roles. First, the photoperiodically induced autumnal diapause functions as an adaptation to the regular seasonal changes in the habitat. In addition, the photoperiod- and food-mediated aestival diapause adapts the biotype to unpredictable, irregularly occurring temporal changes that characterize many drought-prone areas in the western United States. Because prey availability is variable, the prey-mediated aestival diapause allows the *mohave* biotype to exploit abundant prey without suffering devastating mortality during periods when prey are unavailable.

Genetic Basis for Variability in Ecophysiological Mechanisms

To assess the evolutionary flexibility of adaptations to temporal variation, it is essential to determine the genetic variation underlying the traits and also to elucidate the genetic control and expressivity of the various traits. We addressed this challenge for the *C. carnea* species complex with techniques from both Mendelian and simple quantitative genetics. First, we performed a set of selection experiments (1) to determine the amenability of the all-or-none responses to artificial selection and (2) to describe the actual and hidden variation underlying the seasonal reproductive cycle (Tauber and Tauber, 1986b). Second, we conducted an extensive series of hybridization experiments with four very different populations to determine (1) the mode of inheritance of responses to both photoperiod and food, (2) the genetic relationship between the two sets of responses, and (3) the homology of phenotypically similar responses in diverse geographic populations (Tauber and Tauber, 1987).

Our selection studies demonstrated considerable genetic variability in the seasonal cycles of the *C. carnea* (*mohave* strain). The all-or-none response patterns that control the seasonal cycles are amenable to alteration by artificial selection, and both actual and hidden genetic variation is important in the evolution of the seasonal cycles (Tauber and Tauber, 1986b). Our hybridization studies indicated that the variation in photoperiodic requirements for reproduction is largely based on allelic differences at two unlinked, autosomal loci. The long-day reproduction of the *carnea* strain is dominant; the short-day/long-day requirement found in *C. downesi* and the *mohave* strain is recessive. Our studies also revealed that, although the short-day/long-day requirement is controlled by homologous genes in eastern and western populations, the expression of dominance by the alleles at the two loci may vary among geographic populations.

In the *mohave* strain, the genes that determine the photoperiodic reactions also act as suppressors of the genes that govern responsiveness to prey. Variation in the response to prey is controlled by an independent polygenic system with a threshold for the expression of diapause (Tauber and Tauber, 1987).

In summary, the great diversity in the seasonal cycles of the various biotypes in the *C. carnea* species complex appears to be based on variability in a small number of ecophysiological responses and on genetic variation at a small number of major loci. Furthermore, these response patterns and their genetic bases are identifiable through standard genetic analysis.

Trade-offs Among Life-History Traits

Our experimental results suggest that, in the *Chrysoperla* species complex, the responses to photoperiod and the responses to prey are controlled by different genetic systems (Tauber and Tauber, 1986a, 1986b, 1987). Thus, each of the two sets of responses may be able to evolve independently. This raises the question of why all populations do not possess a reproductive response to prey when such a response pattern can synchronize reproduction with both the spatial and the temporal occurrence of food? Is there a constraint on the evolution of the response to prey? Do trade-offs exist?

In examining the relationship between the response to prey and other life-history traits, we identified a consistent phenotypic correlation between the reproductive response to prey and the length of the nondiapause preoviposition period. In all cases, the ability to enter a prey-mediated diapause correlates positively with a relatively long preoviposition period, even under the most favorable conditions for reproduction (an increase in day length and abundant prey) (Tauber and Tauber, 1986a). The average nondiapause preoviposition period of populations that show a response to prey ranges from 8 to 21 days. This period is considerably longer than the 3 to 5 days for populations in which prey do not mediate reproduction.

The ability to enter a prey-mediated diapause seems to be physiologically (and perhaps also genetically) linked with a long preoviposition period. The delay in reproduction is distinct from reproductive diapause; it appears to be a period during which adults "assess" the environment for the presence or absence of prey. This correlation between diapause and the preoviposition period may constrain the evolution of responsiveness to prey unless there is very strong selection pressure in its favor (e.g., advantages that are sufficient to overcome the negative effects of a long nondiapause preoviposition period). Such a situation apparently occurs in areas of western North America, where there is very unpredictable variation in prey availability during the summer.

DISCUSSION AND CONCLUSION

Implications for IPM and Genetic Improvement

The ability to predict and manage insect invasion and colonization represents one of the foremost challenges for the protection of food, fiber, recreational areas, as well as the health of humans, domestic animals, and wildlife. Many of our major insect pests are exotic species that have invaded and spread throughout large areas; most other major pests are native species that have greatly expanded their ranges with the introduction of cultivated crops and domestic animals. Thus, the efficacy and efficiency of pest management requires reliable methods for assessing the degree to which introduced or native species can adapt to new areas and to new cultural practices (Branson and Krysan, 1981).

Similarly, understanding and forecasting colonization is particularly relevant for biological control. An important reason suggested for the failure of natural enemies to achieve control is the introduction of poorly adapted biotypes (van den Bosch, 1971; van den Bosch et al., 1979). For example, natural enemy biotypes that are not well synchronized with their hosts or other requisites are not effective biological control agents. Similarly, biotypes that prefer habitats other than the targeted ones are not effective. Thus, if biological control is to be a dependable, central tactic within the framework of integrated pest management, natural enemies must be matched to the temporal and spatial characteristics of the pest and crop situation. Among the factors to be considered are whether the environment is temporary or stable, whether the cropping system is a monoculture or polyculture, and whether alternate hosts or prey are available. Among the vital characteristics of natural enemies are their habitat preference, host suitability, seasonality, and voltinism, as well as the degree of variability in these traits (e.g., Ruberson et al., 1989).

These practical considerations return us to our original problem of defining the limits to the alterability of insect life cycles. Does information about the ecology, ecophysiological mechanisms, and genetic variability of temporal and spatial adaptations allow us to predict the suitability of different biotypes for specific agronomic situations? Can it help us solve problems related to the collection and importation of natural enemies, for example, whether to introduce a single biotype or multiple biotypes? Furthermore, is this knowledge useful for improving natural enemies through genetic manipulation?

We approached these questions for the *C. carnea* species group from two integral perspectives: adaptation to temporal variation (the seasons) and adaptation to spatial variation (the habitat). First, there are several choices for matching *Chrysoperla* seasonal cycles to specific pest management situations (Fig. 5.5). *C. carnea* (*carnea* strain) is multivoltine; it reproduces throughout spring and summer. *C. downesi* is univoltine; reproduction is confined to late spring. The western *C. carnea* (*mohave* strain) is polymorphic in its reproductive cycle. Generally, part of each *mohave* population is multivoltine, another part is univoltine, and still another is variable, depending on prey abundance. Therefore, we have three choices of seasonal cycle to match with those of the target pests: multivoltine, univoltine, or mixed (polymorphic).

Similarly, the *Chrysoperla* biotypes exhibit considerable variability in habitat preference and adult coloration (Fig. 5.6). During its reproductive phase, *C. carnea* (*carnea* strain) occurs largely in fields and meadows, and the grass green coloration of the adults matches its habitat closely. In contrast, the polymorphic *C. carnea* (*mohave* strain) is variable in both habitat association and adult coloration, but it is most commonly found in broadleaf trees. The dark green *C. downesi* is largely restricted

SEASONAL REPRODUCTIVE CYCLE

		Spring	Summer	Autumn	Winter
Multivoltine					
C. carnea (carnea)		Rep	Rep	Dor	Dor
Univoltine					
C. downesi		Rep	Dor	Dor	Dor
Polymorphic					
C. carnea (mohave)	a)	Rep	Rep	Dor	Dor
	b)	Rep	Dor	Dor	Dor

Rep = Reproduction
Dor = Dormancy

Figure 5.5. Naturally occurring seasonal cycles of the various biotypes of the *Chrysoperla carnea* species complex

to coniferous forests. Thus, in selecting an appropriate biotype for use in particular pest management situations, we have a variety of characteristics to consider.

Based on our knowledge of their temporal and spatial adaptations, we can propose hypotheses to be tested for the various *Chrysoperla* biotypes in biological control programs (Fig. 5.7). For example, we hypothesize that *C. carnea* (*carnea* strain) from the eastern United States generally would be the most effective in row and field crops and vineyards (Tauber and Tauber, 1975; see also Messenger, 1975; Ehler, 1977). This multivoltine biotype produces larvae all summer, and it has the appropri-

SEASONAL HABITAT PREFERENCE

		SPRING	SUMMER	AUTUMN	WINTER
C. carnea (carnea strain):		---Fields and Meadows (light green)	→	Forest Edges (reddish brown)	---
C. downesi:		----------	Conifers (dark green)	----------	
C. carnea (mohave strain):	a)	----------	Conifers (green)	----------	
	b)	----------	Broad Leaved Trees (green, reddish)	----------	
	c)	---Fields and Meadows (green)	→	Conifers (green)	

Figure 5.6. Seasonal habitat preferences of the various biotypes of the *Chrysoperla carnea* species complex

POTENTIAL USE IN PEST MANAGEMENT

SITUATION	CURRENTLY AVAILABLE BIOTYPE
Row and Field Crops	*C. carnea* (*carnea* strain) from eastern U. S.
Fruit Tree Orchards	None
Vineyards	*C. carnea* (*carnea* strain) from eastern U. S.
Urban plantings:	
Eastern U. S.	*C. carnea* (*carnea* or *mohave* strain)
Western U. S.	*C. carnea* (*mohave* or *carnea* strain)
Forests:	
Coniferous	*C. downesi* (eastern & western)
Deciduous	*C. carnea* (*mohave* or *carnea* strain)

Figure 5.7. Matching of *Chrysoperla carnea* biotypes to specific pest management situations

ate habitat preference and adult coloration. Its preoviposition period is short, and, because it has a very low level of genetic variability for the preferred traits, it can be mass-reared efficiently. Moreover, it can be manipulated in the field with artificial food (Hagen et al., 1971), and it can be selected for resistance to certain pesticides (Grafton-Cardwell and Hoy, 1985, 1986). Thus, it appears to be well adapted for use against pests in these agricultural situations.

Our proposals for choosing biotypes for other horticultural situations are also based on their temporal and spatial adaptations (Fig. 5.7). However, because of the variability in the habitat preferences of the *mohave* strain originating from western North America, these recommendations are tentative. Moreover, techniques for mass rearing the *mohave* strain have not been developed; this presents problems for inundative and augmentative releases.

A consideration of the preceding issues and problems leads us to an intriguing question. What would be the result if we introduced a natural enemy biotype that was less than ideal for a given agronomic situation (see van den Bosch, 1971)? Will natural selection produce a better adapted biotype? Our studies have demonstrated that, indeed, the characteristics that adapt biotypes to temporal and spatial variability in the environment are genetically variable and alterable by artificial selection. In addition, the biotypes are hybridizable, and, thus, if one were introduced into an area that contained another biotype, it might not be subject to competitive exclusion in the release area (see Ehler and Hall, 1982). Indeed, because of the ability to hybridize, the introduction of a new biotype might increase the genetic variability that is available to natural selection. If this genetic variability involves traits for the efficient use of resources (e.g., available prey), natural selection might produce a better adapted, more effective biotype. Thus, we can see no long-term detriments, but some possible advantages, to introducing multiple biotypes of the *C. carnea* species complex.

However, our results also demonstrate that there are serious constraints on the evolution of temporal adaptations. First, there appear to be limits to genetic variability. For example, we have no evidence for the expression of prolonged diapause (diapause of more than one year) in the *C. carnea* species complex. Nor has prolonged diapause been reported for other species in the genus. Assuming this situation does not reflect a bias in sampling, we suggest that prolonged diapause is unlikely to evolve even in areas where unfavorable conditions prevail for more than one year. At present, the *C. carnea* species complex does not appear to have the genetic variability for the evolution of this trait, and probably the only way to introduce it would be to use techniques from genetic engineering.

Second, the evolution of certain traits appears to be physiologically or genetically constrained. As we demonstrated earlier, the evolution of prey-mediated reproductive diapause appears to have an associated cost: it is accompanied by a relatively long nondiapause preoviposition period. Thus, we conclude that natural selection can adapt biotypes to new situations but that the evolution of adaptation has limitations. With sufficient knowledge of the genetic variability in specific characteristics and an understanding of the trade-offs between and among these traits, it is possible to outline limits in adaptation to new environments.

This type of knowledge can enable us to produce biotypes with specific sets of temporal and spatial adaptations that do not occur naturally. For example, we propose *C. downesi* for use against pests on conifers in forests, parklands, and urban settings, because its behavior and adult coloration appear to be well adapted to these habitats (Fig. 5.7). However, this biotype is only active in the spring, and pest problems may occur on conifers all summer. Therefore, the situation may require a multivoltine *C. downesi*; unfortunately, such a biotype is not known from nature. Similarly, although mass-reared *C. carnea* (*carnea* strain) are commonly sold for use in commercial fruit orchards, it is not suitable for this habitat. Currently, no biotype has the combination of traits suitable for fruit orchards. For example, the seasonal cycle and adult coloration of the *carnea* strain are appropriate for fruit orchards, but its habitat preference is not. Finally, under some circumstances, it may be desirable to release a biotype that is able to respond reproductively to prey and that also has a short preoviposition period. Again, populations with this combination of traits are unknown from nature. The solution to these problems would be to use our knowledge of the genetic variability in the *Chrysoperla carnea* species complex to develop new biotypes with the desired characteristics through traditional breeding or genetic engineering.

Future Needs

Given the goal of developing new biotypes, what are some of the requirements for genetic improvement? First, we need to understand the ecophysiological and genetic mechanisms underlying the desired traits (Fig. 5.8). Second, we need to quantify the

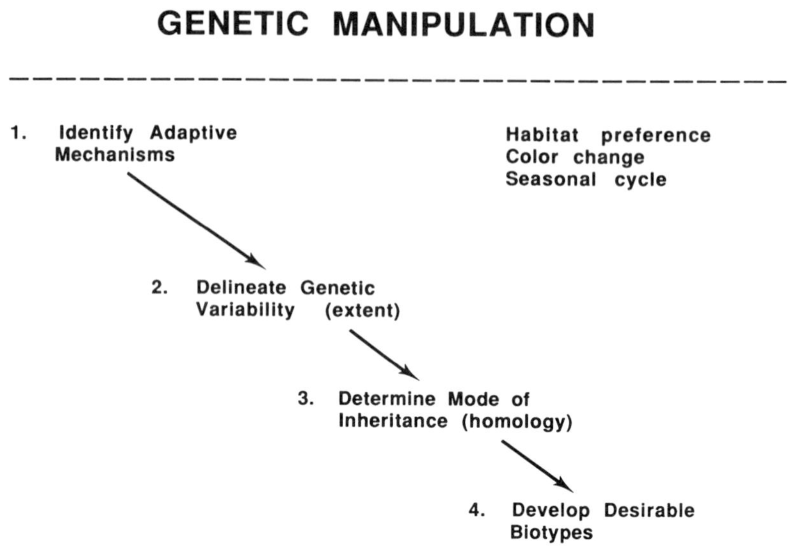

Figure 5.8. Four requirements for developing genetically improved biotypes

extent of genetic variation underlying the major adaptive traits. This can be done through analysis of geographic populations and through quantitative genetic studies. Third, we must work out the mode of inheritance of the traits and determine whether similar responses in different populations are based on homologous genetic systems. With the preceding information, it may be possible to use traditional, sometimes relatively simple, breeding procedures to develop a biotype with the desired traits.

At present, there is sufficient knowledge of the seasonal cycles in *Chrysoperla* to begin such a venture in genetic improvement. However, in looking toward the efficient and effective use of the *C. carnea* biotypes in the future, we should recognize that we lack information in two key areas. Although we know the ecophysiological and genetic bases for the temporal adaptations relatively well, there is insufficient knowledge concerning the mechanisms and genetics of habitat preference. Little experimental work has been done on this crucial aspect (Hagen et al., 1976; Duelli, 1984). The second impediment to genetic improvement is the difficulty in rearing *Chrysoperla* in pure culture for more than five or six generations. There has been insufficient study of nutritional and other developmental requirements of these important natural enemies. These problems must be addressed before biological control can benefit from the full range of temporal and spatial adaptations that the *C. carnea* species complex offers. Only then will this group of insects reach its potential as natural enemies of pests in a variety of agronomic and natural situations. This is true whether naturally occurring, genetically selected, or genetically engineered biotypes are used.

In summary, presently we face an unfinished picture—a picture in which some aspects of temporal and spatial variability are delineated and others are in need of further work. But the projected goals we have considered in this chapter may be achievable within 5 to 10 years. Furthermore, what we learn about one group of organisms, such as *Chrysoperla*, can be applied, in general and sometimes specific ways, to other organisms—both pests and natural enemies. With the appropriate combination of scientific enthusiasm, well-focused experiments, and judicious funding for both traditional biological, as well as new biotechnological studies, there could be dramatic progress in developing the efficient and effective use of these, and other important, natural enemies in IPM systems.

Acknowledgments

We thank G. G. Kennedy, North Carolina State University, for inviting us to participate in the symposium and also K. C. Kim, Pennsylvania State University, for asking us to contribute to the publication. We are grateful to L. E. Ehler and the Department of Entomology, University of California, Davis, for their hospitality during a sabbatical leave (MJT). We thank L. E. Ehler, University of California, Davis, and D. C. Margolies, Kansas State University, for comments on our chapter. Our work has been supported, in part, by the National Science Foundation, the United States Department of Agriculture, and Hatch Project 408, USDA.

REFERENCES

Baker, H. G., and G. L. Stebbins. 1965. *The Genetics of Colonizing Species*. New York: Academic Press.

Beck, S. D. 1980. *Insect Photoperiodism*, 2nd ed. New York: Academic Press.

Branson, T. F., and J. L. Krysan. 1981. Feeding and oviposition behavior and life cycle strategies of *Diabrotica*: An evolutionary view with implications for pest management. *Environ. Entomol.* 10:826–831.

Brown, V. K., and I. Hodek (eds.). 1983. *Diapause and Life Cycle Strategies in Insects*. The Hague, the Netherlands: Junk.

Bush, G. L., and S. R. Diehl. 1982. Host shifts, genetic models of sympatric speciation and the origin of parasitic insect species. Pages 297–305 in Visser, J. H., and A. K. Minks (eds.), *Proc. Fifth Int. Symp. on Insect-Plant Relationships*, PUDOC, Wageningen, Netherlands.

Danilevsky, A. S. 1965. *Photoperiodism and Seasonal Development of Insects* (English translation). London: Oliver & Boyd.

Danks, H. V. 1987. *Insect Dormancy: An Ecological Perspective*. Biol. Survey of Canada (Terrestrial Arthropods), Ottawa.

Darwin. C. 1859. *The Origin of Species*. (A facsimile of the First Edition, with introduction by E. Mayr). Cambridge, Mass.: Harvard Univ. Press.

Denno, R. F. 1983. Tracking variable host plants in space and time. Pages 291–341 in Denno, R. F. and M. S. McClure (eds.), *Variable Plants and Herbivores in Natural and Managed Systems*. New York: Academic Press.

Denno, R. F., and M. S. McClure (eds.). 1983. *Variable Plants and Herbivores in Natural and Managed Systems*. New York: Academic Press.

Diamond, J., and T. J. Case (eds.). 1986. *Community Ecology*. New York: Harper & Row.

Dingle, H. 1972. Migration strategies of insects. *Science* 175:1327–1335.

———. (ed.) 1978. *Evolution of Insect Migration and Diapause*. New York: Springer-Verlag.

———. 1980. Ecology and evolution of migration. Pages 1–101 in Gauthreaux, Jr., S. A. (ed.), *Animal Migration, Orientation, and Navigation*. New York: Academic Press.

———. 1984. Behavior, genes, and life histories: Complex adaptations in uncertain environments. Pages 169–194 in Price, P. W., C. N. Slobodchikoff, and W. S. Gaud (eds.), *A New Ecology*. New York: John Wiley & Sons.

Doutt, R. L., D. P. Annecke, and E. Tremblay. 1976. Biology and host relationships of parasitoids. Pages 143–168 in Huffaker, C. B., and P. S. Messenger (eds.), *Theory and Practice of Biological Control*. New York: Academic Press.

Duelli, P. 1984. Flight, dispersal, migration. Pages 110–116 in Canard, M., Y. Séméria, and T. R. New (eds.), *Biology of Chrysopidae*. The Hague, The Netherlands: Dr. W. Junk Publ.

Ehler, L. E. 1977. Natural enemies of cabbage looper on cotton in the San Joaquin Valley. *Hilgardia* 45:1–106.

Ehler, L. E., and R. W. Hall. 1982. Evidence for competitive exclusion of introduced natural enemies in biological control. *Environ. Entomol.* 11:1–4.

Endler, J. A. 1986. *Natural Selection in the Wild*. Princeton, N. J.: Princeton Univ. Press.

Fleschner, C. A. 1950. Studies on searching capacity of the larvae of three predators of the citrus red mite. *Hilgardia* 20:233–265.

Gould, F. and R. E. Stinner. 1984. Insects in heterogeneous habitats. Pages 427–449 in Huffaker, C. B., and R. L. Rabb (eds.), *Ecological Entomology*. New York: John Wiley & Sons.

Grafton-Cardwell, E. E., and M. A. Hoy. 1985. Intraspecific variability in response to pesticides in the common green lacewing, *Chrysoperla carnea* (Stephens) (Neuroptera: Chrysopidae) *Hilgardia* 53:1–32.

———. 1986. Selection of the common green lacewing for resistance to carbaryl. *Calif. Agric.* Sept. 1986:22–24.

Greenslade, P. J. M. 1983. Adversity selection and the habitat templet. *Am. Nat.* 122:352–365.

Grime, J. P. 1977. Evidence for the existence of three primary strategies in plants and its relevance to ecological and evolutionary theory. *Am. Nat.* 111:1169–1194.

Groves, R. H., and J. J. Burdon (eds.). 1986. *Ecology of Biological Invasions*. Cambridge, Mass.: Cambridge Univ. Press.

Hagen, K. S., P. Greany, E. F. Sawall, Jr., and R. L. Tassan. 1976. Tryptophan in artificial honeydews as a source of an attractant for adult *Chrysopa carnea*. *Environ. Entomol.* 5:458–468.

Hagen, K. S., E. F. Sawall, Jr., and R. L. Tassan. 1971. The use of food sprays to increase effectiveness of entomophagous insects. *Proc. Tall Timbers Conf. Ecol. Anim. Control Habitat Manage.* 2:59–81.

Henry, C. S. 1985. Sibling species, call differences, and speciation in green lacewings (Neuroptera: Chrysopidae). *Evolution* 39:965–984.

Hidaka, T. (ed.). 1977. *Adaptation and Speciation in the Fall Webworm*. Tokyo: Kodansha.

Hinton, H. E. 1960. Cryptobiosis in the larva of *Polypedilum vanderplanki* Hint. (Chironomidae). *J. Insect Physiol.* 5:286–300.

Holm, E. 1988. Environmental restraints and life histories: A habitat templet matrix. *Oecologia* 75:141–145.

Howarth, F. G. 1985. Impacts of alien land arthropods and mollusks on native plants and animals in Hawai'i. Pages 149–179 in Stone, C. P., and J. M. Scott (eds.), *Hawai'i's Terrestrial Ecosystems: Preservation and Management*. Cooperative National Park Resources Study Unit, Univ. of Hawaii, Honolulu.

Ingrisch, S. 1986. The plurennial life cycles of the European Tettigoniidae (Insecta: Orthoptera). 3. The effect of drought and the variable duration of the initial diapause. *Oecologia* 70:624–630.

Istock, C. A. 1984. Boundaries to life history variation and evolution. Pages 143–168 in Price, P. W., C. N. Slobodchikoff, and W. S. Gaud (eds.), *A New Ecology*. New York: John Wiley & Sons.

Jermy, T., E. A. Bernays, and A. Szentesi. 1982. The effect of repeated exposure to feeding deterrents on their acceptability to phytophagous insects. Pages 25–32 in Visser, J. H., and A. K. Minks (eds.), *Proc. Fifth Int. Symp. on Insect–Plant Relationships*. Wageningen, Netherlands: PUDOC.

Kareiva, P. 1986. Patchiness, dispersal, and species interactions: Consequences for communities of herbivorous insects. Pages 192–206 in Diamond, J., and T. G. Case (eds.), *Community Ecology*. New York: Harper & Row.

Keiser, I., and M. Ashraf. 1987. Hardiness, longevity, mortalities, and differences according to sex of adult Oriental fruit flies, Mediterranean fruit flies and melon flies subject to a cold environment and deprivation of food and water. *J. Environ. Sci. Health*, Part A, 22:581–606.

Kennedy, G. G. 1986. Plant—plant-pathogen—insect interactions. Pages 203–216 in Kogan, M. (ed.), *Ecological Theory and Integrated Pest Management Practice*. New York: John Wiley & Sons.

Kennedy, G. G., and D. C. Margolies. 1985. Mobile arthropod pests: Management in diversified agroecosystems. *Bull. Entomol. Soc. Am.* 31:21–27.

Kennedy, J. S. 1961. A turning point in the study of insect migration. *Nature* 189:785–791.

Krysan, J. L., D. E. Foster, T. F. Branson, K. R. Ostlie, and W. S. Cranshaw. 1986. Two years berfore the hatch: Rootworms adapt to crop rotation. *Bull. Entomol. Soc. Am.* 32:250–253.

Lankinen, P., and J. Lumme. 1984. Genetic analysis of geographical variation in photoperiodic diapause and pupal eclosion rhythm in *Drosophila littoralis*.Pages 97–114 in *Photoperiodic Regulation on Insect and Molluscan Hormones*. London (Ciba Foundation Symposium 104): Pitman.

Lee, Jr., R. E., C. P. Chen, and D. L. Denlinger. 1987. A rapid cold-hardening process in insects. *Science* 238:1415–1417.

Lees, A. D. 1955. *The Physiology of Diapause in Arthropods*. London: Cambridge Univ. Press.

Loeschcke, V. (ed.). 1987. *Genetic Constraints on Adaptive Evolution*. Berlin: Springer-Verlag.

Lounibos, L. P., C. van Dover, and G. F. O'Meara. 1982. Fecundity, autogeny, and the larval environment of the pitcher-plant mosquito, *Wyeomyia smithii*. *Oecologia* 55:160–164.

Messenger, P. S. 1975. Parasites, predators, and population dynamics. Pages 201–223 in D. Pimentel (ed.), *Insects, Science, and Society*. New York: Academic Press.

Michod, R. E. 1984. Constraints on adaptation, with special reference to social behavior.

Pages 253–278 in Price, P. W., C. N. Slobodchikoff, and W. S. Gaud (eds.), *A New Ecology*. New York: John Wiley & Sons.

Moeur, J. E., and C. A. Istock. 1980. Ecology and evolution of the pitcher-plant mosquito. IV. Larval influence over adult reproductive performance and longevity. *J. Anim. Ecol.* 49:775–792.

Mooney, H. A., and J. A. Drake. 1986. *Ecology of Biological Invasions of North America and Hawaii*. New York: Springer-Verlag.

Parsons, P. A. 1983. *The Evolutionary Biology of Colonizing Species*. London: Cambridge Univ. Press.

Price, P. W. 1980. *Evolutionary Biology of Parasites*. Princeton, N. J.: Princeton Univ. Press.

Prosser, C. L. 1986. *Adaptational Biology*. New York: John Wiley & Sons.

Rabb, R. G., and G. G. Kennedy (eds.). 1979. *Movement of Highly Mobile Insects: Concepts and Methodology in Research*. Proc. Conf. Raleigh, N.C., 1979. Raleigh, N. C.: University Graphics, North Carolina State Univ.

Roughgarden, J. 1979. *Theory of Population Genetics and Evolutionary Change: An Introduction*. New York: Macmillan.

Ruberson, J. R., Tauber, M. J., and Tauber, C. A. 1989. Intraspecific variability in hymenopteran parasitoids: Comparative studies of two biotypes of the egg parasitoid *Edovum puttleri* (Hymenoptera: Eulophidae). *J. Kansas Entomol. Soc.* 62:189–202.

Saunders, D. S. 1982. *Insect Clocks*, 2nd ed. New York: Pergamon Press.

———. 1984. Introduction: The links between "wet" and "dry" physiology. Pages 2–6 in *Photoperiodic Relation of Insect and Molluscan Hormones*. London (Ciba Foundation Symposium 104): Pitman.

Shapiro, A. M. 1979. The phenology of *Pieris napi microstriata* (Lepidoptera: Pieridae) during and after the 1975–77 California drought, and its evolutionary significance. *Psyche* 86:1–10.

Sibly, R. M., and P. Calow. 1986. *Physiological Ecology of Animals*. Oxford: Blackwell Scientific Publ.

Slansky, Jr., F. 1982. Toward a nutritional ecology of insects. Pages 253–259 in Visser, J. H., and A. K. Minks (eds.), *Proc. Fifth Int. Symp. on Insect–Plant Relationships*. Wageningen, The Netherlands: PUDOC.

Smith-Gill, S. J. 1983. Developmental plasticity: Developmental conversion *versus* phenotypic modulation. *Am. Zool.* 23:47–55.

Solbreck, C. 1985. Insect migration strategies and population dynamics. Pages 641–662 in Rankin, M. A. (ed.), *Migration: Mechanisms and Adaptive Significance*. Univ. Texas Contrib. Marine Sci. Suppl. 27.

Southwood, T. R. E. 1977. Habitat, the templet for ecological strategies? *J. Anim. Ecol.* 46:337–365.

———. 1981. Bionomic strategies and population parameters. Pages 30–52 in May, R. M. (ed.), *Theoretical Ecology*. Oxford: Blackwell Scientific Publs.

———. 1988. Tactics, strategies and templets. *Oikos* 52:3–18.

Stearns, S. C. 1983. Introduction to the symposium: The inter-face of life-history evolution, whole-organism ontogeny, and quantitative genetics. *Am. Zool.* 23:3–4.

Strong, D. R., J. H. Lawton, and R. Southwood. 1984. *Insects on Plants*. Cambridge, Mass.: Harvard Univ. Press.

Tauber, M. J., and C. A. Tauber. 1972. Geographic variation in critical photoperiod and in diapause intensity of *Chrysopa carnea* (Neuroptera). *J. Insect Physiol.* 18:25–29.

———. 1975. Criteria for selecting *Chrysopa carnea* biotypes for biological control: adult dietary requirements. *Can. Entomol.* 107:589–595.

———. 1976a. Environmental control of univoltinism and its evolution in an insect species. *Can. J. Zool.* 54:260–266.

———. 1976b. Developmental requirements of the univoltine species *Chrysopa downesi*: Photoperiodic stimuli and sensitive stages. *J. Insect Physiol.* 22:331–335.

Tauber, C. A., and M. J. Tauber. 1982. Evolution of seasonal adaptations and life history traits in *Chrysopa*: Response to diverse selective pressures. Pages 51–72 in Dingle, H., and J. P. Hegmann (eds.), *Evolution and Genetics of Life Histories*. New York: Springer-Verlag.

———. 1986a. Ecophysiological responses in life-history evolution: Evidence for their importance in a geographically widespread insect species-complex. *Can. J. Zool.* 64:875–884.

———. 1986b. Genetic variation in all-or-none life-history traits of the lacewing *Chrysoperla carnea*. *Can. J. Zool.* 64:1542–1544.

———. 1987. Inheritance of seasonal cycles in *Chrysoperla* (Insecta: Neuroptera). *Genet. Res.* 49:215–223.

———. 1989. Sympatric speciation in insects: Perception and perspective. Pages 307–344 in Otte, D., and J. A. Endler (eds.), *Speciation and its Consequences*. New York: Sinauer.

Tauber, M. J., C. A. Tauber, and S. Masaki, 1986. *Seasonal Adaptations of Insects*. New York: Oxford Univ. Press.

van den Bosch, R. 1971. Biological control of insects. *Annu. Rev. Ecol. Syst.* 2:45–66.

van den Bosch, R., R. Hom, P. Matteson, B. D. Frazer, P. S. Messenger, and C. S. Davis. 1979. Biological control of the walnut aphid in California: Impact of the parasite, *Trioxys pallidus*. *Hilgardia* 47:1–13.

Via, S. 1987. Genetic constraints on the evolution of phenotypic plasticity. Pages 47–71 in Loeschcke, V. (ed.), *Genetic Constraints on Adaptive Evolution*. Berlin: Springer Verlag.

Wellington, W. G., and R. M. Trimble. 1984. Weather. Pages 399–425 in Huffaker, C. B., and R. L. Rabb (eds.), *Ecological Entomology*. New York: John Wiley & Sons.

Wiens, J. A., J. F. Addicott, T. J. Case, and J. Diamond. 1986. Overview: The importance of spatial and temporal scale in ecological investigations. Pages 145–153 in Diamond, J., and T. J. Case (eds.), *Community Ecology*. New York: Harper & Row.

Wiklund, C. 1982. Generalist versus specialist utilization of host plants among butterflies. Pages 181–191 in Visser, J. H., and A. K. Minks (eds.), *Proc. Fifth Int. Symp. on Insect–Plant Relationships*. Wageningen, The Netherlands: PUDOC.

Wilson, E. O. 1971. The plight of taxonomy. *Ecology* 52:741.

———. 1985. Time to revive systematics. *Science* 230:1227.

6

ADAPTATION TO SPATIAL VARIATION IN HABITAT: SPATIAL EFFECTS IN AGROECOSYSTEMS

David C. Margolies

Kansas State University

INTRODUCTION

There is now a well-developed body of theory that suggests that environmental variation can promote genetic variation within a population (Mitter and Futuyma, 1983; Hedrick, 1986). The evidence for this indicates that the degree to which arthropods experience spatial variation in some key environmental factor, such as the host plant, is critical in setting the conditions that maintain genetic variation and evolution (Levins, 1968). While spatial variation in resource distribution is important to explanations of genetic variation, genetic variation in resource utilization is also important to an understanding of adaptation to spatial variation. There is a considerable body of theory and evidence that indicates that an organism's perception of, and response to, patterns of spatial variation can itself evolve (e.g., Harrison, 1980). This chapter will focus on some of the effects of spatial variation among habitats on arthropod adaptations, particularly in agricultural systems where spatial structure is artificially imposed.

Manipulation of the spatial structure of agroecosystems has been suggested as a means of managing arthropod pest populations on crop hosts (Price, 1976; Wellington, 1977; Rabb, 1978; Way, 1979; Browning, 1980; Levins and Wilson, 1980; Cromartie, 1981; Kennedy and Margolies, 1985). Changing spatial structure for management purposes would attempt to reduce pest numbers by altering rates of immigration and emigration, or births and deaths, or some combination of these. This might be achieved by increasing vegetational diversity and hence spatial heterogeneity within an agroecosystem (Risch et al., 1983). On the other hand, reducing diversity might reduce the abundance of certain pests in some systems by minimizing field-to-field movement and by depriving a pest of a sequence of suitable hosts on which its numbers can increase (Way, 1979). One limitation to the successful implementation of management tactics based on manipulating the spatial arrangement of the agricultural environment is our inability to predict pest responses to such changes (Stinner et al., 1982; Hare, 1983; Andow, 1991). Our inability to anticipate pest response weakens pest management strategies.

Pest management almost always imposes selective pressures on pests, because, in most instances, management goals involve reduction of pest fitness (i.e., survival or reproduction) in a crop. When manipulation of the spatial arrangement of an agroecosystem is involved, the nature of the selection pressure will depend on those factors that define spatial structure: the search capacity and host range of the pest, and the relative isolation and suitability of crop and noncrop hosts. Arthropod response to such pressures might involve evolutionary changes in search capacity (e.g., efficiency of host location, rate and distance of dispersal) or host utilization (e.g., efficiency of host use, range of host use). Saks (Chapter 16 in this book) and Krysan (Chapter 18 in this book) present evidence that insect populations have adapted to changes in the structure of their agricultural environments in ways that may contravene pest management. On the other hand, pest arthropods may show no evolutionary response if the population does not contain variability for traits associated with dispersal, search capacity, or host utilization, or if there are genetic correlations that restrict changes in these traits. If patterns of arthropod evolution could be predicted in response to environmental patterns (e.g., Gould, Chapter 3 in this book), agronomists and economic entomologists might be able to impose degrees and arrangements of host variation that would reduce the damaging impact of pest arthropods in agroecosystems. Therefore, the potential for adaptation by arthropods to variation in the spatial distribution of resources should be of major concern in management programs.

In this chapter, I will examine the effect of agriculture on spatial variation in resources used by arthropod pests and the possibility of adaptation by arthropods in response to such practices. This will be done by first defining several components of environmental variation (Southwood, 1977), and then by examining how these might be affected by some proposed manipulations of agroecosystem structure. I will also consider the type of information that may help to predict the outcome of these manipulations and some studies that develop this information. At best, I can only attempt to highlight some general problems and future possibilities in a few selected crops and crop systems because of the limited information available on genetics and adaptation of herbivorous pest species. Andersen and Kareiva (Chapter 12 in this book) discuss some of the implications and effects of spatial variation in agroecosystems on predators and parasitoids.

PATTERNS OF SPATIAL VARIATION AND ADAPTATION

Spatial variation in an environment is the result of the nonuniform distribution of resources and conditions that affect the immediate physiological state or future success of an organism. This nonuniform distribution creates discontinuities in environmental characters, which an organism can perceive or which can affect its survival or reproduction. The space bounded by such a discontinuity is a resource patch (Wiens, 1976). The spatial structure of a particular environment is a function of the magnitude of differences among its component patches relative to an organism's potential performance in each, and the size and isolation of individual patches relative to an organism's mobility (Gillespie, 1974; Addicott et al., 1987). Therefore, the patchiness of an environment is meaningfully defined only in relation to the performance, perceptions, and mobility of individuals or populations in the environment. We can identify patches in terms of the distribution of resource utilization and activity of an organism among habitats (MacArthur and Pianka, 1966; Hassell and Southwood, 1978), and can derive an operational definition of spatial structure from the behavior, survival, and reproduction of a particular population.

The structure of an environment, however, changes over time, as well as through space. The response to temporal variation in a habitat depends on the relationship between the time used for feeding and reproduction and the time the habitat is, or is likely to be, suitable (Hutchinson, 1953; Southwood, 1971, 1977). The stability of a habitat is defined in terms of the historical persistence of resources compared to the generation time of a population. In an environment composed of several habitat patches, the persistence of individual patches is not necessarily synchronized. Therefore, temporal changes within and between patches become a major source of variation in patch quality in space, and the effect of variation in space and time on the survivorship and reproduction of individuals is closely connected.

Dispersal by arthropods is thought to be one adaptive response to this interplay of spatial and temporal variation in the suitability of local environments (Southwood, 1962; Johnson, 1969). I am defining *dispersal* as a process by which individuals abandon one habitat in order to locate another. The process allows individuals and populations to keep pace with the changing location of suitable habitats (Southwood, 1962), where suitability is defined by survival and reproductive performance. *Dormancy* is an alternative response to changing habitat suitability, which has been discussed by Tauber and Tauber (Chapter 5 in this book). The success of both dispersal and dormancy depend on the pattern of appearance and disappearance of suitable habitats and the relative isolation of these habitats. In theory, dispersal should confer greater fitness than dormancy when changes in habitat suitability are asynchronous, that is, when conditions in some patches within dispersal range remain favorable while the local habitat is deteriorating (Solbreck, 1978). Because finding a suitable new habitat entails some risk of failure and death, or at least an interruption of feeding and reproduction, individuals faced with a heterogeneous environment are expected to resist leaving a habitat to some degree (Taylor and Taylor, 1977; Taylor, 1981). However, individuals face risks whether or not they disperse and may benefit if they do disperse. The importance of dispersal behavior and the maintenance of this behavior in a population presumably occur in response to the chances of survival and reproduction in the present habitat, the likelihood that a better habitat is available, and the immediate reduction in survival and reproduction incurred while making the transition.

Several authors have attempted to characterize parameters of environmental variation in a way that would allow us to compare the risks and advantages of dispersal and make testable predictions regarding those conditions under which it is likely to occur. Reddingius and Den Boer (1970) proposed net reproductive rate, R_0, as a measure of habitat suitability and used variability of R_0 among habitats as a biologically meaningful definition of spatial heterogeneity to model "spreading-of-risk." This approach, however, ignores the risk of moving between habitats that differ in suitability, as well as the possibility of change in R_0 over time. Instead of using R_0 as the sole criteria of suitability, Southwood (1977) suggested a classification of environmental variability that emphasized the predictability of the environment as the basis of adaptation to variation. He proposed that the fitness associated with alternative habitats could be best expressed if the rate of increase in a habitat were weighted by the chances of surviving to find and reproduce in a new habitat. Further, evaluation of reproductive success in a habitat should take into account the variance in rate of increase over time. These three components (differences between habitats in potential rate of increase, change in the ability of a habitat to sustain a rate of increase, and probability of successful transition between habitats) form a framework for analyzing the effect of environmental variation on its inhabitants.

Human intervention in agroecosystem structure has affected both the spatial variability and the temporal predictability of the environment. Because most agroecosystems are mosaics of crops and noncrops, and because periodic habitat senescence or destruction is characteristic of agricultural production, changing location of suitable hosts in an agricultural environment is likely to be a significant factor in the population dynamics of arthropod pests in agricultural systems (Southwood, 1971; Denno, 1978). The economic impact, as well as the population dynamics, of many agricultural pests lies in their ability to move among crops and exploit the spatial and temporal variation inherent in this environment (Johnson, 1969; van Steenwyk, 1978; Margolies and Kennedy, 1985). Therefore, the structure of that variation should affect the economic impact of mobile arthropod pests and is a legitimate area of concern in pest management. Because a crop host serves as a habitat as well as a food source for herbivorous arthropod pests, manipulating crop type and placement can affect all three components of spatial variation described by Southwood. Management practices that might alter the spatial structure of the environment are discussed later in relation to the components of spatial variation that they are most likely to affect and potential pest adaptation to the altered environment.

COMPONENTS OF, AND ADAPTATION TO, SPATIAL VARIATION IN AGROECOSYSTEMS

Suitability of Different Habitats

It is possible to reduce pest numbers on, and damage to, specific crops by manipulating the relative suitability of those crops or of other pest hosts for pest reproduction and survival (Rabb et al., 1984). The use of resistant crop varieties and trap crops are examples of this approach. However, considered within the framework of an agroecosystem, changing the suitability of one host relative to others may affect the immediate distribution of pests on hosts throughout the system. For example, even though the use of resistant corn varieties could protect the corn crop in eastern North Carolina from economic damage by *Helicoverpa zea*, other accessible host crops, such as soybean and cotton, might be subject to increased pest pressure, depending on the manner in which *H. zea* performance is altered in corn (e.g., resistance by antibiosis, antixenosis, or tolerance) (Kennedy et al., 1987). In addition, changing the suitability of one host will alter relative pest performance among different hosts. This effect may impose selection on the ability of arthropod pest populations to utilize other crop hosts, depending on the manner in which pest populations perceive the spatial structure of their environment (see the section titled "Patterns of Spatial Variation and Adaptation"). In this section, I will address adaptive changes in host utilization by pest arthropods within the context of an agroecosystem in which alternate hosts are available. Although I will focus on intraspecific variation in performance on different hosts, preference behavior may also be an important determinant of the potential for host use adaptation in a population (Singer, 1983; Futuyma and Peterson, 1985; Thompson, 1988). There are, however, few studies that have tried to correlate genetic variation in performance on different hosts with host preference (Via, 1986; Singer et al., 1988; Jaenike, 1989).

Reproduction and survival of arthropod pests vary within an agricultural environment, even on suitable hosts, in response to both natural conditions and production practices (Stinner, 1979; Rabb et al., 1984). The ability of arthropod pests to exploit these various and varying hosts and the response to imposed variation in host quality depend on the genetic structure of host-related performance in a pest population (Futuyma and Peterson, 1985). The genetic makeup of a pest species that can use several host crops or varieties might be of two types: one could be composed of individuals, each of which is able to utilize efficiently a number of hosts, whereas another could be composed of several populations, each of which is able to use only one host efficiently (Roughgarden, 1972; Fox and Morrow, 1981). Genetic differences in host-associated performance within populations of twospotted spider mite, *Tetranychus urticae* (Gould, 1978; Fry, 1990), vegetable leafminer, *Liriomyza sativae* (Via, 1984a, 1984b), and pea aphid, *Acyrthosiphon pisum* (Via, 1991), all of which utilize a wide range of hosts, as well as evidence of genetic differences among scale insect demes associated with individual host plants (Alstad and Edmunds, 1983) and even individual twigs (Alstad and Corbin, 1990), suggest that polyphagous populations are most likely composed of a variety of specialized genotypes. If genotypes do differ in their performance on different hosts, then selection might operate within a host to improve pest performance on that host (Via, 1984a). The ability of a population to adapt to a new host or to changes in the suitability of an old host may then depend on limitations imposed by specialization and on the opportunity to specialize.

It is widely held that specialization by herbivores on a particular host involves some loss of efficiency on other hosts (Dethier, 1954); that is, there is some "tradeoff" between performance on one host and that on others. Current theories of biotype and host race formation assume that adaptation to one resource reduces the ability to simultaneously perform well on another (Diehl and Bush, 1984). If performance on a host is genetically determined, the existence of trade-offs implies either pleiotropy or linkage between the genes affecting host utilization. Therefore, these trade-offs should be observable as negative genetic correlations between performance on different hosts (Rausher, 1984; Via, 1984b). Such genetic interdependence would effectively slow the rate of adaptation of a population to each host, compared to the rate that could occur if performance on different hosts was independent (Lande, 1982). This is because selection for increased performance on one host would cause a decrease in performance on the other, and neither could approach their respective optima. However, positive genetic correlations or no correlations have been more commonly observed. Gould (1979) found that selection for increased survival of *T. urticae* on cucumber increased survival on potato and tobacco as well as on cucumber, had no effect on survival on plantain, but decreased survival on beans. Using full-sib families, Via (1984b) found positive correlations between pupal weight of *L. sativae* sibs reared on both tomato and cowpea, and no correlation of development time between the same hosts. Hare and Kennedy (1986) found no evidence of correlations between survival of *Leptinotarsa decemlineata* full-sib progeny reared on potato and on horsenettle. Negative genetic correlations between performance on different hosts have rarely been found, even when they have been expected and looked for (but see Gould, 1979), although the absence of negative correlations for some measures of performance among hosts does not deny the potentially important effects of negative correlations in some as yet unexamined measure of performance (Rausher, 1984). These findings, qualified by the preceding caveat, imply that many arthropod populations that utilize several hosts may be able to adapt to each host independently, in which case, pest populations might adapt rapidly to changes in the composition of their environment.

Even if performance on different hosts is genetically independent, the rate of movement between host-associated populations can affect the degree of local differentiation of populations on different hosts. If selection on different hosts favors different genotypes, genetic differentiation can only be maintained if there is low gene flow between populations using the different hosts (Hartl, 1987). Via (1984a) demonstrated the potential for population level host specialization in populations of *L. sativae* taken from tomato and cowpea in adjacent fields because there were significant genotype–host interactions. However, she found little difference between the two populations in average response on the two hosts. Via suggested that population divergence was impeded by yearly crop rotations, which result in massive dispersal during spring colonization and the mixing of genotypes from different hosts. In a study of allozyme frequencies among *Musca domestica* populations on different Iowa farms, Black and Krafsur (1986) also found evidence of divergent, habitat specific adaptation that was interrupted by massive seasonal dispersal. In both cases, the absence of host- or habitat-adapted populations is probably due to frequent dispersal among hosts, which is likely in diverse agroecosystems in which various alternate hosts are available.

Duration of Habitat Suitability

The preceding section suggests possible evolutionary paths open to arthropods faced with spatial variation in host suitability, assuming that alternate hosts are available and that their relative suitability is constant. I now want to consider the more realistic situation in which suitability varies more or less independently among hosts; that is, the suitability of one host may be in decline, but other suitable hosts are available. Arthropod performance on, and preference for, a host varies, in part, in response to host phenology, and relative suitability varies among hosts in response to host specific phenological patterns. These patterns are often asynchronous among different hosts within a season (Stinner, 1979; Margolies and Kennedy, 1985). Where alternate hosts are available, reduction of the relative suitability of a host should increase the probability of dispersal from that host. Understanding the consequences of asynchrony among host patches requires consideration of the relative duration of favorable conditions on a host.

It is obviously advantageous for arthropods to disperse when the local host is deteriorating or soon will be deteriorating and better conditions are likely to occur elsewhere. However, if the duration of present host suitability is uncertain or the appearance of better hosts is variable, then dispersal may not be beneficial (Stearns, 1976). Individuals in a population faced with such unpredictable conditions are likely to differ in their tendency to disperse (Southwood, 1977). Polymorphisms in the tendency of arthropod populations to disperse are well documented (see references in Harrison, 1980). Although immediate environmental factors do influence the occurrence of dispersal polymorphisms, there is ample evidence that intraspecific genetic differences in insect populations contribute to variation in wing length (Vepsalainen, 1978; Dingle, 1982; Roff, 1986; Zera and Rankin, 1989) and dispersal behavior (Dingle, 1968; Caldwell and Hegmann, 1969; Rose, 1972; Harrison, 1980; Bell and Tortorici, 1987; Palmer and Dingle, 1989). The actual proportion of dispersers in these genetically polymorphic populations appears to be correlated with some measures of the predictability or permanence of the environment in natural ecosystems (Caldwell, 1974; Jarvinen and Vepsalainen, 1976; Denno, 1979; Denno and Grissell, 1979; Groeters, 1989).

In agricultural systems, as well as in natural ones, the proportion of an insect or mite population that abandons a host should depend on the duration of suitable conditions in that crop in relation to the reproductive and feeding time of the pest. Whether the pest abandons the crop by means of dispersal or goes into dormancy would then depend on the probable availability of alternate hosts. If we can assume that alternate hosts are available, it may be possible to evaluate the extent to which arthropod pests have adapted to variation in habitat duration by comparing dispersal rates of pest populations from agroecosystems that appear to differ in durational stability. This information may then allow us to estimate the potential for pests to adapt to changes in the agroecosystem. Because polyphagous pest species are likely to have several alternate hosts in diversified agricultural environments, the proportion of dispersers in polyphagous species is more likely to be related to variation in host duration among agroecosystems than to variation in host availability.

My own work with spider mites suggests that the same is true in agroecosystems. The twospotted spider mite (TSM), *Tetranychus urticae* Koch, is a highly polyphagous, multivoltine pest. Long-distance aerial dispersal between hosts has been shown to be an important component of mite life history and pest status in orchards and row crops (Hoy et al., 1985; Margolies and Kennedy, 1985). A specific aerial dispersal behavior has been identified that enhances the probability that mites will be carried aloft as aerial plankton (Smitley and Kennedy, 1985) and therefore appears to be the primary long-distance dispersal mechanism. Measuring the proportion of a population that engages in this behavior is thus likely to indicate the frequency with which individuals abandon one habitat for another.

I examined the frequency of aerial dispersal behavior in four populations of TSM with the objective of relating dispersal in each population to the stability of the environment from which they were taken. The four populations were collected from the primary crop in three different agroecosystems: row crop (on corn), orchard (on apples), and greenhouse (on lima beans). These agroecosystems appeared to represent a gradient from less to more durable. Mite populations persisted in the corn agroecosystem by moving between, and utilizing a series of, temporarily suitable hosts (Margolies and Kennedy, 1985), so this system may be classified as relatively unstable. Mite populations in orchards move between tree hosts and weedy hosts on a more seasonal basis (Georgala, 1955; Tedders et al., 1984); therefore, this system may have greater durational stability. Because the greenhouse population was maintained as a laboratory culture, lima bean plants were constantly being replenished, and this habitat was highly stable and long lasting. More mites from the corn populations exhibited the aerial dispersal response than did mites from the apple orchard, which in turn exhibited the aerial dispersal response in greater proportion than mites from the greenhouse population. This pattern is consistent with the hypothesis that the corn agroecosystem is less durable than the apple agroecosystem and that the apple agroecosystem is less durable than the greenhouse environment.

Results from reciprocal crosses between corn and apple populations indicate that the difference in dispersal response between these mites is genetic and that the response is a quantitative character. There is, in addition to the preceding evidence for genetic variation in dispersal behavior between mite populations, evidence that there is additive genetic variation in aerial dispersal within a mite population (Li and Margolies, 1993). This indicates a potential for mite populations to evolve in response to changes in the agroecosystem, which differentially affect dispersers and nondispersers. The increasing evidence of genetically based dispersal polymorphisms in populations from natural systems (Harrison, 1980) suggests that further investigation will find the same in other economically important arthropods.

Accessibility of New Habitats

In the previous section, the effect of variable host duration on arthropod adaptation was examined under the assumption that alternate hosts were relatively accessible. The successful survival and reproduction of an individual that disperses from a host depends on accessibility of alternate hosts. However, suitable hosts may not be equally accessible in different environments or at different times. The accessibility of a host depends on the ability to locate the host, as well as the distribution of a host in an environment. If accessibility of other hosts is not consistent or if locating other hosts entails high mortality, then traits other than dispersal may also be adaptive. Thus, the effects of accessibility can interact with the influences of relative suitability and duration of different hosts to influence host use. In this section, I will consider host accessibility in agroecosystems in terms of its effect on host use and movement.

The spatial structure of an ecosystem may render crop hosts relatively inaccessible to dispersing arthropods even if suitable hosts are available. Distribution of hosts in an ecosystem determines the probability that a dispersing individual will locate a suitable new host. This probability can be affected by the size and distribution of host patches in a diverse ecosystem, both of which affect the rate with which herbivores move into a host patch (Kareiva, 1983; Saks, Chapter 16 in this book). For example, it is possible for a host to be available but relatively inaccessible if host patches are small or sparse (see references in Kareiva, 1983).

Mixed cropping systems or polycultures may reduce pest impact, because increased spatial diversity may reduce host location by dispersers. In their review of the effect of spatial diversity of crops on pest abundance, Risch et al. (1983) concluded that herbivore movement in annual cropping systems was the most important factor governing herbivore numbers in diverse agroecosystems. Among the studies they reviewed that examined mechanisms, the differences in pest abundance between simple and diverse systems were often tied to the host-finding ability of herbivores and the availability of resources. Rausher (1981), for example, demonstrated that susceptibility of hosts to the pipevine swallowtail butterfly, *Battus philenor*, was reduced in diverse vegetation because host discovery was inhibited. And Bach and Tabashnik (1990) concluded that tomato neighbors affected long-range host finding by diamondback moths, *Plutella xylostella*, looking for cabbage, as well as possibly reducing early larval survival on cabbage. Whether reduced insect numbers on hosts intercropped with nonhosts occurs because the host is somehow lost against a nonhost background or because search time increases and, with it, mortality (Cromartie, 1981), the end result in many cases is a reduction in the abundance of arthropod pests on a host and in the agroecosystem.

Reduction in number is especially likely to be true of herbivorous pests that have a narrow range of acceptable hosts (Risch et al., 1983; Kareiva, 1983). The range of host use can affect host finding by arthropod pests because, in a diverse agroecosystem, a generalist feeder may be more likely to encounter a suitable host than a more specialized feeder. For example, the twospotted spider mite, *T. urticae*, has a broad host range, including both grasses and broadleaf plants, while the Banks grass mite, *Oligonychus pratensis*, utilizes a small range of grasses. The proportion of aerial dispersers is about 60% in populations of twospotted spider mites from a corn agroecosystem but only 15–20% in Banks grass mite populations from the same system (Margolies, 1987). This difference in their tendency to disperse may be related to prevalence of alternate hosts and therefore to the likelihood of each species encountering a new host. Risch (1981) investigated the numerical response of six chrysomelid species to varying combinations of beans, squash, and corn in artificial agroecosystems. Some of these beetles were able to feed on only one of the three crops, while other species could feed on more than one. Risch found that the abundance of beetle species in polycultures that included one or more nonhost crops was less than the abundance of the same species in pure cultures of their host crop (based on per unit weight of host plant available). Beetles that fed on both crops in a diculture, however, showed an opposite numerical response; they were more abundant in the diculture than in any of the host monocultures. These studies suggest that the range of hosts utilized influences the perception of, and response to, the spatial structure of the ecosystem in terms of both dispersal and host finding.

Dispersal appears to be an obvious advantage for arthropods on crop hosts that are faced with the likelihood of local resource depletion or destruction. However, dispersal is unlikely to be adaptive if new hosts are not available within the range of movement of an arthropod. In such a situation, dormancy may be a more appropriate response if unfavorable conditions are likely to revert back to favorable ones, as happens over winters in temperate climates (Tauber and Tauber, Chapter 5 in this book). The response of northern corn rootworm beetles, *Diabrotica barberi*, to annual crop rotation (Krysan, Chapter 18 in this book) illustrates possible consequences of changing the probability of host location in a seasonal environment. Management of the northern corn rootworm is based on crop rotation practices that remove the corn host from the beetles' overwintering sites in previous corn plantings. Postdiapausing eggs hatch into larvae that cannot move very far before feeding. Therefore, moving the host reduces colonization of the new corn crop and subsequent pest numbers in the host. However, Krysan et al. (1984, 1986) and Levine et al. (1992) have documented extended diapause, ranging from one to four years, in some percentage of beetle eggs found in cornfields. Beetles with a two-year or longer diapause may emerge when a suitable host (i.e., corn) has been rotated back into a field. Thus, beetle populations may adapt to rotation schedules by having some genotypes that can "wait" for the return of corn to a field. Similarly, local populations of the cabbage root fly, *Delia radicum*, may have evolved delayed emergence from diapause in response to availability of crop hosts (Hawkes et al., 1987). For these insects with prolonged diapause, agricultural crops provide a more permanent rather than a temporary habitat.

Even if the ability of an arthropod to locate and move among host patches is large with respect to the spatial arrangement of hosts, host accessibility will depend in part on the proportion of an ecosystem composed of suitable hosts. Because the use of antibiotic or antixenotic crop varieties to manage pests effectively renders the crop inaccessible, this practice may impose strong selective pressure for genotypes that can overcome host plant resistance or that selectively use other crop hosts. This problem was examined by Lampert et al. (1983) when a resistant wheat was deployed to control cereal leaf beetle, *Oulema melanopus* (L.). Resistant wheat was introduced into a Michigan agroecosystem containing at least two of the beetles' hosts, wheat and oats. Presence of adult beetles, eggs, and larvae were monitored in oats, susceptible wheat, and resistant wheat for five years, during which time the proportion of small-grain acreage planted in resistant wheat increased. As the percentage of total small-grain acreage sown in resistant wheat went up to approximately 80%, adult ovipositional activity and the proportional abundance of larvae shifted to oats and susceptible wheat, but above 80%, beetle activity abruptly declined in oats and increased in resistant wheat. Wellso (1979) had earlier found that a relatively high percentage of beetles ($> 7\%$) were capable of sustaining feeding and oviposition when restricted to resistant wheat by taking advantage of within- and between-plant variation in the pubescence that caused the resistance. Lampert et al. (1983) suggested that the response seen in beetle host use when resistant wheat became widely deployed was caused by an increase in the portion of the beetle population composed of this phenotype, although this hypothesis was not tested. Thus, evidence from the field, as well as from theoretical models (e.g., Gould, 1986), supports the possibility that adaptation of target pests to the widespread deployment of resistant varieties will occur with decreased accessibility of nonresistant hosts.

CONCLUDING REMARKS

The central theme of this book is that arthropods live in a variable environment and can adapt to this variable environment. To evaluate, and possibly predict, the impact of habitat variation on the evolution of arthropods pests, we must first be able to define this variation in terms of the selective pressure it imposes. The definition of environmental structure proposed by Southwood (1977) and used in this chapter allows us to do this by partitioning spatial structure into quantifiable components. Reproductive rate and survival on a host, host duration, dispersal capabilities of an arthropod, and availability of, and distance between, hosts are difficult but not impossible factors to measure. However, the nature of arthropod response to spatial variability also depends on the specific morphology, physiology, behavior, and life history of a population, and on the genetic basis of, and variability in, those traits. Our ability to make generalizations regarding adaptation to spatial variation is limited by lack of information about variability of resource use and life-history traits within populations.

In agroecosystems, human inputs are dynamic and integral components of the spatial and temporal variability that arthropod pests must face. In this chapter, I have attempted to describe the impact of agriculture on the spatial environment of arthropod pests, particularly as agriculture affects resource diversity and accessibility. I have also argued that the relative quality and accessibility of different crop hosts impose selective pressure on pest arthropods. At present, the evidence for specific responses to such selection is sparse and inferential, and I may have overemphasized the importance or generality of some selective factors. Nonetheless, the overall picture that these examples provide may prove useful in directing future research. There may be certain rules that apply to the evolution of pest population in any environment and that constrain insect adaptation, and only a detailed study of variation in resource use and behavior in relation to the spatial structure of its resources will reveal them. When we finally have a better understanding of the genetic mechanisms involved in the interaction of arthropods and their spatial environment, it may then be possible to make useful predictions about the consequences of changing environmental structure.

Acknowledgments

I am grateful to George Kennedy and Fred Gould for their suggestions and support on this project. I wish to thank Jim Krysan, Ralph Howard, Marion Harris, and Sheila Hochhauser for their helpful comments on an earlier version of this manuscript. This is contribution No. 88-406-B from the Kansas Agricultural Experiment Station, Kansas State University, Manhattan.

REFERENCES

Addicott, J. F., J. M. Aho, M. F. Antolin, D. K. Padilla, J. S. Richardson, and D. A. Soluk. 1987. Ecological neighborhoods: Scaling environmental patterns. *Oikos* 49:340–349.

Alstad, D. N., and K. W. Corbin. 1990. Scale insect allozyme differentiation within and between host trees. *Evol. Ecol.* 4:43–56.

Alstad, D. N., and G. F. Edmunds, Jr. 1983. Adaptation, host specificity, and gene flow in the black pineleaf scale. Pages 413–426 in Denno, R. F., and M. S. McClure (eds.), *Variable Plants and Herbivores in Natural and Managed Environments.* New York: Academic Press.

Andow, D. A. 1991. Vegetational diversity and arthropod population response. *Ann. Rev. Entomol.* 36:561–586.

Bach, C. E., and B. E. Tabashnik. 1990. Effects on nonhost plant neighbors on population densities and parasitism rates of the diamondback moth (Lepidoptera: Plutellidae). *Environ. Entomol.* 19:987–994.

Bell, W. J., and C. Tortorici. 1987. Genetic and non-genetic control of search duration in adults of two morphs of *Drosophila melanogaster. J. Insect Physiol.* 33:51–54.

Black, W. C. IV, and E. S. Krafsur. 1986. Temporal and spatial trends in allozyme frequencies in house fly populations, *Musca domestica* L. *Theor. Appl. Genet.* 71:673–681.

Browning, J. A. 1980. Genetic protective mechanisms of plant-pathogen populations: Their coevolution and use in breeding for resistance. Pages 52–75 in Harris, M. K. (ed.), *Biology and Breeding for Resistance to Arthropods and Pathogens in Agricultural Plants.* Texas Agric. Exp. Stn., Texas A&M Univ., College Station, Texas.

Caldwell, R. L. 1974. A comparison of the migratory strategies of two milkweed bugs, *Oncopeltus fasciatius* and *Lygaeus kalmii.* Pages 304–316 in Barton-Browne, L. (ed.), *Experimental Analysis of Insect Behavior.* Berlin: Springer-Verlag.

Caldwell, R. L., and J. P. Hegmann. 1969. Heritability of flight duration in the milkweed bug *Lygaeus kalmii. Nature* 223:91–92.

Cromartie, W. J., Jr. 1981. The environmental control of insects using crop diversity. Pages 223–251 in Pimentel, D. (ed.), *CRC Handbook of Pest Management in Agriculture, Vol. 1.* Boca Raton, Fla.: CRC Press.

Denno, R. F. 1978. The optimum population strategy for planthoppers (Homoptera: Delphacidae) in stable marsh habitats. *Can. Ent.* 110:135–142.

———. 1979. The relation between habitat stability and the migratory tactics of planthoppers (Homoptera: Delphacidae). *Misc. Publ. Entomol. Soc. Am.* 11:41–49.

———. 1983. Tracking variable host plants in space and time. Pages 291–341 in Denno, R. F., and M. S. McClure (eds.), *Variable Plants and Herbivores in Natural and Managed Environments.* New York: Academic Press.

Denno, R. F., and E. E. Grissell. 1979. The adaptiveness of wing dimorphism in the salt marsh-inhabiting planthopper, *Prokelisia marginata* (Homoptera: Delphacidae). *Ecology* 60:221–236.

Dethier, V. G. 1954. Evolution of feeding preference in phytophagous insects. *Evolution* 8:33–54.

Diehl, S. R., and G. L. Bush. 1984. An evolutionary and applied perspective of insect biotypes. *Ann. Rev. Entomol.* 29:471–504.

Dingle, H. 1968. The influence of environment and heredity on flight activity in the milkweed bug *Oncopeltus*. *J. Exp. Biol.* 48:175–184.

———. 1982. Function of migration in the seasonal synchronization of insects. *Entomol. Exp. Appl.* 31:36–48.

Fox, L. R., and P. A. Morrow. 1981. Specialization: Species property or local phenomenon. *Science* 211:887–893.

Fry, J. D. 1990. Trade-offs in fitness on different hosts: Evidence from a selection experiment with a phytophagous mite. *Am. Nat.* 136:569–580.

Futuyma, D. J., and S. C. Peterson. 1985. Genetic variation in the use of resources by insects. *Ann. Rev. Entomol.* 30:217–238.

Georgala, M. B. 1955. The biology of orchard mites in the Western Cape Province. *Union S. Afr., Dept. Agric. Tech. Serv. Sci. Bull.* 360.

Gillespie, J. 1974. The role of environmental grain in the maintenance of genetic variation. *Am. Nat.* 108:831–836.

Gould, F. 1978. Predicting the future resistance of crop varieties to pest populations: A case study of mites and cucumbers. *Environ. Entomol.* 7:622–626.

———. 1979. Rapid host evolution in a population of the phytophagous mite *Tetranychus urticae* Koch. *Evolution* 33:791–802.

———. 1986. Simulation models for predicting durability of insect-resistant germ plasm: Hessian fly (Diptera: Cecidomyiidae)-resistant winter wheat. *Environ. Entomol.* 15:11–23.

Groeters, F. R. 1989. Geographic and clonal variation in the milkweed-oleander aphid, *Aphis nerii* (Homoptera: Aphididae), for winged morph production, life history, and morphology in relation to host plant permanence. *Evol. Ecol.* 3:327–341.

Hare, J. D. 1983. Manipulation of host suitability for herbivore pest management. Pages 655–680 in Denno, R. F., and M. S. McClure (eds.), *Variable Plants and Herbivores in Natural and Managed Environments*. New York: Academic Press.

Hare, J. D., and G. G. Kennedy. 1986. Genetic variation in plant-insect associations: Survival of *Leptinotarsa decemlineata* populations on *Solanum carolinense*. *Evolution* 40:1031–1043.

Harrison, R. G. 1980. Dispersal polymorphisms in insects. *Ann. Rev. Ecol. Syst.* 11:95–118.

Hartl, D. L. 1987. *A Primer of Population Genetics*. Sunderland, Mass.: Sinauer Press.

Hassell, M. P., and T. R. E. Southwood. 1978. Foraging strategies of insects. *Ann. Rev. Ecol. Syst.* 9:75–98.

Hawkes, C., R. Kowalski, and G. L. Brindle. 1987. An alternative to migration in agricultural insects? *Funct. Ecol.* 1:165–167.

Hedrick, P. W. 1986. Genetic polymorphism in heterogeneous environments: A decade later. *Ann. Rev. Ecol. Syst.* 17:535–566.

Hoy, M. A., J. J. R. Groot, and H. E. van de Baan. 1985. Influence of aerial dispersal on persistence and spread of pesticide-resistant *Metaseiulus occidentalis* in California almond orchards. *Entomol. Exp. Appl.* 37:17–31.

Hutchinson, G. E. 1953. The concept of pattern in ecology. *Proc. Acad. Nat. Sci. Philadelphia* 105:1–12.

Jaenike, J. 1989. Genetic population structure of *Drosophila tripunctata*: Patterns of variation and covariation of traits affecting resource use. *Evolution* 43:1467–1482.

Jarvinen, O., and K. Vepsalainen. 1976. Wing dimorphism as an adaptive strategy in water striders (*Gerris*). *Hereditas* 84:61–68.

Johnson, C. G. 1969. *Migration and Dispersal of Insects by Flight*. London: Methuen.

Kareiva, P. 1983. Influence of vegetation texture on herbivore populations: Resource concentration and herbivore movement. Pages 259–289 in Denno, R. F., and M. S. McClure (eds.), *Variable Plants and Herbivores in Natural and Managed Systems*. New York: Academic Press.

Kennedy, G. G., F. Gould, O. M. B. dePonti, and R. E. Stinner. 1987. Ecological, agricultural, genetic, and commercial considerations in the deployment of insect-resistant germ plasm. *Environ. Entomol.* 16:327–338.

Kennedy, G. G., and D. C. Margolies. 1985. Mobile arthropod pests: Management in diversified agroecosystems. *Bull. Entomol. Soc. Am.* 31(3):21–27.

Krysan, J. L., D. E. Foster, T. F. Branson, K. R. Ostlie, and W. S. Cranshaw. 1986. Two years before the hatch: Rootworms adapt to crop rotation. *Bull. Entomol. Soc. Am.* 32:250–253.

Krysan, J. L., J. J. Jackson, and A. C. Lew. 1984. Field termination of egg diapause in *Diabrotica* with new evidence of extended diapause in *D. barberi* (Coleoptera: Chrysomelidae). *Environ. Entomol.* 13:1237–1240.

Lampert, E. P., D. L. Haynes, A. J. Sawyer, D. P. Jokinen, S. G. Wellso, R. L. Gallun, and J. J. Roberts. 1983. Effects of regional release of resistant wheats on the population dynamics of the cereal leaf beetle (Coleoptera: Chrysomelidae). *Ann. Entomol. Soc. Am.* 76:972–980.

Lande, R. 1982. A quantitative genetic theory of life-history evolution. *Ecology* 63:607–615.

Levine, E., H. Oloumi-Sadeghi, and J. R. Fisher. 1992. Discovery of multiyear diapause in Illinois and South Dakota northern corn rootworm (Coleoptera: Chrysomelidae) eggs and incidence of the prolonged diapause trait in Illinois. *J. Econ. Entomol.* 85:262–267.

Levins, R. 1968. *Evolution in Changing Environments*. Princeton, N. J.: Princeton Univ. Press.

Levins, R., and M. Wilson. 1980. Ecological theory and pest management. *Ann. Rev. Entomol.* 25:287–308.

Li, J., and D. C. Margolies. 1993. Quantitative genetics of aerial dispersal behaviour and life history traits in *Tetranychus urticae. Heredity* 70:(in press).

MacArthur, R. H., and E. R. Pianka. 1966. On optimal use of a patchy environment. *Am. Nat.* 100:603–609.

Margolies, D. 1987. Conditions eliciting aerial dispersal behavior in Banks grass mite, *Oligonychus pratensis* (Acari: Tetranychidae). *Environ. Entomol.* 16:928–932.

Margolies, D. C., and G. G. Kennedy. 1985. Movement of the two-spotted spider mite, *Tetranychus urticae*, among hosts in a corn-peanut agroecosystem. *Entomol. Exp. Appl.* 37:55–61.

Mitter, C., and D. J. Futuyma. 1983. An evolutionary-genetic view of host-plant utilization. Pages 427–459 in Denno, R. F., and M. S. McClure (eds.), *Variable Plants and Herbivores in Natural and Managed Environments*. New York: Academic Press.

Palmer, J. O., and H. Dingle. 1989. Responses to selection on flight behavior in a migratory population of milkweed bug (*Oncopeltus fasciatus*). *Evolution* 43:1805–1808.

Price, P. W. 1976. Colonization of crops by arthropods: Non-equilibrium communities in soybean fields. *Environ. Entomol.* 5:605–611.

Rabb, R. L. 1978. A sharp focus in insect populations and pest management from a wide area view. *Bull. Entomol. Soc. Am.* 24:55–61.

Rabb, R. L., G. K. DeFoliart, and G. G. Kennedy. 1984. An ecological approach to managing insect populations. Pages 697–728 in Huffaker, C. B., and R. L. Rabb (eds.), *Ecological Entomology.* New York: John Wiley and Sons.

Rausher, M. D. 1981. The effect of native vegetation on the susceptibility of *Aristolochia reticulata* (Aristolochiaceae) to herbivore attack. *Ecology* 62:1187–1195.

Rausher, M. D. 1984. Trade-offs in performance on different hosts: Evidence from within- and between-site variation in the beetle *Deloyala guttata. Evolution* 38:582–595.

Reddingius, J., and P. J. Den Boer. 1970. Simulation experiments illustrating stabilization of animal numbers by spreading of risk. *Oecologia* 5:240–284.

Risch, S. J. 1981. Insect herbivore abundance in tropical monocultures and polycultures: An experimental test of hypotheses. *Ecology* 62:1325–1340.

Risch, S. J., D. Andow, and M. A. Altieri. 1983. Agroecosystem diversity and pest control: Data, tentative conclusions, and new research directions. *Environ. Entomol.* 12:625–629.

Roff, D. A. 1986. The evolution of wing dimorphism in insects. *Evolution* 40:1009–1020.

Rose, D. J. W. 1972. Dispersal and quality in populations of *Cicadulina* species (Cicadellidae). *J. Anim. Ecol.* 41:589–609.

Roughgarden, J. 1972. Evolution of niche width. *Am. Nat.* 106:683–718.

Singer, M. 1983. Determinants of multiple host use by a phytophagous insect population. *Evolution* 37:389–403.

Singer, M. C., D. Ng, and C. D. Thomas. 1988. Heritability of oviposition preference and its relationship to offspring performance within a single insect population. *Evolution* 42:977–985.

Smitley, D. R., and G. G. Kennedy. 1985. Photo-oriented aerial-dispersal behavior of *Tetranychus urticae* (Acari: Tetranychidae) enhances escape from the leaf surface. *Ann. Entomol. Soc. Am.* 78:609–614.

Solbreck, C. 1978. Migration, diapause, and direct development as alternative life histories in a seed bug, *Neacoryphus bicrucis*. Pages 195–217 in Dingle, H. (ed.), *The Evolution of Insect Migration and Diapause.* New York: Springer-Verlag.

Southwood, T. R. E. 1962. Migration of terrestrial arthropods in relation to habitat. *Biol. Rev.* 37:171–214.

———. 1971. The role and measurement of migration in the population system of an insect pest. *Trop. Sci.* 13:275–278.

———. 1977. Habitat, the template for ecological strategies? *J. Anim. Ecol.* 46:337–365.

Stearns, S. C. 1976. Life history tactics: A review of the ideas. *Quart. Rev. Biol.* 51:3–49.

Stinner, R. E. 1979. Biological monitoring essentials in studying wide-area moth movement. Pages 199–211 in Rabb, R. L., and G. G. Kennedy (eds.), *Movement of Highly Mobile Insects: Concepts and Methodology in Research.* Raleigh, N. C.: North Carolina State Univ.

Stinner, R. E., J. Regniere, and K. Wilson. 1982. Differential effects of agroecosystem structure on dynamics of three soybean herbivores. *Environ. Entomol.* 11:538–543.

Taylor, L. R., and R. A. J. Taylor. 1977. Aggregation, migration and population mechanics. *Nature* 265:415–421.

Taylor, R. A. J. 1981. The behavioral basis of redistribution I. The Δ-model concept. *J. Anim. Ecol.* 50:573–586.

Tedders, W. L., J. A. Payne, and J. Inman. 1984. A migration of *Tetranychus urticae* (Acari: Tetranychidae) from clover into pecan trees. *J. G. Entomol. Soc.* 19:498–502.

Thompson, J. N. 1988. Evolutionary ecology of the relationship between oviposition preference and performance of offspring in polyphagous insects. *Entomol. Exp. Appl.* 47:3–14.

Vepsalainen, K. 1978. Wing dimorphisms and diapause in *Gerris*: Determination and adaptive significance. Pages 218–253 in Dingle, H. (ed.), *Evolution of Insect Migration and Diapause.* New York: Springer-Verlag.

Via, S. 1984a. The quantitative genetics of polyphagy in an insect herbivore. I. Genotype-environment interaction in larval performance on different host plants. *Evolution* 38:881–895.

———. 1984b. The quantitative genetics of polyphagy in an insect herbivore. II. Genetic correlations in larval performance within and among host plants. *Evolution* 8:896–905.

———. 1986. Genetic covariance between oviposition preference and larval performance in an insect herbivore. *Evolution* 40:778–785.

———. 1991. The genetic structure of host plant adaptation in a spatial patchwork: Demographic variability among reciprocally transplanted pea aphid clones. *Evolution* 45:827–852.

Van Steenwyk, R. 1978. Short-range movement of major agricultural pests. Pages 17–22 in Vaughn, C. R., W. Wolf, and W. Klassen (eds.), *Radar, Insect Population Ecology and Pest Management.* NASA Conf. Publ. 2070.

Way, M. J. 1979. Significance of diversity in agroecosystems. Pages 9–12 in *Proc. IX Intnl. Congr. Crop Prot.*

Wellington, W. G. 1977. Returning the insect to insect ecology: Some consequences for pest management. *Environ. Entomol.* 6:1–8.

Wellso, S. G. 1979. Cereal leaf beetle interaction with and ovipositional adaptation to a resistant wheat. *Environ. Entomol.* 8:454–457.

Wiens, J. A. 1976. Population response to patchy environments. *Ann. Rev. Ecol. Syst.* 7:81–120.

Zera, A., and M. A. Rankin. 1989. Wing dimorphism in *Gryllus rubens*: Genetic basis of morph determination and fertility differences between morphs. *Oecologia* 80:249–255.

7

HABITAT PATTERNS AND PESTICIDE RESISTANCE

B. A. Croft and J. E. Dunley

Oregon State University

INTRODUCTION

Pesticide resistance is a complex evolutionary phenomenon. It takes place when a population develops the ability to tolerate doses of a toxicant that are lethal to the majority of normal individuals of the species (Anonymous, 1957). Resistance occurs in a wide diversity of arthropod species, including many insect and mite pests of importance to agricultural and natural resource management (Georghiou, 1986; Roush and Croft, 1986; Roush and McKenzie, 1987). Resistance evolution is influenced by an organism's genetic features, including the number, dominance, and fitness associated with its resistance genes; aspects of its life history, including its mode and rate of reproduction; and its population dynamics and migration characteristics (Georghiou and Taylor, 1986; Roush and McKenzie, 1987).

At the genetic level, habitat patterns influence the development of resistance through the promotion or retardation of gene flow. Gene flow is the effect of immigration and acts to alter the frequency of a gene within a population (Falconer, 1981). In agricultural systems, gene flow is generally between demes or subpopulations that may be located in treated or untreated habitats. The rates of emigration and immigration of these demes ultimately change the frequency of the resistant individuals within them, raising or lowering the resistance of the population. Although these phenomena fundamentally follow the principles of population genetics, very little experimental work has been done, primarily due to the lack of information on resistance at the genetic level (Levin et al., 1986); it is very difficult to monitor the movement of one gene throughout a population.

Ecologically, gene flow between populations occurring in treated habitat and surrounding vegetation (or species pools) determines the rate of resistance evolution and its stability over time. Movement of susceptible organisms into a treated habitat to interbreed with a resistant population can slow the rate of pesticide resistance development (see Comins, 1977; Tabashnik and Croft, 1982; Georghiou and Taylor, 1986) (Fig. 7.1). There also may be outward emigration of resistant individuals from a treated management site into the surrounding environment to interbreed with susceptible strains (Fig 7.1). At high densities and high levels of movement of resistant organisms, the net effect of outward emigration is the increase of resistant organisms in the refugia, and thus an area-wide or regional resistance pattern may occur (e.g., see later examples of this pattern in pear psylla populations). Clearly, the community ecology and population dynamics or genetics of susceptible and resistant pests in treated and surrounding untreated habitats are of great interest to those seeking to minimize or modulate the evolution of pesticide resistance in the field (Whalon and Croft, 1985; Liss et al., 1986; Croft and van de Baan, 1988).

As noted, it is very difficult to measure experimentally the influence of species pool habitats and migration of susceptible and resistant variants on rates of resistance evolution. This is for several reasons. First, documenting the rates of migration of highly mobile species, such as arthropod pests or their natural enemies, is technically difficult (Rabb and Kennedy, 1979; MacKenzie et al., 1986). Also, since levels of population and community interaction occur over large spatial scales (often on the order of many acres or square miles of habitat), it is almost impossible to establish regional plots that have treated and untreated controls for comparison. In commercial agriculture, for example, few growers are willing to leave their crops unsprayed to allow for experimentation on resistance development.

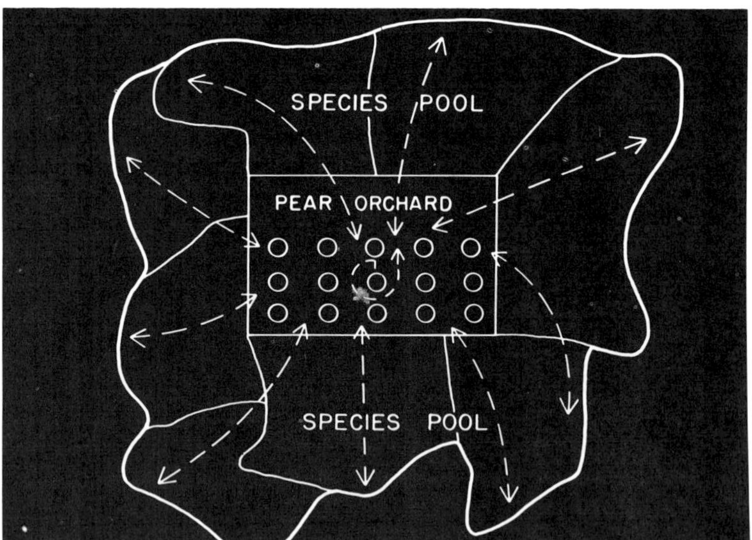

Figure 7.1. Generalized representation of species pool habitat surrounding a tree-fruit orchard.

Because of difficulties in experimentally evaluating the effects of habitat and species immigration on resistance evolution, most of our understanding of their influence comes from models of pesticide resistance (see above) or from descriptive scientific studies. Insight may come by following a series of resistance episodes in the field, especially when a population or an ecological perspective is maintained. It is especially useful if experience with several resistance episodes can be had in a common environment or if resistance in the same organism can be followed in a number of different environmental settings. In almost all cases, our experience has been in following population trends and resistance levels, rather than on following gene flow.

Because of our research experience in studying resistance evolution in arthropod pests and natural enemies of deciduous tree fruit and berry crops, the case histories discussed in this chapter focus on these agroecosystems. These examples point out patterns observed between habitat conditions and the evolution of resistance in individual species populations and later in communities of species. In most cases, our perspectives were obtained from studies made to monitor or limit resistance in pests or to exploit resistance in beneficial natural enemies (i.e., to practice resistance management) and achieve selective pesticide use in integrated pest management (IPM) programs.

CASE HISTORIES

Spider Mite Resistance on California Strawberries

A simple illustration of the influence of surrounding habitat on the evolution of pesticide resistance in a treated environment is represented by patterns of acaricide resistance that evolved in populations of the two-spotted spider mite, *Tetranychus urticae* Koch, on strawberry in the Pajaro Valley of California (Miller et al., 1985). Here, the effects of early season immigration of this spider mite on rates of resistance development to the acaricides cyhexatin and formetanate were evaluated over a two-year period in 1983–1984.

In this study, strawberry leaves from four commercial field sites were collected to assess rates of immigrant mites colonizing plants and resistance levels of spider mites from the time transplants were first put out in early December until after the first few acaricide sprays were applied in March and April. Similarly, rates of spider mite buildup and resistance development were measured in overwintered mite populations that were occurring on surrounding habitat plants at field borders in early spring.

Tests indicated that, initially, some acaricide-susceptible mites had overwintered in storage warehouses and were present at similar densities at all four study sites on newly transplanted strawberry plants early in the growing season (Table 7.1). However, at three sites, acaricide-resistant spider mites rapidly immigrated into fields from surrounding wild plant hosts and orchard trees as the season progressed. Subsequently, the susceptible spider mite variants were diluted out (Lewis, Bachan, and Eaton fields, Table 7.1). (*Note*: Susceptible mites were especially reduced later in the

Table 7.1. Mean Density and Bioassay Results of Samples of *Tetranychus urticae* Collected from Nursery Plants, Strawberry Field Sites, and Surrounding Vegetation in the Pajaro Valley, California (adapted from Miller et al., 1985)

Field Site	Nursery Mite Density/LC_{50}	Border Mite LC_{50}	Immigration Index Near/Far	Field Mites Density LC_{50}
Lewis	0.005/0.005[a]	0.036	19.8/0.16	1.460/0.033
Bachan	0.018/0.004[b]	0.143	4.4/0.17	1.020/0.075
Eaton	0.006/0.003	0.050	1.2/0.17	0.203/0.124
Smith	0.005/0.008	0.040	0.03/0.0	0.312/0.0035

[a]Estimate based on average LC_{50} values from nursery mite values.
[b]Value based on 1983 estimates due to early immigration of resistance mites into field plots before bioassays were made in 1984.

growing season after selection by sprays had occurred, but immigrant resistant mites that had colonized these three fields were less severely affected.) At one site (Smith field, Table 7.1), isolated from surrounding habitat sources of weeds and orchard trees, immigration of resistant mites was very low in early season. Only susceptible mites were present at this site for several months beyond the time that resistance problems had caused control failures at the other sites.

It was concluded that rates of resistance development in Pajaro Valley strawberry planting were directly related to the levels and proximity of resistant populations developing on surrounding weed and orchard plants and their immigration rates into newly planted fields. Miller et al. (1985) suggested modification of crop production practices that might limit seasonal development and immigration of resistant mites into newly planted fields and thereby extend the effectiveness of acaricides. This included greater use of biological controls and limiting the immigration of resistant mites by planting fields in isolated locations or by carefully controlling surrounding weed hosts and managing orchard trees and mites occurring on them. The latter tactic was deemed the most feasible solution under current crop production constraints.

Acaricide Resistance in Spider Mites in Rogue River Valley, Oregon Pear

Similar to the case with California strawberries, *T. urticae* cause severe damage to southern Oregon pears when they cannot be controlled because of acaricide resistance (Croft et al., 1987; Flexner, 1988). Monitoring studies has demonstrated that, while native wild vegetation and crops surrounding orchards often maintain susceptible populations of *T. urticae* Koch, the rate of immigration of these mites into commercial orchards is too low to dilute out resistant mites. However, studies of mite populations on weeds and ground cover that occur in orchards have shown that susceptible mites are maintained on these plants in large numbers. Furthermore, these mites readily immigrate up into trees during the growing season and apparently hybridize or dilute out resistant mite populations in trees. By encouraging this dilution effect through both ground cover management and pesticide use practices, resistance development can be modulated. These measures are effective to the extent that cyhexatin can be used seasonally in moderation, provided that ground cover mite populations are not similarly exposed to these treatments (Flexner, 1988).

One additional aspect of resistance in this environment is that resistance characteristics of *T. urticae* in both orchards and the surrounding habitat seem to differ from region to region in the Rogue River Valley (Flexner, 1988). For example, in the central part of the valley (where orchards predominate, mingled with a few other untreated fields and wild plant hosts, such as weeds), resistance levels in mites from both treated and untreated environments are relatively high when compared with other regions (Fig 7.2). The main migration trends in the region are from treated sites out to untreated sites, so populations in the surrounding habitat tend to become as resistant as treated orchard populations. In contrast, on the periphery of the valley (where orchards are less dense and untreated fields and wild plants are more common), resistance levels in orchards and in untreated habitats are lower than in other areas. This indicates that the inflow of susceptible mites from the surrounding habi-

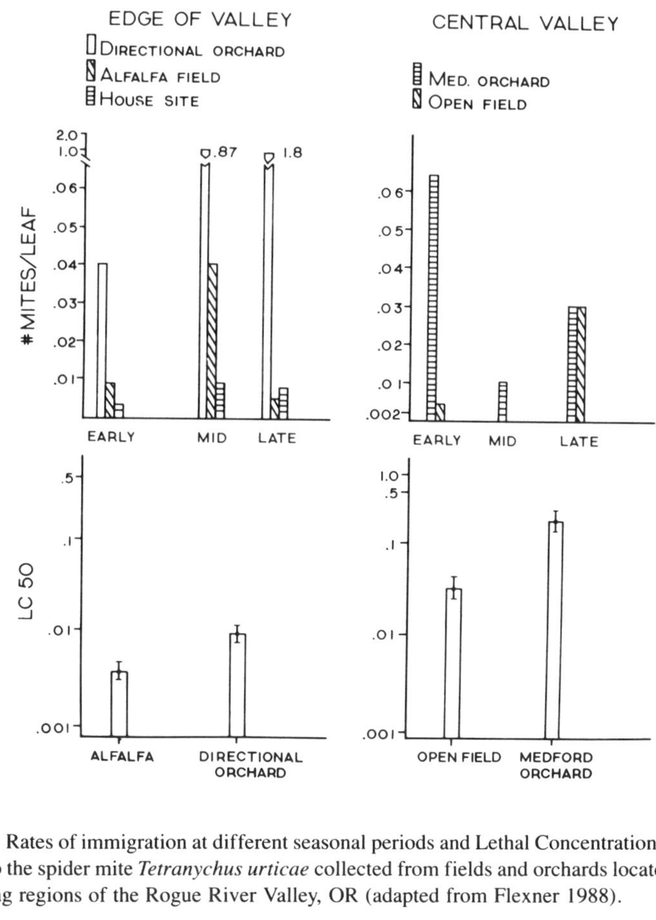

Figure 7.2. Rates of immigration at different seasonal periods and Lethal Concentration (LC_{50}) values of cyhexatin to the spider mite *Tetranychus urticae* collected from fields and orchards located in two general fruit growing regions of the Rogue River Valley, OR (adapted from Flexner 1988).

tat is probably greater than the outflow of resistant forms in treated orchards and that refugia for untreated susceptible pests are contributing significantly to the lower levels of resistance in the region (Fig 7.2).

Pear Psylla Resistance Patterns in Three Oregon Valleys

Regional resistance patterns in populations of *Psylla pyricola* Foerster from Oregon pear orchards are even more reflective of the influence of surrounding habitat types on resistance development than are the patterns for spider mites (see discussion earlier). One reason is that this alate insect species is much more mobile than are the passively wind-dispersed spider mites. Also, in many areas of western North America, the dense pattern of pear production favors the area-wide exposure of psylla populations. The host specificity and reproductive attributes of many generations per season, rapid developmental rate, and so on of this pest also make it highly prone to develop resistant populations (Riedl et al., 1981; Follett et al., 1985).

Follett et al. (1985) bioassayed populations of pear psylla from intensively sprayed, minimally sprayed (IPM), recently abandoned, and long-abandoned pear orchards in the Hood River Valley, the Rogue River Valley and the Willamette Valley for resistance development during 1982–1983 (Table 7.2). They found resistance patterns within orchards that did not correlate with orchard treatment histories in the Hood River and Rogue River Valleys, where orchard densities per unit area are high.

Table 7.2. Resistance Levels for *Psylla pyricola* Collected from Western Oregon, Pear Orchards (adapted from Follett et al., 1985)

			Compounds			
			Azinphosmethyl[a]		Endosulfan	
Region	Orchard	Orchard Type[c]	LC_{50}[a]	Fold-R[b]	LC_{50}[a]	Fold-R[b]
Hood River Valley	McCarty	C	4.21	14	0.63	5
	Facteau	U	6.42	21	1.73	12
	Merz	I	5.47	18	1.26	9
	Tamiyasu	C	3.60	12	1.55	11
Rogue River Valley	Rogue	U	6.30	20	0.66	5
	Cherry ln.	U	10.20	33	0.54	4
	Carpenter	U	12.64	41	0.71	5
	Hanley	E	6.84	22	0.64	5
Willamette Valley	Corvallis	C	6.90	22	0.56	4
	Wiley	C	4.52	15	0.27	2
	OSU Ent. Fm.	E	0.31	0	0.14	0
	OSU Bot. Fm.	E	5.18	17	0.38	3

[a]LC_{50} values based on either entire probit lines or diagnostic dosages.
[b]Resistance values calculated in relation to OSU Entomology Farm strain.
[c]C = Commercial, U = Unsprayed, I = IPM, E = Experimental Orchards.

However, there was good association between treatment histories and resistance levels in the more widely scattered pear orchards of the Willamette Valley.

A regional hypothesis was proposed to explain the different patterns of resistance observed in the three valley areas. In the more arid, locally dense orchard situations of the Hood River and Rogue River Valleys, regional resistance patterns were present, because populations of this crop specific pest are almost entirely exposed to pesticides during the course of the growing season. Even the few long-abandoned orchards in the area harbored resistant psylla, which apparently immigrate there from treated orchards (Table 7.2). It is well known that abandoned orchards do not produce high populations of psylla due to the limited nutrition and lowered vigor of the trees, whereas managed orchards are nutritious and fertile habitats for development of this pest (Follett et al., 1985). In the Willamette Valley, the ratio of sprayed to unsprayed orchards is more even, and abandoned orchards survive better than in the other two areas due to greater amounts of rainfall. In the Willamette Valley, apparently the mix of orchards and the distance from orchard to orchard is enough so that only locally specific patterns of pesticide resistance develop and are maintained over time.

The biological attributes and regional patterns of resistance development of pear psylla indicate that maintaining systems of resistance management for this species will be very difficult. Manipulation of susceptible populations in surrounding alternate host plants is not an alternative for allowing appreciable reversion of resistant populations. Instead, preventive programs of resistance management must be based on moderation of pesticide use and careful monitoring of resistance condition. Programs using these principles have been implemented with some success in recent years to stem resistance evolution to the synthetic pyrethroid insecticides (Riedl et al., 1981; van de Baan et al., 1989). In contrast, such resistance management pratices have not been widely employed in areas of Western Europe and the eastern United States. This has allowed resistance to the synthetic pyrethroids to develop rapidly in these areas, and these compounds alone are no longer useful for controlling this pest (van de Baan et al., 1989). [*Note*: In some cases, use of synergists with synthetic pyrethroids has provided short-term control of *Psylla sp.*, especially in Western Europe (van de Baan et al., 1989).]

Resistance in Predaceous Phytoseiid Mites from the Willamette and Hood River Valleys, Oregon

Mites of the family Phytoseiidae are the most prone of all groups of arthropod natural enemies to develop traits of insecticide resistance (Croft and Strickler, 1983; Hoy, 1985). These traits allow for selective use of pesticides to favor these biological control agents over their spider mite prey. Resistant phytoseiids can develop from natural selection in the field, artificial selection under laboratory conditions, or both. Genetically improved strains of phytoseiid mites are widely used in biological control or IPM programs on several commercial crops (Croft and Strickler, 1983; Hoy, 1985).

Two species that have shown some of the widest ranges and highest levels of pesticide resistance of all natural enemy species are *Metaseiulus occidentalis* (Nesbitt) and *Typhlodromus pyri* Scheuten. These two species co-occur in apple orchards in the Willamette and Hood River Valleys, where they provide excellent biological control of tree fruit pest mites, including *T. urticae* Koch and *Panonychus ulmi* (Koch) (Hadam et al., 1986; Croft, unpublished data). Preliminary studies of the emigration, immigration, and colonization habits of these mites have demonstrated that *M. occidentalis* immigrates more rapidly into new sites and over longer distances than does *T. pyri* (Dunley, unpublished data). Also, these species show different patterns of occurrence in wild and unsprayed vegetation in both valley areas. Generally, *T. pyri* is much more common in these areas than is *M. occidentalis*. The latter species seems to be more commonly associated with pesticide-treated habitats (although it is found readily at low densities on unsprayed wild host plants), whereas *T. pyri* is much more cosmopolitan and is readily found on many commercial crops and wild plant species (Hadam et al., 1986; Croft, unpublished data).

While the different distributions and dispersal abilities of predatory mites influence their abilities to move into new orchards (or trees that are temporarily rid of predaceous mites by nonselective pesticide use) and control pest mites, they also seem to affect patterns of pesticide resistance evolution in these species. For example, *T. pyri* collected from orchards typically show locally specific patterns of pesticide resistance. With *M. occidentalis*, pesticide resistance patterns are more regional in nature, being more uniform, regardless of whether they are collected from sprayed orchards or wild unsprayed habitats (e.g., Croft and AliNiazee, 1983; Hadam et al., 1986; Croft, unpublished data). One possible explanation for these trends is the tendency for *M. occidentalis* to occur primarily in treated crops and for *T. pyri* to be commonly present in both wild and crop habitats. Another possibility is the greater mobility of *M. occidentalis* when compared with *T. pyri*.

The different population dynamics and occurrence in different plant habitats of *M. occidentalis* and *T. pyri* have ramifications for managing their levels of pesticide resistance in the field. First, insofar as releasing genetically improved strains is concerned, these features in *M. occidentalis* could easily be diluted out by native immigrant mites moving into a release orchard or study site. The likelihood that the released resistance factor would remain local would be much better in the case of *T. pyri*. On the other hand, a naturally developed resistance factor would spread faster from orchard to orchard in populations of *M. occidentalis* than for *T. pyri*. In some cases, growers have compensated for this lack of dispersion in *T. pryi*. In several orchard production areas where this species occurs, growers often transfer resistant *T. pyri* from orchard to orchard by taking branches and leaf cuttings from tree to tree. This practice is seldom necessary for the more rapidly dispersing *M. occidentalis*.

Tortricoid Movement and Pesticide Resistance

Evolution and Management The incidence of resistance in tortricid species infesting deciduous tree fruits, especially to organophosphate pesticides, is on the rise globally and is becoming increasingly important to IPM programs (Croft and

Hull, 1991; Croft and Riedl, 1991). For example, the number of cases of resistance reported in the literature for this group has gone from one during 1965–1970 to two during 1970–1975 and back to one during 1975–1980, but it jumped to six during 1980–1985. The recent jump in resistance cases is critically important, because these species often are key pests that attack the fruit directly and must be controlled at low population levels (Croft and Hull, 1991).

While many secondary pests of deciduous fruit crops and their natural enemies develop resistance to organophosphate insecticides much earlier than do tortricoids (Croft, 1982), these pesticides continue to be used in IPM programs, because they still control most key pests of these crops (e.g., in North America, they still control codling moth, apple maggot, plum curculio, oriental fruit moth, and others). The tortricoids are a pivital group insofar as resistance development is concerned. If these lepidopteran pests develop resistance to the OPs, growers will switch to other pesticide classes, such as synthetic pyrethroids, which are highly effective on key pests but not so selective on the natural enemies of secondary pests of these crops as are organophosphates (Croft and Whalon, 1982; Croft and Riedl, 1991; see further discussion of resistance management in deciduous tree fruit species complexes in the following section).

Recent studies of tortricoid species have increased our understanding of their dynamic features of regional movement and pesticide resistance. They suggest novel ways to manage resistance among these pests in the field. For example, Knight and Croft (1986) recently reported regional patterns of movement in populations of *Argyrotaenia citrana* (Fernald) based on pheromone monitoring of adult males and direct larval sampling in caneberry fields of the northwestern United States. They observed a period of rapid, extensive movement of males across a large and diverse area of habitat in the Willamette Valley, following the principal mating period of females (Fig 7.3). Earlier in the season, males were much more local in their flight patterns, remaining near caneberry fields and sites of female development and calling. Knight (1986) felt that the rapid dispersal of male moths and the intermixing with wild populations during these periods may contribute to the lack of developed resistance observed in the area.

Another interesting attempt to manage resistance in tortricoid pests of deciduous tree fruits has been reported for New Zealand populations of the light brown apple moth, *Austrotortrix postvittana* (Walker). Suckling and coworkers (Suckling et al., 1984a, 1984b, 1985) summarized their plans for achieving resistance management of this species in an isolated peninsular area of the south island near Nelson. Their survey of the distribution of populations showed a restricted area of apparently stable resistance to azinphosmethyl in apple orchards. These resistant moths were surrounded by a large population of susceptible moths on alternate host plants that was potentially capable of diluting out the genes for resistance in commercial orchards (Fig 7.4). There was strong evidence that dilution would be a major factor in helping to prevent the spread of the OP resistance trait.

The research needed to develop a program of resistance management for this species involved maintaining the susceptible population as a source for dilution, as well as improving the dilution ratio by reducing the resistant population to a very low

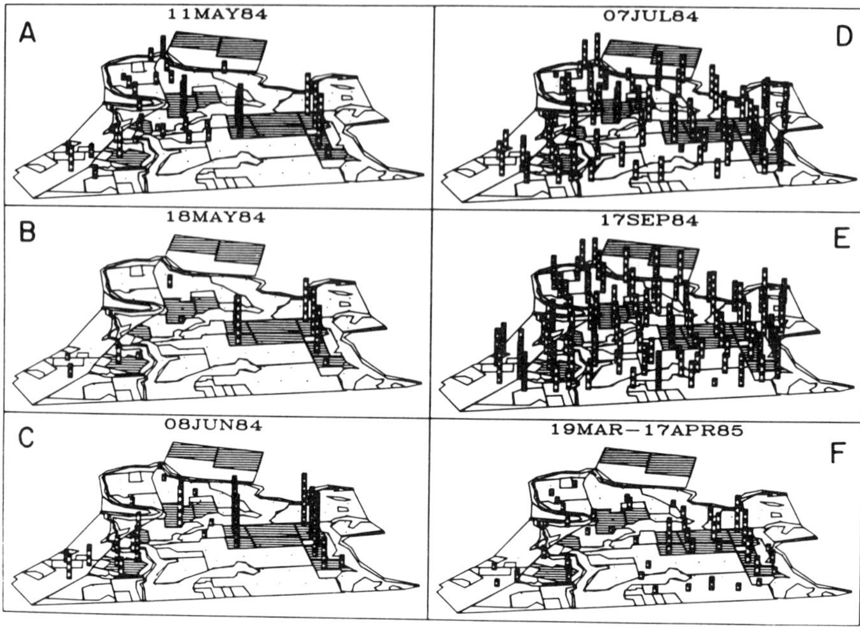

Figure 7.3. Spatial distribution of pheromone trap catch, in the regional study site near Woodburn, OR, during (A) a week in spring 1984 with warm temperatures; (B) a week in spring 1984 with cool temperature; (C) the week of peak catch in 1984; (D) the week of peak number of traps catching moths of *Argyrotaenia citrana* during the overwintering generation flight in spring 1984; (E) the last week of study in 1984; and (F) early spring 1985. Hatched areas are caneberry fields. Height of pillars is an approximate measure of weekly catch (Adapted from Knight & Croft 1986).

level by using a non-cross-resistant pesticide (e.g., a synthetic pyrethroid). In addition, orchardists were asked to eliminate light-brown apple moth host plants within the resistance management zone. While the resistance management program in New Zealand continues to evolve, a recent plan included the following features as part of an IPM program:

1. Close monitoring of the level and distribution of resistance using pheromone-collected males
2. Destruction of wild host plant habitats containing resistant individuals near apple-producing areas
3. Spraying chlorpyrifos or pyrethroids to reduce resistant moth populations to low levels in orchards and nearby wild habitats
4. Conservation of azinphosmethyl-susceptible moth reservoirs in the nearby bush
5. Enhanced immigration of susceptible males into the resistance area by using female sex phermones

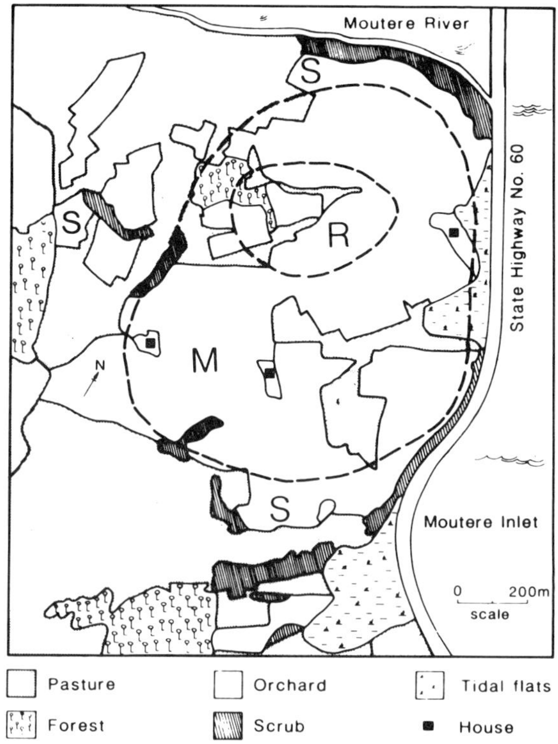

Figure 7.4. Schematic representation of the distribution of azinphos-methyl resistance in light brown apple moth (*Austrotortrix postvittana*) at Mariri in April 1983, after a 'calender' schedule of azinphos-methyl sprays. Zone codes: R = resistant individuals; M = mixed resistant and susceptible individuals; S = susceptible individuals (Adapted from Suckling et al, 1984b).

This use of pheromones to manipulate immigration rates of male moths and thereby flood out resistant moths is an innovative illustration of how susceptible species pool populations may be managed for practical benefit.

Immigration and Residency Effects on Resistance Evolution in a Pest and Natural Enemy Complex

The influence of immigration and residency factors on the evolution of pesticide resistance in a larger complex of orchard pests and their natural enemies has been studied by our group over the past decade (see earlier summaries in Croft, 1982; Croft and van de Baan, 1988). These studies have included resistance monitoring, modeling, and experimental investigations.

In a simulation study, Tabashnik and Croft (1985) compared relative rates of resistance development in a 24-species complex of apple pests and natural enemies that occur in midwestern and eastern North America. Factors such as generations per year, immigration rates, initial and overwintering population sizes, fecundity developmental rates, sex ratio, pesticide exposure levels, and percent of time spent in orchards by adults were included as variable factors in the simulation. The influence of each of the biological or pesticide-related factors were evaluated to estimate their impact on patterns of resistance observed in the field. As shown in Fig 7.5, the time of resistance development in the simulation for plant-feeding pests was very similar to that observed in nature. In natural enemies, the sequence of species developing resistance was similar to that observed in nature; however, the rate of resistance development in the model was considerably faster than that observed in the field (Fig 7.5). As pointed out by Tabashnik and Croft (1985) and Tabashnik (1986), one factor not represented in the model that may explain the faster estimated predictions of resistance was the differential influence of pesticides on the food supply of pests versus natural enemies following application in the field. Whereas a pest usually has an unlimited food supply for the resistant variant to use following selection, the resistant natural enemy has had its food supply decimated by the same applica-

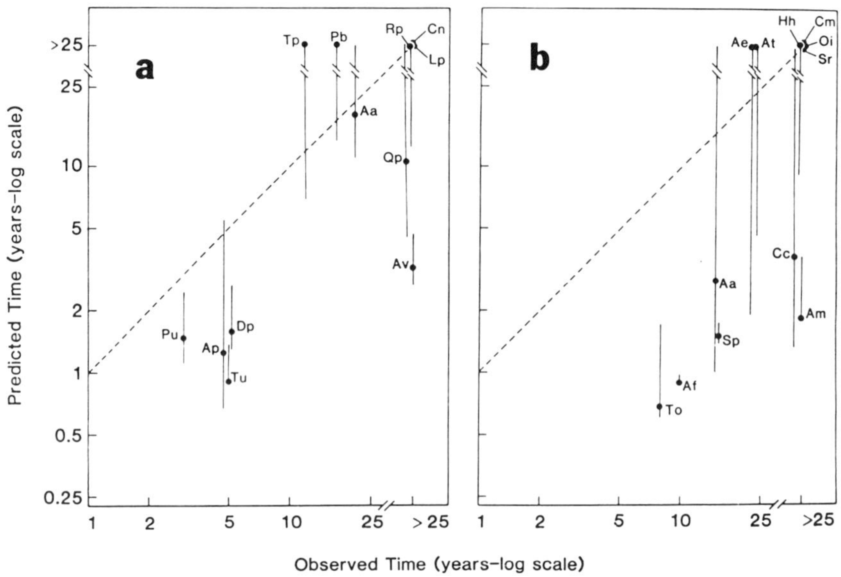

Figure 7.5. Predicted vs. observed times to evolve resistance to azinphosmethyl for apple arthropods. Predicted time (•) = simulated time to evolve resistance using means of estimates of population ecology parameters. Observed time = years after 1955 (1st widespread use of azinphosmethyl) to 1st report of resistance. Vertical bars show range of predicted times from sensitivity analysis. Predicted = observed along the dashed lines. a. Pests. n = 12. Spearman's rank correlation coefficient, r_s = .652, $p < 0.05$. Aa = *Archips argyrospilus*, Ap = *Aphis pomi*, Av = *Argyrotaenia velutinana*, Cn = *Conotrachelus nenuphar*, Dp = *Dysaphis plantaginea*, Lp = *Laspeyresia pomonella*, Pb = *Phyllonorcyter blancardella*, Pu = *Panonychus ulmi*, Qp = *Quadraspidiotus perniciousus*, Rp = *Rhagoletis pomonella*, Tp = *Typhlocyba pomaria*, Tu = *Tetranychus urticae*. b. Natural enemies (same assumptions as pests) n = 12. r_s = .692, $p < 0.025$. Aa = *Aphidoletes aphidimyza*, Ae = *Anagrus epos*, Af = *Amblyseius fallacis*, Am = *Aphelinus mali*, At = *Aphelopus typhlocyba*, Cc = *Chrysopa carnea*, Cm = *Coleomegilla maculata lingi*, Hh = *Hyaliodes harti*, Oi = *Orius insidiosus*, Sp = *Stethorus punctum*, Sr = *Syrphus ribesii*, To = *Typhlodromus occidentalis* (Adapted from Tabashnik & Croft, 1985).

tion. Therefore, it must find alternate food resources or, as usually happens, it either starves or migrates out of the treated area to interbreed with susceptible strains of the species. When the latter happens, the resistance factor is diluted out. In almost all cases of field-developed resistance among natural enemies observed to date, a resistance factor has usually developed first in its host or prey, but the chemical continues to be used for some other pest species, and thus the natural enemy continues to be selected in the presence of reasonable populations levels of its food supply (Croft and Strickler, 1983).

The major point from the data in Fig 7.5 is that one of the main attributes that contributes to the rapid development and persistence of resistance in orchard-inhabiting species is their tendency to reside in treated habitats and to have limited immigration from outside environments. This may be expressed as the ratio of susceptible immigrants moving into the treated habitat per generation relative to the number of residents present in the treated population per generation. In the simulation study, this ratio was highly correlated with the tendency for different species to develop resistant strains.

The influence of immigration of susceptible forms of pests and natural enemies in tree fruit crops on resistance evolution also has been studied by monitoring immigration in the field (Whalon and Croft, 1985; Croft and van de Baan, 1988). In Fig. 7.5, results from immigration studies are presented as an index for both (1) pests and (2) natural enemies. Time to develop resistance for individual species is plotted against their features of reproduction and immigration and residency. Reproduction estimates are from the expert opinions of researchers who work extensively with these species (see Croft, 1982; Tabashnik and Croft, 1985, for the specific details of how these estimates were collected). The reproductive index was calculated from the equation: 1 female (eggs/female $\times$ % survival to adult $\times$ female adults/male + female adults) to the exponent of number of generations/season $\times$ % overwintering survival. The immigration and residency ratio was estimated as indicated.

As shown in Fig. 7.6a, b, there was a strong positive correlation between both reproduction and residency estimates and time to resistance development. The higher the rate of increase was and the greater the number of residents to immigrants was, the faster or more likely is the species to develop resistant populations. The most rapid rates of resistance development for pests and natural enemies were for the spider mites (*T. urticae, Panonychus ulmi*) and phytoseiid mite predators (*M. occidentalis, Amblyseius fallacis*), respectively. Rapidly developing pests like aphids, leafhoppers, and leaf miners, along with the natural enemies *Stethorous* and *Aphidoletes sp.*, were next most likely to develop resistant strains in the field rapidly. Notice that, whereas both pests and natural enemy species have similar ranges in reproductive characteristics, there are major differences between the two groups in immigration and residency attributes. Natural enemies (usually as susceptible forms) tend to come more often from outside the treated habitat as compared to pests and therefore would dilute out any resistance already present within orchards. In contrast, pests tend to reside as resistant forms in treated habitats and are therefore less diluted by susceptible immigrants from surrounding species pool habitats.

Another way to illustrate the differences in the immigration features of apple

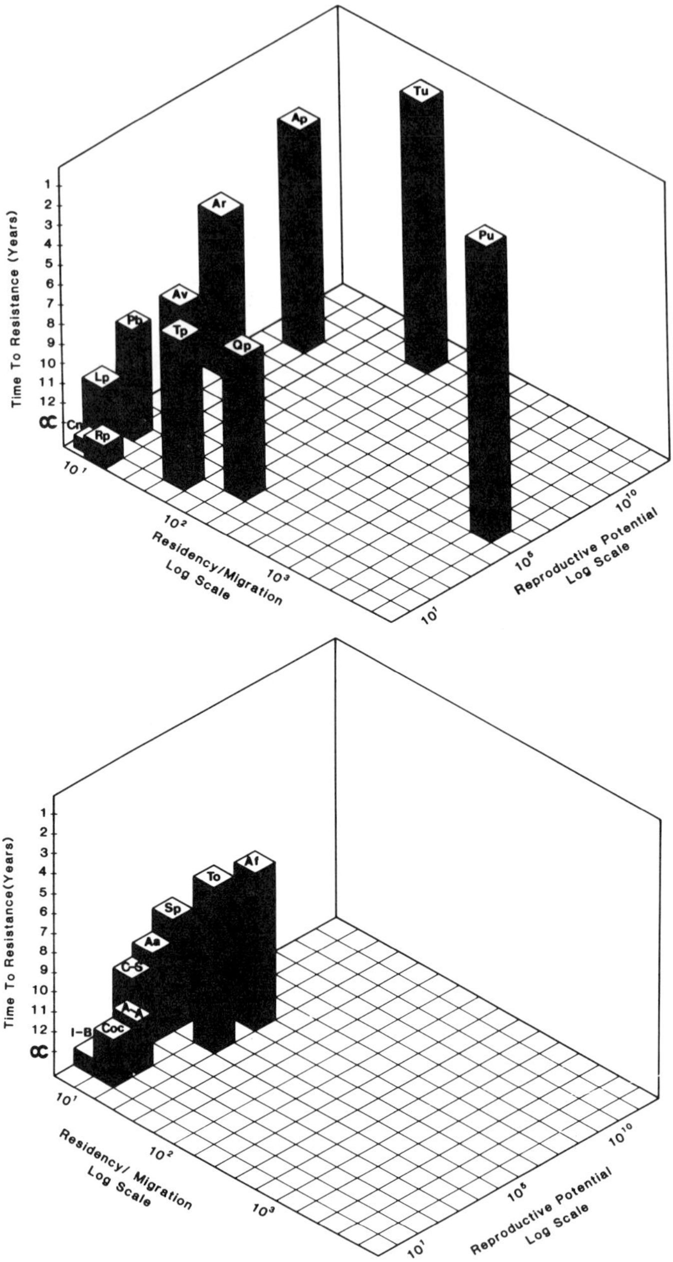

Figure 7.6. Estimated time to evolve resistance in 11 species of arthropod pests (a) and 8 species or groups of species of arthropod natural enemies (b) associated with deciduous tree fruit crops. Time to evolve resistance is shown as a function of the residency-to-immigration ratio and reproductive potential. (See Figure 7.5 for key to species designations; additional designations: C-S–Chrysopid-Syrphids, A-A–*Aphelinus mali-Aphelopus typhlocyba,* Coc–Coccinellids, I-B–Ichneumonid-Braconids) (from Croft and van de Baan 1988).

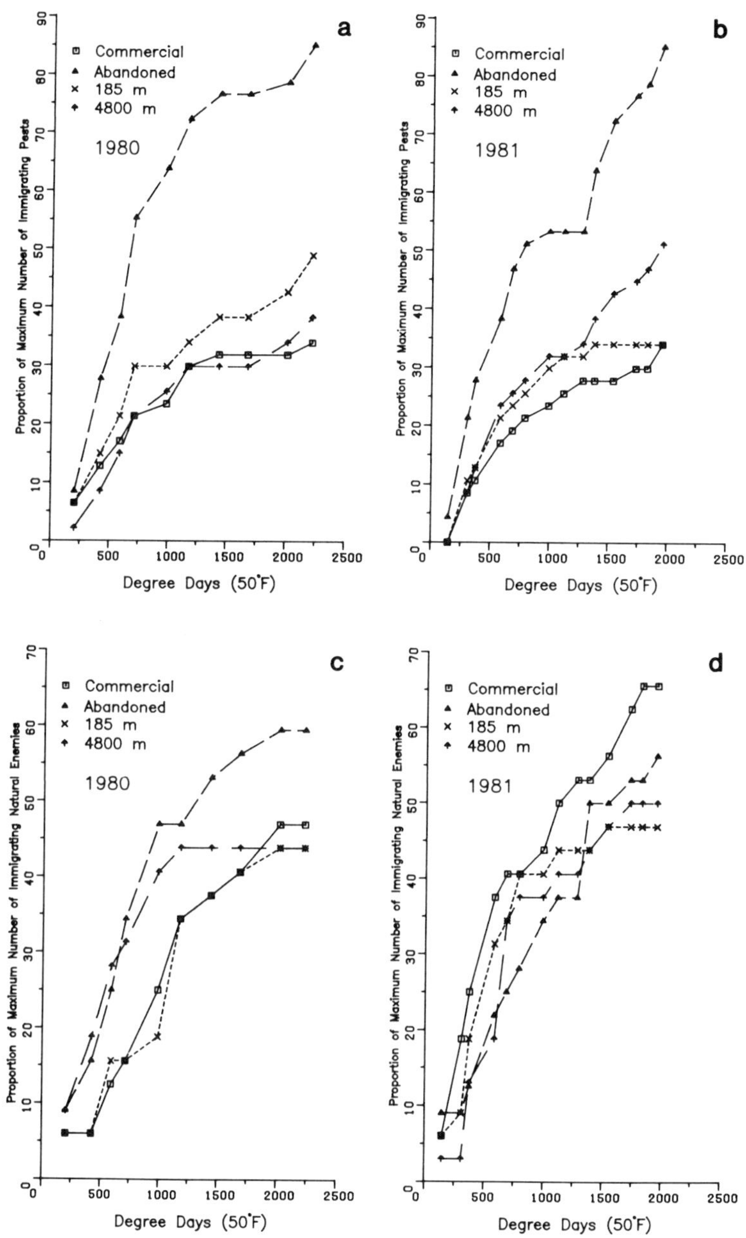

Figure 7.7. The percent of the maximum number of immigrating pest (a and b) and natural enemy (c and d) species into portable apple orchards placed in and around apple orchards in apple-growing regions of Michigan. Portable orchards were established in 1980 in areas of intense apple production, while in 1981 portable orchards were placed in less intense apple-growing regions (from Whalon and Croft 1985).

pests and their generalist natural enemies can be seen in a plot of the immigration rates of these two classes of organisms observed in the studies of Whalon and Croft (1985) (Fig. 7.7). Portable apple trees that had been eradicated of pests or natural enemies were set out in two regional species pool environments (i.e., dense apple production versus less dense apple production) and under one of four sets of local species pool conditions: (1) within 20 meters of an unsprayed apple orchard, (2) within 20 meters of a sprayed apple orchard, (3) within 185 meters of an unsprayed apple orchard, and (4) within 4,800 meters of an unsprayed apple orchard. For pests, the trees set out near abandoned orchards had the largest number of species and the fastest rate of colonization. However, there were no significant differences among immigration rates for natural enemies across the four treatments. These results reflect that specialist apple pests tend to move from apple to apple or, more likely, from one treated system to another, whereas generalist predators seem to come from sources other than orchards and seem to be almost ubiquitous in any environment. They therefore would tend to move from unsprayed systems into treated systems and dilute out those few organisms that may be residents in the treated environment. Also, because the area that is treated usually represents only a small portion of a generalist predator's habitat, a relatively small percentage of predators are ever directly exposed to the pesticides, further reducing the likeihood of maintaining a resistant population.

Although there are many exceptions to the broad generalizations made earlier, the tendency for pests to reside in treated systems and for natural enemies to immigrate into these managed systems may be of primary significance in explaining the high frequency of resistance among pest arthropods and the paucity of cases among beneficial natural enemies (Croft and Strickler, 1983; Founier et al., 1985). In several cases, the understanding gained about these influences has affected practical programs of pesticide resistance management implemented for both the pests and the natural enemies occurring on these crops (Croft, 1982; Croft and van de Baan, 1988). However, much additional research is needed to evaluate the specific effects on resistance of these general patterns of movement and to devise better means to take advantage of them in implementing programs of resistance management.

REFERENCES

Anonymous. 1957. World Health Expert Committee on Insecticides, 7th Report. *WHO Tech. Rep. Ser.* No. 125.

Comins, N. H. 1977. The development of insecticide resistance in the presence of migration. *J. Theor. Biol.* 64:177–197.

Croft, B. A. 1982. Developed resistance to insecticides in apple arthropods: A key to pest control failures and successes in North America. *Entomol. Exp. Appl.* 31:88–110.

———. 1990. Unpublished data.

Croft, B. A., and M. T. AliNiazee. 1983. Differential tolerance or resistance to insecticides in *Typhlodromus arboreus* and associated phytoseiid mites from apple in the Willamette Valley, Oregon. *J. Econ. Entomol.* 12:1420–1423.

Croft, B. A., S. C. Hoyt, and P. H. Westigard. 1987. Spider mite management on pome fruit revisited: Organotin and acaricide resistance management. *J. Econ. Entomol.* 80:304–311.

Croft, B. A., and L. A. Hull. 1991. Chemical control and resistance in tortricoid pests of pome and stone fruits. Pages 473–483 Chapters 2, 3, 4 in van der Geest, L. R. S., and H. H. Evenhuis (eds.), *Tortricoid Pests*. Amsterdam: Elsevier.

Croft, B. A., and H. W. Riedl. 1991. Chemical control and resistance to pesticides in the codling moth, *Cydia pomonella*. Pages 371–383 Chapters 2, 1, 6 in van der Geest, L. R. S., and H. H. Evenhuis (eds.), *Tortricoid Pests*. Amsterdam: Elsevier.

Croft, B. A., and K. Strickler. 1983. Natural enemy resistance to pesticides: Documentation, characterization, theory, and application. Pages 669–702 in Georghiou, G. P., and T. Saito (eds.), *Pest Resistance to Pesticides*. New York: Plenum. 809 pp.

Croft, B. A., and M. E. Whalon. 1982. Selective toxicity of pyrethroid insecticides to arthropod natural enemies and pests of agricultural crops. *Entomophaga* 27:3–21.

Croft, B. A., and H. E. van de Baan. 1988. Ecological and genetic factors influencing evolution of pesticide resistance in tetranychic and phytoseiid mites. *Exper. Appl. Acarol.* 4:277–300.

Dunley, J. 1990. Unpublished data.

Falconer, D. S. 1981. *Introduction to Quantitative Genetics*, 2nd ed. London: Longman.

Flexner, J. L. 1988. Organotin resistance in *Tetranychus urticae* on pear: Components and their integration for resistance management. Ph.D. Thesis. Oregon St. Univ. Corvallis. 103 pages.

Follett, P., B. A. Croft, and P. H. Westigard. 1985. Regional patterns of insecticide resistance in pear psylla (*Psylla pyricola*). *Can. Entomol.* 117:565–573.

Founier, D., M. Pralavorio, J. B. Berge, and A. Cuany. 1985. Pesticide resistance in Phytoseiidae. Pages 423–430 in Helle, W., and M. W. Sabelis (eds.), *Spider Mites: Their Biology, Natural Enemies and Control*. Amsterdam: Elsevier.

Georghiou, G. P. 1986. Magnitude of the resistance problem. Pages 14–43 in *Pesticide Resistance: Strategies and Tatics for Management*. Washington, D.C.: USNAS/NRC Press.

Georghiou, G. P., and C. E. Taylor. 1986. Factors influencing the evolution of resistance. Pages 157–169 in *Pesticide Resistance: Strategies and Tactics for Management*. Washington, D.C.: USNAS/NRC Press.

Hadam, J. J., M. T. AliNiazee, and B. A. Croft. 1986. Phytoseiid mites of major crops in the Willamette Valley, OR and pesticide resistance in *Typhlodromus pyri* Schulten. *Environ. Entomol.* 15:1255–1262.

Hoy, M. A. 1985. Recent advances in the genetics and genetic improvement of the Phytoseiidae. *Annu. Rev. Entomol.* 30:345–370.

Knight, A. L. 1986. Regional population dynamics of *Argyrotaenia citrana* (Lepidoptera: Tortricidae) in caneberry: Phenology, overwintering survivorship, and movement as measured by pheromone trap catches and larval sampling. Ph.D Thesis., Oregon St. Univ. Corvallis. 175 pp.

Knight, A. L., and B. A. Croft. 1986. Regional population dynamics and seasonal spatial patterns of *Agryrocaenia citrana* (Lepidoptera: Tortricidae) as measured by a pheromone trap grid and larval sampling. *Envrion. Entomol.* 16:59–67.

Levin, B. R., et al. (14 other authors). 1986. Population biology of pesticide resistance: Bridging the gap between theory and practical application. Pages 143–156 in *Pesticide*

Resistance: Strategies and Tactics for Management. Washington, D.C.: USNAS/NRC Press.

Liss, W. J., L. J. Gut, P. H. Westigard, and C. E. Warren. 1986. Perspectives on arthropod community structure, organization and development in agricultural crops. *Annu. Rev. Entomol.* 31:455–478

MacKenzie, D. R., C. S. Barfield, G. G. Kennedy, and R. D. Berger (eds.). 1986. *The Movement and Dispersal of Agriculturally Important Biotic Pests.* Baton Rouge, La.: Claitor Pub.

Miller, R. W., B. A. Croft, and R. D. Nelson. 1985. Effects of early-season immigration on cyhexatin and formetanate resistance in *Tetranychus urticae* on strawberry in California. *J. Econ. Entomol.* 78:1379–1388.

Rabb, R. L., and G. G. Kennedy (eds.). 1979. *Movement of Highly Mobile Insects: Concepts and Methodology in Research.* North Carolina: N. C. St. Univ. Press.

Riedl, H., P. H. Westigard, R. S. Bethell, and J. E. DeTar. 1981. Problems with chemical control of pear psylla. *Calif. Agric.* 35:7–9.

Roush, R. T., and B. A. Croft. 1986. Experimental population genetics and ecology studies of resistance in arthropods. Pages 257–270 in *Pesticide Resistance: Strategies and Tactics for Management.* Washington, D.C.: USNAS/NRC Press.

Roush, R. T., and J. McKenzie. 1987. Ecological genetics of insecticide and acaricide resistance. *Annu. Rev. Entomol.* 32:361–380.

Suckling, D. M., R. B. Chapman, and D. R. Penman. 1984a. Insecticide resistance in the light-brown apple moth (*Epiphyas postvittana* Walker) (Lepidoptera: Tortricidae): larval responses to azinphosmethyl. *J. Econ. Entomol.* 77:579–582.

Suckling, D. M., C. H. Wearing, and W. P. Thomas. 1984b. Insecticide resistance in the light-brown apple moth: A case for resistance management. *Proc. 37th N. Z. Weed Pest Cont. Conf.*, 248–252.

Suckling, D. M., D. R. Penman, R. B. Chapman, and C. H. Wearing. 1985. Pheromone use in insecticide resistance surveys of light-brown apple moths (Lepidoptera: Tortricidae). *J. Econ. Entomol.* 78:204–207.

Tabashnik, B. E. 1986. Evolution of pesticide resistance in predator-prey systems. *Bull. Entomol. Soc. Am.* 32:156–161.

Tabashnik, B. E., and B. A. Croft. 1982. Managing pesticide resistance in crop-arthropod complexes: Interactions between biological and operational factors. *Environ. Entomol.* 11:1137–1144.

Tabashnik, B. E., and B. A. Croft. 1985. Evolution of pesticide resistance in apple pests and their natural enemies. *Entomophaga* 30:37–49.

van de Baan, H. E., P. H. Westigard, E. C. Burts, and B. A. Croft. 1989. Seasonal susceptibility to insecticides in insecticide-resistant pear psylla, *Psylla pyricola* (Homoptera: Psyllidae). *Crop Protect.* 8:122–126.

Whalon, M. E., and B. A. Croft. 1985. Dispersal of apple pests and natural enemies in Michigan. *Mich. St. Univ. Res. Rept.* 467. 23 pages.

SECTION III

ECOLOGY OF INVADING POPULATIONS

8

HABITAT FACTORS IN NEW PEST INVASIONS

David Pimentel

Cornell University

INTRODUCTION

Since the United States was first colonized in 1620, many insect, plant pathogen, and weed species have invaded U.S. agricultural, livestock, grassland, forest, and other ecosystems. Once established in their new habitat, some of these pests significantly reduce crop and other biomass resources, despite all efforts to control them with pesticides and other means (Pimentel and Levitan, 1986).

The damage inflicted by pests on crops, forages, livestock, and forests seriously affects the potential yields of agricultural and forest systems and represents a major loss of food and fiber products in the marketplace (Pimentel, 1986a). Before proceeding, I need to clarify what is meant by "new pest invasions." Often new pest invasions implies that a species was newly introduced into the United States and attacked crops, livestock, or forests. However, another definition also exists. For example, when we introduced new crop, livestock, and tree species into the United States and then native insect, plant pathogen, and weed species attacked the introductions, these pests were also considered "new pest invaders"—the newly introduced crop was invaded by native pest species. These two views of "new pest invasions" will be discussed later; however, the prime interest in this chapter is pest invaders introduced from outside the United States.

For the present, in analyzing agricultural and forestry losses to pests, I will consider both new pest invaders from outside the United States and native species that became pests of crops and livestock as one group. Current crop losses attributed to all pests in the United States (insects, plant pathogens, and weeds) are estimated to be 37% (Pimentel, 1981, 1986a). Of this, 13% is due to insects, 12% to plant pathogens, and 12% to weeds. These losses occur, despite the annual use of about 500,000 kilograms (kg) of pesticides plus various biological and other nonchemical controls (Pimentel, 1981).

In forests, losses due to insects, other animals, and plant pathogens are estimated to be about 25% (Dasmann, 1966). Of this, 14% is due to insects, 8% to plant pathogens, and 3% to other animals. Interestingly, only about 10% of the total pesticides used in the United States is dispersed in forests (Pimentel and Levitan, 1986). This number is small partly because the economic return on investment per hectare in forest production is relatively low and has to be spread over 20 to 100 years. Pesticide treatments are thus expensive and are rarely used in forest production.

Annual livestock losses due to pests are considered substantial and are estimated to be about 13% of the total production (Drummond et al., 1981). These losses are caused by both diseases and insects. Insects are vectors of many livestock diseases, including various viral, bacterial, rickettsial, and protozoan diseases.

Translating crop, forest, and livestock losses into dollars gives a better appreciation of the losses that all pests cause each year in the United States. For crops, the loss of potential production is calculated to be $51 billion; for forests, $2 billion; and for livestock, $13 billion (Pimentel, 1986a). This totals $66 billion—a substantial economic loss for the United States.

The exact number of major pests in the United States is not known, but it is estimated to be about 2,000 species. This is about 1% of the total number of all species (200,000) that exist in the United States (Pimentel et al., 1980). If all major and minor pests were included, the number of pests might be increased about five fold (to about 10,000). The prime difficulty is how to define a pest and what the factors are that affect the severity of attack.

Numerous habitat factors, including the plants, animals, microbes, climate, and various other physical and chemical factors, influence the potential establishment and severity of attack of new invading pest species. In the United States, most crop plants have been introduced from many regions of the world (Pimentel, 1986b). For example, corn was introduced from Central America, soybeans from China, wheat from the Middle East, and potatoes from Bolivia and Peru. In fact, the majority of annual grain, vegetable, and fruit crops are from the tropics. In contrast, most trees in commercial forests in the United States are native species. Similar to crops, most livestock species have also been introduced, and several of these are of tropical origin (e.g., swine and chickens).

As mentioned, all of the introduced crops, trees, and livestock are attacked by a combination of introduced and native insect, plant pathogen, and weed species. Some of these pest invaders are from the native habitat of the crops (e.g., cabbage white butterfly), forests, and livestock, whereas others have been brought in from habitats that are foreign to the crops, forests, and livestock. In addition, many of the pest insects and plant pathogen invaders moved from native U.S. plant and animal hosts and have become serious pests of the introduced annual tropical crops (e.g., Colorado potato beetle) (Pimentel, 1977). The introduced crop, tree, or livestock species has never coevolved with these potential pests and thus has little or no resistance to attack from the native pests (Pimentel, 1977). The lack of evolution through genetic feedback to establish a balance in food demand by the feeding parasites and supply of food resources by the crops, forests, and livestock often has resulted in severe outbreaks of the pests on the various imported crops (Pimentel, 1961, 1988).

In contrast to crops, most U.S. forests contain native tree species that have evolved genetic balance with their pests. This is one factor that reduces the severity of insect herbivore and plant pathogen attacks on forest trees compared with those in crops. However, when such insects as the gypsy moth and such plant pathogens as the chestnut blight have been introduced from foreign regions into U.S. forests, severe outbreaks have resulted (Elton, 1958; Pimentel, 1961). In both cases, the associations between the introduced species and the native tree hosts were new, no balance had evolved, and severe damage was inflicted on the native tree species.

The habitats of crops, livestock, and forests have certain attributes that make them favorable environments for invading pest species. In this chapter, I will analyze the ecological origin of the major plant and animal invaders of agricultural, livestock, and forestry habitats of the United States and what habitat factors favored the success of these invaders.

DISTURBED HABITATS

Generally, it is easier for a new species of plant or animal to invade a new habitat if the habitat is disturbed (Allee et al., 1949; Odum, 1971; Fox and Fox, 1986; Williamson and Brown, 1986; Joenje, 1987; van Lenteren et al., 1987). The argument is made that "environmental resistance" is much less in a disturbed habitat compared with a climax community in a stable habitat. One of the best examples of a disturbed habitat that is considered more susceptible to invasion is crop habitats. Most of the natural vegetation and associated animals are destroyed before a selected crop is planted in the newly disturbed habitat. This modified habitat is thus susceptible to invasion by other plant species.

The newly planted crop also provides a simplified and ideal habitat for some new invaders, including insects, plant pathogens, and weeds. For insects and plant pathogens, food is abundant. The pests can move with ease from one host crop plant to another. In addition to this abundance of food, often there are few parasites and predators to limit the growth in the invading pest populations (Pimentel, 1977; Williamson and Brown, 1986; Groves and Burdon, 1986). Often parasite and predator populations are scarce in the disturbed habitats, because the natural enemies cannot increase until their food hosts and prey are present. This gives the pest population an advantage in time for their populations to become well established and grow rapidly.

In addition to having a disturbed habitat planted to a single plant species, most often the crop habitat is planted to an introduced crop plant (Pimentel, 1986b). Most of the crops planted in the United States are tropical annuals. Nearly all of the perennial crops, including fruit and forage crops, are also introduced species. Thus, the disturbed agricultural habitats also include a foreign species of crop plant that is often susceptible to native pest invaders (Pimentel, 1961, 1977).

When the pest invaders in U.S. crops are examined, it is interesting to note that only 40% of the insect species that are pests in crop habitats are foreign-introduced species (Table 8.1). Most of the foreign introduced invaders are from Europe. Some were brought in with the crops, and others were probably in the ships' ballast. In the

Table 8.1. Major Insect Pest Species in U.S. Crops and the Number of Invaders and Their Origins (Pimentel, 1986b)

Crop	Total Number of Major Insect Pest Species	Number of Biological Invaders	Number of Native Insect Species
Corn	16	3	13
Rice	2	2	0
Other small grains	11	6	5
Cotton	15	4	11
Sugar cane and sugar beets	4	0	4
Tobacco	6	1	5
Vegetables[a]	41	15	26
Fruits[b]	40	16	24
Alfalfa/clover and other legumes	13	10	3
Total	148	57	91

	Number of Species
Europe	25
Africa	1
Asia	4
Tropical America	12
Oceania	1
Multiple regions	14
Total	57

[a]Asparagus, beans, potato, cabbage, onion, peas, cowpeas, cucurbits, and carrots.
[b]Apples, citrus, and pears.

Sources: Metcalf et al. (1962); USDA (1974); Davidson and Lyon (1979).

seventeenth and eighteenth centuries, ships that arrived in America were relatively empty of goods but were filled with soil ballast. To make room for American products, the ballast, which contained insects, weeds, plant pathogens, and other organisms, was emptied on U.S. shores (Elton, 1958; Sailer, 1983).

Several factors in crop habitats have affected the ability of foreign insect invaders to become established in the modified habitats. First, most of the annual grain, vegetable, and fruit crops grown in the United States are tropical and subtropical in origin (Pimentel, 1984). The result is that few insect and plant pathogen invaders from the tropics could survive the cold winter climates of North America. Most tropical insect species, for instance, lack the ability to hibernate or enter diapause in order to survive the freezing conditions typical of the northern United States.

The other 60% of the insect pests attacking crops in the United States are also invaders, but they are native insects that moved from native vegetation to feed on the introduced crops cultured in the United States (Table 8.1). The crop plants introduced from the tropics and subtropics lack genetic feedback-evolved resistance to native insect herbivores, because they had never coevolved (Pimentel, 1977, 1986b). Thus, a large number of native insect herbivores have increased their food resource by utilizing the susceptible crop plants as a food resource.

It is typical throughout the world that most insect pests on crops are native species that moved onto introduced crops. For example, in most regions where sugarcane was introduced and established as a crop, a unique set of pest species has become associated with it. Of 1,645 known pest species of sugarcane, "959 occur only in a single region and 156 in two regions. Only 18 pest species occur in more than 10 regions" (Strong et al., 1977). These data suggest that sugarcane pest groups develop primarily from native arthropods that have moved from local plants and not from arthropods introduced into the region from the native region of sugarcane. This example also indicates that the limitation of foreign insects feeding in U.S. crop habitats appears to have little to do with the freezing climatic conditions in North America. Sugarcane is a tropical and subtropical plant that can only be cultured under nonfreezing conditions. Note that nearly 60% of the arthropod pests of sugarcane are, in fact, tropical native herbivores that have moved onto sugarcane in that habitat in which sugarcane was introduced.

An example of a native U.S. herbivore moving onto an introduced crop is the invasion of the white potato by the so-called Colorado potato beetle. The potato originated in Bolivia and Peru (Hawkes, 1944). The Colorado potato beetle is native to the United States and feeds primarily on wild sandbur, *Solanum rostratum* (Elton, 1958). When the potato was introduced into the United States, the plant became exposed to the beetle. The introduced potato plant lacked any natural resistance to the beetle, and the beetle became the most serious insect pest of potatoes here and in many parts of the world, including Europe and Eurasia.

A total of 28 major arthropod pests attack livestock that were introduced into the United States, including cattle, horses, sheep, goats, hogs, and poultry (Table 8.2). Approximately 64% of these pests have a worldwide distribution, and most were probably introduced along with the livestock. Only two insect species, the horn fly and the face fly of cattle, were introduced independent of cattle.

Several ecological characteristics of both the habitat and the insect invaders helped the invaders become established in the United States (Lattin and Oman, 1983). First, food was ample in the crop habitats. Second, the invaders were able to

Table 8.2. Major Arthropod Pests of Livestock and the Number of Invaders and Their Origins (Pimentel, 1986b)

Total Number of Major Arthropod Pest Species	Number of Biological Invaders	Number of Native Arthropod Pest Species	Number of Worldwide Arthropod Pest Species
28	2[a]	8	18

[a]Horn fly and face fly of cattle introduced from Europe.

Sources: Metcalf et al. (1962); Davidson and Prentice (1967); Davidson and Lyon (1979).

tolerate the climatic conditions and have their life history synchronized with the climate of the region (Groves and Burdon, 1986; Williamson and Brown, 1986). In addition, the new pest invaders were able to resist the attack of natural enemies that were present in the new crop habitats. As mentioned, however, the disturbed habitat of crops would generally have fewer natural enemies than in natural U.S. habitats (Pimentel, 1977).

The patterns of the invasions of foreign plant pathogens in vegetable and grain crops in the United States are distinctly different from those of insect invaders. Almost 40% of the 155 major plant pathogens in vegetables that were investigated had a worldwide distribution (Table 8.3). Often, the pathogenic microbes were brought in when the crop host was introduced into the United States. Thus, these introductions might, in a sense, be classed as worldwide distributions. Frequently, the pathogens were brought in with the seeds of the crops themselves.

The number of plant pathogens introduced into U.S. vegetable crop habitats, other than those having a worldwide distribution, was 31% compared to 37% for insects (Table 8.3). Also, in contrast to insects, most of the introduced plant pathogens came from tropical America and secondarily from Europe.

Still, the number of native plant pathogens that moved from native vegetation to attack crops was substantial, or about 30% (Table 8.3). This trend also agrees favorably with those of pest insects.

All of the plant pathogens associated with grain crops, except one, have a worldwide distribution (Table 8.4). Only one pathogen, that associated with corn, is of native origin. Most of the grain pathogens were probably spread with the transport of grain (seeds) worldwide.

In agriculture, all plants other than the crops planted in the disturbed habitat are considered weeds. Many of the weeds were introduced from various regions of the world, whereas some of the weed species are native to the United States. About 73%

Table 8.3. Major Plant Pathogens of U.S. Vegetable Crops and the Number of Invaders and Their Origins (Pimentel, 1986b)

Total Number of Major Plant Pathogen Species	Number of Biological Invaders	Number of Native Pathogen Species	Number of Worldwide Pathogen Species
155	48	47	60

Origins of Pathogen Invaders	Number of Species
Europe	19
Tropical America	13
Asia	1
Multiple Regions	18
Total	51

Sources: Chupp and Sherf (1960); Purseglove (1972, 1977); CMI (1983).

Table 8.4. Major Plant Pathogens of U.S. Grain Crops and the Number of Invaders and Their Origins (Pimentel, 1986b)

Total Number of Major Plant Pathogen Species	Number of Biological Invaders	Number of Native Pathogen Species	Number of Worldwide Pathogen Species
74	0	1	73

Sources: Leonard and Martin (1963); Purseglove (1972); CMI (1983).

of the 80 major U.S. weed species originally came from regions outside the United States (Table 8.5). Approximately 67% of the weed invaders came from Europe and Eurasia, because of the long-term selection and the intensive early trade between the United States and Europe and Eurasia. Several of these arrived on ballast from ships, as mentioned earlier. Some of these weed invaders eventually found their way into U.S. crop habitats.

The mixing in of weed seeds with crop seeds when harvested was probably the most important means of introducing weeds into the United States. Today, effective methods exist for removing most of the weed seeds from crop seeds, and importers are very concerned about the dangers that exist in importing weed seeds into the United States. However, before 1900, the production of pure crop seed was difficult, and sufficient care was not exercised.

For various reasons, the pattern of the invasion of weeds in U.S. pastures is quite different from that associated with crops. Only 41% of the pasture weed invaders came from regions outside the United States, whereas 73% of the crop weed invaders were introduced species (Tables 8.5 and 8.6). Similar to cultivated croplands, most (78%) of the weed species that were introduced into the United States and invaded pastures and hayfields also came from Europe and Eurasia (Tables 8.5 and

Table 8.5. Major Weed Species in U.S. Cultivated Crops and the Number of Invaders and Their Origins (Pimentel, 1986b)

Total Number of Major Weed Species	Number of Biological Invaders	Number of Native Weed Species
80	58	22
Origins of Weed Invaders		**Number of Species**
Europe		22
Eurasia		17
Africa		1
Asia		4
Tropical America		6
Mediterranean		1
Multiple regions		7
Total		58

Table 8.6. Major Weed Species in U.S. Pastures and the Number of Invaders and Their Origins (Pimentel, 1986b)

Total Number of Major Weed Species	Number of Biological Invaders	Number of Native Weed Species
110	45	65

Origins of Weed Invaders	Number of Species
Europe	17
Eurasia	18
Africa	0
Asia	4
Tropical America	3
Mediterranean	2
Multiple regions	1
Total	45

Source: USDA (1971).

8.6). In addition to being carried in soil and ballast, some weeds probably were transported in straw and hay used in packaging goods and materials that were shipped to the United States. Although there was a greater opportunity for the introduction of weed seeds into pastures than crops, probably fewer introduced weeds became established in planted pastures and hayfields. The pasture habitats were more resistant, because they were planted to more stable perennial-plant communities. The pasture habitats were seldom plowed and disturbed, as is typical when culturing annual crops.

Unfortunately, no data exist on the precise number of weed species that were introduced into the United States and the number that actually became established. Overall, however, fewer weed species appear to have been introduced and established in natural ecosystems than in the managed agricultural ecosystems. Harper (1965) reported that "it is extremely questionable whether there is any single proven case of the extinction or even the decline in abundance of a native species which can be directly attributed to aggression by an alien species that was introduced." There is no question, however, that the invaders may have significant effects on native species (Hadfield and Mountain, 1981; Howarth, 1983; Pimentel et al., 1989).

NATURAL HABITATS INVADED

Based on ecological principles, one would expect fewer invaders in natural habitats than in disturbed habitats. This supports Harper's analysis and generally confirms that interspecific competition from new pest invaders seldom results in the displacement and extinction of a native species by an introduced species. This is confirmed by an analysis of the numbers and extent of invaders in natural habitats, such as forests. In U.S. forests, only 27% of the insect pest species are invaders, and the

Table 8.7. Major Insect Pest Species in U.S. Forests and the Number of Invaders and Their Origins (Pimentel, 1986b)

Origin of Major Pests	Number of Species
Native	51
Europe	15
Other introduced	4
Total	70

Sources: Metcalf et al. (1962); Davidson and Prentice (1967); Davidson and Lyon (1979).

remainder are native pests (Table 8.7). However, some of the most serious insect pests in forests are introduced insects (Elton, 1958).

Exactly which species are considered serious forest pests depends as much on the extent of economic damage caused by the pest as the size of the pest population. For example, the white pine weevil, which is relatively rare in terms of numbers in white pine forests, is one of the most serious pests of pine, because it attacks, and thereby kills, the growing tip of the pine tree. When this happens, another branch becomes the growing tip, altering the shape of the tree and diminishing its value as saw timber. No precise data exist on the number of plant pathogens that are invaders in forests versus those that are native in forests.

It appears that few introduced pest species have caused the extinction of native vegetation (Harper, 1965; Simberloff, 1981; Joenje, 1987). In U.S. forests, most of the weed tree species, such as white poplar, black locust, and hawthorn, are native species that do not have characteristics suitable for their use as saw timber and pulpwood. No major species of weed tree has been introduced into the United States.

ESSENTIAL HABITAT REQUIREMENTS OF NEW PEST INVADERS

In general, the essential habitat requirements of new pest invaders fit into three major categories: food, climate, and natural enemies (biological control). These factors and their ecological relationship to the success of pest invaders are discussed below.

Food Resources

All insect and plant pathogen pest invaders, except for a few, like dermestid insects and termites, feed on living matter, crops, trees, and livestock. Weed plants are not dependent on living organisms like pest insects and plant pathogens, because they are producers. Most insect pests and plant pathogens are classed either as herbivores, parasites, or predators. Thus, they are typical of about one-half of all species, which obtain their food resources directly from living hosts (Pimentel, 1968; Price, 1975). Most species have from 10 to 2,000 species of herbivore and parasite species feeding on them (Pimentel, 1988).

That most pest species feed as parasites emphasizes the importance of the food economy to insect and plant pathogen pests. Certainly, this relationship plays a major role in determining the success of insect and plant pathogen species in invading new habitats. The food economy is significant to organisms in their natural habitat as well as in their new habitat. It was Elton (1927) who first pointed out that "the whole structure and activities of the community are dependent upon the questions of food supply." The essential balanced supply–demand economy in herbivore and parasite host systems appears to be maintained often by genetic-feedback evolution that is achieved through natural selection (Pimentel, 1961).

All pest insect and plant pathogen species require certain essential nutrients (including particular amino acids, B-complex vitamins, vitamin C, lipogenic factors, fatty acids, sterols, fat-soluble vitamins, and several minerals) to be present and in proper balance. If these requirements are not met, the fitness and survival of the herbivore and parasite are reduced (Auclair, 1957; Dadd, 1973; White, 1976; Onuf et al., 1977; Mattson, 1980; Friend and Dadd, 1982).

In the native habitat of insect pests and plant pathogen species, it has been proposed that genetic-feedback evolution plays an important role in limiting the intensity of feeding pressure on host populations to some level of "harvestable" energy (Pimentel, 1961, 1988). Because most of the essential resources necessary for growth, maintenance, and reproduction account for most (about 90%) of the food host's resources, only a small portion (less than 10%) of the host's resources are available to the exploiter species (Pimentel, 1988).

In a recent analysis of 92 cases of natural herbivore and parasite populations feeding on plant hosts, the primary productivity harvested from various plant hosts averaged only 7% (Pimentel, 1988). In contrast, the lack of defenses in plant hosts has frequently resulted in outbreaks of the new pest invaders feeding on the new plant hosts. This has been typical, for example, of the Eurasian gypsy moth introduced into Massachusetts; the Japanese beetle introduced into the eastern United States; and the Dutch elm disease and European bark beetle that have caused major reduction in American elms (Elton, 1958). In all of these associations, the herbivore and other parasitic organisms increased rapidly on their susceptible hosts, which lacked defenses against their new parasites. Host numbers often declined because of the intense feeding pressure of the parasitic organisms.

This lack of balance between new pest invaders and their host plants in the United States is well documented. For example, of 212 major pest insect species introduced into the United States and established here, 139 (65%) were not considered pests in their native ecosystems (Calkins, 1983).

Precisely how many insect species have been introduced into the United States is not known; however, an estimated 1,500 new insect species have become established in the U.S. habitat (Figure 8.1). Although earlier the "ballast idea" was presented as an explanation for the introduction of many small organisms previous to 1900, note that most insect introductions occurred after 1900. This is probably due to more rapid transportation (e.g., aircraft and ships) and a higher volume of goods and people moved today than earlier in U.S. history. The 1,500 species of new insect invaders confirms two facts: (1) the relative ease with which insect species can invade

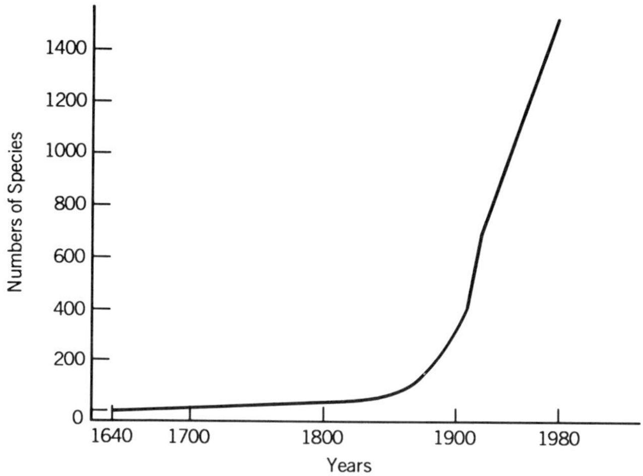

Figure 8.1. Number of species of insects and mites introduced into the 48 contiguous states from 1640 to 1980 (after Sailer, 1983)

the United States, and (2) that not all niches are full. Some controversy surrounds the question of whether niches are full or not in a habitat, but the problem centers primarily on the definition of *niche* and not whether new species can invade new habitats (Williamson and Brown, 1986). Clearly, a great many introduced species can find suitable food and other resources for their survival in the United States (Pimentel, 1985).

Climatic Conditions

Because the United States has primarily a temperate climate, restrictions are placed on some species invading this environment. Insects, microbes, and plants from the tropics and subtropics have difficulty invading temperate-climatic conditions like those in the United States and other regions (Allee et al., 1949; Pimentel, 1977; Groves and Burdon, 1986; Williamson and Brown, 1986). From data collected in the United Kingdom, it appears that insects from a cool climate are more than 10 times as likely to become established than if the species were from a tropical rain habitat (Table 8.8). Plants from cool temperate regions are about twice as likely to become established as those from a tropical rain climate (Table 8.8). No data exist for microbes, but it is assumed that they would react similarly to the invertebrates.

Moisture can also be an important limitation to many new pest invaders, but, in general, it does not appear to be as severe a limitation as temperature. This is, in part, confirmed by the data assembled in the United Kingdom on invaders. Note that insects from a similar moist, cool, temperate environment were more than twice as likely to become established than those from a dry climate (Table 8.8). Interestingly, plants from a dry habitat did not survive as well in the moist United Kingdom environment as those from a moist, cool environment.

Table 8.8. Percentage of Invasive Species Becoming Established at More Than One Locality in the United Kingdom (Williamson and Brown, 1986)

Climate	Vertebrates	Plants	Insects
Tropical, moist	44	38	3.75
Dry	53	44	21.6
Warm, temperate	66	56	35.1
Cool, temperate	33	66	50

It should be noted that, for vertebrate species invaders, neither temperature nor moisture affected the percentage of invasive species becoming established in the United Kingdom (Table 8.8). Since many of these species were selected for introduction and were not accidental introductions, it is assumed that they were selected because it was expected that they could survive in the United Kingdom habitat.

Natural Enemies and Biological Control of New Pest Invaders

A wide array of parasite and predator species present in the U.S. habitat where the new pest invader is introduced may prevent the establishment of the new invader species. If the invader becomes established, its population may be highly susceptible to attack by U.S. native parasites and predators (Pimentel, 1963; Groves and Burdon, 1986; Williamson and Brown, 1986). This is especially true of small populations that can be exterminated easily by the attack of relatively few predators and parasites. The population of the new invader is small and might become extinct because of the minimal population effect (Allee et al., 1949).

Parasite and predator pressure, of course, are well known for limiting the abundance and distribution of host animals and plants (Huffaker, 1980; Pimentel, 1986c). Numerous programs in biological control have demonstrated the capacity of natural enemies in limiting the number of pest populations.

In principle, the new pest invaders should be more susceptible to the attack of U.S. native parasite and predator species (Pimentel, 1961, 1963). The reason for this is the ecology of new parasite and predator host population associations. The new pest invader has never interacted with the native parasites and predators, and, therefore, they should lack defenses to the attack from the new natural enemies that they encounter in the new habitat. Thus, the new pest invaders should be more susceptible to attack from the native parasitic and predaceous species and biological control in the new habitat (Pimentel, 1963, 1984; Hokkanen and Pimentel, 1984).

That new associations between parasite and host result in the parasite having a greater impact on the host than native parasite and native host has been documented. Pimentel (1963, 1986c) and Hokkanen and Pimentel (1984) found, for example, that the success of introducing natural enemies for biological control was about 75% higher when new parasite–host (predator–prey) associations were employed than those based on long-evolved, native parasite–host systems. The lack of evolved interspecific balance in new species associations explains the high success rate of new associations in biological control.

Exactly how many potential new pest invaders have been prevented from becoming established or were kept relatively rare in the United States because of native parasites and predators attacking the new invaders is unknown. However, numerous records exist of native U.S. natural enemies moving to attack the new invaders as their hosts and prey. For instance, the Japanese beetle was introduced from Asia and was later attacked by a native bacterium (*Bacillus popillae*) that appears to have come from one of the native scarab beetles (Pimentel, 1963). This bacterium that causes milky white disease has been cultured and spread for the biological control of the Japanese beetle in the United States.

The gypsy moth is another new pest invader that was introduced from Europe (Elton, 1958). A total of 61 natural enemies were brought to the United States for biological control of the gypsy moth (Coulson et al., 1986). Of the natural enemies that were introduced, 10 became established in the United States and are providing partial control of this moth (Nichols, 1961). These 10 introduced biocontrol species, however, make up only a small percentage (about 10%) of the total number of natural enemies attacking this moth. It is reported that a total of about 95 parasitic and predaceous species attack the gypsy moth in the United States (Nichols, 1961; Campbell and Podgwaite, 1971; Podgwaite and Campbell, 1972; Campbell, 1974; Leonard, 1974). Thus, most are native natural enemies attacking the gypsy moth invader.

Exactly how much genetic change has taken place in the natural enemies of the gypsy moth, the gypsy moth itself, and the trees on which the moth feeds is unknown. However, with the degree of selective pressure by the parasites and predators on the gypsy moth and moth on host trees, some genetic change is strongly suspected in all populations. What all these probable genetic changes mean to the ecology of these systems and other biological systems, and new invader plant host systems needs to be investigated.

Pimentel (1961, 1963, 1968), Levin (1972), Levin and Pimentel (1981), Franz (1986), and others have proposed that genetic changes take place in these interacting populations. However, this does not necessarily mean that economic biological control will fail (Pimentel, 1986c). Genetic balance between interacting populations and economic biological control lead to entirely different outcomes. The sole aim of biological control is to keep the target pest economically low. Most past assessments contained no quantitative population data or genetic information and were merely observations that the pest remained under control.

CONCLUSION

Numerous factors in the habitat appear to influence the potential success of a new invader establishing itself. The habitat factors that appear to be most important for establishment are food, climate, natural enemies, genetics, and other factors, including disturbing and simplifying a habitat. Food is essential for all organisms and thus is probably the number one factor in the total habitat system. Because new invading insects and plant pathogens attack plant hosts that lack defenses, the new invaded plant host is often highly susceptible to attack from the new invaders.

Another dimension of this problem is that the new invading species also lacks defenses to attack from natural enemies that are native in the U.S. habitat. This tends to make the new invader highly susceptible to attack by native natural enemies. The exact role that native parasites and predators have played in limiting or preventing the establishment of new invaders still remains unknown. However, based on our knowledge of the success of biological controls, these natural enemies clearly play an important role in limiting new invading species.

The percentage of new invading species that become pests is unknown and in large part depends on how one defines a pest. If the focus is only on those species that are classed as major pests or those organisms that cause significant economic losses, then the following conclusion can be drawn regarding insects: of the 1,500 insects that have been introduced, approximately 235, or about 16%, have become pests (Sailer, 1983).

The exact proportion of the introduced microbes and plants that have become pests is unknown. For vertebrate species, probably less than 5% have become major pests in the United States.

In an investigation of new invaders in the United Kingdom, approximately 10% of the invaders were estimated as having become pests. Their definition of *pest* was not clear, but this generally agrees with the situation in the United States, if all invading species are assessed as a whole.

It is still not possible accurately to predict which species will successfully establish themselves if introduced. However, in general, each new invader will need food, few or no natural enemies, a tolerable climate, and other tolerable physical and chemical factors. Without suitable food and/or if a tropical species is introduced into the United States, there is little or no chance that it will become established. This is especially true of insects and microbes introduced from the tropics.

In summary, numerous habitat factors, including food, climate, natural enemies, genetics, disturbances, and other factors operating individually and/or in combination can determine the potential of a new invader establishing itself in the United States. More research is needed on the subject of habitats and invaders before it will be possible to make accurate predictions on the potential invasion of a new species. Because of the ecological and economic importance of this question, it is hoped that many new investigations will be initiated.

REFERENCES

Allee, W. C., A. E. Emerson, O. Park, T. Park, and K. P. Schmidt. 1949. *Principles of Animal Ecology.* Philadelphia: Saunders.

Auclair, J. L. 1957. Developments of resistance of plants to insects. *Proc. Entomol. Soc. Ont.* 88:7–17.

Calkins, C. O. 1983. Research on exotic insects. Pages 321–359 in Wilson, C. L., and C. L.

Graham (eds.), *Exotic Plant Pests and North American Agriculture.* New York: Academic Press.

Campbell, R. W. 1974. The gypsy moth and its natural enemies. U.S. Dep. Agric., *For. Serv. Agric. Info. Bull.* 381.

Campbell, R. W., and J. D. Podgwaite. 1971. The disease complex of the gypsy moth. I. Major components. *J. Invertebr. Pathol.* 18:101–107.

Chupp, C., and A. F. Sherf. 1960. *Vegetable Diseases and Their Control.* New York: Ronald Press.

Commonwealth Mycological Institute (CMI). 1983. *Distribution Maps of Plant Diseases.* Maps 1–554 and indexes. Kew, Surrey, England: Commonwealth Agricultural Bureaux.

Coulson, J. R., R. W. Fuester, P. W. Schaefer, L. R. Ertle, J. S. Kelleher, and L. D. Rhoads. 1986. Exploration for and importation of natural enemies of the gypsy moth, *Lymantria dispar* (L.) (Lepidoptera: Lymantriidae), in North-America—An update. *Proc. Entomol. Soc. Wash.* 88:461–75.

Dadd, R. H. 1973. Insect nutrition: Current developments and metabolic implications. *Annu. Rev. Entomol.* 18:381–420.

Dasmann, R. F. 1966. *Environmental Conservation.* New York: John Wiley & Sons.

Davidson, A. G., and R. M. Prentice (eds.). 1967. *Important Forest Insects and Diseases of Mutual Concern to Canada, the United States, and Mexico.* Ottawa, Canada: Queens Printer and Controller of Stationery.

Davidson, R. H., and W. F. Lyon. 1979. *Insect Pests of Farm, Garden and Orchard,* 7th ed. New York: John Wiley & Sons.

Drummond, R. O., G. Lambert, H. E. Smalley, and C. E. Terrill. 1981. Estimated losses of livestock to pests. Pages 111–128 in Pimentel, D. (ed.), *Handbook of Pest Management in Agriculture,* Vol. 1. Boca Raton, Fla.: CRC Press.

Elton, C. S. 1927. *Animal Ecology.* London: Sidgwick and Jackson.

———. 1958. *The Ecology of Invasions by Animals and Plants.* London: Methuen.

Fox, M. D., and B. J. Fox. 1986. The susceptibility of natural communities to invasion. Pages 57–66 in Groves, R. H., and J. J. Burdon (eds.), *Ecology of Biological Invasions.* Cambridge, Mass.: Cambridge Univ. Press.

Franz, J. M. (ed.). 1986. *Biological Plant and Health Protection.* Gustav Fischer Verlag, Stuttgart: Fortschr. Zool. 32.

Friend, W. G., and R. H. Dadd. 1982. Insect nutrition: A comparative perspective. Pages 205–247 in Draper, H. H. (ed.), *Advances in Nutritional Research.* New York: Plenum.

Groves, R. H., and J. J. Burdon (eds.). 1986. *Ecology of Biological Invasions.* Cambridge, Mass.: Cambridge Univ. Press.

Hadfield, M. G., and B. S. Mountain. 1981. A field study of a vanishing species, *Achatinella mustelina* (Gastropoda, Pulmonata) in the Waianae Mountains of Oahu. *Pac. Sci.* (1980) 34:345–358.

Harper, J. L. 1965. Establishment, aggression, and cohabitation in weedy species. Pages 243–268 in Baker, H. G., and G. L. Stebbins, *The Genetics of Colonizing Species.* New York: Academic Press.

Hawkes, J. G. 1944. Potato collecting expeditions in Mexico and South America. Imperial Bureau of Plant Breeding and Genetics. Cambridge, England: School of Agriculture.

Hokkanen, H., and D. Pimentel. 1984. New approach for selecting biological control agents. *Can. Entomol.* 116:1109–1121.

Howarth, F. B. 1983. Classical biocontrol: Panacea or Pandora's box. *Proc. Hawaii. Entomol. Soc.* 24:239–244.

Huffaker, C. B. 1980. *New Technology of Pest Control.* New York: John Wiley & Sons.

Joenje, W. 1987. The SCOPE programme on the ecology of biological invasions: An account of the Dutch contribution. *Proc. K. Ned. Akad. Wet. Ser. C Biol. Med. Sci.* 90:3–13.

Lattin, J. D., and P. Oman. 1983. Where are the exotic insect threats? Pages 93–137 in Wilson, C. L., and C. L. Graham (eds.), *Exotic Plant Pests and North American Agriculture.* New York: Academic Press.

Leonard, D. E. 1974. Recent developments in ecology and control of the gypsy moth. *Annu. Rev. Entomol.* 19:197–229.

Leonard, W. H., and J. H. Martin. 1963. *Cereal Crops.* New York: Macmillan.

Levin, S. A. 1972. A mathematical analysis of the genetic feedback mechanism. *Am. Nat.* 106:145–164. (Erratum 1973. 104:320).

Levin, S., and D. Pimentel. 1981. Selection of intermediate rates of increase in parasite-host systems. *Am. Nat.* 117:308–315.

Mattson, W. J. 1980. Herbivory in relation to plant nitrogen content. *Annu. Rev. Ecol. Syst.* 11:119–161.

Metcalf, C. L., W. P. Flint, and R. L. Metcalf. 1962. *Destructive and Useful Insects: Their Habits and Control,* 4th ed. New York: McGraw-Hill.

Nichols, J. O. 1961. The gypsy moth in Pennsylvania—Its history and eradication. *Pa. Dep. Agric. Misc. Bull.* No. 4404.

Odum, P. 1971. *Fundamentals of Ecology,* 3rd ed. Philadelphia: Saunders.

Onuf, C. P., J. M. Teal, and I. Valiela. 1977. Interactions of nutrients, plant growth, and herbivory in a mangrove ecosystem. *Ecology* 58:514–526.

Pimentel, D. 1961. Animal population regulation by the genetic feedback mechanism. *Am. Nat.* 95:65–79.

———. 1963. Introducing parasites and predators to control native pests. *Can. Entomol.* 95:785–792.

———. 1968. Population regulation and genetic feedback. *Science* 159:1432–1437.

———. 1977. Ecological basis of insect pest, pathogen and weed problems. Pages 3–31 in Cherrett, J. M., and G. R. Sagar (eds.), *The Origins of Pest, Parasite, Disease and Weed Problems.* Oxford, England: Blackwell Scientific Publ.

———. (ed.) 1981. *Handbook of Pest Management in Agriculture,* Vols. I-III. Boca Raton, Fla.: CRC Press.

———. 1984. Perennial grasses on the prairies: New crops for the future? Pages 124–137 in *1985 Yearbook of Science and the Future.* Chicago: Encyclopaedia Britannica.

———. 1985. Using genetic engineering for biological control: Reducing ecological risks. Pages 129–140 in Halvorson, H. O., D. Pramer, and M. Rogul (eds.), *Engineered Organisms in the Environment: Scientific Issues.* Washington, D.C.: American Society for Microbiology.

———. 1986a. Agroecology and economics. Pages 299–319 in Kogan, M. (ed.), *Ecological Theory and Integrated Pest Management Practice.* New York: John Wiley & Sons.

———. 1986b. Biological invasions of plants and animals in agriculture and forestry. Pages 149–162 in Mooney, H. A., and J. A. Drake (eds.), *Ecology of Biological Invasions of North America and Hawaii.* New York: Springer-Verlag.

———. 1986c. Population dynamics and the importance of evolution in successful biological control. Pages 3–18 in Franz, J. M. (ed.), *Biological Plant and Health Protection*. Gustav Fischer Verlag, Stuttgart: Fortschr. Zool. 32.

———. 1988. Herbivore population feeding pressure on plant hosts: Feedback evolution and host conservation. *Oikos* 53:289–302.

Pimentel, D., and L. Levitan. 1986. Pesticides: Amounts applied and amounts reaching pests. *BioScience* 36:86–91.

Pimentel, D., E. Garnick, A. Berkowitz, S. Jacobson, S. Napolitano, P. Black, S. Valdes-Cogliano, B. Vinzant, E. Hudes, and S. Littman. 1980. Environmental quality and natural biota. *BioScience* 30:750–755.

Pimentel, D., M. S. Hunter, J. A. LaGro, R. A. Efroymson, J. C. Landers, F. T. Mervis, C. A. McCarthy, and A. E. Boyd. 1989. Benefits and risks of genetic engineering in agriculture. *Bioscience* 39:606–614.

Podgwaite, J. D., and R. W. Campbell. 1972. The disease complex of the gypsy moth. II. Aerobic bacterial pathogens. *J. Invertebr. Pathol.* 20:303–308.

Price, P. W. 1975. *Evolutionary Strategies of Parasitic Insects and Mites*. New York: Plenum.

Purseglove, J. W. 1972. *Tropical Crops: Monocotyledons*. New York: Halsted Press, John Wiley & Sons.

———. 1977. *Tropical Crops: Dicotyledons*. London: Longman Group Limited.

Sailer, R. I. 1983. History of insect introductions. Pages 15–38 in Wilson, C. L., and C. L. Graham (eds.), *Exotic Plant Pests and North American Agriculture*. New York: Academic Press.

Simberloff, D. 1981. Community effects of introduced species. Pages 79–107 in Nitecki, M. H. (ed.), *Biotic Crises in Ecological and Evolutionary Time*. New York: Academic Press.

Strong, D. R., E. P. McCoy, and J. R. Rey. 1977. Time and the number of herbivore species. The pests of sugarcane. *Ecology* 58:167–175.

U.S. Dept. of Agriculture (USDA). 1971. Common weeds of the United States. New York: Agr. Res. Serv., Dover Publications.

———. 1974. Guidelines for the use of insecticides. U.S Dep. Agric. Handb. No. 542.

van Lenteren, J. C., J. Woets, P. Grijpma, S. A. Ulenberg, and O. P. J. M. Minkenberg. 1987. Invasions of pest and beneficial insects in the Netherlands. *Proc. K. Ned. Akad. Wet. Ser. C Biol. Med. Sci.* 90:51–80.

White, T. C. R. 1976. Weather, food and plagues of locusts. *Oecologia* 22:119–134.

Williamson, M. H., and K. C. Brown. 1986. The analysis and modelling of British invasions. *Philos. Trans. R. Soc. Lond. Ser. B Biol. Sci.* 3114:505–522.

9

LIFE-HISTORY STRATEGIES OF IMMIGRANT PLANTHOPPERS

Robert F. Denno

University of Maryland

INTRODUCTION

An immigrant or colonizing species is one that easily crosses barriers, rapidly establishes itself, and then expands its numbers and range in the new habitat (Ehrlich, 1986). Thus, there are three components to successful invasion: (1) immigration and arrival in the new habitat, (2) establishment involving mating and oviposition, and (3) population increase and spread. Several ecologists have suggested that successful invasion ought to be associated with an integrated suite of characteristics involving dispersal, reproduction, and often diapause (Southwood, 1977; Dingle, 1978, 1985; Denno and Dingle, 1981; Denno, 1983, 1985).

The invasion process itself poses a significant dilemma for many colonizing species. Given a limited energy budget for organisms, choices must be made as to specific allocations. For example, energy spent on migration may not be available for reproduction, and vice versa. The lower fecundity of the migratory compared to the nonmigratory forms of aphids, crickets, bugs, and other insects attests to the existence of this trade-off (Dixon, 1972; Utida, 1972; Roff, 1984, 1986; Zera, 1984). Thus, traits like high dispersability, which facilitates crossing barriers and colonizing new resources, and high fecundity, which promotes rapid establishment, are antagonistic—at least to some extent.

The advantage of possessing both high dispersability and high reproductive potential has most certainly played a role in the evolution of flight polymorphisms (Roff, 1984, 1986; Denno et al., 1985, 1989). For wing polymorphic insects, dispersal and reproductive functions are in part partitioned between different morphs (Roff, 1986). The result is an efficient dispersing morph with reduced fecundity or an efficient reproductive form with restricted dispersal ability. Such life-history flexibility may provide the answer for many organisms in their chase of patchy and changing resources and may increase their chances of successfully invading distant habitats.

Many authors, including Baker and Stebbins (1965), Dingle (1985), and Ehrlich (1986), have categorized the expected life-history traits and ecological attributes of a successful invading species. Ehrlich's list represents what might be called an approximation of a consensus of trained ecologists, or, as he calls it, the "folk knowledge" of the field. The proposed characteristics of a successful invader are as follows:

1. Abundance in the original range increases the source pool of potential colonists.
2. Polyphagy increases the probability of finding suitable food in the new habitat.
3. High fecundity and early age to first reproduction favor rapid establishment.
4. High genetic variability increases the probability for successful establishment under, and adjustment to, novel conditions.
5. Mating prior to migration in the source area allows a fertilized female to colonize alone and precludes mate finding under low-density conditions in the new habitat.
6. Larger individuals should be able to fly further and longer, and possess greater reproductive potential, thus increasing the chances for colonization of distant resources.
7. Organisms associated with humans have a greater chance of accidental or, as the case with biological control agents, intentional introduction into new habitats.
8. The ability of a organism to function over a wide range of biotic and physical conditions will promote tolerance of new situations.

My approach in this chapter is to use this list of characteristics as a set of predictions to be tested against data regarding the life-history traits and ecological attributes of a single group of insects, the planthoppers. Planthoppers are ideal for this exercise, because there are both migratory and sedentary species (Denno, 1978; Denno et al., 1987, 1989). From this analysis will come a set of characteristics that describe colonizing (migratory) planthoppers and perhaps might be used to assess the likelihood of a species becoming a successful invader. A second objective will be to consider the generality of the results that emerge from planthopper case studies.

NATURAL HISTORY AND CASE STUDIES OF PLANTHOPPERS

The planthoppers in this study are homopterans in the family Delphacidae. Their primary host plants are grasses, including rice, corn, sugarcane, and cereals (Denno, 1978, 1980; Mochida and Okada, 1979; O'Brien and Wilson, 1985). Consequently, many are severe agricultural pests, either because they directly damage crops (Sogawa and Cheng, 1979) or because they transmit plant pathogens while feeding in the phloem (Purcell, 1985). Females have swordlike ovipositors and insert their eggs in various plant tissues (Denno et al., 1981; O'Brien and Wilson, 1985). The eggs hatch into nymphs that pass through five instars. They can be extremely abundant, and the excessive feeding that results from heavy infestations often kills plants or reduces yield (Sogawa and Cheng, 1979).

Most planthopper populations are composed of two discrete wing forms (Denno, 1976, 1983, 1985; Denno et al., 1985, 1987, 1989). There are macropterous individuals with fully developed wings and brachypterous adults with reduced wings that cannot fly. Macropters are capable of long-distance flight, whereas brachypters can move only short distances (Kisimoto, 1976, 1979; Denno and Grissell, 1979; Denno et al., 1980, 1985, 1989; Roderick, 1987). Nevertheless, brachypters have a reproductive advantage because they reproduce earlier and lay more eggs (Denno, 1985; Denno et al., 1989). There is a strong environmental component to wing form determination, although this differs between species and sometimes between the sexes (Kisimoto, 1956, 1965; Denno, 1976, 1983; Denno et al., 1985; Roderick, 1987). Individuals have the capability of molting into either a brachypter or a macropter, depending on the conditions they experience as nymphs; various environmental cues trigger a developmental switch that determines wing form. Usually, crowded conditions on poor-quality hosts result in the production of macropterous individuals that can migrate to better habitats elsewhere (Denno et al., 1985). The wing form composition of planthopper populations in the field reflects the migratory ability of a species (Denno et al., 1987). For this report, I categorize those species with > 50% macropterous forms as migratory species and those species with < 50% macropterous forms (often much less than 50%) as sedentary species.

Two case studies underscore the flexible life histories of planthoppers and how wing polymorphism promotes the effective tracking of changing host plants. These studies also serve to emphasize that co-occurring species answer problems of resource tracking with different life-history solutions; that is, life-history traits (e.g., migration and reproduction) may be integrated differently among species that successfully exploit the same host plant. The two case studies involve (1) planthoppers that feed on rice in Asia and (2) planthoppers that occur on intertidal grasses in North America. First, let us focus on planthoppers that feed on rice.

Two planthoppers, *Nilaparvata lugens* Stål (Brown Planthopper) and *Sogatella furcifera* (Horvath) (White-Backed Planthopper) are major pests of rice throughout most of their range in Southeast Asia and the temperate Orient (Dyck and Thomas, 1979). However, the nature of the host plant and the population dynamics of these planthoppers differ markedly between tropical and temperate Asia. For example, in tropical Asia, there is continuous cropping with staggered planting and harvest times, resulting in a patchy resource that is available year round (Dyck et al., 1979; Kenmore, 1980; Kenmore et al., 1984). Figure 9.1a shows the phenology of *N. lugens* in tropical Asia, where there are often two generations after colonizing rice paddies. Macropters colonize over a long period of time, and their colonizing density is relatively high. The offspring of colonists are either brachypters, which remain in the field, or macropters, which emigrate. However, the rapid buildup of

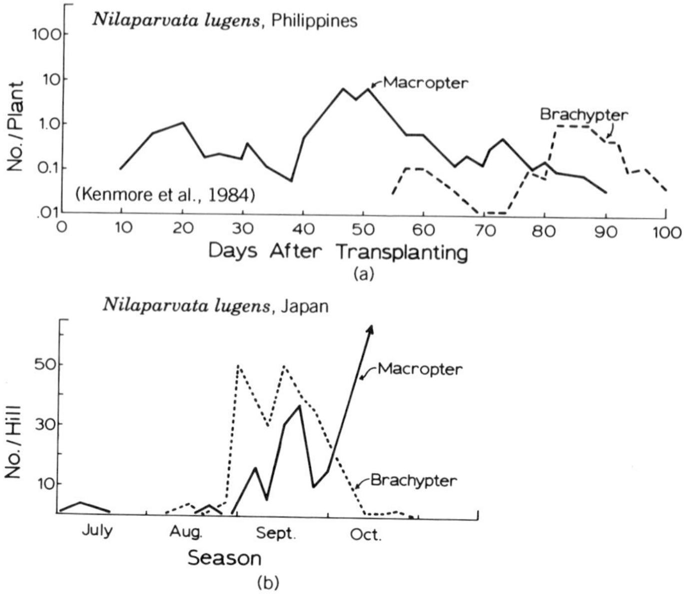

Figure 9.1. (a) Phenology of the macropterous and brachypterous adults of *Nilaparvata lugens* in a tropical rice paddy in the Philippines. Data are replotted from Kenmore et al. (1984). (b) Phenology of the macropterous and brachypterous adults of *N. lugens* in a temperate rice paddy in Japan. Data are replotted from Kisimoto (1965). Note the narrow window of colonization by macropters during July in Japan, compared with the continuous colonization of rice in the Philippines.

planthoppers in the paddy is largely attributable to the more fecund brachypterous form, while movement between paddies is only possible by macropters (Kisimoto, 1977; Hirao, 1979; Kenmore, 1980). The phenology of *S. furcifera* in tropical rice is similar to that for *N. lugens*, and the production of macropterous forms is density dependent in the females of both species (Cook and Perfect, 1985).

In temperate areas like Japan, the situation is very different. Here, neither *N. lugens* nor *S. furcifera* survive the winter season, and recolonization occurs every year by immigrants originating from central China (Kisimoto, 1976, 1979). Typically, migrations of planthoppers proceed eastward across the East China Sea to Japan, a distance of 2,000 kilometers (km). Following colonization, the general phenology of both planthopper species in Japanese paddies is similar but differs markedly from the pattern in tropical Asia because there is a discrete period of colonization by macropters in July and their initial density is very low (Fig. 9.1b). However, like tropical Asia, heavy end-of-the-season infestations are attributable to the midseason occurrence of brachypters (Kisimoto, 1977; Hirao, 1979). Despite similarities in their phenology in rice, there are several important differences in migration and reproduction between *N. lugens* and *S. furcifera*.

Life-history differences between *N. lugens* and *S. furcifera* are elucidated by the studies of Kisimoto (1979), who, one day in July, sampled the settling immigrant macropters of both species in paddies at different distances from the westernmost

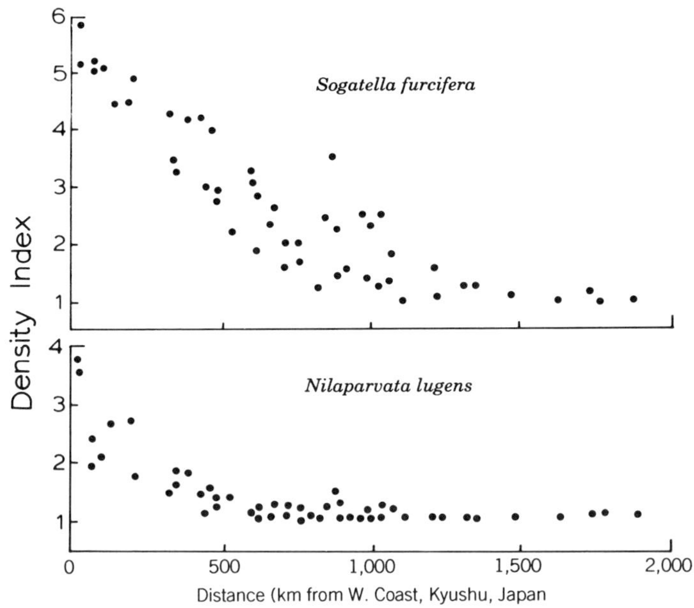

Figure 9.2. Density of settling immigrants (macropters) of *Sogatella furcifera* and *Nilaparvata lugens* at different distances from the West Coast of Kyushu, Japan. The migration was from west to east, or from left to right in this graph, and originated in central China. Note that the density of settlers for both species decreased along the eastward course of the migration but that the initial colonizing density of *S. furcifera* was much higher than that for *N. lugens*. Density index: 1 = no planthoppers; 2 = 1–5; 3 = 6–25; 4 = 26–125; 5 = 126–625; 6 = more than 625 (from Kisimoto, 1979).

coast of Japan (Fig. 9.2) The direction of the migration was west to east, or from left to right in Figure 9.2. It is clear that the density of both species decreased linearly with distance from the West Coast. From his data, Kisimoto (1979) calculated that the founding population of *N. lugens* was about 1/100 of that for *S. furcifera*. During the preceding week, west of Japan on the East China Sea, Mochida (1974) sampled the same migration and caught very similar numbers of both species, about 2,700 of each. Thus, *S. furcifera* appears to be a much better long-distance colonizer than *N. lugens*.

Even though the initial density of *S. furcifera* was one order of magnitude higher than that for *N. lugens*, populations of *Nilaparvata* grew much faster after colonization (see Fig. 9.3). The population of *N. lugens* increased 513 times from its initial density, compared with only a fourfold increase in *S. furcifera*. Kuno (1979) attributes the rapid population increase of *N. lugens* primarily to its high fecundity compared with *S. furcifera*. Raised under the same conditions in the laboratory, macropters of *N. lugens* lay 543 eggs, compared with only 385 for *S. furcifera* (Kisimoto, 1965). Thus, we see evidence for two alternative colonizing strategies and a difference in the way reproduction and migration are integrated. *S. furcifera* will locate distant resources better, but *N. lugens* will leave more offspring if it arrives at a favorable habitat.

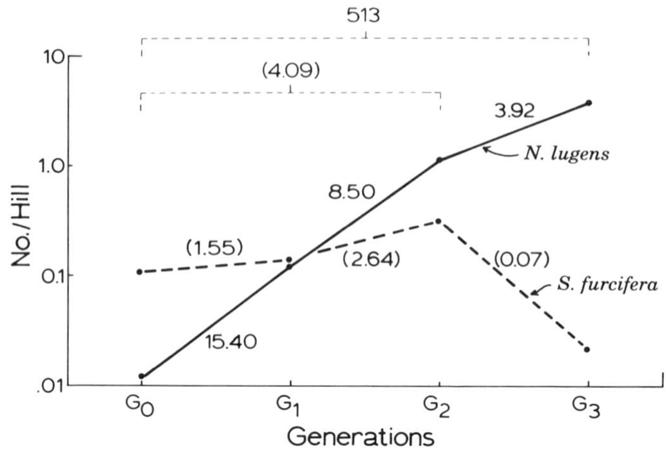

Figure 9.3. Population growth of *Nilaparvata lugens* and *Sogatella furcifera* in a Japanese rice paddy. Numbers on graph are replacement rates of adults from one generation to the next. A value of 1.0 is equivalent to exact replacement and no population growth. The initial density of colonists is much higher for *S. furcifera*, but populations of *N. lugens* increase more rapidly. Data replotted from Kuno (1979).

Now I would like to turn briefly to the second case study in a natural rather than agricultural system. The host plant is cord grass, *Spartina alterniflora* Loisel, which occurs abundantly along the Atlantic and Gulf coasts of North America in the tidal marshes (Denno, 1983). Two species of planthoppers, *Prokelisia marginata* (Van Duzee) and *Prokelisia dolus* Wilson feed and reproduce exclusively on this grass (Denno et al., 1987). There is, however, a dramatic difference in life-style between these two congeners. *P. marginata* is a highly mobile species, capable of long-distance dispersal (Denno, 1983; Denno et al., 1985, 1987; Antolin and Strong, 1987). Macropters have been trapped on oil platforms as far as 32 km offshore in the Gulf of Mexico (Sparks et al., 1986). In Atlantic Coast populations, *P. marginata* undergoes interhabitat migrations between optimal overwintering and feeding sites (Denno and Grissell, 1979; Denno, 1983). The best sites for overwintering occur on the high marsh in short-form *Spartina*, where nymphs can nestle into the protective thatch. The most favorable sites for feeding and development are located in tall-form *Spartina* along creeksides some distance away. Tall-form *Spartina* offers no protection during winter, because creek banks are swept clean of vegetation (Denno and Grissell, 1979). Thus, tall-form *Spartina* is a temporary habitat that can be colonized in spring and escaped from in fall only by macropters. Brachypters of *P. marginata* remain on the high marsh associated with local patches of nutritious plants (Denno and Grissell, 1979; Denno, unpubl. data). The wing form composition of *P. marginata* populations, which are 90% macropterous, reflects its migratory life-style (Denno, 1985; Denno et al., 1985, 1987). By contrast, only about 20% of the adults of *P. dolus* are macropterous (Wilson, 1982; Denno, 1985; Denno et al., 1985). *P. dolus* is a very sedentary species. It overwinters and reproduces only on the high marsh and never effectively colonizes tall-form *Spartina* (Denno, unpubl. data). The difference in the proportion of macropters or migrants between populations of *P. marginata* and *P. dolus* is in large part attributable to differences in their wing form response to crowding. The density threshold that triggers an individual to molt to a macropter is much lower for *P. marginata* than for *P. dolus*. For example, in laboratory studies, it takes a density of only 5 nymphs per cage to trigger 75% of *P. marginata* females to molt to macropters, whereas in *P. dolus* females, 75% macroptery is produced at a density of 40 under the same caged conditions. Furthermore, males of *P. marginata* are long-winged, even when raised at very low densities, while macroptery in *P. dolus* males is density dependent and identical to the female response (Denno et al., 1991). Differences in the life-styles of these two co-occurring planthoppers may well reflect a different ability to tolerate crowding and nutritionally inferior host plants.

TEST OF PREDICTIONS FOR INVADERS

Now I would like to refer back to the original list of proposed characteristics for colonizing species and see how planthoppers compare. First, we may ask if invading species of planthoppers are abundant in their original range. This appears to be true; all long-distance migrating species that I have investigated, including *N. lugens, S. furcifera, Laodelphax striatellus* (Fallen), *Javesella pellucida* (F.), and *P. marginata*, are abundant in their source areas (Raatikainen and Vasarainen, 1976; Denno, 1977; Dyck et al., 1979; Kenmore et al., 1984; Denno et al., 1985; Cook and Perfect, 1985). For example, it is common for *P. marginata* to reach densities in excess of 1,000 adults per meter (m), and it is one of the most abundant planthoppers in *Spartina* (Denno et al., 1987). Furthermore, *J. pellucida*, a migrant pest of oats in Europe, is the most abundant sap-feeding herbivore in the system (Raatikainen and Vasarainen, 1976). Admittedly, it is difficult to compare the absolute abundances of species in their source areas due to differences in sampling, but abundance results in host plant deterioration and the production of migrant forms that are able to colonize new habitats. Thus, abundance and mobility are related, but not perfectly. Not all abundant species are migratory. For example, *P. dolus* is as abundant as *P. marginata*, but it is relatively sedentary (Denno et al., 1987). Thus, abundance per se cannot be used to predict a species's potential to migrate. Other factors must be considered, such as sensitivity to changes in host plant nutrition (see Prestidge and McNeill, 1982).

Second, are invaders polyphagous? Here I define *polyphagy* as feeding on many genera of grasses, *oligophagy* as feeding on two or three genera of grasses, and *monophagy* as feeding within a single grass genus. There appears to be little relationship between invasion potential and diet breadth (see Table 9.1). For example, of the migratory planthoppers, *N. lugens* and *P. marginata* are monophagous, whereas *S. furcifera* and *J. pellucida* are very polyphagous (Mochida and Okada, 1979; Sogawa, 1982; Denno et al., 1987; Wilson and O'Brien, 1987). At least one case of oligophagy appears to be explained by the addition of sugarcane as a host for the

Table 9.1. Diet Breadth for Migratory Species of Planthoppers

Species	Diet	Host Plants	Reference
Javesella pellucida	Polyphagous	Many grasses	Wilson and O'Brien, 1987
Laodelphax striatellus	Polyphagous	Many grasses	Wilson and O'Brien, 1987
Nilaparvata lugens	Monophagous	Rice	Sogawa, 1982
Peregrinus maidis	Oligophagous	Corn-sorghum	Fisk et al., 1981
Perkinsiella saccharicida	Monophagous	Sugarcane	Wilson and O'Brien, 1987
Prokelisia marginata	Monophagous	*Spartina*	Denno et al., 1987
Saccharosydne saccarivora	Oligophagous	Sugarcane-*Andropogon*	Metcalf, 1969
Sogatella furcifera	Polyphagous	Many grasses	Wilson and O'Brien, 1987
Sogatodes oryzicola	Oligophagous	Rice-*Echinochloa*	King and Saunders, 1984

Andropogon-feeding *S. saccharivora* (Metcalf, 1969). Thus, diet breadth is a poor predictor of colonizing ability for planthoppers.

Third, we can ask if invasive planthoppers have high fecundities and early reproduction. Figure 9.4a shows the relationship between the total fecundity of the brachypter and the fecundity of the macropter for several migratory and sedentary species of planthoppers. The line represents a projection of equal fecundity for the two wing forms. Two patterns are apparent. First, most points fall above the line, documenting that brachypters are generally more fecund than macropters. Second, migratory species do not appear to be more fecund than sedentary species; both are scattered throughout the distribution. In fact, for the *Prokelisia* species, *P. marginata*, the migrator, has a significantly lower fecundity than *P. dolus* (Denno et al., 1989). However, among migratory species, those with greater fecundity (e.g., *N. lugens*) may establish more effectively in new habitats, as Kuno (1979) suggests.

Early reproduction is, in part, influenced by preoviposition period, or the number of days spent as an adult prior to first oviposition. Fig. 9.4b shows the relationship between the length of the preoviposition period of the brachypter and macropter for migratory and sedentary species of planthoppers. All points fall below the unity line, documenting that macropters reproduce later in life when compared with brachypters. However, all six migratory species lie closer to the origin than the three sedentary species, suggesting that selection has favored earlier reproduction in migratory species. The extent to which this pattern is upheld as more data become available remains to be seen.

Fourth, do invading planthoppers differ genetically from more sedentary species? Here the answer is, perhaps; there are no data for comparison, despite the fact that this issue has attracted considerable attention from evolutionary biologists (e.g., Baker and Stebbins, 1965; Lewontin, 1965, Parsons, 1983). It seems obvious that a reasonable degree of genetic variability would be helpful to invaders, but, presently,

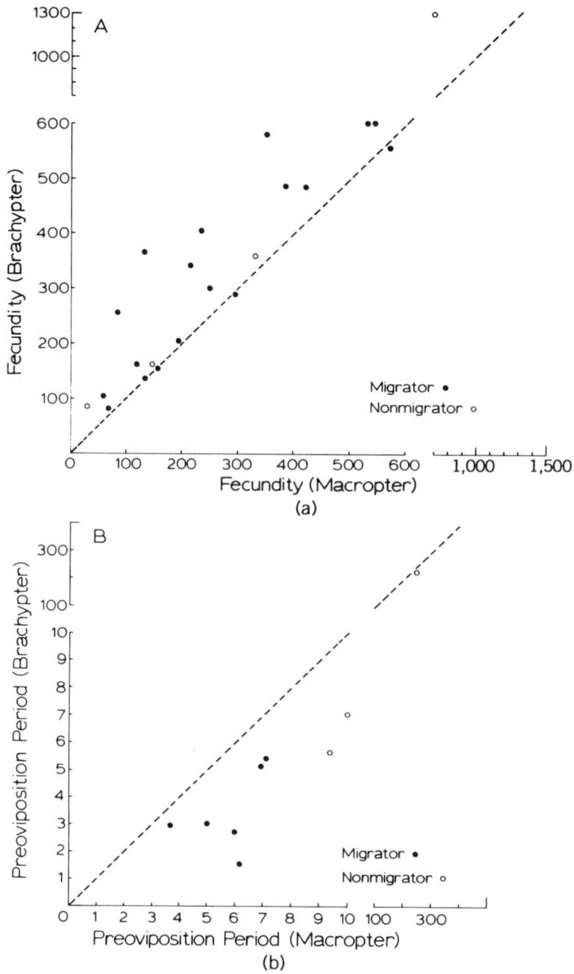

Figure 9.4. (a) Relationship between the total fecundity of the brachypterous and macropterous females of migratory and nonmigratory species of planthoppers. Most points fall above the line of equal fecundity, indicating that brachypters are more fecund. There is no tendency for migratory species to be more or less fecund than sedentary species. (b) Relationship between the preovipositional period of the brachypterous and macropterous females of migratory and nonmigratory species of planthoppers. All points fall below the unity line, documenting that brachypterous females reproduce earlier in life than macropters. Also, points for migratory species are clustered closer to the origin, suggesting that there is a general tendency for migratory species to reproduce earlier in life when compared with nonmigratory species. Data are extracted from Denno (1985) and Denno et al. (1989).

Ehrlich (1986) suggests that we cannot list in detail genetic attributes, present in some species but absent in close relatives, that have provided advantages to the former in novel environments. Ehrlich (1986) goes on to posit that successful invaders may be more likely to originate from marginal populations, which, at least in *Drosophila* (Brussard, 1984), have more genic heterozygosity and a greater tendency to disperse. Furthermore, both positive and negative correlations between life-history traits do occur in planthoppers (Roderick, 1987), but we do not know the extent to which these differ between populations or species. Suffice it to say, however, that the *Prokelisia-Spartina* system is ideal for pursuing issues concerning the genetics of colonization, because there is extensive intra- and interspecific variation in dispersal characters (Denno and Grissell, 1979; Denno, 1983; Strong and Stiling, 1983; Denno et al., 1985, 1987; Antolin and Strong, 1987).

Fifth, are invaders fertilized females that are able to colonize alone? At least for *N. lugens* and *S. furcifera* the answer is, no. Kisimoto (1976) sampled immigrating planthoppers in Japan and after dissection found that almost all females were unmated (there were no spermataphores) and that their reproductive systems were underdeveloped. Mating occurs after arrival in the new habitat (Kuno, 1979). This is probably true for *P. marginata* as well, because mating is frequently observed after migrants have just arrived in tall-form *Spartina* (Denno, unpubl. data).

For all planthoppers studied, there is an oogenesis flight syndrome, in which the female reproductive system fails to develop until after migration (Denno, 1985; Denno et al., 1989). Thus, there is a physiological basis for the separation of migratory and reproductive behavior. Not only does this phenomenon explain the reproductive delay characteristic of migratory forms (Fig. 9.4b), but it may also explain why macropterous females are not receptive to males for mating until after arrival in the new habitat.

There is a dramatic density wing form pattern that appears to be unique to long-distance migrators. In sedentary species like *P. dolus* the production of migratory forms is density dependent in both sexes (Denno et al., 1991). In migratory species like *P. marginata*, only females show a density dependent wing form response. Males remain long-winged even when raised at the lowest densities (Denno et al., 1985; Roderick, 1987). This pattern holds true for most of the mobile species that I have examined. *P. marginata*, (Denno et al., 1985), *N. lugens* (Fig. 9.5a; Kisimoto, 1956, 1965), *S. furcifera* (Kisimoto, 1965,) and *Sogatodes orizicola* (Muir) (King and Saunders, 1984) all show a lack of density dependence in male response. Consistent with the lack of density dependent wing form response in the males of *P. marginata* is the fact that wing form is under genetic control in males but is determined environmentally in females (Roderick, 1987). *Javesella pellucida* (Fab.) (Fig. 9.5b; Mochida, 1973), is somewhat exceptional in that it is quite mobile yet shows density dependent macroptery in males as well as females. *Muellerianella fairmairei* (Perris) (Fig. 9.5c; Drosopoulos, 1977) and *M. brevipennis* (Boheman) (Drosopoulos, 1977), two very immobile species like *P. dolus*, respond as predicted and show density dependent macroptery in both sexes.

Colonizing species often face very low densities after arrival in new habitats. There are numerous statements to this effect in the planthopper literature, as well as to the difficulty that planthoppers have finding mates under these conditions (e.g., Kuno, 1979). I suggest that macroptery in males is an adaptation that facilitates mate finding in migratory species that frequently face low-density situations. Observations on mate location in *N. lugens* are consistent with this argument. The sexes of planthoppers communicate with each other by substrate-transmitted high-frequency

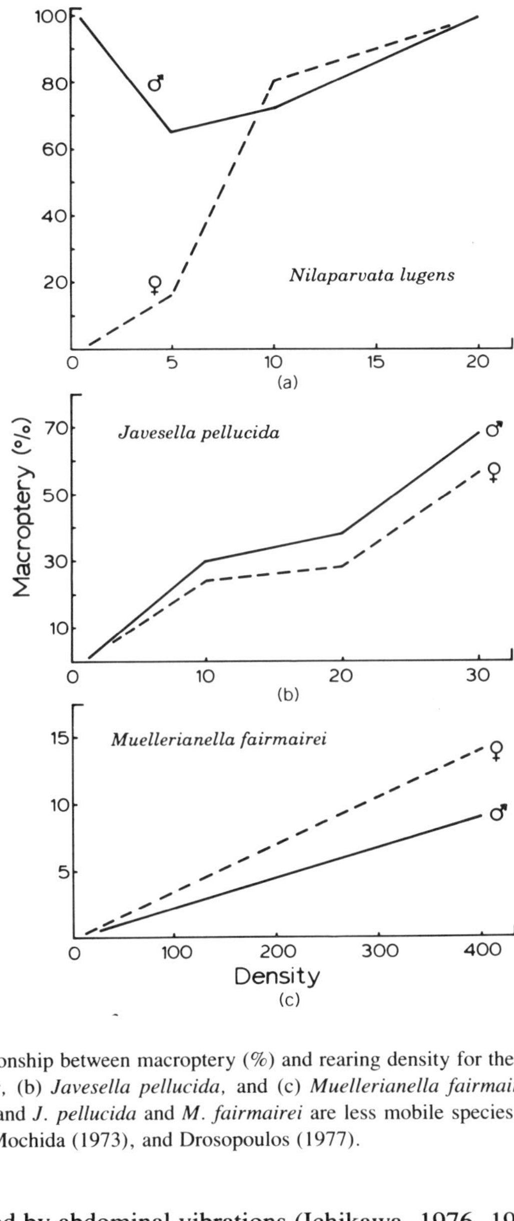

Figure 9.5. Relationship between macroptery (%) and rearing density for the males and females of (a) *Nilaparvata lugens*, (b) *Javesella pellucida*, and (c) *Muellerianella fairmairei*. *N. lugens* is a long-distance migrator, and *J. pellucida* and *M. fairmairei* are less mobile species. Data are replotted from Kisimoto (1965), Mochida (1973), and Drosopoulos (1977).

sounds produced by abdominal vibrations (Ichikawa, 1976, 1977; Claridge, 1985). "Calls" are species specific and function in mate location and reproductive isolation (Claridge et al., 1984; Claridge, 1985). Even one generation after colonization, the density of *N. lugens* adults in rice paddies can be as low as two adults per 10 hills of rice (Kuno, 1979). Brachypterous females remain stationary, "calling" males while males fly among the hills (Ichikawa, 1977). When both sexes appear on the same plant, acoustic exchanges ensue, and males move actively about until contact and mating occur (Claridge, 1985). Thus, wings allow males to locate females under very low-density conditions. Sedentary species that exploit more persistent host plants may not face such low densities, and, as a consequence, selective pressures associated with remaining in the habitat and exploiting current resources may select against wings in both sexes except at very high densities.

I would like to continue now with point six and ask whether invading planthoppers are larger than their relatives. This appears to be true, although the sample size is small (Fig. 9.6). In a survey of three genera containing migratory and sedentary species, migrators were the largest species in *Nilaparvata* and *Sogatella*. Within the genus *Prokelisia*, although *P. marginata* was not the largest species, it was larger than *P. dolus*, its sedentary cohabitor. Apparently, selection has favored large body size in migratory planthoppers, perhaps for reasons concerning powerful flight and/or reduced desiccation.

Seventh, is there evidence that humans have been responsible for planthopper introductions? Trivially, the answer is, yes. The sugarcane planthopper, *Perkinsnella saccharivora* (Westwood), was introduced to Hawaii in cane cuttings from Australia (Swezey, 1936). The habit of inserting their eggs into plant tissue predisposes them to accidental introductions in transported plant material.

Last, we can query if invaders function over a wide range of biotic or physical conditions. Although we have no explicit information on this point for planthoppers, there is some circumstantial evidence suggesting that mobile species may, in fact, be

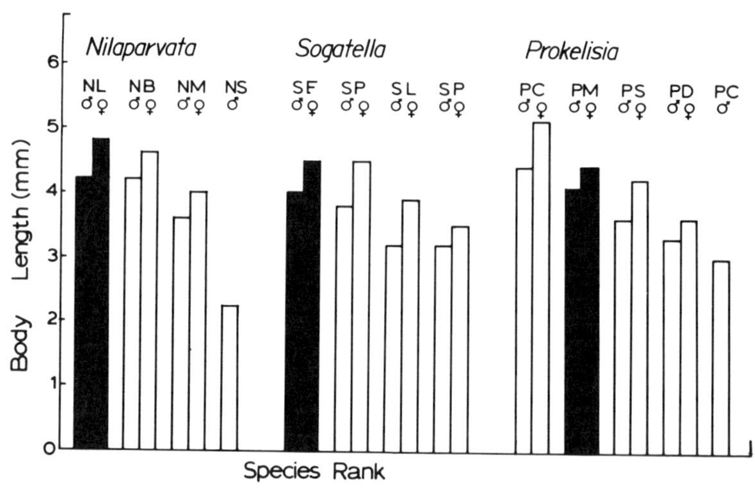

Figure 9.6. Ranked body lengths in millimeters (mm) of the macropterous males and females of various species of *Nilaparvata*, *Sogatella* and *Prokelisia*. Note that migrating species (black bars) tend to be the largest species in each genus. Open bars are less mobile species. NL = *Nilaparvata lugens*, NB = *N. bakeri*, NM = *N. muiri*, NS = *N. seminula*, SF = *Sogatella furcifera*, SP = *S. panicicola*, SL = *S. longifurcifera*. SP = *S. pusana*, PC = *Prokelisia crocea*, PM = *P. marginata*, PS = *P. salina*, PD = *P. dolus*, PC = *P. carolae*. Body lengths taken from Okada (1977), Mochida and Okada (1979), and Wilson (1982).

very intolerant of fluctuating levels of amino nitrogen in their host plants. Prestidge and McNeill (1982) studied a variety of grassland-inhabiting planthoppers (Delphacidae) and leafhoppers (Cicadellidae). They found that the leafhopper complex showed a variety of responses to changes in amino nitrogen, with a tendency for most leafhoppers to be polyphagous in terms of host species but specialists in terms of nitrogen requirements. This, in turn, meant that many planthoppers were highly mobile, moving from host to host within and between seasons, tracking spatial changes in the most suitable plants. A few much less mobile species, including the brachypterous planthopper, *Dicranotropis hamata* (Boheman), were host plant specialists, but nitrogen generalists. As a result, they were much less dependent on available nitrogen levels for reproduction.

This situation, at least superficially, appears to be comparable with the behavior and dynamics of the *Prokelisia* planthoppers on their *Spartina* host. There is much evidence suggesting that *P. marginata* may be a "nitrogen specialist": (1) adults feed on the most nutritious plant parts (Denno et al., 1980); (2) in laboratory choice tests, females preferentially feed and oviposit on plants containing the most soluble nitrogen (Denno, 1985); (3) development on more nutritious plants results in larger adults that are more fecund (Denno and McCloud, 1985; Denno et al., 1986); (4) in the field, macropterous adults accumulate in fertilized stands of *Spartina* (Denno, 1983); and (5) interhabitat migrations are associated with tracking spatial changes in host plant nitrogen (Denno, 1983; Denno et al., 1985, 1987). Much of this kind of data is not available for the predominantly brachypterous *P. dolus*, but it is known that this species does not disperse to low-marsh habitat, where much more nutritious host plants grow (Denno et al., 1987, unpubl. data).

If this scenario were true, it could mean that colonizing planthoppers are intolerant of variation in host plant nitrogen. In fact, host plant nitrogen is implicated in the evolution and maintenance of migration in *P. marginata* (Denno, 1983, 1985; Denno et al., 1985). It may be that such "nitrogen specialists" are, in fact, preadapted for long-distance dispersal and colonization insomuch as mobility and host location are integral features of their life-styles.

The extent to which invading planthoppers are tolerant of a wide range of physical conditions is not known. Data are available for *N. lugens* on the hatchability of eggs, nymphal survival, embryonic development (Suenaga, 1963), and the length of the preoviposition period (Mochida, 1974) at different temperatures. Again, there is no comparable data for closely related sedentary species. Planthoppers in the genus *Prokelisia* are ideally suited for such a test.

CONCLUSIONS

Predictions from Ehrlich's original list that seem to be upheld by immigrant (migratory) planthopper species are as follows: (1) immigrants tend to be abundant in their original range; (2) invasive species have shorter preoviposition periods, compared with more sedentary species; (3) immigrant species tend to be larger than their more sessile relatives; and (4) association with humans does seem to increase the chances for introduction to foreign lands, as was the case for the sugarcane planthopper. However, contrary to Ehrlich's list of predictions, (1) immigrant planthoppers are *not* all polyphagous (some are monophagous); (2) fecundity is variable and *not* consistently higher for immigrant species; and (3) immigrant species do *not* migrate as fertilized females. They colonize as virgins and must be mated after arrival in the new habitat. However, the males of all the migratory species examined lacked a density dependent wing form response, because, even under low-density conditions, long-winged forms result. This response probably facilitates mate location for colonizing species that often face low-density conditions in the new habitat.

Predictions that remain untested due to a paucity of data are those involving genetics. Namely, immigrants should show a high degree of genetic variability, and they should be able to function under a wide range of physical or biological conditions. However, if nitrogen specialization turns out to be generally associated with mobility, then intolerance of variable conditions may predispose certain planthoppers for invasion. Although the genetics of plant resistance in planthoppers has received much attention (e.g., Pathak and Khush, 1979), workers are just now beginning to analyze the genetics of life-history traits that directly influence the invasion process (e.g., Roderick, 1987). Of particular interest should be an investigation of genetic correlations between life-history traits for fugitive and sedentary species. Such studies of genetic substructure will elucidate the extent to which selection for dispersal characters may influence other life-history traits, such as fecundity, age to first reproduction, or development time. For example, a negative correlation between dispersal and fecundity in *S. furcifera* may explain why it is so effective at locating new habitats but relatively poor at exploiting them, as evidenced by reduced fecundity and slow population growth.

We may ask if these findings on planthoppers might be useful in predicting the invasion potential of other insect groups. The answer is, probably not very useful, at least not in any specific way. The suite of traits that has evolved in invading planthoppers is likely shaped in part by phylogeny. Until we have a better idea of the limits that phylogeny sets on life-history evolution, we will be unable to use rigorously the traits of one taxon as measures of optimality for another. Furthermore, similar environmental problems may be solved with equivalently adaptive but different combinations of traits, making prediction even more difficult (Denno and Dingle, 1981; Tallamy and Denno, 1981).

Finally and probably foremost is the fact that polymorphic insects, such as planthoppers, are thorns in the side of existing life-history theory. This is so because, rather than a species or a population being characterized by a suite of traits, each morph must be treated as a different coordinated life-history syndrome. Life-history theory needs to deal better with the challenge of polymorphisms, because there is growing evidence that they might be the rule rather than the exception.

Acknowledgments

This research was supported by National Science Foundation Grants BSR–8206603 and BSR-8614561. This is Scientific Article Number A-4830, Contribution Number 7856 of the Maryland Agricultural Experiment Station, Department of Entomology.

REFERENCES

Antolin, M. F., and D. R. Strong. 1987. Long-distance dispersal by a parasitoid (*Anagrus delicatus*, Mymaridae) and its host. *Oecologia* 73:288–292.

Baker, H. G., and G. L. Stebbins (eds.). 1965. *The Genetics of Colonizing Species*. New York: Academic Press.

Brussard, P. F. 1984. Geographic patterns and environmental gradients: The "central-marginal" model in *Drosophila* revisited. *Annu. Rev. Ecol. Syst.* 15:25–64.

Claridge, M. F. 1985. Acoustic behavior of leafhoppers and planthoppers: Species problems and speciation. Pages 103–125 in Nault, L. R. and J. G. Rodriguez (eds.), *The Leafhoppers and Planthoppers*. New York: John Wiley & Sons.

Claridge, M. F., J. Den Hollander, and J. C. Morgan. 1984. Specificity of acoustic signals and mate choice in the brown planthopper *Nilaparvata lugens* (Stal). *Entomol. Exp. Appl.* 35:221–226.

Cook, A. G., and T. J. Perfect. 1985. Seasonal abundance of macropterous *Nilaparvata lugens* and *Sogatella furcifera* based on presumptive macroptery in fifth-instar nymphs. *Ecol. Entomol.* 10:249–258.

Denno, R. F. 1976. Ecological significance of wing-polymorphism in Fulgoroidea which inhabit salt marshes. *Ecol. Entomol.* 1:257–266.

———. 1977. Comparison of the assemblages of sap-feeding insects (Homoptera-Hemiptera) inhabiting two structurally different salt marsh grasses in the genus *Spartina*. *Environ. Entomol.* 6:359–372.

———. 1978. The optimum population strategy for planthoppers (Homoptera-Delphacidae) in stable marsh habitats. *Can. Entomol.* 110:135–142.

———. 1980. Ecotope differentiation in a guild of sap-feeding insects on the salt marsh grasses, *Spartina patens*. *Ecology* 61:702–714.

———. 1983. Tracking variable host plants in space and time. Pages 291–341, in Denno, R. F., and M. S. McClure (eds.) *Variable Plants and Herbivores in Natural and Managed Systems*. New York: Academic Press.

———. 1985. Fitness, population dynamics, and migration in planthoppers: The role of host plants. Pages 623–640 in Rankin, M. A., (ed.), *Migration: Mechanisms and Adaptive Significance. Contrib. Mar. Sci.* 27, Supplement.

Denno, R. F., and H. Dingle. 1981. Considerations for the development of a more general life history theory. Page 1–6 in Denno, R. F., and H. Dingle (eds.), *Insect Life History Patterns: Habitat and Geographic Variation*. New York: Springer-Verlag.

Denno, R. F., and E. E. Grissell. 1979. The adaptiveness of wing-dimorphism in the salt marsh-inhabiting planthopper, *Prokelisia marginata* (Homoptera: Delphacidae). *Ecology* 60:221–236.

Denno, R. F., and E. S. McCloud. 1985. Predicting fecundity from body size in the planthopper, *Prokelisia marginata* (Homoptera: Delphacidae). *Environ. Entomol.* 14:846–849.

Denno, R. F., M. J. Raupp, D. W. Tallamy, and C. F. Reichelderfer. 1980. Migration in heterogeneous environments: Differences in habitat selection between the wing-forms of the dimorphic planthopper, *Prokelisia marginata* (Homoptera: Delphacidae). *Ecology* 61:859–867.

Denno, R. F., M. J. Raupp, and D. W. Tallamy. 1981. Organization of a guild of sap-feeding insects: Equilibrium vs. nonequilibrium coexistence. Pages 151–181 in Denno, R. F., and

H. Dingle (eds.), *Insect Life History Patterns: Habitat and Geographic Variation*. New York: Springer-Verlag.

Denno, R. F., L. W. Douglass, and D. Jacobs. 1985. Crowding and host plant nutrition: Environmental determinants of wing-form in *Prokelisia marginata*. *Ecology* 66:1588–1596.

Denno, R. F., L. W. Douglass, and D. Jacobs. 1986. Effects of crowding and host plant nutrition on a wing-dimorphic planthopper. *Ecology* 67:116–123.

Denno, R. F., M. E. Schauff, S. W. Wilson, and K. L. Olmstead. 1987. Practical diagnosis and natural history of two sibling salt marsh-inhabiting planthoppers in the genus *Prokelisia* (Homoptera: Delphacidae). *Proc. Entomol. Soc. Wash.* 89:687–700.

Denno, R. F., K. L. Olmstead, and E. S. McCloud. 1989. Reproductive cost of flight capability: A comparison of life history traits in wing dimorphic planthoppers. *Ecol. Entomol.* 14:31–44.

Denno, R. F., G. K. Roderick, K. L. Olmstead and H. G. Döbel. 1991. Density-related migration in planthoppers (Homoptera: Delphacidae): The role of habitat persistence. *Amer. Natur.* 138:1513–1541.

Dingle, H. 1978. Migration and diapause in tropical, temperate, and island milkweed bugs. Pages 254–276 in Dingle, H., (ed.), *The Evolution of Insect Migration and Diapause*. New York: Springer-Verlag.

———. 1985. Migration. Pages 375–415 in Kerkut, G. A., and L. I. Gilbert (eds.), *Comprehensive Insect Physiology, Biochemistry and Pharmacology, Vol. 9, Behavior.* New York: Pergamon Press.

Dixon, A. F. G. 1972. Fecundity of brachypterous and macropterous alatae in *Drepanosipum dixoni* (Callaphididae, Aphididae). *Entomol. Exp. Appl.* 15:335–340.

Drosopoulos, S. 1977. Biosystematic studies on the *Muellerianella complex* (Delphacidae, Homoptera, Auchenorrhyncha). *Meded. Landbouwhogesch.* Wageningen 284:77–114.

Dyck, V. A., and B. Thomas. 1979. The brown planthopper problem. Pages 3–17 in *Brown Planthopper: Threat to Asian Rice Production*. Los Banos, Philippines: IRRI.

Dyck, V. A., B. C. Misra, S. Alam, C. N. Chen, C. Y. Hsieh, and R. S. Rejesus. 1979. Ecology of the brown planthopper in the tropics. Pages 61–98 in *Brown Planthopper: Threat to Asian Rice Production*. Los Banos, Philippines: IRRI.

Ehrlich, P. R. 1986. Which animal will invade? Pages 79–95 in Mooney, H. A., and J. A. Drake (eds), *Ecology of Bological Invasions of North America and Hawaii*. New York: Springer-Verlag.

Fisk, J., E. A. Bernays, R. F. Chapman, and S. Woodhead. 1981. Report of studies on the feeding biology of *Peregrinus maidis*. *Cent. Overseas Pest Res./Int. Crops Res. Inst.* for Semi-Arid Tropics, London.

Hirao, J. 1979. Forecasting brown planthopper outbreaks in Japan. Pages 101–112 in *Brown Planthopper: Threat to Asian Rice Production*. Los Banos, Philippines: II-RRI.

Ichikawa, T. 1976. Mutual communication by substrate vibrations in the mating behavior of planthoppers (Homoptera: Delphacidae). *Appl. Entomol. Zool.* 11:8–21.

———. 1977. Sexual communications of planthoppers: The rice brown planthopper. Pages 84–94 *In Proc. Int. Sem. Vol. 1.* Taiwan, Republic of China: Food and Fertilizer Technology Center for the Asian and Pacific Region.

Kenmore, P. E. 1980. Ecology and outbreaks of a tropical insect pest of the green revolution, the rice brown planthopper, *Nilaparvata lugens* (Stal). Ph.D. Dissertation. Berkeley, Calif.: Univ. of California.

Kenmore, P. E., F. O. Carino, C. A. Perez, V. A. Dyck, and A. P. Gutierrez. 1984. Population regulation of the rice brown planthopper *Nilaparvata lugens* Stal) within rice fields in the Philippines. *J. Plant Prot. Trop.* 1:19–37.

King, A. B. S., and J. L. Saunders. 1984. The invertebrate pests of annual food crops in Central America. London: Overseas Development Administration.

Kisimoto, R. 1956. Factors determining the wing-form of adult, with special reference to the effect of crowding during the larval period of the brown planthopper, *Nilaparvata lugens* Stal. Studies of the polymorphism in the planthoppers (Homoptera: Araeopidae). I. *Oyo-Kontyu* 12:105–111.

———. 1965. Studies on the polymorphism and its role playing in the population growth of the brown planthopper, *Nilaparvata lugens* Stal. *Bull.* Shikoku *Agric. Exp. Stn.* 13:1–106.

———. 1976. Synoptic weather conditions inducing long-distance immigration of planthoppers, *Sogatella furcifera* Horvath and *Nilaparvata lugens* Stal. *Ecol. Entomol.* 1:95–109.

———. 1977. Bionomics, forecasting of outbreaks and injury caused by the rice brown planthopper. Pages 27–41 in *Proc. Int. Sem., Vol 1.* Taiwan, Republic of China: Food and Fertilizer Technology Center for the Asian and Pacific Region.

———. 1979. Brown planthopper migration. Pages 113–124 in *Brown Planthopper: Threat to Asian Rice Production* Los Banos, Philippines: IRRI.

Kuno, E. 1979. Ecology of the brown planthopper in temperate regions. Pages 45–60 in *Brown Planthopper: Threat to Asian Rice Production.* Los Banos, Philippines: IRRI.

Lewontin, R. C. 1965. Selection for colonizing ability. Pages 77–94 in Baker, H. G., and G. L. Stebbins (eds.), *The Genetics of Colonizing Species.* New York: Academic Press.

Metcalf, J. R. 1969. Studies on the biology of the sugarcane pest *Saccharosydne saccharivora* (Westw.) (Hom., Delphacidae). *Bull. Entomol. Res.* 59:393–408.

Mochida, O. 1964. On oviposition in the brown planthopper, *Nilaparvata lugens* (Stal) (Hom., Auchenorrhyneha). II. *Jap. J. Appl. Entomol Zool.* 8:14–142.

———. 1973. The characters of the two wing-forms of *Javesella pellucida* (F.) (Homoptera: Delphacidae), with special reference to reproduction. *Trans. R. Entomol. Soc. Lond.* 125:177–225.

———. 1974. Long-distance movement of *Sogatella furcifera* and *Nilaparvata lugens* (Homoptera: Delphacidae) across the East China Sea. *Rice Entomol. Newsletter* 1:18–22.

Mochida, O., and T. Okada. 1979. Taxonomy and biology of *Nilaparvata lugens* (Hom., Delphacidae). Pages 21–43 in *Brown Planthopper: Threat to Asian Production.* Los Banos, Philippines: IRRI.

O'Brien, L. B., and S. W. Wilson. 1985. Planthopper systematics and external morphology. Pages 61–102 in Nault, L. R., and J. G. Rodriguez (eds.), *The Leafhoppers and Planthoppers.* New York: John Wiley & Sons.

Okada, T. 1977. Taxonomic characters for the identification of the rice brown planthopper (*Nilaparvata lugens*) and related species in the Asian and Pacific Region. Pages 1–26 in *Proc. Int. Sem. Vol 1.* Taiwan, Republic of China: Food and Fertilizer Technology Center for the Asian and Pacific Region.

Parsons, P. A. 1983. The Evolutionary Biology of Colonizing Species. Cambridge, Mass.: Cambridge Univ. Press.

Pathak, M. D., and G. S. Khush. 1979. Studies of varietal resistance in rice to the brown planthopper at the International Rice Research Institute. Pages 285–302 in *Brown Planthopper: Threat to Asian Rice Production.* Los Banos, Philippines: IRRI.

Prestidge, R. A., and S. McNeill. 1982. The role of nitrogen in the ecology of grassland Auchenorryncha. Pages 257–281, in Lee, J. A., I. H. Rorison, and S. McNeill (eds.), *Nitrogen as an Ecological Factor. Symp. Br. Ecol. Soc.* 22. Oxford, England: Blackwell Sci. Pubs.

Purcell, A. H. 1985. The ecology of bacterial and mycoplasma plant diseases spread by leafhoppers and planthoppers. Pages 351–380 in Nault, L. R., and J G. Rodriguez (eds.), *The Leafhoppers and Planthoppers.* New York: John Wiley & Sons.

Raatikainen, M., and A. Vasarainen. 1976. Composition, zonation and origin of the leafhopper fauna of oatfields in Finland. *Ann. Zool. Fenn.* 13:1–24.

Roderick, G. K. 1987. Ecology and evolution of dispersal in California populations of a salt marsh insect, *Prokelisia marginata.* Ph.D. Dissertation. Berkeley, Calif.: Univ. of California.

Roff, D. A. 1984. The cost of being able to fly: A study of wing polymorphism in two species of crickets. *Oecologia* 63:30–37.

———. 1986. The evolution of wing dimorphism in insects. *Evolution* 40:1009–1020.

Sogawa, K. 1982. The rice brown planthopper: Feeding physiology and host plant interactions. *Annu. Rev. Entomol.* 27:49–73.

Sogawa, K., and C. H. Cheng. 1979. Economic thresholds, nature of damage, and losses caused by the brown planthopper. Pages 125–142 in *Brown Planthopper: Threat to Asian Rice Production.* Los Banos, Philippines: IRRI.

Southwood, T. R. E. 1977. Habitat, the templet for ecological strategies? *J. Anim. Ecol.* 46:337–465.

Sparks, A. N., R. D. Jackson, J. E. Carpenter, and R. A. Muller. 1986. Insects captured in light traps in the Gulf of Mexico. *Ann. Entomol. Soc. Am.* 79:132–139.

Strong, D. R., and P. D. Stiling. 1983. Wing dimorphism changed by experimental density manipulation in a planthopper (*Prokelisia marginata*, Delphacidae). *Ecology* 64:206–209.

Suenaga, H. 1963. Analytical studies on the ecology of two species of planthoppers, the white-backed planthopper (*Sogata furcifera* Horvath) and the brown planthopper (*Nilaparvata lugens* Stal), with special reference to their outbreaks. *Bull. Kyushu Agric. Exp. Stn.* 8:1–152.

Swezey, O. H. 1936. Biological control of the sugarcane leafhopper in Hawaii. *Bull. Exp. Stn. Hawaii. Sugar Planters Assoc.* (Ent. 21) 40:57–101.

Tallamy, D. W., and R. F. Denno. 1981. Alternative life history patterns in risky environments: An example from lacebugs. Pages 129–147 in Denno, R. F., and H. Dingle (eds.), *Insect Life History Patterns: Habitat and Geographic Variation.* New York: Springer-Verlag.

Utida, S. 1972. Density dependent polymorphism in the adult of *Callosobrachus maculatus* (Coleoptera, Bruchidae). *J. Stored Prod. Res.* 9:111–126.

Wilson, S. W. 1982. The planthopper genus *Prokelisia* in the United States (Homoptera: Fulgoroidea: Delphacidae). *J. Kans. Entomol Soc.* 55:532–546.

Wilson, S. W., and L. B. O'Brien. 1987. A survey of planthopper pests of economically important plants (Homoptera: Fulgoroidea). Pages 343–360 in Wilson, M. R., and L. R.

Nault (eds.), Proc. 2nd Int. Workshop on Leafhoppers and Planthoppers of Economic Importance. London: CIE.

Zera, A. J. 1984. Differences in survivorship, development rate and fertility between the long-winged and wingless morphs of the waterstrider, *Limnoporus coniculatus*. *Evolution* 38:1023–1032.

10

POPULATION DYNAMICS OF INVADING PESTS: FACTORS GOVERNING SUCCESS

M. W. Brown

U.S. Department of Agriculture, Agricultural Research Service, Appalachian Fruit Research Station

INTRODUCTION

The life of a newly introduced species is not always an easy one. Many obstacles must be overcome before the new species can gain a foothold in its new environment and become established. There are biotic obstacles in dealing with other organisms that may feed on, compete with, or be fed upon by the introduced species. There are also abiotic obstacles by which the introduced species encounters physical conditions that may be different from those of its native habitat. The obstacles to establishment can be briefly summarized as struggles to create a niche in a community in which the member species have evolved with each other and the environment. However, if the introduced species becomes established in its new habitat, it may have a relatively easy existence, released from the biotic restraints of its native habitat in a new environment with relatively underutilized resources. At this stage, it is the invaded community that now must adapt to the new species. The histories of the gypsy moth, *Lymantria dispar* (L.) (Forbush and Fernald, 1896; Doane and McManus, 1981) and Japanese beetle (*Popillia japonica* Newman) (Smith and Hadley, 1926) in North America demonstrate the level of success that an introduced species may be able to attain.

Several books published recently on the ecology of exotic pests (Wilson and Graham, 1983; Groves and Burdon, 1986; Mooney and Drake, 1986) discuss the biological and ecological attributes of invading species and the habitats they invade. However, little data exist on the population dynamics of the early stages of a newly introduced species's adaptation to a new habitat. Success or failure of the invasion is often determined in the earliest stages of establishment, but it is also difficult to obtain data on these stages. During early stages of invasion, population density is so small that, if detected, adequate sampling could eradicate the species. Some exotic species are detected early, but these are generally known pests. Natural population processes are disrupted by attempts at control, quarantine, or eradication of these pest species. More likely, the species would go undetected until it was established. Once this happens, its population dynamics is not likely to be different from that of native species (Turnbull, 1967). At this point, it has become an integral part of the community, and its population is affected by the same factors that affect other species in the community. As such, population processes are too variable to generalize.

Because of the paucity of data on population dynamics of invading species and the variability of possible scenarios of population dynamics, I can only present some of the principles of population dynamics as they apply to these species. Through the course of invasion, different population processes become important at different times. Not all of the processes discussed are important to all species as they invade, but they all may act to some degree. For one species, a certain process may be so restrictive that the invasion fails, whereas for another species that same process may be unimportant.

POPULATION PROCESSES

Population dynamics is a complex topic concerned with changes in numbers of individuals, factors that cause those changes, and processes that moderate the effect of these factors. All of these population processes and the factors that affect them can be discussed within the framework of the logistic equation

$$N_t = N_o e^{(b - d)t} - E_t + I_t + \Theta$$

This equation reads as follows: the number of individuals in the population at some point in time (N_t) is the result of the number of individuals in the population at some previous time (N_o) multiplied by an exponential growth rate that is comprised of births (b) minus deaths (d) over the time period (t) of interest. The number emigrating (E_t) must be subtracted from the population, and the number immigrating (I_t) must be added. Random effects (Θ) affect the population with unpredictable intensity and direction. From this equation, it can be seen that all population processes and external factors involved in population dynamics fall into five separate, but not mutually exclusive, categories: initial population, birth processes, death processes, movement, and random events.

Initial Population

The initial, or founding, population is very important for the future population dynamics of invading species. This initial population, which may be as small as a single fertilized or parthenogenetic individual or as large as a boat load of infested products, contains all the genetic variability available for adaptation to the new environment. More genetic variability in the initial population may reduce the time required for establishment (Scorza, 1983; Barrett and Richardson, 1986). Although the gene pool may be augmented by future waves of invasion and mutation, the genetic content of the initial population can determine the fate of the introduced population. As an extreme example, if all individuals in the initial population are male, the invasion is doomed to failure. A similar outcome results if the initial population lacks genetic resources to adapt to novel food sources, interspecific interactions, or abiotic conditions.

The gene pool from which invading individuals originated is also important in determining invasion success and future population dynamics (Remington, 1968). Remington's hypothesis is that invaders from the core of a species's native habitat may be highly heterozygous but may possess poor adaptability and may be burdened by a large genetic load. Invaders from a fringe area may have high levels of homozygosity and be well adapted to extreme conditions. Remington concluded that small numbers of individuals from core populations should have the lowest potential for successful invasion. Larger numbers of individuals from a core population should have the greatest potential for success but the lowest probability of subsequent speciation. If the initial invading population is comprised of small numbers of individuals, those from the fringe of the species's native range should have the highest probability of success, as well as the highest probability of subsequent speciation.

Birth Processes

Birth processes are inherent properties of the species, modified by environmental conditions. Fertility, the realized reproduction per female, is a more important property for the dynamics of an invading population than fecundity, the maximum potential reproduction per female. Fertility is a result of the species's fecundity interacting with environmental conditions, such as climate, nutritional quality of the food, and any other stress that may affect reproductive success. Environmental stresses can be especially important to invading populations adapting to a new environment. Energy otherwise available for reproduction must be spent locating suitable resources in an unfamiliar setting (unfamiliar in the sense that the species is exposed to conditions for which it may not have been selected). There may be a substantial, though possibly temporary, depression of fertility through increases in physiological stress because of different nutritive quality or digestibility of the food. Novel food and other resources could be superior, however, to native conditions, and fertility may actually be greater than in the native habitat. Competition with native species could also cause a decrease in fertility. Energy diverted from reproduction to competition for resources and less than optimal amounts of food as a result of competition may decrease reproductive ability.

The type of reproduction, sexual versus asexual, can be very important to the population dynamics of a species during the initial stages of invasion. If the species is parthenogenetic, only one individual needs to survive to reproductive maturity in order to perpetuate the population. Thus, parthenogenetic species have an advantage over sexually reproducing species that must maintain a minimum population density to ensure mating. The critical population size depends upon the type of mate-finding mechanisms employed by the species. With highly developed mate-finding abilities, a low minimum population density is required. However, if mate finding is through random encounters, a relatively high population density may be needed to ensure successful mating and maintain the invasion.

Death Processes

Mortality can affect any stage of the insect's life cycle and can be from either biotic or abiotic factors. Temperature, moisture, and light (intensity, photoperiod), alone or in combination, may cause mortality, even if other habitat conditions are favorable for successful establishment. Not only can extremes of these factors cause mortality, but also suboptimal conditions can kill some individuals in the population and stress others. Physiological stress inflicted by suboptimal abiotic conditions may increase susceptibility to biotic agents, particularly pathogens (Tanada, 1973), or reduce reproductive potential. Although abiotic mortality factors generally act in a density independent manner and cannot regulate populations in the classical sense (Nicholson, 1958), they may be important in determining overall long-term abundance by mediating biotic interactions.

Biotic mortality agents, which can be density dependent or density independent, are generally not as important as abiotic agents in the early stages of colonization and establishment (Goeden and Louda, 1976). An invading species is removed from predators, parasitoids, parasites, and disease organisms that limit population growth in its native habitat. Generalist natural enemies may attack the new species, but the host/prey population will be too small for significant mortality to be likely. As the invading species becomes more abundant, more natural enemy species will adapt to the new host/prey and become more significant in the species's population dynamics. Also, for species that become pests, natural enemies are likely to be introduced to add further to the impact of native natural enemies.

A potentially more important biotic mortality factor for an invading species is interspecific competition. This may result in direct mortality or act indirectly, reducing fertility by limiting resource availability. Depending on the similarity of resource requirements to that of native species, abundance of potential competitors, and availability of limiting shared resources, competition could range from prohibitive to nonexistent for an invading species. Once an invading species becomes successfully established, competition is not likely to be a serious mortality factor (Connell, 1980). A population's niche requirements tend to be plastic and to shift over time in response to other species's niche requirements and resource availability—a process termed *character displacement* (Brown and Wilson, 1956), *character divergence* (Pearson and Mury, 1979) or *niche shifts* (Diamond, 1978). Character displacement, though commonly thought to be a response to competitive interactions, can also take place in response to an absence of competition (Schluter, 1988). In cases such as this, the invading species may evolve a niche that would be broader than in its native habitat or may lead to rapid speciation through adaptive radiation.

Movement

Movement can have a negative (emigration) or positive (immigration) effect on population size (Hughes, 1979). The population of any invading species is dependent upon at least one initial immigration event from its native habitat to its new habitat. Further immigration may increase genetic variability. Movement is unlikely to be important in most invading species's population dynamics during the early stages of invasion. Low population density will preclude any density dependent dispersal. The species must be capable, however, of local dispersion to find suitable food and habitat, as well as to expand the area of infestation (Southwood, 1962). Excessive emigration could become a problem for sexually reproducing species if it reduces density below the minimum required for effective mate finding. Emigration is risky and only beneficial if the probability of finding a new suitable habitat is greater than the probability of survival in the original habitat (Southwood, 1977). Otherwise, emigration will be a net loss by depleting the local population without enhancing the regional population.

Random Events

Random events, or chance, often play a significant role in population dynamics (May, 1974a). This is especially true for introduced species in the early stages of invasion. Populations are small, local, and may be stressed as they adapt to the new habitat. Such populations are susceptible to extinction, or conversely, may benefit greatly from the action of random events. In fact, introduction of the species is often the result of one or more random events. Chance is neither a population parameter nor an attribute of the habitat, but it can act through any of the factors previously discussed. Often it is triggered by climatic or human-induced stochasticity.

Chance can completely reverse the trend of establishment. An otherwise successful invasion may be eradicated by random events like climatic extremes. Human-made disturbances, such as an insecticide application for another species or major construction that alters the habitat, are also chance events with respect to a species's population dynamics. Such events may cause direct mortality to the population or indirect mortality through major disruptions of the community. On the other hand, chance disruptions of the habitat may aid the establishment of the introduced species. Unusually conducive climatic conditions or low populations of potential natural enemies or competitors can provide conditions suitable for establishment, whereas normal conditions may prevent it. Random disruptions may actually increase suitability of the habitat for an invading species. Establishment of introduced species commonly occurs in disturbed habitats (Elton, 1958; Swincer, 1986).

VACANT NICHES

Many authors (e.g., Turnbull, 1967; Simberloff, 1986) discuss, explicitly or implicitly, the potential success of invading species in terms of vacant niches. By this they mean that in natural communities there are niches, functional roles (Elton, 1927), that are not occupied. By examining a community for vacant niches, one could predict which species could successfully invade that community. Although this concept may be useful, it is flawed in theory (Herbold and Moyle, 1986). If vacant niches do exist, transition should be easy for an invading species. The new species could easily enter the new community, because its niche would be present. It could invade the community with little or no disturbance to other resident species and immediately become an integral part of the community. Some introduced species illustrate such a scenario (Simberloff, 1981), but there is also ample evidence that many introduced species do alter the invaded community (Elton, 1958; Turnbull, 1967; Herbold and Moyle, 1986; Vitousek, 1986). Even such successful invaders of apparently vacant niches as the Japanese beetle and the gypsy moth required ca. 5 (Smith and Hadley, 1926) and 20 (Forbush and Fernald 1896) years, respectively, of adjusting to the new community before their presence was noticed (i.e., became adapted to the point where their population increased to noticeable levels). This time lag is contrary to expectations from the theory of vacant niches. Also, biological control efforts against imported weed species is often unsuccessful (Goeden and Louda, 1976), further evidence against the concept of vacant niches. An imported weed would be expected to bring with it many vacant niches that would be only incompletely filled by native herbivores. Thus, an exotic herbivore could be introduced to fill the same niche as it had in the native habitat. That such introductions are not often successful weakens the vacant-niche hypothesis.

The niche is neither a property of the community (Clapham, 1973) nor of the species (Patten and Auble, 1981); rather, it is a property of a specific population within a specific community (Whittaker et al., 1973). A single species can have different niches in different communities, depending upon the available resources and other species present. A population's niche only exists within the context of the community the population inhabits. One of the more accepted definitions of *niche* is the range of all environmental resources used by the species (Hutchinson, 1944). Taken out of its natural environment, a species can exist on a much broader range of resources than that which it utilized in its native habitat, because of the limitations imposed by the community (i.e., other species using the same resources) and physical resource limitations. Within a community, there are many species that overlap in their requirements for at least some resources. Therefore, each species in the community must interact with other species to acquire enough resources to survive. This may entail narrowing the resource base in some dimensions and shifting in others. The niche is a result of dynamic evolution between the species and the community (Hutchinson, 1978).

A newly arrived species must evolve its own niche among the matrix of existing niches. Resident species have evolved their own niches over long periods of time in association with the environment and other species that share the same resources. It is into this evolved system of niches that the new species must exist. The need to evolve a niche among previously evolved niches in a community is one explanation for the time lag between invasion and successful establishment, such as has been reported for the gypsy moth (Forbush and Fernald, 1896). Once a new species gains a foothold in the invaded community, it must adapt to the new environment before its population can expand significantly. This adaptation need not be as long as reported for the gypsy moth, however. It has been shown that species can adapt to novel food sources in as short as 10 generations (Gould, 1979; Claridge and den Hollander, 1980; Wasserman and Futuyma, 1981). Species such as the spotted alfalfa aphid, *Therioaphis maculata* (Buckton) (Metcalf et al., 1962), and recently the Russian wheat aphid, *Diuraphis noxia* (Mordvilko) (Stoetzel, 1987), have made remarkably rapid adaptation and subsequent range expansion in the southwest United States.

The requirement of evolving a niche is another reason why disturbed habitats are more easily invaded than natural habitats. All niches are relatively new and evolving in a disturbed habitat, so the invading species may be close to the same stage of adaptation as are native species. In undisturbed habitats, native species may be close to an equilibrium in niche space with the other species in the community. In these undisturbed habitats, invading species must alter this equilibrium and generate a new equilibrium with itself as a member of the community.

STAGES OF INVASION

The invasion and subsequent establishment of an introduced species can be divided into three stages of population behavior: colonization, establishment, and naturalization. The duration, population dynamics, and role of the factors in each stage varies from species to species. However, some generalizations about each stage can be made.

With the help of a conceptual model, one possible form of a species invasion is described, and generalizations are made about the interaction of factors previously discussed. The sequence of colonization, establishment, and naturalization is not a one-time event but, rather, is repeated continually as the species expands its range. As population density increases at the invasion site, individuals disperse to surrounding areas. The population at the site of invasion proceeds to the naturalization stage while new sites are colonized in a widening area of infestation. This continually expanding range of infestation in roughly concentric zones, to the extent that environmental conditions allow, is apparent on time series maps of infestations for many introduced species (Elton, 1958; Swincer, 1986).

A Conceptual Model

A generalized model of the population dynamics of an invading species is given in Figure 10.1. The model was developed and simulated using the modeling software Stella (High Performance Systems, Lyme, New Hampshire) (see Costanza, 1986). In the flow chart of Figure 10.1, the rectangle symbolizes the state variable population, representing the invading species. The double-line arrows represent a flow of the state variable; "clouds" at the end or origin of arrows indicate that the flow is into, or from, regions or states outside of the realm of the model. Circles are converters, representing factors that influence the flow of the state variable. Connectors, the single-line arrows, indicate the direction of the interaction of two

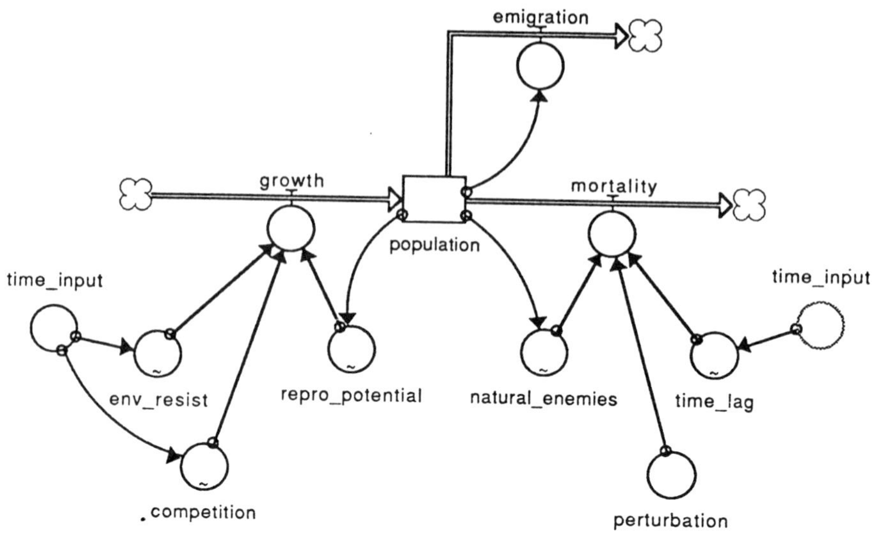

Figure 10.1. Flow chart of the conceptual model of population dynamics of an invading species using the simulation program Stella.

converters or a converter and a state variable. A wavy line inside a circle means that the factor was modeled graphically; otherwise, it was with an algebraic expression. There is a "ghost" variable (time-input, on the right side of Figure 10.1), indicating that it is the same as the time-input variable on the left side of the flowchart, drawn as a ghost so as to minimize clutter in the diagram.

The differential equation described by the flowchart (Figure 10.1) is composed of one additive component, growth, and two subtractive components, emigration and mortality. Figure 10.2a presents the population density over the duration of the simulation (100 generations), and Figure 10.2b growth and mortality. Population growth is a function of three factors: reproductive potential, competition, and environmental resistance (Figure 10.2c). Reproductive potential is a decreasing sigmoid function of density, modeling the effect of crowding on fertility of the population. Competition is modeled as a time dependent factor, initially low, rising to a high at 20 generations, and decreasing exponentially thereafter (Figure 10.2c). Realistically, competition is not time dependent, but the overall effect in this scenario is a reasonable, although by no means only, interaction while maintaining a relatively simple structure to the model. Competition is modeled to reduce reproductive potential rather than as a direct mortality factor, making the assumption that competition results in a reduction of available resources to below optimal levels but not below the threshold for development. Environmental resistance is a decreasing sigmoid function of time (Figure 10.2c) This factor is comprised of all of the elements previously discussed that reduce fertility while the invading species adapts to the new habitat: novel food resources, food location among different environmental cues, new interspecific interactions, different climatic conditions, and so forth.

Emigration is a density dependent loss to the population. Above a threshold of

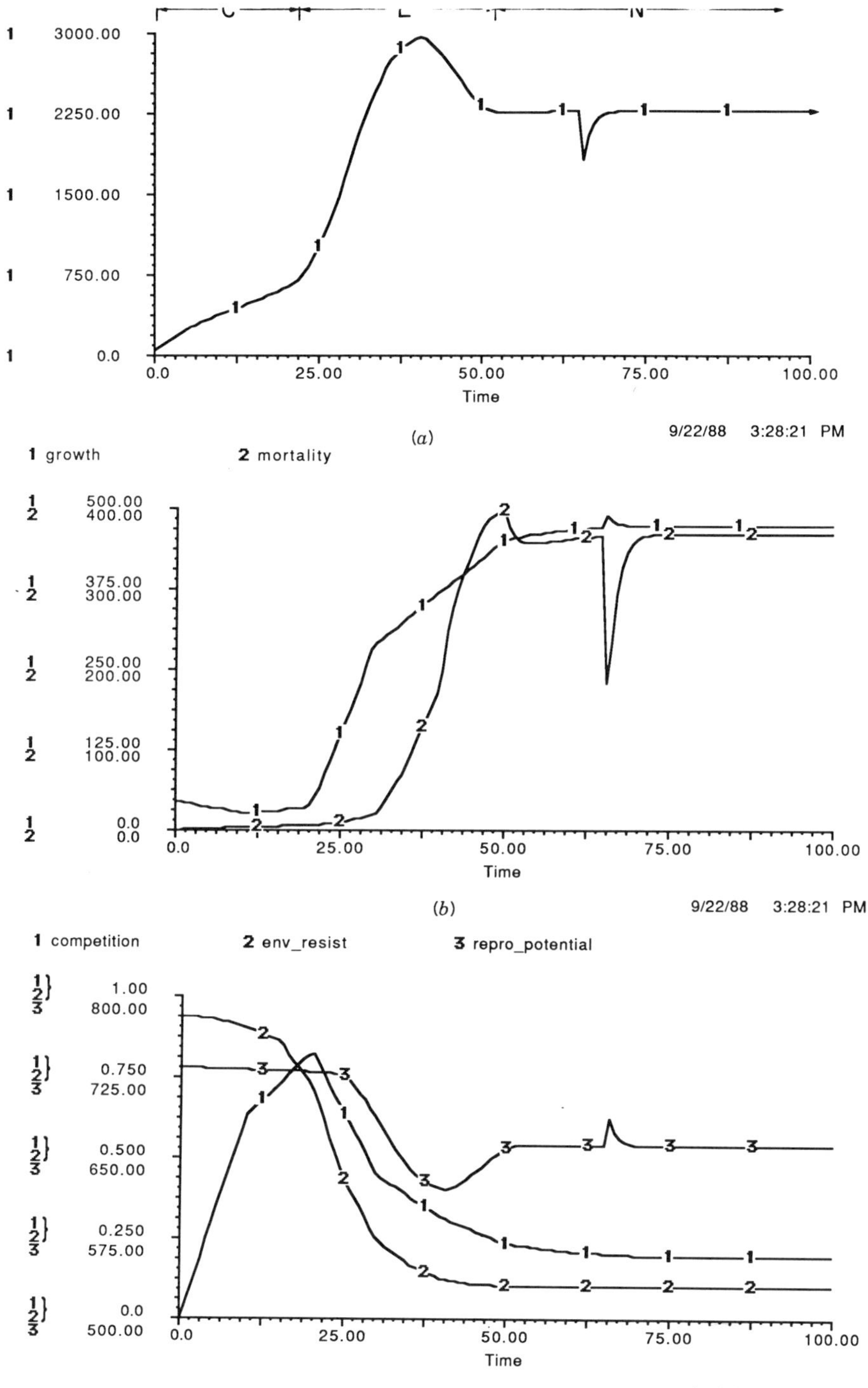

Figure 10.2. One hundred generation simulation output of the conceptual model: (a), population density of an invading species as modeled indicating the stages of infestation (C = colonization, E = establishment, N = naturalization); (b), overall population growth and mortality; (c), components of population growth (competition, environmental resistance, reproductive potential); and (d), components of mortality (natural enemies, time lag, emigration).

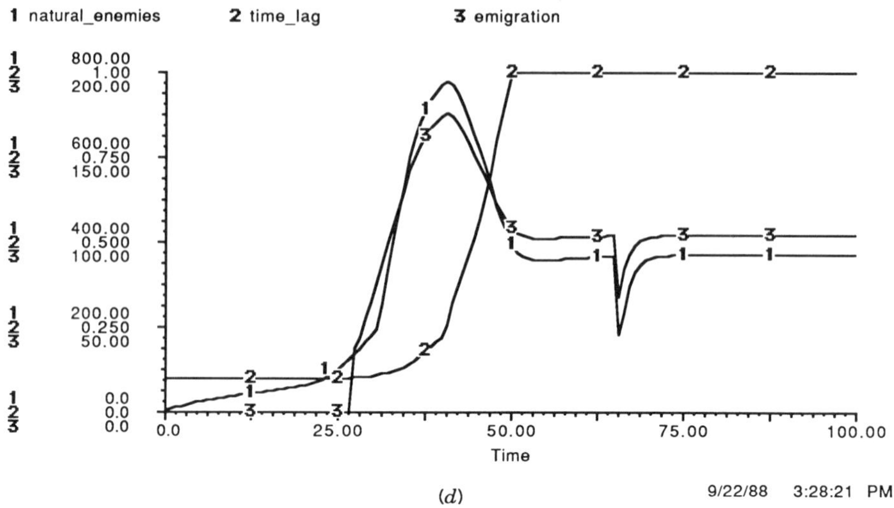

(d)

Figure 10.2. ***(Continued)***

1,250 individuals, there is a sigmoid increase in emigration to a high of 20% of the population. Mortality is a function of three factors: natural enemies, time lag, and perturbation. Mortality from natural enemies is an increasing sigmoid function of host density. Time lag is a time dependent, multiplicative increase in natural enemy mortality (Figure 10.2d). The time lag models a gradual adaptation of natural enemies to the introduced species an/or the introduction of exotic natural enemies between generations 35 and 50. A perturbation to the model is added at generation 65 in the form of a single increase in mortality in order to simulate the effects of a random event on the population dynamics of the invading species during the naturalization stage of invasion.

The model represents just one possible scenario of the dynamics of an invading species's population. The model is deterministic; therefore, the role of chance, while important to invading populations, is not included for the sake of simplicity and to show more clearly the effects of other factors in the model. The behavior of the model is dominated by density dependent factors. This is not meant to imply that density independent factors are unimportant. Density independent factors are often stochastic (May, 1974a), and, as described earlier, the model is deterministic. Density independent factors can have a major influence over all of the interactions incorporated into the model. They modify the action of density dependent factors and may govern overall population density (Nicholson, 1954; Henson, 1968) and even success or failure of the invasion.

The simulation of this model (Figure 10.2a) demonstrates one of the many possible population behaviors of an invading species. The numbers in the simulation are relative; the important aspects of the model are the behavior of the population and the role played by the various factors modeled into the system. The three stages of infestation are indicated in Figure 10.2a. Population behavior and the role of each factor in each stage are discussed next.

Colonization

The initial stage in the population dynamics of invading species is colonization (Figure 10.2a). This is the most critical stage in determining the fate of invasion. In this stage, population density is low, and, therefore, the probability of extinction as a result of chance is high. The species must adjust to the environmental conditions encountered and must establish a niche in spite of a small gene pool. Throughout the adaptation process, extinction may come as a result of any number of factors: insufficient population density to ensure mating, climatic conditions beyond the species's tolerance, natural enemies, competition, or one of many random factors.

The pattern of resource distribution and how it affects the species's ability to locate resources also may be a critical factor. Some or all of the required resources may be patchily distributed (Wiens, 1976; Southwood, 1977). The invading species must be able to locate resources within a foreign environment with a multitude of biotic, chemical, and environmental cues that are different from those in which its resource-locating abilities have evolved. The species may be able successfully to colonize a suitable patch of habitat, but, if it cannot locate other patches within the matrix of the changing environment, the invasion is doomed (Taylor and Taylor, 1977).

The duration of the colonization stage depends on the genetic variability of the founding population and the closeness of the invaded environment, biotic and abiotic, to its native habitat. Colonization can last from as short as a few generations, as in the case of the Japanese beetle (Smith and Hadley, 1926), to as long as 20 generations, as in the case of the gypsy moth (Forbush and Fernald, 1896), or more. Actual duration of colonization is difficult to determine, since date of first invasion often is unknown and the species is noticed only once it has successfully colonized and become more abundant. The colonization stage ends when the species becomes adapted to the invaded environment and has established a niche for itself. Only then can the population density begin to increase significantly.

For the hypothetical invading species in the model, the colonization stage is marked by a gradual increase in population density as the population begins to adapt to its new habitat (Figure 10.2a). Both population growth and mortality are low (Figure 10.2b) because it is adapting to new environmental conditions. The dominant factor affecting growth is environmental resistance (Figure 10.2c). Reproductive potential is consistently high (Figure 10.2c) because of low population density, but net growth is low because of environmental resistance. During colonization, competition increases as population density of the invading species increases. The new species is at a competitive disadvantage, because it is still in the process of evolving a niche for itself. Therefore, it may be in competition with more species than it would within its eventual niche. Competition begins to taper off as the species adapts and becomes a better competitor in the new habitat. Mortality from natural enemies is of minor importance during colonization, because host/prey density is low and native natural enemies are not adapted to the new species as a food source. Emigration is negligible in this scenario, due to low population density. There will be some expansion of the area infested as population density slowly grows but no significant emigration to areas discontinuous with the original infested area.

Many invading species never progress beyond the colonization stage and may remain at low density for long periods of time. In these cases, the initial invading population may have possessed enough genetic variation to survive but not enough to overcome the environmental and biotic resistance to exploit the new habitat fully. Such species may still overcome environmental resistance and achieve pest status. New immigrants could arrive with more suitable genetic variations, or conditions may change to be more amenable to the species. As long as these species remain extant, there is some potential for attaining pest status.

Establishment

In the establishment stage, population density rises from the low level of colonization to a higher level (Figure 10.2a). In many species, this rise in abundance may be very rapid, overshoot the eventual equilibrium density, and then crash. This outbreak is often the first time the species is detected, but it is often too late for easy eradication, because the species has adapted to its new environment. As the population increases, the area of infestation also increases as more individuals are produced than can survive in the local habitat (Taylor and Taylor, 1977). During establishment, there is little chance of extinction, unless population density increases so rapidly that it renders the habitat unsuitable. The establishment stage should generally last only a few generations at any one site before unrestrained growth is reversed by density dependent mortality.

In the scenario modeled in Figures 10.1 and 10.2 establishment is marked by a rapid population increase, followed by a rapid decline (Figure 10.2a). This type of population behavior during establishment has been documented for the alfalfa blotch leaf miner, *Agromyza frontella* (Rondani) (Harcourt et al., 1987). There could be a more gradual increase to a higher general equilibrium density or repeated population explosions and collapses in this stage of invasion. The primary reason for the population increase is a reduction of environmental resistance (Figure 10.2c) as the invading species becomes adapted and utilizes the resources more efficiently. Competition also decreases (Figure 10.2c) as the species evolves into a niche that minimizes competitive interactions with other species (Diamond, 1978). Reproductive potential also decreases (Figure 10.2c), due to inverse density dependence on population size. The overall trend, however, is population growth, as the reduction in environmental resistance and competition is greater than the reduction in reproductive potential. Lagging behind population growth is an increase in mortality (Figure 10.2b). The lag in mortality is due partially to classical delay in density dependent mortality (Nicholson and Bailey, 1935) and partially to the adaptation and/or introduction of natural enemies (Figure 10.2d). Emigration also increases with the increase in population density, being a mortality factor for the local population of the invading species (Figure 10.2d).

Naturalization

During naturalization, the introduced species and biotic components of the habitat begin adapting to each other in the same way in which native species continuously adapt to their environment. The introduced species has developed a niche and is an interacting member of the community, reacting to changes in abundance of other species and influencing them in turn. Other trophic levels, both above and below the introduced species, also interact with the new species. Hosts, predators, parasitoids, parasites, and disease organisms also adapt to the introduced species. In short, a new community structure and a new set of interspecific interactions arise with an additional species. Since the invaded species is becoming naturalized during this stage, its population dynamics become similar to those of native species (Turnbull, 1967).

Population behavior during naturalization may resemble that of native species after a perturbation: damped oscillations, stable oscillations, steady-state equilibrium, multiple equilibria, or chaos (May, 1974b). Although abiotic factors are still important in governing the population dynamics during this stage, biotic factors largely determine the pattern of population behavior. Food quality, with respect to the nutritional needs of the introduced species, influences fertility. The degree to which native natural enemies adapt to the species as a host and the intensity of competitive interactions with other species at the same trophic level also can have an important impact on population behavior. Chance events may have a dramatic effect on populations during naturalization but are only temporary disruptions, unlikely to cause area-wide extinction.

In the model simulation, population density and intensity of all inputs stabilize. Being a deterministic model driven largely by density dependent factors, the model population will remain in equilibrium until some chance event occurs to disrupt the system. Such a perturbation is put into the model at generation 65. After a brief disruption, the equilibrium is re-established at the same density as prior to the chance event (Figure 10.2).

In summary, the population dynamics of invading species is a poorly studied phenomenon. It is an important subject for regulatory officials, such as those in the Animal and Plant Health Inspection Service (APHIS) and similar state agencies, but it is very difficult to collect information on a population when density is low during the initial stages of invasion. In this chapter, I have tried to present my ideas on what factors are important to the population dynamics of invading species. The conceptual model presented demonstrates just one possible sequence of events. The model is useful in explaining interactions of the various factors and the population behavior of a species as it progresses through the stages of invasion and becomes adapted to the new habitat. I am sure that many variations on the theme presented have taken place, and exceptions to all of the generalizations I have made can be cited. My hope is that this chapter will stimulate discussion of, and research into, the important subject of population dynamics of the early stages of species invasions.

Acknowledgments

I which to thank Robert L. Pienkowski, Jeffrey J. Schmitt, Ralph Scorza, L. Claire Stuart, Charles L. Wilson, and an anonymous reviewer for valuable comments on an earlier version of this manuscript, and E. Richard Hoebeke and Thomas E. Wallenmaier for initiating my interest in this subject.

REFERENCES

Barrett, S. C. H., and B. J. Richardson. 1986. Genetic attributes of invading species. Pages 21–33 in R. H. Groves, and J. J. Burdon (eds.). *Ecology of Biological Invasions*. New York: Cambridge Univ. Press.

Brown, W. L., and E. O. Wilson. 1956. Character displacement. *Syst. Zool.* 5:49–64.

Clapham, W. B. 1973. *Natural Ecosystems*. New York: Macmillan.

Claridge, M. F., and J. den Hollander. 1980. The "biotypes" of the rice brown planthopper, *Nilaparvata lugens*. *Entomol. Exp. Appl.* 27:23–30.

Connell , J. H. 1980. Diversity and the coevolution of competitors, or the ghost of competition past. *Oikos* 35:131–138.

Costanza, R. 1986. Simulation modeling on the Macintosh using Stella. *ISEM* 8:75–88.

Diamond, J. M. 1978. Niche shifts and the rediscovery of interspecific competition. *Am. Sci.* 66:322–331.

Doane, C. C., and M. L. McManus (eds.). 1981. The Gypsy Moth: Research Toward Integrated Pest Management. U. S. For. Serv. Tech. Bull. 1584.

Elton. C. S. 1927. *Animal Ecology*. London: Sidgwick and Jackson.

———. 1958. *The Ecology of Invasions by Animals and Plants*. London: Methuen.

Forbush, E. H., and C. H. Fernald. 1896. *The Gypsy Moth: Porthetria dispar* (Linn.). Boston: Wright and Potter Printing Co.

Goeden, R. D., and S. M. Louda. 1976. Biotic interference with insects imported for weed control. *Annu. Rev. Entomol.* 21:325–342.

Gould, S. J. 1979. Rapid host range evolution in a population of a phytophagous mite *Tetranychus urticae* Koch. *Evolution* 33:791–802.

Groves, R. H., and J. J. Burdon (eds.). 1986. *Ecology Of Biological Invasions*. New York: Cambridge Univ. Press.

Harcourt, D. G., J. C. Guppy, J. Drolet, and J. N. McNeil. 1987. Population dynamics of alfalfa blotch leaf miner, *Agromyza frontella* (Diptera: Agromyzidae), in eastern Ontario: Analysis of numerical change during the colonization phase. *Environ. Entomol.* 16:145–153.

Henson, W. R. 1968. Some recent changes in the approach to studies of climatic effects on insect populations. Pages 37–46 in Southwood, T. R. E. (ed.), *Insect Abundance*. Symp. R. Entomol. Soc. London. No. 4. Oxford, England: Blackwell Scientific.

Herbold, B., and P. B. Moyle. 1986. Introduced species and vacant niches. Am. Nat. 128:751–760.

Hughes, R. D. 1979. Movement in population dynamics. Pages 14–32 in R. L., Rabb, and G. G. Kenedy (eds). *Movement of Highly Mobile Insects: Concepts and Methodology In Research*. North Carolina State Univ., Raleigh: University Graphics.

Hutchinson, G. E. 1944. Limnological studies in Connecticut. VII. A critical examination of the supposed relationship between phytoplankton periodicity and chemical changes in lake water. *Ecology* 25:3–26.

———. 1978. *An Introduction to Population Ecology*. New Haven: Yale Univ. Press.

May, R. M. 1974a. *Stability and Complexity in Model Ecosystems*. Monographs in Population Biology, No. 6. Princeton: Princeton Univ. Press.

———. 1974b. Biological populations with nonoverlapping generations: Stable points, stable cycles, and chaos. *Science* 186:645–647.

Metcalf, C. L., W. P. Flint, and R. L. Mctcalf. 1962. *Destructive and Useful Insects*. New York: McGraw-Hill.

Mooney, H. A., and J. A. Drake (eds.). 1986. *Ecology of Biological Invasions of North America and Hawaii*. New York: Springer-Verlag.

Nicholson, A. J. 1954. An outline of the dynamics of animal populations. *Aust. J. Zool.* 2:9–65.

———. 1958. Dynamics of insect populations. *Annu. Rev. Entomol.* 3:107–136.

Nicholson, A. J., and V. A. Bailey. 1935. The balance of animal populations. *Proc. Zool. Soc. Lond.* 3:551–598.

Patten, B. C., and G. T. Auble. 1981. System theory of the ecological niche. *Am. Nat.* 117:893–922.

Pearson, D. L., and E. J. Mury. 1979. Character divergence and convergence among tiger beetles (Coleoptera: Cicindelidae). *Ecology* 60:557–566.

Remington, C. L. 1968. The population genetics of insect introduction. *Annu. Rev. Entomol.* 13:415–426.

Schluter, D. 1988. Character displacement and the adaptive divergence of finches on islands and continents. *Am. Nat.* 131:799–824.

Scorza, R. 1983. Ecology and genetics of exotics. Pages 219–238 in Wilson C. L., and C. L. Graham (eds.), *Exotic Plant Pests and North American Agriculture*. New York: Academic Press.

Simberloff, D. 1981. Community effects of introduced species. Pages 53–81 in Nitecki, T. H. (ed.), *Biotic Crises in Ecological and Evolutionary Time*. New York: Academic Press.

———. 1986. Introduced insects: A biogeographic and systematic perspective. Pages 3–26 in Mooney, H. A., and J. A. Drake (eds.), *Ecology of Biological Invasions of North America and Hawaii*. New York: Springer-Verlag.

Smith, L. B., and C. H. Hadley. 1926. The Japanese Beetle. *U. S. Dept. Agric. Circ.* 363: 1–66.

Southwood, T. R. E. 1962. Migration of terrestrial arthropods in relation to habitat. Biol. Rev. 37:171–214.

———. 1977. Habitat, the templet for ecological strategies? *J. Anim. Ecol.* 46:337–365.

Stoetzel, M. B. 1987. Information on and identification of *Diuraphis noxia* (Homoptera: Aphididae) and other aphid species colonizing leaves of wheat and barley in the United States. *J. Econ. Entomol.* 80:696–704.

Swincer, D. E. 1986. Physical characteristics of sites in relation to invasions. Pages 67–76 in Groves, R. H., and J. J. Burdon (eds.), *Ecology of Biological Invasions*. New York: Cambridge Univ. Press.

Tanada, Y. 1973. Environmental factors external to the host. *Ann. N.Y. Acad. Sci.* 217:120–130.

Taylor, L. R., and R. A. J. Taylor. 1977. Aggregation, migration and population mechanics. *Nature* 265:415–421.

Turnbull, A. L. 1967. Population dynamics of exotic pests. *Bull. Entomol. Soc. Am.* 13:333–337.

Vitousek, P. M. 1986. Biological invasions and ecosystem properties: Can species make a difference? Pages 163–176 in Mooney, H. A., and J. A. Drake (eds.), *Ecology of Biological Invasions of North America and Hawaii*. New York: Springer-Verlag.

Wasserman, S. S., and D. J. Futuyma. 1981. Evolution of host plant utilization in laboratory populations of the southern cowpea weevil, *Callosobruchus maculatus* Fabricius (Coeoptera: Bruchidae). *Evolution* 35:605–617.

Whittaker, R. H., S. A. Levin, and R. B. Root. 1973. Niche, habitat, and ecotope. *Am. Nat.* 107:321–338.

Wiens, J. A. 1976. Population responses to patchy environments. *Annu. Rev. Ecol. Syst.* 7:81–120.

Wilson, C. L., and C. L. Graham (eds.). 1983. *Exotic Plant Pests and North American Agriculture*. New York: Academic Press.

11

SPREAD OF INVADING ORGANISMS: PATTERNS OF SPREAD

D. A. Andow

University of Minnesota

P. M. Kareiva

University of Washington

S. A. Levin

Cornell University

A. Okubo

State University of New York

INTRODUCTION

The invasion of new habitats by organisms is a fundamental ecological process. Invading species have had tremendous ecological and economic impact (Elton, 1958; Wilson and Graham, 1983; Mooney and Drake, 1986), and understanding the invasion process could greatly affect efforts to conserve species, regulate deliberate species introductions, control pests, and enforce quarantines.

Once an invading species establishes a population, it may spread and expand its geographic range. At times this seems like an inexorable march, as for the boll weevil across the southern United States or the Colorado potato beetle across the United States and Europe; but sometimes it occurs at a snail's pace, as for the Japanese beetle in the eastern United States. In this chapter, we examine the *spread* of invading organisms, that is, the process of range expansion. In particular, to develop the theoretical basis for generalities about spread, we evaluated several methods to describe patterns of range expansion in relation to simple mathematical models of spread. We focus our analysis on muskrat (*Ondatra zibethica*), cereal leaf beetle (*Oulema melanopus*), rice water weevil (*Lissorhoptrus oryzophilus*), and small cabbage white butterfly (*Pieris rapae*). We chose muskrat, because considerable work has focused on this species (Skellam, 1951; Okubo, 1980), and we chose the other species, because they are economically important and because excellent records of geographic expansion exist. We compare the predictions of the simple mathematical models to the patterns of range expansion of the invaders in order to expose the limitations of the simple analysis and to learn more about the ecology of those four major invasions. Complementary investigations for these four species are reported in Andow et al. (1990). Similar analyses have been carried out by Lubina and Levin (1988) for the California sea otter.

THEORY

Although Brownlee (1911) had developed equations to describe the geographic spread of epidemics, the first important mathematical models of spread originated with Fisher (1937), Kolmogorov et al. (1937), and Kendall (1948). In the ecological literature, the major early contributions were made by Skellam (1951) and Kierstead and Slobodkin (1953). Most of these investigators considered deterministic approximations for the dynamics of the expected population size or density of a stochastic spread model, and most of these deterministic approximations were based on particular cases of the two-dimensional linear diffusion and growth equation

$$\frac{\partial N}{\partial t} = f(N) + \frac{\partial^2}{\partial x^2}[D_x(x,y)N] + \frac{\partial^2}{\partial y^2}[D_y(x,y)N] \tag{11.1}$$

in which N denotes the expected density of the species in question at position x,y at time t, D represents a diffusion or dispersal coefficient dependent on x,y, and f is a function describing net population change due to birth and death. These diffusion models have the advantages of being testable and easily quantifiable (Kareiva, 1982; Banks et al., 1985), and in many cases provide remarkably good descriptions of animal movement (Kareiva, 1983). Mollison (1977), however, has questioned the relevance of equation (11.1) to the issue of ecological spread, because it does not account explicitly for a dispersing life stage and assumes that dispersal distances are normally distributed.

One way of using equation (11.1) to investigate spread is to consider a colonizing population of initial size N_o, with all individuals at position $(x,y) = (0,0)$, and then to determine how rapidly the growing population will spread from (0,0). A simple version of equation (11.1), to which this analysis can be applied, assumes a homogeneous, unstructured population growing exponentially and spreading in a homogeneous two-dimensional environment:

$$\frac{\partial N}{\partial t} = D\left[\frac{\partial^2 N}{\partial x^2} + \frac{\partial^2 N}{\partial y^2}\right] + \alpha N \tag{11.2}$$

where α is the intrinsic rate of population growth.

The solution to equation (11.2), given the initial conditions mentioned above, is

$$N(r,t) = \frac{N_o}{4\pi Dt} \exp [\alpha t - r^2/4Dt] \quad (11.3)$$

where $r^2 = x^2 + y^2$. This solution does not form a true front for the spreading populations. Indeed, equation (11.3) implies that, at arbitrarily small positive t, the population N will be positive at arbitrarily large distances away from the point of introduction; in other words, the population will have an infinite rate of spread. An identical problem arises in the diffusion theory of heat. In both cases, the conclusion is the same: the diffusion approximation is not good at distances far from the release point. In practice, the problem of infinite rate of spread is treated by recognizing that there is a threshold level, N_F, below which the population cannot be detected. Thus, we can calculate the rate of spread of the "front" where the theoretical distribution in equation (11.3) intersects the threshold level $N = N_F$ (Skellam, 1951). We calculate the average rate of spread, V_F, of this advancing front, located at R_F at time t, from equation (11.3) (Okubo, 1980):

$$V_F = R_F/t = \{ 4\alpha D + (4D/t) \ln [N_o/t(4\pi D)^{1/2} N_F] \}^{1/2} \quad (11.4)$$

provided the bracketed term is positive; $V_F = 0$ otherwise. For long time, V_F asymptotically approaches

$$V_F = (4 \alpha D)^{1/2} \quad (11.5)$$

provided that α is positive. Remarkably, equation (11.5) is independent of the actual value of the threshold, N_F.

Equation (11.2) can be modified to include many nonlinear forms of $f(N)$, and the prediction in equation (11.5) remains robust. Kolmogorov et al. (1937) and Fisher (1937) were among the first of many to examine the one-dimensional equation

$$\frac{\partial N}{\partial t} = D \frac{\partial^2 N}{\partial x^2} + \alpha N (1 - N) \quad (11.6)$$

and various more general cases:

$$\frac{\partial N}{\partial t} = D \frac{\partial^2 N}{\partial x^2} + F(N) \quad (11.7)$$

They showed that the general case admits asymptotic fronts of range expansion with asymptotic speed V_F given in equation (11.5), with $\alpha = F'(0)$. Bramson (1983) has specified the necessary and sufficient conditions on the initial distribution of organisms, $N(x,0)$, for convergence of the spread rate to the asymptotic speed V_F and has shown that convergence occurs on the order of log t.

Mollison (1977) began with a different approach and suggested that, if juveniles are the only dispersers, the biologically more accurate formulation of spread stems from an approximation of the stochastic one-dimensional linear-contact birth process

$$\frac{\partial N}{\partial t} = \alpha \bar{N}, \tag{11.8}$$

where

$$\bar{N} = \int_u N(x - u)\, dV(u) \tag{11.9}$$

is a convolution integral, representing the addition of offspring to location x from reproducing organisms at all other locations, $N(x - u)$. The quantity $V(u)$ is the contact distribution function and is the probability distribution for the distance of individual moves, and $dV(u)$ is the probability density that an individual will move u units. Equation (11.8) can be approximated by assuming that V is a symmetric distribution and by taking a Taylor expansion of $N(x - u)$ around x. Equation (11.8) becomes

$$\frac{\partial N}{\partial t} = \alpha \left[N \int dV - \frac{\partial N}{\partial x}\int u dV + \frac{\partial^2 N}{\partial x^2}\int \frac{u^2}{2}\, dV - \right.$$
$$\left. \frac{\partial^3 N}{\partial x^3}\int \frac{u^3}{6}\, dV + \frac{\partial^4 N}{\partial x^4}\int \frac{u^4}{24}\, dV + \ldots\right] \tag{11.10}$$

This can be simplified, because the total probability $\int$ dV $= 1$ and because symmetry of V sets all odd powers to zero. Setting $\int u^2 dV = \sigma^2$, the variance of V, and assuming that terms of u^4 or higher may be neglected, we obtain

$$\frac{\partial N}{\delta t} = \alpha N + \alpha \frac{\sigma_v^2}{2} \frac{\partial^2 N}{\partial x^2} \tag{11.11}$$

Under this formulation,

$$V_F = \alpha\, (2\sigma_v^{\,2})^{1/2} \tag{11.12}$$

This agrees with equation (11.5) if D is identified with $(\alpha\, \sigma^2)/2$, which has the same units as D.

For the density dependent case, Mollison (1977) suggested the form

$$\frac{\partial N}{\partial t} = \alpha \bar{N}(1 - N) \tag{11.13}$$

By the same reasoning used to derive equation (11.11), we approximate equation (11.13) by

$$\frac{\partial N}{\partial t} = \alpha \frac{\sigma^2{}_v}{2}(1 - N)\frac{\partial^2 N}{\partial x^2} + \alpha N (1 - N) \qquad (11.14)$$

which differs from equation (11.6) in that the diffusion coefficient is density dependent.

Mollison (1977) summarizes a number of investigations that determine the asymptotic rate of spread of a front of invading organisms for equation (11.8). We have reproduced part of Mollison's (1977) summary in Table 11.1, for different contact distributions, V. An important qualitative prediction that emerges from all of these models is that a species range should tend asymptotically to increase linearly with time in a homogeneous habitat. There remain substantial differences between the two theoretical approaches, but we shall not discuss these further in this chapter. Instead, we examine methods for quantifying spread in rate by four organisms and for evaluating the prediction of asymptotic linearity in spread rate.

Table 11.1. Theoretical Constant Coefficients, k, for the Time Asymptotic Velocities of Spread, $V_F = k\,\alpha\sigma_v$ for Spread of Organisms in One and Two Dimensions (from Mollison, 1977); from equation (11.12)

Model	k
One Dimension	
Linear approximation [like equation (11.11)]	1.414
V [equation (11.8)]	
Nearest neighbor	1.509
Continuous uniform	1.568
Normal	1.649
Geometric ($\sigma_v = 2$)	1.851
Exponential	1.834
Two Dimension	
Linear approximation [equation (11.11)]	1.000
V [like equation (11.8)]	
Nearest neighbor	
Square lattice—axial	1.114
—45° direction	1.067
Exponential	1.297

QUANTIFYING SPREAD

Skellam (1951) suggested that spread rate could be quantified by measurement of the change in the square root of the area, A, of the geographic distribution of the expanding population. More generally, for a population with a convex geographic range,

$$A = \frac{1}{2} \int_{0}^{2\pi} r_\theta^2 \, d\theta$$

where the position of the front is described in polar coordinates, (r_θ, θ), with the origin being the point of release of the organisms. Thus, $A^{1/2} = \pi^{1/2}\overline{r}_R$, where $\overline{r}_R$ is the root mean-squared radius of the distribution.

In addition to $\overline{r}_R$, we consider other measures, such as $\overline{r}_A$ (the arithmetic mean radius), $\overline{r}_G$ (the geometric mean radius), and $\overline{r}_H$ (the hyperbolic mean radius). The theoretical basis for choosing among these different measures is ambiguous. Some statistical properties of these measures, however, can be recognized. Clearly, for circular distributions, $\overline{r}_R = \overline{r}_A = \overline{r}_G = \overline{r}_H$, and, for irregularly shaped geographic distributions, $\overline{r}_R \geqslant \overline{r}_A \geqslant \overline{r}_G \geqslant \overline{r}_H$. To span these possibilities, we focus our analysis on $\overline{r}_R$, $\overline{r}_A$, and $\overline{r}_H$. Since $\overline{r}_R$ will tend to give greater weight to long radii and $\overline{r}_H$ will tend to give greater weight to short radii, differences in these measures will be largest for highly asymmetrical geographic distributions. We compared these measures for a reasonably symmetrical range expansion pattern (muskrat from Prague) and a highly asymmetrical range expansion pattern (cereal leaf beetle from southwest Michigan) by measuring the radius of the geographic distribution in eight directions and calculating the appropriate averages.

Theoretically, the rate of spread depends on phenomena occurring in the neighborhood of the expanding front (Mollison, 1977; Bramson, 1983), so measurement can be confined to that neighborhood. In addition, neighborhood measurements of range expansion have several practical advantages (Fig. 11.1). Spread frequently is impeded in one or more directions by some geographic barrier, and, because of habitat variation, spread may occur faster in some regions than others or may be directed into a corridor. In addition, the actual distribution of an organism in a given year rarely is known completely, so neighborhood measurement of spread could give a more meaningful description.

While there are good theoretical and practical reasons for neighborhood measurement of spread, the theoretical behavior of neighborhood spread measures is unknown and theoretical definitions for neighborhoods are unspecified; that is, in heterogeneous environments, estimating theoretical asymptotic spread rates by measurements of spread that are determined in local neighborhoods can only be an approximation because of the time it takes to relax to an asymptotic front. Nonetheless, we rely on these neighborhood estimates of range expansion to describe the change in geographic distributions over time. We denote by $\overline{\Delta r}$ the average change in geographic range in a local neighborhood over the time Δt. We approximate $\overline{\Delta r}$ by the average displacement over the entire geographic boundary,

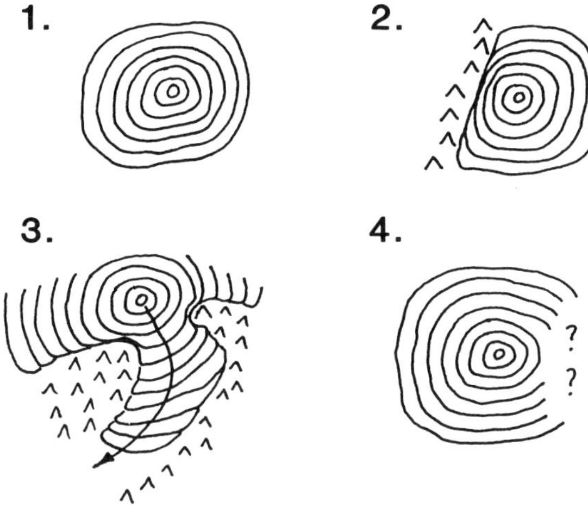

Figure 11.1. Practical complications in the study of spread of invading organisms: (1) idealized spread, (2) geographic barrier impedes spread, (3) geographic features direct spread down a corridor, (4) poor information on the distribution of the spreading organism

$$\overline{\Delta r} = \frac{1}{L_{\Omega_n}} \int_{\Omega_n} \min_{x} \| (x,y) \| \, d\Omega_n \tag{11.15}$$

where Ω_n is the new geographic boundary, y is on Ω_n, x is an arbitrary point on the old boundary, $\|(x, y)\|$ is the Euclidean distance between x and y, and L_{Ω_n} is the length of Ω_n (Fig. 11.2). The distance min $\|(x, y)\|$ is the minimum distance that an organism had to move to get from the old range, which is delimited by Ω_o, to the point y on the boundary of the new range, Ω_n. This measure of Δr averages these distances over the length of the new boundary, L_{Ω_n}. In this chapter, we measure the longest and shortest of these distances, denoted

$$\Delta r_{\max} = \max_{y} \left(\min_{x} \| (x,y) \| \right)$$

and

$$\Delta r_{\min} = \min_{y} \left(\min_{x} \| (x,y) \| \right) \tag{11.16}$$

and estimated $\overline{\Delta r}$ by

$$\overline{\Delta r_R} = \left[(\Delta r^2_{\max} + \Delta r^2_{\min})/2 \right]^{1/2} \tag{11.17}$$

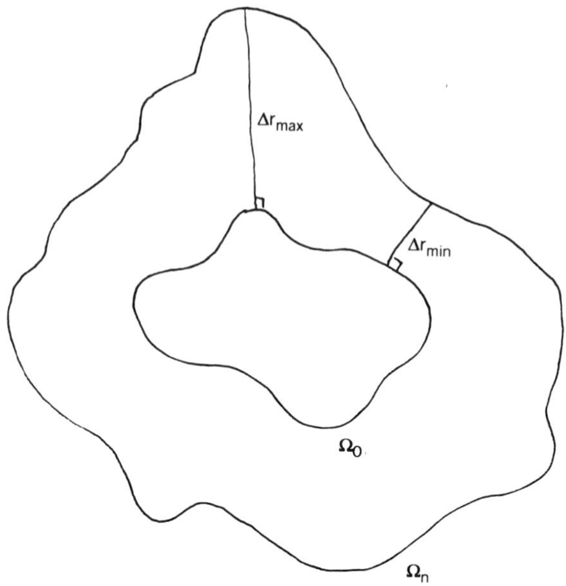

Figure 11.2. Neighborhood measure of spread

which is our neighborhood estimate of spread increments. We calculated $v(t) = \overline{\Delta r}/\Delta t$, where $v(t)$ is the velocity or rate of spread for each sequential pair of geographic range boundaries and t is the time after invasion.

To test the hypothesis that spread is linear or that the asymptotic rate of spread is constant, as predicted by equations (11.5), (11.12) and Table 11.1, we estimated the regression,

$$v(t) = \alpha_0 + \alpha_1 t + \epsilon \tag{11.18}$$

Under the null hypothesis that spread rate is constant, $\alpha_1 = 0$, the graph of $v(t)$ versus t should be a flat line with zero slope.

THE INVADERS

Muskrat

Muskrat, *Ondatra zibethica*, was released near Prague in 1905. This represents one of the best known examples of spread, having been analyzed by Skellam (1951) and cited in numerous textbooks. The map by Ulbrich (1930, reproduced by Elton, 1958) and updated maps (Hoffman, 1958; Frank and Härle, 1964, 1967) describe the spread of muskrat from Prague. Muskrat spread has been observed also within three localities in France (Eure, Belfort, and Ardennes) (Dorst and Giban, 1954) and four localities in Finland (northern Pohjanma, southern Kokemäenjoki, Pieksämäki, and coastal Unsimaa) (Artimo, 1960).

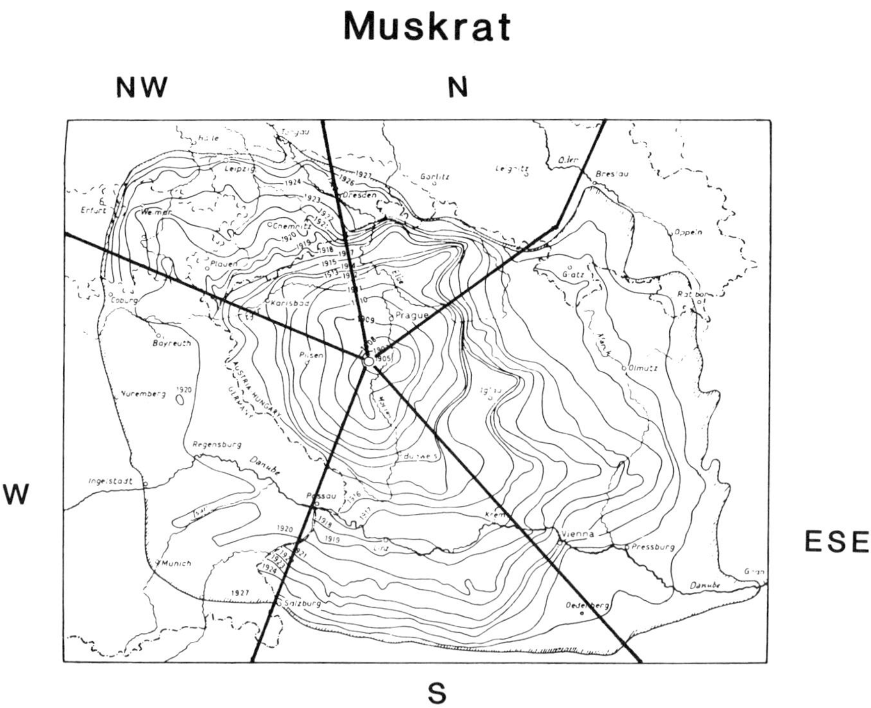

Figure 11.3. Spread of muskrat from Prague (reproduced from Elton, 1958)

For our analysis, we divided muskrat spread from Prague into five sectors (Fig. 11.3): west (W), northwest (NW), north (N), east southeast (ESE), and south (S). The W sector was drawn to reflect a zone of poor information. The ESE sector represents movement across the Hungarian plain. The N sector corresponds to movement down the Elbe and its tributaries, whereas the NW sector reflects slower movement northward without these major rivers as corridors. The S sector simply reflects movement due south, but not across plains. Muskrat spread in France and Finland was not divided into sectors: instead, for each focus, an estimated radius was calculated as $\overline{r}_R$.

Cereal Leaf Beetle

Cereal leaf beetle, *Oulema melanopus*, was first discovered in southwestern Michigan in 1958 but is believed to have invaded about 1948 (Haynes and Gage, 1981). At the time of discovery, it was feared that *O. melanopus* would become an important pest of wheat; therefore, its spread was monitored carefully by the USDA Cooperative States Research Service. (Fig. 11.4, Anonymous, 1963, 1965a, 1965b, 1967, 1968, 1969, 1970, 1971, 1973, 1974, 1976; Wilson and Treece, 1968; Wilson et al., 1969, 1970, 1972; Wilson, 1974). Early in the invasion, investigators sampled approximately 20 fields in each township with sweep nets; approximately 1,000

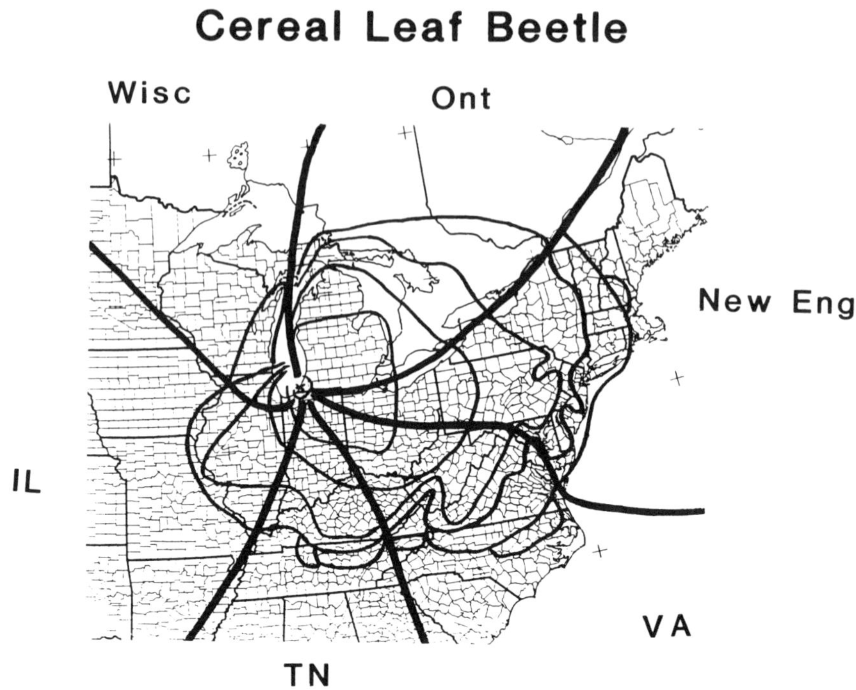

Figure 11.4. Spread of cereal leaf beetle from southwest Michigan

sweeps per field were made (Haynes and Gage, 1981). With this sampling intensity, average densities of 5 larvae per acre could be detected with 90% probability (Ruesink and Haynes, 1973) within the entire township. In 1967, the methods were changed and larval damage was examined (Wilson and Treece, 1968). Based on the published data, we calculated that, with the new sampling effort, beetle densities of approximately 10 larvae per acre would probably be detectable.

We divided cereal leaf beetle spread into six sectors (Fig. 11.4): east to New England (New Eng or NE), northeast through Ontario (Ont), southeast through Virginia and North Carolina (VA), south to Tennessee (TN), southwest through Illinois and Missouri (IL), and west, then north through Wisconsin (Wisc). The Great Lakes and St. Lawrence River formed the Ont/New Eng boundary, and Lake Michigan, the Ont/Wisc boundary. The remaining boundaries were somewhat arbitrary, but they reflect variation in the spread pattern.

Rice Water Weevil

Rice water weevil, *Lissorhoptrus oryzophilus*, was first discovered in Japan in 1976 (Iwata, 1979). The initial invasion occurred in Aichi Prefecture near Nagoya (Tsuzuki and Isogawa, 1976). It is considered to be the greatest insect threat to rice production in Japan, and recently it has colonized the Asian mainland. Its distribution and spread have been carefully monitored by the Japanese, and spread maps have been published by Iwata (1979) and Kiritani (1984). Sampling is detailed in Tsuzuki et al. (1977, 1979).

We divided rice water weevil spread (Fig. 11.5) into three sectors: north to the Japan Sea (N), northeast and north through northern Japan (NE), and west through western Japan, Shikoku, and Kyushu (W). Spread in N included northern Aichi, Gifu, eastern Fukui, Ishikawa, and western Toyama Prefectures.

Small Cabbage White Butterfly

Small cabbage white butterfly, *Pieris rapae*, has successfully invaded North America several times. The first recorded successful invasion was in 1860 near Quebec City, and secondary invasions occurred in New York City in 1868, in Charleston in 1873, and in the Florida Panhandle in 1874 (Scudder, 1887). All of these appear to have been immigrants from Europe. Shapiro (1974) argued that the imported cabbageworm also invaded the West Coast from an Asian population sometime before 1867. We focus on the East Coast invasions, which were documented by Scudder (1887) and are based on correspondence and collection records of professional and amateur entomologists.

We divided small cabbage white butterfly spread into six sectors (Fig. 11.6): east through the Maritime Provinces (Que), directly west from Quebec City across northern Ontario (Ont), southwest along the St. Lawrence River and along the Great Lakes (Lakes), directly south to Rhode Island (New Eng), and southwest along the southern shore of the St. Lawrence River and then either east through Ohio and the Great Plains (Plains), and continuing southwest through West Virginia, Alabama, Louisiana, and Texas (South). The Ont/Lakes boundary reflected the lack of information in the Ont sector, and the Lakes/Plains boundary was the Great Lakes. The New Eng/South boundary corresponded to a line from Quebec City to New York City. The other borders reflected variation in the spread map. We also analyzed secondary invasions starting at New York City, Charleston, and the Florida Panhandle. In addition, the disjunct colonies at Chicago and in Indiana were analyzed.

EVALUATING MEASURES OF SPREAD

When spread of an organism occurred at an apparent constant rate, all measures of spread showed constant rates of spread (Fig. 11.7). The rate of spread, however, varied, depending on the measure, and $\overline{\Delta r_R}$ measured spread as being more variable in time than $\overline{r}_R$, $\overline{r}_A$, and $\overline{r}_H$. The estimated asymptotic rate of spread was different for each measure.

A more detailed comparison of $\overline{r}_R$, $\overline{r}_A$ with $\overline{\Delta r_R}$ revealed some differences. For muskrat, $\overline{r}_R$, $\overline{r}_A$, and $\overline{r}_H$ agreed very closely with each other, because spread was uniform in all directions during most of the 22-year period following initial colonization

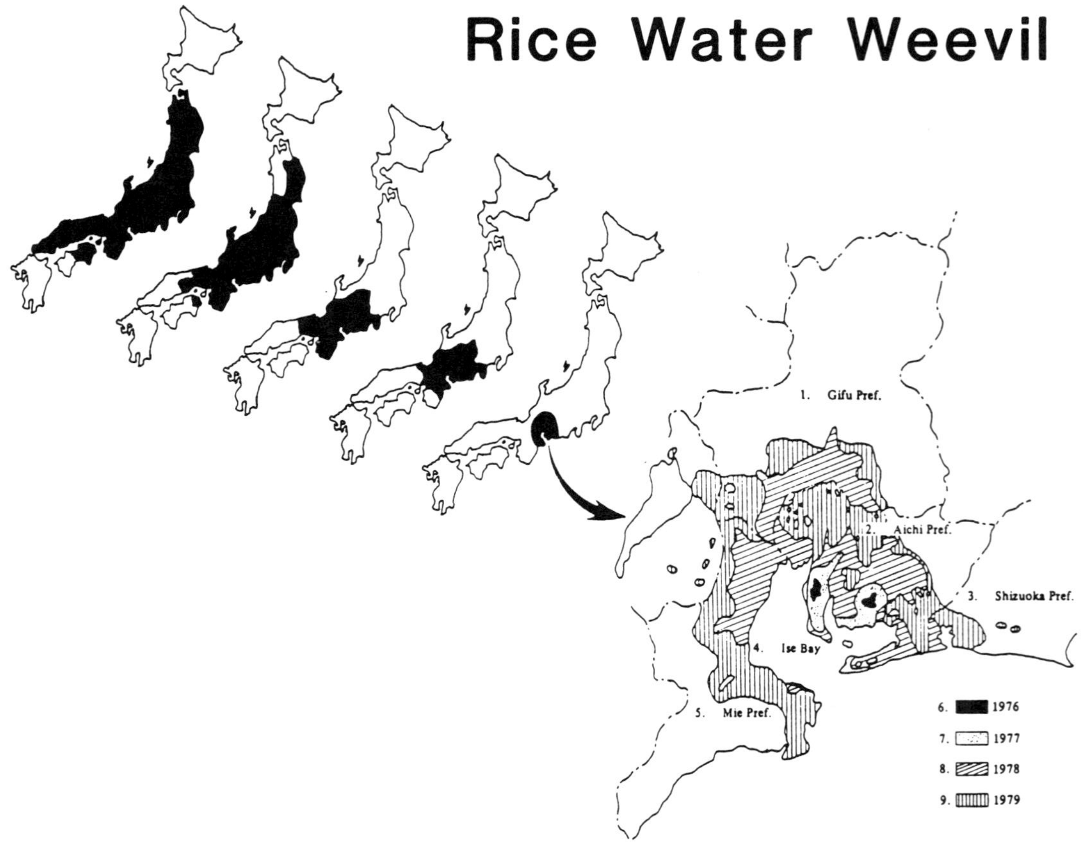

Figure 11.5. Spread of rice water weevil from southern Aichi Prefecture (reproduced from Iwata, 1979)

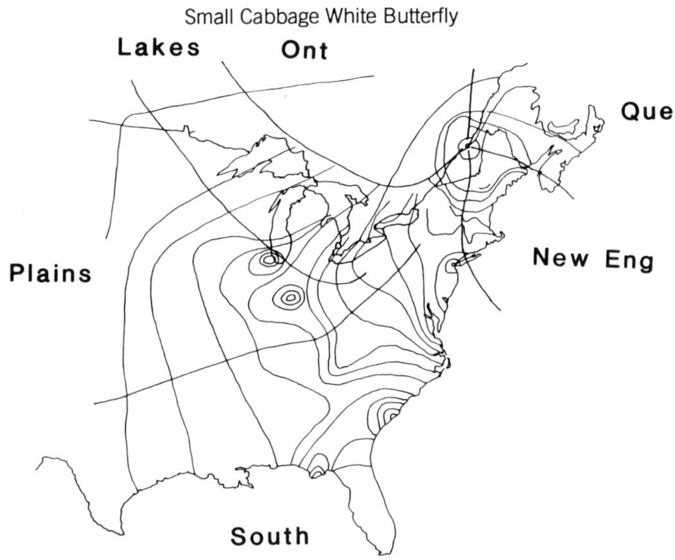

Figure 11.6. Spread of small cabbage white butterfly in eastern North America (after Scudder, 1887)

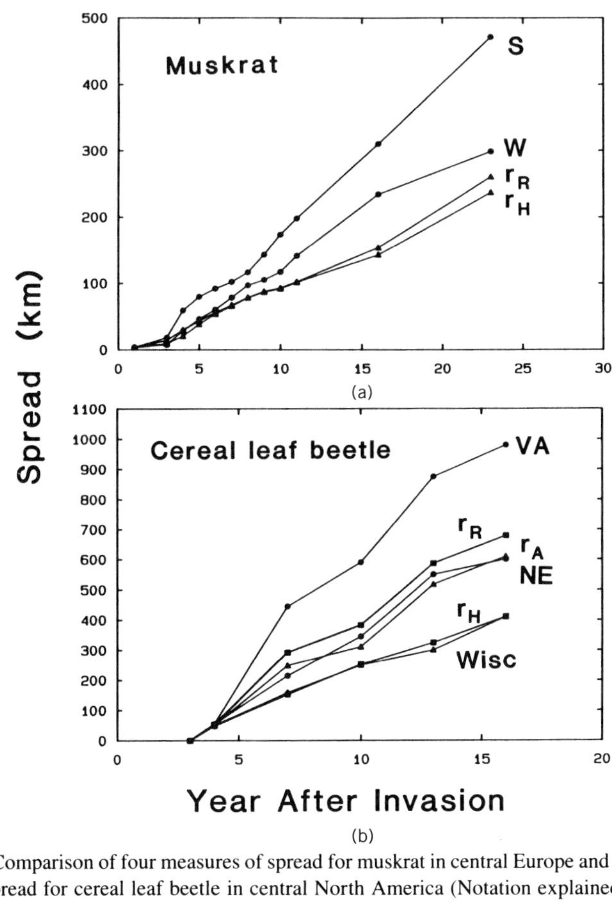

Figure 11.7. Comparison of four measures of spread for muskrat in central Europe and comparison of six measures of spread for cereal leaf beetle in central North America (Notation explained in text.)

in 1905 (Fig. 11.7a, $\overline{r}_A$ is not shown but is between $\overline{r}_R$ and $\overline{r}_H$). By contrast, the neighborhood measure, $\overline{\Delta r_R}$, shown here for the S and W sectors, led to a different asymptotic rate (Fig. 11.7a). Spread in S and ESE showed relatively constant rates, but the spread rates fluctuated, and the fluctuations were not coincident (Fig. 11.8a). Clearly, spread was unequal in the different sectors.

For cereal leaf beetle, $\overline{r}_R$, $\overline{r}_A$, and $\overline{r}_H$ were different from each other because of the asymmetrical spread pattern of the beetle. The results for three of the neighborhood measures, shown here for the NE, VA, and Wisc sectors, showed even greater variation (Fig. 11.7b). Spread in NE, VA, was rapid; spread in Wisc was slow; VA spread oscillated, and NE spread was fairly constant (Fig. 11.8b). For cereal leaf beetle, as for muskrat, the rate of spread was unequal in the different sectors.

Clearly, neighborhood measures of spread reveal greater detail about the spread of invading organisms. Moreover, they are theoretically justified and have several

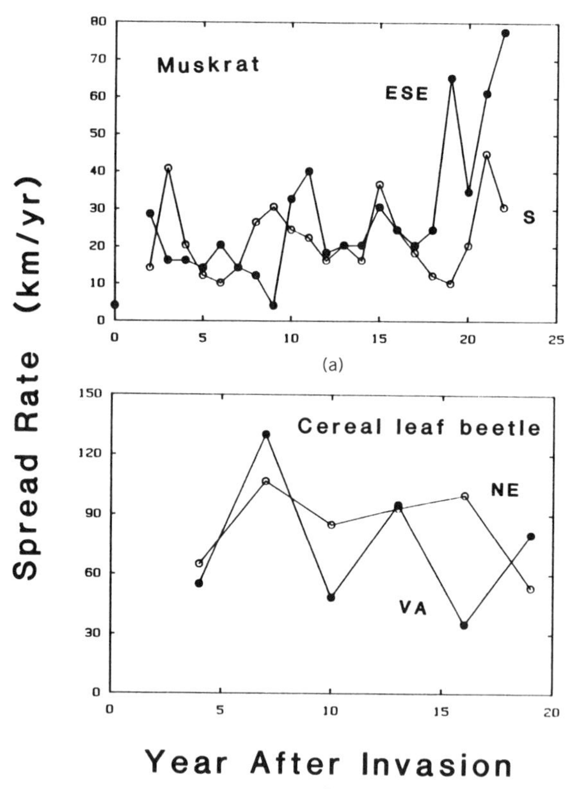

Figure 11.8. Spread rate of muskrat and cereal leaf beetle in selected sectors (see text for notation), showing fluctuations in spread rate in some of the sectors

practical advantages over the global summary measures. Therefore, in the rest of this chapter, we discuss spread as measured by our neighborhood measure in equation (11.17).

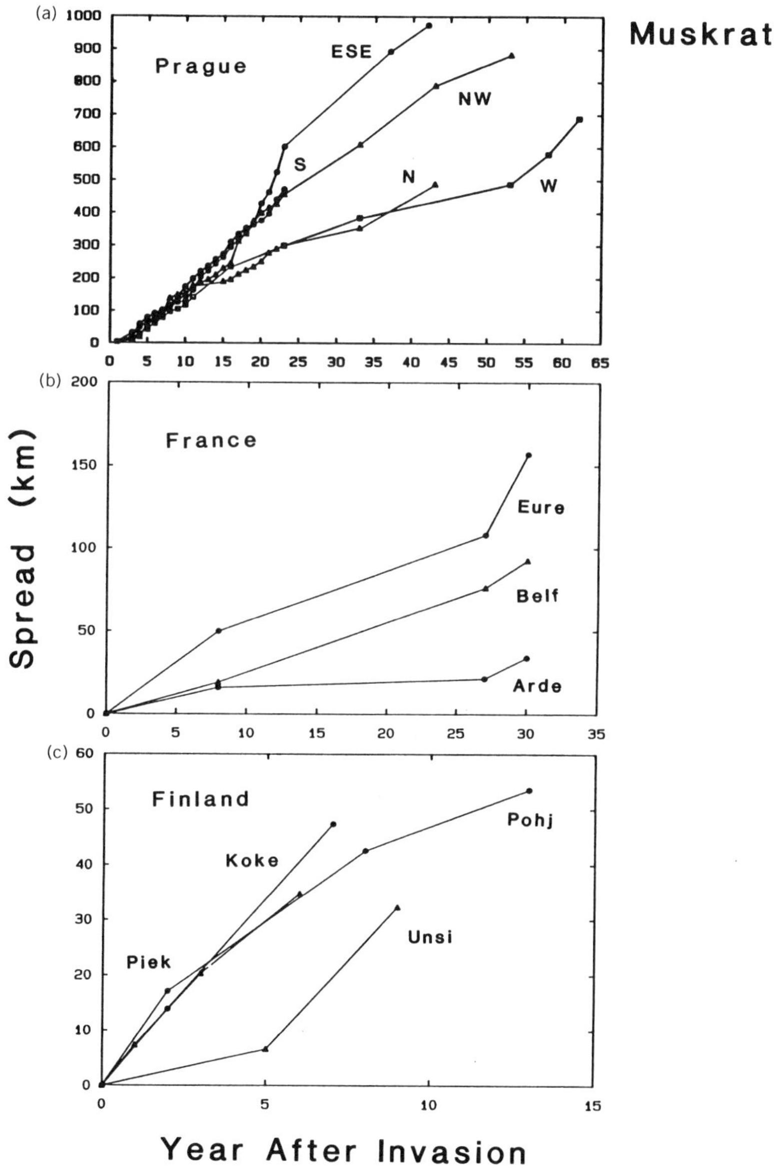

Figure 11.9. Radial spread of muskrat (a) in central Europe from Prague in five sectors, (b) in France, and (c) in Finland

RADIAL SPREAD

Muskrat

The radial spread of muskrat is summarized in Figure 11.9. The slope of these plots measures the average rate of spread from Prague and from various foci in France and Finland. Average rates of spread varied considerably, ranging from 10.3 kilometers

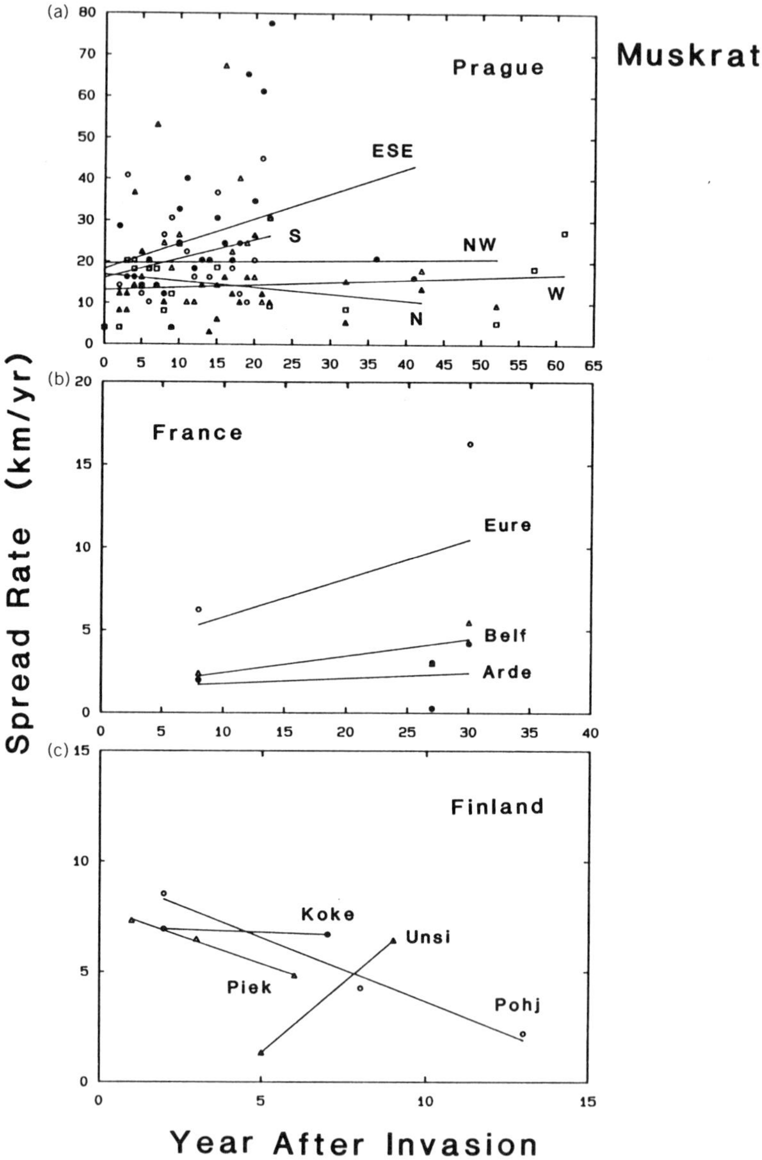

Figure 11.10. Spread rate of muskrat from Prague, in France, and in Finland. Fitted regression lines for equation (11.18) are shown, and test for constancy in spread rate. Cases correspond with Figure 11.9.

per year (km/yr) (W) to 25.4 km/yr (ESE) from Prague, from 0.9 to 4.6 km/yr in France, and from 3.5 to 6.7 km/yr in Finland. For the entire Central European invasion, the spread rate was about 11.4 km/yr (restricted only to Ulbrich's 1930 data). Eastern Europe seemed to be more conducive to muskrat range expansion than western and northern Europe. Although average rates clearly differed among regions, spread rates generally were constant in time (Fig. 11.10). Spread rates did vary considerably in space from less than 1 km/yr to nearly 80 km/yr, and in one case, Pieksämäki, Finland, decelerated significantly.

Cereal Leaf Beetle

After an initial lag of at least four years, the radial spread of cereal leaf beetle was also strikingly linear (Fig. 11.11a). That there was a time lag before range expansion began is consistent with our simple theoretical models. Indeed, for small α, or large

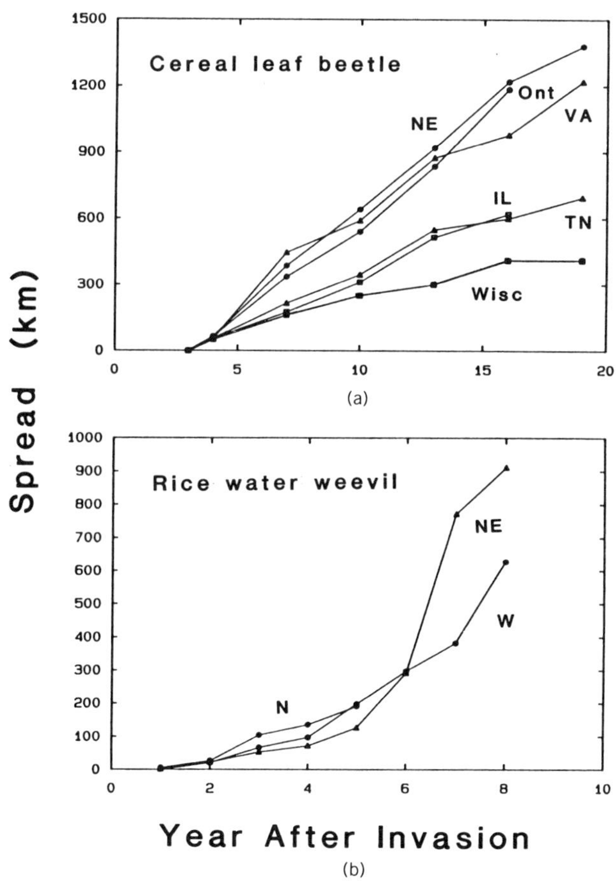

Figure 11.11. Radial spread of cereal leaf beetle in central North America in six sectors and of rice water weevil in Japan in three sectors

D or large N_F/N_o, equation (11.4) predicts a time lag before range expansion begins. Average rates of spread varied greatly, from a low of 26.5 km/yr (Wisc) to a high of 89.5 km/yr (New Eng); for the entire invasion, the rate of spread averaged about 52.0 km/yr. Expansion into Wisconsin was extremely erratic, possibly because Chicago was a poor habitat and the westerly winds inhibited beetle movement (Shade and Wilson, 1964). Cereal leaf beetle spread rates generally were constant in time (Fig. 11.12a), but, similar to muskrat, the rate of spread varied considerably, ranging from 0 km/yr to about 130 km/yr.

Rice Water Weevil

In contrast to the spread of muskrat and cereal leaf beetle, spread of rice water weevil showed no indication of increasing linearly with time (Fig. 11.11b, 11.12b). This beetle is difficult to observe in the field, but its damage is characteristic. In 1976, the

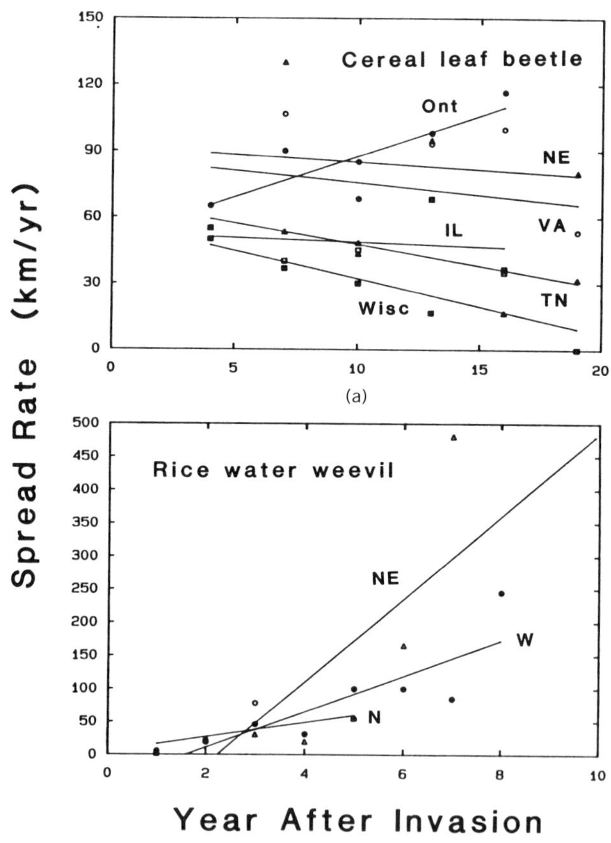

Figure 11.12. Spread rate of cereal leaf beetle and rice water weevil. Fitted regression lines for equation (11.18) are shown, as well as test for constancy in spread rate. Cases correspond with Figure 11.11.

year of its discovery, 730 hectares were found to be infested (Iwata, 1979), so the beetle probably had invaded Japan several years earlier. If the year of invasion had actually been several years before 1976, then spread would have been even more strongly nonlinear. Indeed, spread rates accelerated dramatically. The spread rate was initially between 28 km/yr (Toohoku) and 47 km/yr (Kyushu), and increased to about 274 km/yr (Toohoku) and 113 km/yr (Kyushu). Average acceleration was about 62 km/yr^2 (Toohoku) and 28 km/yr^2 (Kyushu).

In this species, acceleration of spread rates could be related to its dispersal behavior. Rice water weevil disperses in two different ways: first by crawling and swimming from paddy to paddy, and second by flying. If it crawls or swims, new infestations are near old ones and they are more readily detected. If it flies and colonizes a new paddy, it is more difficult to detect, because the beetles are hard to find and they are recognized only when they become abundant enough to cause damage to the rice. Moreover, if relatively few beetles fly long distances, then early observations of range expansion will be related to the short distances that the majority of beetles move, whereas later observations will be related to the long distances that the few beetles move, because these populations will become large enough to detect.

Small Cabbage White Butterfly

Range expansion of small cabbage white butterfly was complicated by multiple invasions, which occurred in New York City, Charleston, and the Florida Panhandle, and by great leaps forward to Chicago and Indiana. Range expansion was mainly toward the south (Fig. 11.6).

Overall, spread was nonlinear (Fig. 11.13 and 11.14), especially in the very long sectors (Ont, Lakes, S, Plains). It was also nonlinear in the shorter New Eng sector, although not as obviously so. Spread from the disjunct centers (Indiana and Chicago) and secondary invasions (New York, South Carolina, Florida) was generally linear, although some cases appear nonlinear (Fig. 11.13). In the main invasion from Quebec City, spread rates accelerated; from an initial value of 28 km/yr (Lakes, S), they increased to 103 (Lakes) to 235 (S) km/yr. Average acceleration of spread ranged between 6.9 to 18.5 km/yr^2.

Acceleration of spread rates could have been related to the southward expansion of this species. In Quebec, *P. rapae* completes three generations per year, whereas, in Missouri, it completes six to seven generations per year. Thus, as it moved south, population growth rate and dispersal rate would increase, accelerating spread rates.

DISCUSSION

All of the simple diffusion and spatial contact models with realistic biological dispersal predict that the spread rate of invading organisms will tend to a limit and that, asymptotically, the geographic range will expand linearly with time.

By plotting radial spread of four species, we observed approximately linear rates of spread for muskrat and cereal leaf beetle, accelerating rates of spread for rice

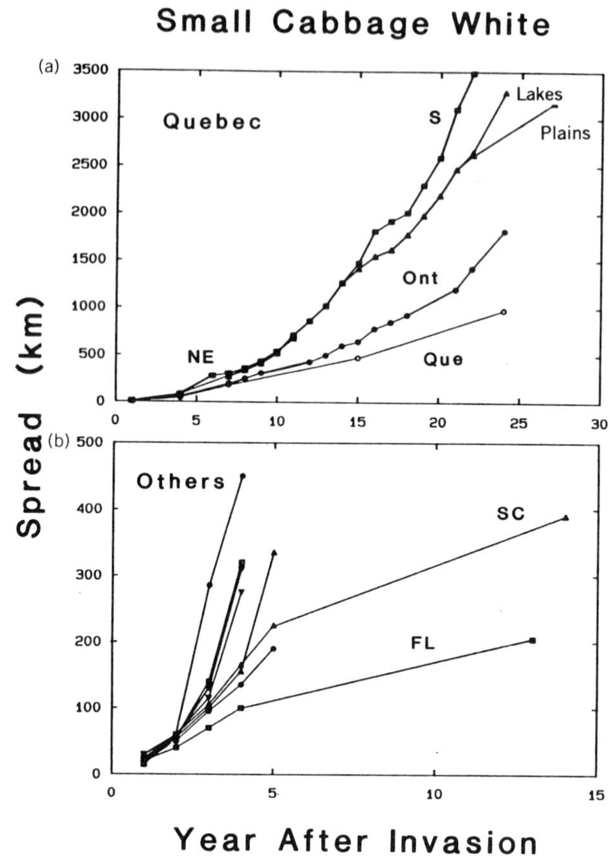

Figure 11.13. Radial spread of small cabbage white butterfly (a) in eastern North America in six sectors and (b) three secondary invasions in six directions and two disjunct populations. Only the eastern spread from the Florida Panhandle and the southern spread from South Carolina are labeled in (b).

water weevil and small cabbage white butterfly, great leaps forward (Mollison, 1977) in small cabbage white butterfly, and pronounced geographic variation in all species. The simple models do not predict the accelerated spread observed for rice water weevil and imported cabbageworm. These simple models, however, provide an essential reference point for examining the spread of invading organisms. While the models are not valid over the entire range of a spreading species, they make qualitative predictions about the pattern of spread and provide an important null hypothesis in examining the spread of invading organisms.

Two critical needs remain. First, a theoretical treatment of our measures of spread is needed. What is the appropriate measure that will reflect theoretical predictions accurately? This is needed before a quantitative comparison of theory and observation can be affected. Second, theory can be tested quantitatively by estimating spread rates from independent estimates of α and D (or σ_v) and by comparing these estimates with observed rates of spread. Do theoretical predictions correspond with

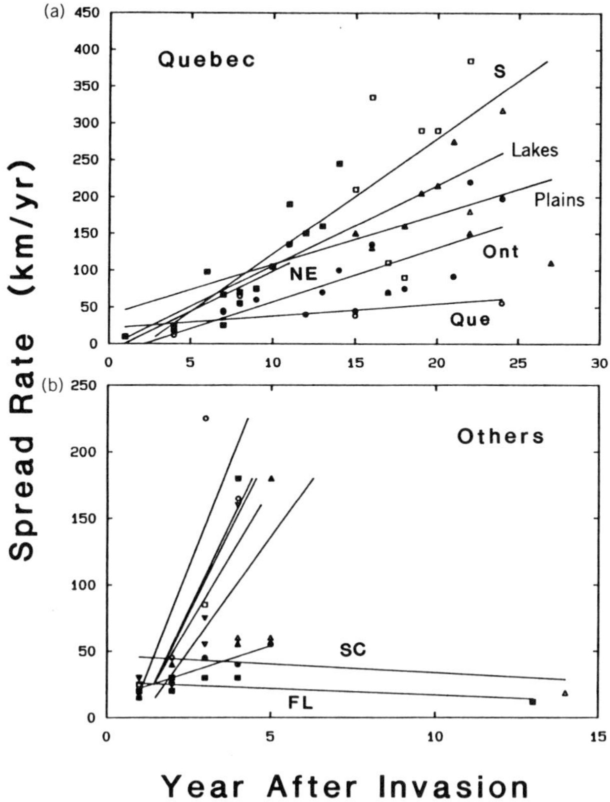

Figure 11.14. Spread rate of small cabbage white butterfly in eastern North America. Fitted regression lines for equation (11.18) are shown, as well as test for constancy in spread rate. Cases correspond with Figure 11.13.

observations? What is the appropriate spatial scale for measuring α and D? Can predictions of the diffusion approximation be distinguished from predictions of the various spatial contact models (Table 11.1)? Clearly, there are several outstanding theoretical and practical problems related to the spread of invading organisms. We hope this chapter stimulates further work along these lines.

Acknowledgments

D. A. Andow acknowledges parital support by USDA-NC-105, USDA-NC-180, and U.S. EPA (to SAL). S. A. Levin acknowledges partial support by U.S. EPA Cooperative Agreement CR812685, NSF Grant No. DMS-8406472, and McIntire Stennis Grant NYC-183568. S. A. Levin thanks All Souls College and the Centre for Mathematical Biology, University of Oxford, for the opportunity to think about mathematical aspects of the problem of biological invasions. A. Okubo would like to acknowledge support from the Centre for Mathematical Biology, University of Oxford, and the Science and Engineering Research Council of Great Britain (Grant No. GR/D/13573).

This is publication ERC-178 of the Ecosystems Research Center, Cornell University and 20,207 of the Minnesota Agricultural Experiment Station. The work and conclusions published herein represent the views of the authors and do not necessarily represent the opinions, policies, or recommendations of the sponsoring agencies and institutions.

REFERENCES

Andow, D. A., P. M. Kareiva, S. A. Levin, and A. Okubo. 1990. Spread of invading organisms. Landscape Ecol. 4:177–188.

Anonymous. 1963. Cereal leaf beetle. *Coop. Econ. Insect Rep.* 13:1031.

———. 1965a. Cereal leaf beetle. *Coop. Econ. Insect Rep.* 15:138–139.

———. 1965b. Cereal leaf beetle. *Coop. Econ. Insect Rep.* 15:1288.

———. 1967. Spread of cereal leaf beetle by year. *Coop. Econ. Insect Rep.* 17:720.

———. 1968. Cereal leaf beetle spread. *Coop. Econ. Insect Rep.* 18:289.

———. 1969. Spread of cereal leaf beetle. *Coop. Econ. Insect Rep.* 19:534.

———. 1970. Distribution of cereal leaf beetle. *Coop. Econ. Insect Rep.* 20:86.

———. 1971. Distribution of cereal leaf beetle. *Coop. Econ. Insect Rep.* 21:115.

———. 1973. Distribution of cereal leaf beetle. *Coop. Econ. Insect Rep.* 23:189.

———. 1974. Distribution of cereal leaf beetle. *Coop. Econ. Insect Rep.* 24:53.

———. 1976. Distribution of cereal leaf beetle. *Coop. Plant Pest Rep.* 1:848.

Artimo, A. 1960. The dispersal and acclimatization of the muskrat, *Ondatra zibethica* (L.), in Finland. *Paper Game Res.* 21. Helsinki: Finnish Game Foundation.

Banks, H. T., P. M. Kareiva, and P. Lamm. 1985. Estimation of temporally and spatially varying coefficients in models for insect dispersal. *J. Math. Biol.* 22:258–277.

Bramson, M. 1983. Convergence of solutions of the Kolmogorov equation to travelling waves. *Mem. Am. Math. Soc. Vol. 285*. Providence, R. I.: American Mathematical Society.

Brownlee, J. 1911. The mathematical theory of random migration and epidemic distribution. *Proc. R. Soc. Edin.* 31:262–289.

Dorst, J., and J. Giban. 1954. Les mammiferes acclimates en France detuis un siecle. *Terre Vie* 1954:217–229.

Elton, C. S. 1958. *The Ecology of Invasions by Animals and Plants*. London: Methuen.

Frank, V. F., and A. Härle. 1964. Die Entwicklung des Bisam befalls (*Ondatra zibethica*) in der Bundesrepublik Deutschland von 1957 bis 1963. *Nachrichtenbl. Dtsch. Pflanzenschutzdienstes (Braunschw).* 16:145–147.

———. 1967. Der zeitiger Stand und voraussichtliche Entwicklung des Bisam befalls (*Ondatra zibethica*) in der Bundesrepublik Deutschland (nach den Meldungen der Pflanzenschutzanter der Bundeslander). *Nachrichtenbl. Dtsch. Pflanzenschutzdienstes (Braunschw).* 19:123–125.

Fisher, R. A. 1937. The wave of advance of advantageous genes. *Ann. Eugenics* 7:355–369.

Haynes, D. L., and S. H. Gage. 1981. The cereal leaf beetle in North America. *Annu. Rev. Entomol.* 26:259–287.

Hoffman, M. 1958. *Die Bisamratte*. Akademische Verlagsgesellschaft. Leipzig: Geest & Portig K-G.

Iwata, T. 1979. Invasion of the rice water weevil, *Lissorhoptrus oryzae* Kuschel, into Japan, spread of its distribution and abstract of the research experiments conducted in Japan. *Jap. Pestic. Info.* 36:14–21.

Kareiva, P. M. 1982. Experimental and mathematical analyses of herbivore movement: Quantifying the influence of plant spacing on foraging discrimination. *Ecol. Monogr.* 52:261–282.

———. 1983. Local movement in herbivorous insects: Applying a passive diffusion model to mark recapture field experiments. *Oecologia* 57:322–327.

Kendall, D. G. 1948. A form of wave propagation associated with the equation of heat conduction. *Proc. Camb. Philos. Soc.* 13:11–66.

Kierstead, H., and L. B. Slobodkin. 1953. The size of water masses containing plankton bloom. *J. Mar. Res.* 12:141–147.

Kiritani, K. 1984. Colonizing insects—(Colonist as a member of native insect community). *Insekutariumu (Insectarium)* 9:248–262.

Kolmogorov, A., I. Petrovsky, and N. Piscounov. 1937. Étude de l'équation de la diffusion avec croissance de la quantité de matière et son application à un problèma bilogigue. *Moscow Univ. Bull. Ser. Int. Sect. A.* 1:1–25.

Lubina, J. A., and S. A. Levin. 1988. The spread of a reinvading organism: Range expansion of the California sea otter. *Am. Nat.* 131:526–543.

Mollison, D. 1977. Spatial contact model for ecological and epidemic spread. *J. R. Stat. Soc. B.* 39:283–326.

Mooney, H. A., and J. A. Drake (eds.). 1986. *Ecology of Biological Invasions of North America and Hawaii*. New York: Springer-Verlag.

Okubo, A. 1980. *Diffusion and Ecological Problems: Mathematical Models*. Berlin: Springer-Verlag.

Ruesink, W. G., and D. L. Haynes. 1973. Sweepnet sampling for the cereal leaf beetle, *Oulema melanopus*. *Environ. Entomol.* 2:161–172.

Scudder, S. H. 1887. The introduction and spread of *Pieris rapae* in North America, 1860–1886. *Mem. Boston Soc. Nat. Hist.* 4:53–69 plus 1 plate.

Shade, R. E., and M. C. Wilson. 1964. Population buildup of the cereal leaf beetle and the apparent influence of wind on dispersion. *Purdue Univ. Agric. Exp. Stn. Res. Prog. Rep.* 98:1–7.

Shapiro, A. M. 1974. Butterflies and skippers of New York State. *Search Agric. (Geneva, N. Y.)* 4(3):1–60.

Skellam, J. G. 1951. Random dispersal in theoretical populations. *Biometrika* 38:196–218.

Tsuzuki, H., and Y. Isogawa. 1976. (The first record of the rice water weevil, tentatively identified as *Lissorhoptrus oryzophilus*, in Aiti prefecture, Japan). *Syokubutu Boeki* 30:341 plus 1 plate (in Japanese).

Tsuzuki, H., T. Asayama, Y. Morita, and T. Iwata. 1977. (The recent situation of the rice water weevil and its control measures). *Syokubutu Boeki* 31:483–486 (in Japanese).

Tsuzuki, H., T. Asayama, J. Kayumi, T. Sakashita, H. Yasuda, T. Shimohara, S. Takeshima, H. Kawata, Y. Hasegawa, and T. Tanaka. 1979. (The occurrence of the rice water weevil in five prefectures in 1979 and its control measures). *Syokubutu Boeki* 33:530–535 (in Japanese).

Ulbrich, J. 1930. *Die Bisamratte: Lebensweise Gang ihrer Ausbreitung in Europa, wirtschaftliche Bedeutung und Bekampfung* Dresden: Verlag und Druck von C. Heinrich.

Wilson, C. L., and C. L. Graham (eds.). 1983. *Exotic Plant Pests and North American Agriculture*. New York: Academic Press.

Wilson, M. C. 1974. 1973 report of infestation of oats by the cereal leaf beetle. *Coop. Econ. Insect Rep.* 24:443–447.

Wilson, M. C., and R. E. Treece. 1968. 1967 cereal leaf beetle infestation and oats crop loss survey. *Coop. Econ. Insect Rep.* 18:343–348.

Wilson, M. C., R. E. Treece, and R. E. Shade. 1969. 1968 cereal leaf beetle infestation and oats crop loss survey. *Coop. Econ. Insect Rep.* 19:408–417.

———. 1970. 1969 cereal leaf beetle infestation and oats crop loss survey. *Coop. Econ. Insect Rep.* 20:457–468.

———. 1972. Infestation of oats by the cereal leaf beetle in 1970 and 1971. *Coop. Econ. Insect Rep.* 22:371–378.

12

INTERACTIONS BETWEEN IMPORTED PREDATORS AND THEIR PREY IN PATCHY ENVIRONMENTS

Mark Andersen

University of California at Irvine

Peter M. Kareiva

University of Washington

INTRODUCTION

In order to control introduced pests, applied entomologists have been importing exotic natural enemies into North America since the beginning of the twentieth century (Turnbull and Chant, 1961). The majority of these deliberate introductions fail (Hall and Ehler, 1979), even though the natural enemies are typically screened to ensure that their new habitat has a suitable climate. Many reasons have been given for the unsuccessful establishment of invading populations, including low reproductive rates (Crawley, 1986), competition from resident species (Moulton and Pimm, 1986), the ravages of diseases and predators (Crawley, 1986), and genetic inflexibility (Baker and Stebbins, 1965). Populations of imported beneficial insects may also fail to establish, for a very different reason: perhaps these predators and parasitoids become extinct because they cannot find food. In particular, although biological control agents usually feed on pest insects, these "pests" are rarely so abundant that they are "easy to find."

Our ideas about the difficulty that a biocontrol agent may experience in locating its prey have been developed from our own studies of the ladybug, *Coccinella septempunctata* Linnaeus. This generalist predator has been released by the U.S. Department of Agriculture (USDA) in several states for the control of aphid pests (Angelet et al., 1979) and is reputed to be a great success (Obrycki et al., 1982). In fact, *C. septempunctata*'s reputation for aphid control is so great that the USDA has been collecting and mass rearing these beetles for widespread release in the western United States (Comis and Heppner, 1986). While our own experiments indicate that *C. septempunctata* is potentially a tremendously effective predator (Kareiva, 1985), we have also found that this ladybug can have difficulty finding prey that appear to be locally abundant. We think that the problems experienced by *C. septempunctata* in hunting for prey are problems that plague all predators. Furthermore, studies of biological control agents tend to neglect these problems. Our goal in this chapter is to alert the reader to a strong selective pressure that plays a major role in determining the fate of individuals or populations of predators and parasitoids, even when their prey are so abundant as to be called pests. Nothing that we say is particularly new; it has long been known that the act of searching for prey often requires special adaptations and strategies (Curio, 1976). Our contribution is to pinpoint simple measurable features of predator searching that can be translated into predictions about the expected success of a predator seeking patchily distributed prey. While our specific discussion invariably centers on *C. septempunctata* feeding on aphids, our approach and message should extend well beyond coccinellid-aphid interactions.

PATCHY PEST POPULATIONS AND SOME LIMITATIONS OF FIELD AND LABORATORY STUDIES OF PREDATOR EFFICIENCY

Pests are not abundant everywhere, and, even where they are abundant, their distribution tends to be very spotty (Hassell, 1982; Trumble et al., 1983; Taylor, 1984). Consequently, a predator must often search vast stretches of habitat to find highly localized pest outbreaks. Upon finding a high-density patch of prey, a predator should remain in the vicinity of those prey until they have been consumed. In other words, predators should aggregate at clusters or outbreaks of their prey; aggregation both enhances a predator's success and increases its effectiveness as a biological control agent. Not surprisingly then, many predators, including ladybugs, can be found aggregated where their prey are most abundant (Turchin and Kareiva, 1989; Curio, 1976). However, simply observing that predators are clustered where their prey are clustered does not tell us what fraction of the predator population is without prey or what fraction of its life the average predator spends amidst plentiful patches of prey. In fact, static observations of the spatial coincidence of predators and prey cannot adequately characterize the foraging of predators (Heads and Lawton, 1983; Kareiva, 1985). A more dynamic examination of predator–prey interactions is needed in order to assess predator efficacy.

Quantitative studies of predator efficiency have been a standard feature of research programs whose aim is the evaluation of potential biological control agents (Luck et al., 1988). One severe limitation of these studies is that they typically quantify the effectiveness of predators in the best possible situations. For example, a predator's killing rate may be measured by observing it in a small container in which the environment is so simple that encounters with prey are virtually unavoidable (Sundby, 1966). Similarly, many so-called field experiments on predator efficiency confine predators inside cages; a predator so confined cannot accidentally wander away from its prey (which themselves are often at unusually high densities) and will be forced to search the same area over and over again (Shands and Simpson, 1972; Shands et al., 1972). Even the most realistic studies typically place predators directly on plants known to harbor prey and then record the predator's performance (Tamaki and Weeks, 1972). All of these experimental procedures make it too easy for a predator to find its prey. In less contrived settings, predators must immigrate into fields with patches of prey and then find and, it is hoped, remain at those prey hot spots. Rarely have population ecologists examined this chain of events, and most of what are called "predator efficiency studies" measure how fast predators capture and kill prey that have been found for them. But, in nature, capturing and killing prey is not the problem; finding the prey is the problem! For example, our own studies of pea aphids in commercial pea fields indicate that, at the time when fields are typically sprayed for aphid control, aphids are abundant only in a few isolated patches (Fig. 12.1). If ladybugs were to replace insecticides in such a system, they would need to be effective at finding scarce incipient aphid outbreaks *before* the aphid populations built up to harmful levels. Our acknowledgment of the problems that predators have in finding prey explains our emphasis on predator hunting and aggregation.

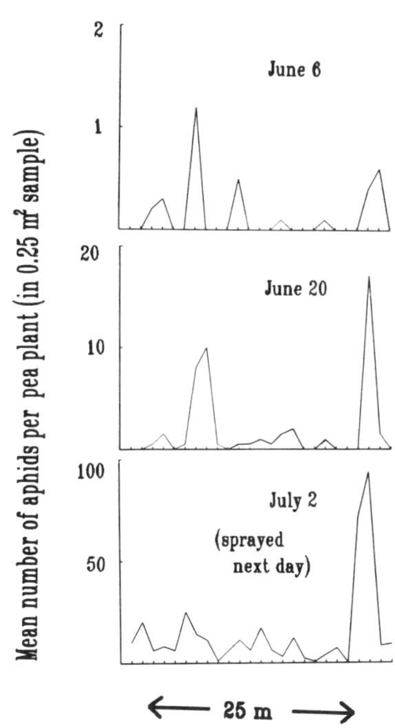

Figure 12.1. Naturally occuring aphid patchiness on "homogeneous" commercial pea fields in Skagit County, Washington. The data are from a 25 m transect through the field, with 0.25 m2 quadrats placed at 1-meter intervals.

Another important aspect of predator evaluation is the construction of fecundity and mortality schedules so that potential rates of population growth can be calculated (Huffaker and Rabb, 1984). The possibility that prey may be hard to find also has implications for the design of these demographic studies. If predators often face prey shortages, then we should pay close attention to the survival and reproduction of nearly starved predators.

In summary, we think that, to understand the success or failure of introduced biological control agents, more research is needed on their ability to find scarce patches of prey and to endure prolonged periods of starvation. Current approaches to the evaluation of biological control agents produce overly optimistic portraits, because they quantify predator performance in unrealistically favorable situations. In the next section, we discuss a model that more appropriately directs attention to predators seeking prey that are possibly scarce and certainly patchy.

A THEORETICAL FRAMEWORK IN WHICH TO EXAMINE PREDATOR HUNTING AND AGGREGATION

On the assumption that aggregative behavior can contribute to the success of a predator, we have developed a model that translates easily measured features of a predator's movement behavior into aggregation rates (Kareiva and Odell, 1987). The model is composed of two partial differential equations,

$$\frac{\partial p}{\partial t} = \left[I + E(V) + \frac{\partial}{\partial x}\left(D(V)\frac{\partial P}{\partial x} - A(V)P\frac{\partial V}{\partial x}\right)\right] + F(V,P) \qquad (12.1a)$$

and

$$\frac{\partial V}{\partial t} = \Psi \frac{\partial^2 V}{\partial x^2} + G(V,P) \qquad (12.1b)$$

where V is prey density, P is predator density, and x is position along a single spatial dimension; the model can be generalized to two spatial dimensions (i.e., x-y coordinates). Equation (12.1b) describes how prey density changes in space and time as a result of two components: random motion at a constant rate "Ψ," and local birth and death processes represented by the function $G(V,P)$; G typically includes a logistic function for population growth in the absence of predators and an additional term that accounts for losses due to predation. Equation (12.1a) describes how predator density changes in space and time, again both due to movement and to local births and deaths. In the predator equation, movement is represented by the bracketed terms, and local births and deaths are accounted for by the function $F(V,P)$. Elsewhere, we discuss the implications of this model for predator–prey dynamics (Kareiva and Odell, 1987). Here, we want to focus on the aspects of the model that relate to a predator's hunting behavior, which means the terms inside the brackets in equation (12.1a). I is a constant immigration rate, implying that predators arrive in a field without regard for prey availability; $E(V)$ is a function describing long-range emigration, presumably at a rate that increases as prey density declines. These two model components can be thought of in an agricultural setting as representing the movement of predators between fields. By themselves, $E(V)$ and I can produce a pattern of predator aggregation in regions of high-prey density. However, most predators respond to variations in prey density at a smaller spatial scale through local movement. In equation (12.1a), this local movement is depicted by $D(V)$ and $A(V)$; $D(V)$ is random motion at a rate that depends on prey density, whereas $A(V)$ is oriented or directed movement, again at a rate that depends on prey density. $A(V)$ is the aggregation term that causes predators to move toward regions of increasing prey density at a rate that is proportional to the magnitude of change in prey density per unit change in position (i.e., proportional to $\partial V/\partial x$).

The functions $D(V)$, $E(V)$, and $A(V)$, and the parameters associated with those functions, can be obtained for any predator by observing that predator's flight, velocity, and turning behavior at a variety of prey densities (details are given in Kareiva and Odell, 1987). It is important to realize that the pattern of predator aggregation that one should expect to observe depends not only on a predator's intrinsic behavioral program but also on the spatial arrangement of prey. This dependence is explicitly handled by having emigration, random motion, and oriented motion depend either on V or $\partial V/\partial x$ in equation (12.1a). One value of equation (12.1a) is that it can be used to translate short-term observations of individual predators into predictions about the extent to which populations of predators will aggregate at different sorts of prey clusters. We have tested these predictions in replicated field experiments and have found that the model performs remarkably well for *C. septempunctata* (Kareiva and Odell, 1987). Here, we want to discuss the model from a different perspective, using it as a framework for investigating the failures and successes of predators that must hunt for patchily distributed prey.

Equations (12.1a) and (12.1b) represent a major departure from earlier models of the influence of dispersal on predator–prey dynamics. Previous models either concentrated on within-field movement of natural enemies (Beddington et al., 1978; Hassell and May, 1974, 1985; Hassell, 1978) or dealt with between-field dispersal to the exclusion of local dynamics (Vandermeer, 1973; Hastings, 1977; Renshaw, 1982). Our model includes both aspects of predator behavior, and, in particular, treats within-field movements in mechanistic detail. The advantage of our mechanistic approach is that the parameters that determine predator success are estimated from direct measurements of the behavior of individual predators, rather than observations of the static spatial pattern of predators and their prey. We reiterate that characterizations of a predator–prey spatial pattern are not sufficient to reveal the underlying movement behavior that generates the pattern. We also emphasize that, although our model accounts for between-field movement, we have not given it the same detailed treatment accorded within-field movement. Nevertheless, long-range dispersal clearly must contribute to the regional persistence of a biocontrol agent, and more theoretical and empirical attention need to be given to predator between-field movement.

Using equations (12.1a) and (12.1b) as a theoretical basis for evaluating predator effectiveness, numerical simulations reveal that the key to a predator's success is a large value for $A(V)$ in equation (12.1a). Fortunately, $A(V)$ can be related to a few quantifiable features of predator foraging behavior. In particular, predators will aggregate most rapidly when they

1. Change plants frequently or move rapidly through a field in the absence of prey
2. Tend to move in straight lines and infrequently reverse directions in the absence of prey
3. Spend a large fraction of their time actually searching plants
4. Reduce their velocity and turn more frequently in the midst of a prey patch

Prediction (3) is obvious; predictions (1) and (2) arise because predators must search widely for prey patches if they are to aggregate at those patches and must not search the same area over and over again; prediction (4) arises because, for predators to aggregate, an individual predator entering a prey patch must alter its movement behavior so as to remain in that patch. One of the ways in which a model such as equation (12.1a) can be used is to make precise statements about how $A(V)$ should change for a given change in velocity or turning behavior and, in turn, what this change in $A(V)$ means for aggregation. These predictions can be used to compare the aggregative behavior and hence the effectiveness of different biocontrol agents. In the next section, we report field studies of ladybug aggregation and relate our findings to the previous predictions regarding $A(V)$ and foraging behavior. Later in this chapter, we will use equations (12.1a) and (12.1b) to compute numerical solutions corresponding to different ladybug behaviors.

TRYING TO UNDERSTAND GOOD AND POOR AGGREGATION IN *COCCINELLA SEPTEMPUNCTATA*: HABITAT EFFECTS, PLANT EFFECTS, AND INDIVIDUAL DIFFERENCES

From 1981 to 1988, we have been studying the foraging ecology of the seven-spotted ladybug, *Coccinella septempunctata*. This ladybug was initially released by the USDA in New Jersey and subsequently in several other states. It has become widely established and is reputed to be an extremely effective predator of aphids (Obrycki et al., 1982). By experimentally removing ladybugs from goldenrod fields, we have verified that these predators can control aphid populations (Kareiva, 1985). However, the ability of *C. septempunctata* to suppress aphid populations varies with vegetation and habitat structure. For instance, in large continuous stands of goldenrod, *Coccinella* generally hold aphid populations at low densities; but, when the goldenrod is experimentally fragmented into patches, the aphids frequently escape ladybug control and erupt in local high-density outbreaks (Kareiva, 1987). Habitat fragmentation appears to disrupt the ability of ladybugs to control aphids, because it slows down the rate at which ladybugs aggregate at incipient aphid outbreaks (Kareiva, 1985, 1987). But, even though ladybugs aggregate more slowly in fragmented goldenrod fields, *Coccinella* can still be found clumped together at aphid clusters. For this reason, we initially thought that aggregation would always be an important feature of *Coccinella* foraging behavior, albeit at different rates, depending on habitat structure (Kareiva, 1987; Kareiva and Odell, 1987).

With a firm belief in ladybug aggregation, we embarked on an evaluation of the effectiveness of different ladybugs as biocontrol agents, using rate of aggregation as a focal measure of each species's promise. We began in 1986 by releasing *Coccinella septempunctata* in commercial pea fields in order to quantify the rate at which they aggregated at patches of the pea aphid, *Acyrthosiphon pisum* (Harris). To our surprise, the ladybugs revealed absolutely no tendency to aggregate in those parts of pea fields that harbored elevated densities of pea aphids, even though identical experiments had yielded striking evidence of aggregation in goldenrod fields (Fig. 12.2). We then tested the ability of two other ladybug species, *Hippodamia convergens* Guérin-Méneville and *Hippodamia variegata* (Goeze), to aggregate at aphid clusters in pea fields and found at best very weak aggregation. Initially, these findings made us question our emphasis on an "aggregation model." However, examination of the individual behavior of ladybugs revealed that the absence of a strong aggregative response in pea fields was, in a sense, a vindication of our model, because the model predicts just such a failure, given the ladybug behaviors that we observed in pea foliage. In particular, whereas ladybugs move rapidly in straight paths and rarely reverse direction when they are searching goldenrod stands that lack aphids, they move slowly and reverse direction often when they are searching pea fields, even if the plants are entirely without aphids (Fig. 12.3). Furthermore, ladybugs fall off pea plants so frequently that in pea fields they spend one-quarter of their foraging time on the ground (Fig. 12.4); in contrast, *Coccinella* spends virtually no time on the ground when it is hunting through goldenrod. In summary, ladybugs should not be expected to aggregate at aphid clusters in pea fields, because they are ineffective at moving from plant to plant and because their behavior does not markedly change when they do find aphids on peas.

Although our model explains, in one sense, why ladybugs do not aggregate in pea fields, we still need to ask what it is about the pea system that produces "nonaggregating" movement patterns. Our hypothesis is that pea plant morphology (especially slippery leaves) and architecture (i.e., tangled, twisted foliage) interfere with ladybug movement. Furthermore, we suspect that plant architecture in general may constrain the ability of predators to aggregate at prey clusters, just as habitat patchiness has been shown to limit the ability of ladybugs to aggregate at aphid outbreaks (Kareiva, 1987). We have been examining this hypothesis by quantifying the searching behavior of ladybugs on plants that differ markedly in their architectures, for instance, peas versus beans, and leafless peas versus normal peas. Preliminary results indicate a striking effect of plant architecture on ladybug searching—an effect that varies among species (Kareiva and Doak, 1990). For instance, whereas *C. septempunctata* is inevitably frustrated when it is searching through peas because it frequently falls to the ground or gets hung up on the underside of a leaf, *Hippodamia*

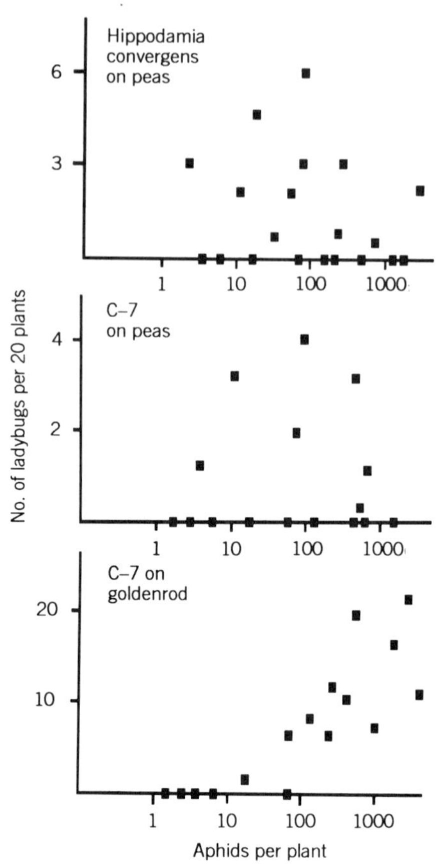

Figure 12.2. Aggregation of released ladybugs in fields containing naturally occurring patches of aphids (C-7 = *Cocinella septempuctata*). The experiments involved the release of adult ladybugs, followed by sampling of numerous meter-square quadrats for both ladybugs and aphids. An increase in ladybug density with increasing aphid density can be taken as evidence of aggregation by the ladybugs. The samples were taken two days after the ladybugs were released, and the initial position of the ladybugs was random with respect to the aphid patches.

variegata seems to negotiate pea foliage relatively well (Kareiva and Doak, 1990). These species specific differences may arise because of differences in the size or morphology of the species, as well as of differences in behavior. The important point is that predators may be successful only in certain types of vegetation, because they can move effectively only in those types of vegetation. Furthermore, different ladybug species may, by virtue of their morphology and behavioral inclinations, be differentially adapted to searching particular vegetation. Although we are certainly not the first to notice the effect of plant structure on predator efficiency, previous discussions of the influence of plant morphology on insect predators have concentrated almost exclusively on within-plant searching (Shah, 1982; Belcher and Thurston, 1982; Rabb and Bradley, 1968; Rotheray, 1986; Quilici and Iperti, 1986; Obrycki

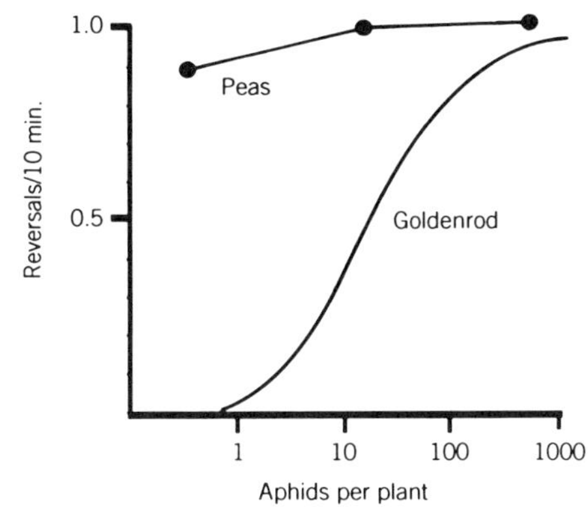

Figure 12.3. The reversal frequency of *Coccinella septempunctata* when observed foraging on rows of goldenrod plants and pea plants harboring various densities of aphids. Only three points are shown for peas, since observations have been made at only three aphid densities (20 ladybugs observed at each prey density); the goldenrod curve is shown as a continuous line, because it has been well characterized by two years of experimentation (see Kareiva and Odell, 1987). A reversal is recorded whenever a ladybug moving in one direction down the row of plants turns around to move in the opposite direction.

and Tauber, 1984). In contrast, we want to emphasize that plant morphology also alters predator efficiency, because it influences movement between plants.

Having found that different predator species respond differently to plant architecture, it is natural to ask whether individuals within a species exhibit any consistent behavioral idiosyncrasies in the way in which they move through foliage. To address this question, we observed different ladybugs repeatedly under controlled field conditions. Two of the parameters we measured were (1) percentage of time spent foraging and (2) mean rate of displacement per unit time spent foraging. To minimize complicating factors, we recorded these parameters for individuals foraging in the absence of prey, a simplification that also makes sense, because behavior when no

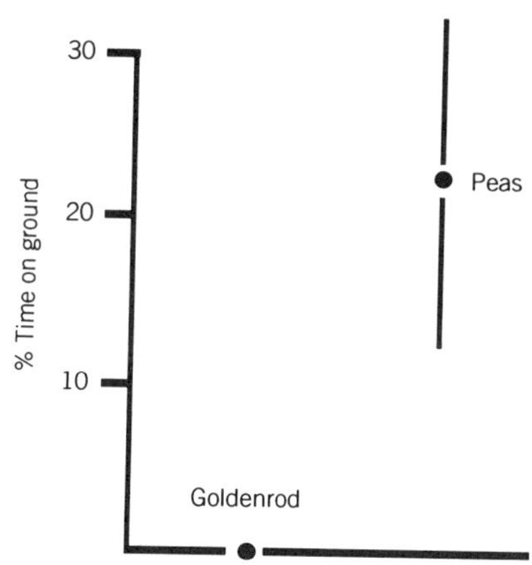

Figure 12.4. The fraction of foraging time *Coccinella septempunctata* spends wandering on the ground as opposed to searching plants. The goldenrod data represent > 50 individuals, and the pea data represent 20 individuals, observed for periods of from 10 to 30 minutes.

prey are present is key to the aggregation response. Our analyses include only ovipositing females of the same age, reared and maintained under identical laboratory conditions. The ladybugs were observed foraging in the same common plot of commercial pea or bean plants and at the same times of day. When they were not being used in foraging experiments, our ladybug gladiators were held in small cardboard containers with unlimited sugar water and an *ad libitum* supply of aphids (except on the experimental days, on which all individuals were starved for two hours prior to the observations). Individuals differed strikingly in their foraging styles: some ladybugs would frequently move from pea plant to pea plant, whereas others would be easily halted by an inconveniently placed leaf; some ladybugs would fall often to the ground, whereas others seemed adept at staying on the plants. Since these observations were done in plots from which aphids had been removed, encounters with prey cannot explain the differences that we observed. In addition, each ladybug was observed on at least five different days, and the differences that we observed appeared to be quite consistent. Indeed, in an analysis of variance, the between-individual factor explained 60–80% of the variance, depending on the behavioral parameter under consideration (see Table 12.1). Of course, these experiments do not indicate genetic differences among ladybugs with respect to foraging behavior, but they do show that phenotypic differences can be striking and consistent, and they suggest that it would be worthwhile to ask whether the differences are heritable. This is particularly true because the parameters that varied among individuals figure prominently in a ladybug's ability to find and aggregate at clusters of prey. More precisely, the behavioral parameters that differed among individuals play an important role in determining $A(V)$ and $D(V)$ in equation (12.1a).

One final interesting feature of individual variability is the possibility that different individuals might consistently forage in such a way that they do best on different plant species. For example, one beetle may do comparatively well on peas, whereas another beetle may do best on beans. These sorts of individual plant interactions did appear in our data (Fig. 12.5), but our sample sizes are still too small to examine the statistical significance of such interaction terms.

Table 12.1. Analysis of Variance: Partitioning Variation in Percent Time on the Ground into Within-Individual and Between-Individual Components[a]

Percent Time on the Ground				
Source	df	MS	F	P
Between	11	0.28	3.1	0.003
Within	50	0.09		(n.s.)

[a]These data represent *Coccinella septempunctata* foraging in a row of peas during 10-minute bouts of observation. Note that most of the behavioral variation appears to reflect consistent differences between individuals. The same trend holds for other behavioral parameters, such as reversal frequency and foraging velocity.

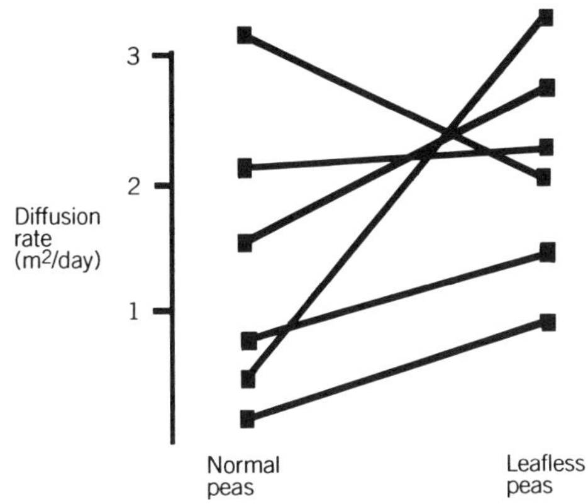

Figure 12.5. Using diffusion rate to summarize how different individual *Coccinella septempunctata* respond to the morphological variation in the plants they search for prey. In these experiments, individuals were observed on both normal and leafless peas in succession and returned to the laboratory. The pea plants were empty of aphids. Diffusion rate is calculated as mean-squared displacement divided by twice the time period of observation, which was 20 minutes for these observations.

USING THE AGGREGATION MODEL TO EXPLORE EVOLUTIONARY CHALLENGES FACING INVADING BIOLOGICAL CONTROL AGENTS: SPECULATIONS ABOUT THE CONSEQUENCES OF MINOR CHANGES IN BEHAVIOR

Traditionally, predator–prey models, such as equations (12.1a) and (12.1b), are used to address questions about population dynamics (Murdoch et al., 1985). However, because the behavior that corresponds to local movement in equation (12.1a) has a strong stochastic element, it might be better to interpret the numerical solutions of equation (12.1a) as an expectation for an individually foraging ladybug. In other words, we can imagine holding the prey dispersion constant and then finding the expected spatial distribution of activity for a single foraging predator. Using the model in this way, we can compare the advantages to a predator of different mixes of random movement and directed movement [i.e., different forms of $D(V)$ and $A(V)$ in equation (12.1a)] when presented with different arrangements of prey. Viewed in this context, the better a predator is at aggregating at prey clusters, the more food on average the predator will consume. The relative magnitudes of the aggregation and diffusion components of predator movement $A(V)$ and $D(V)$, respectively, in equation (12.1a) are determined by the way in which the turning frequency of a foraging predator depends on local prey density through the influence of local prey density on predator gut fullness (see Karieva and Odell, 1987, for details). By altering the sensitivity of turning behavior to local prey density, we can adjust the relative contributions of random and directed motion to predator movement. In Fig. 12.6, we display the results of numerical solutions to equation (12.1a) for different prey dispersions and styles of predator movement, by which we mean combinations of $A(V)$ and

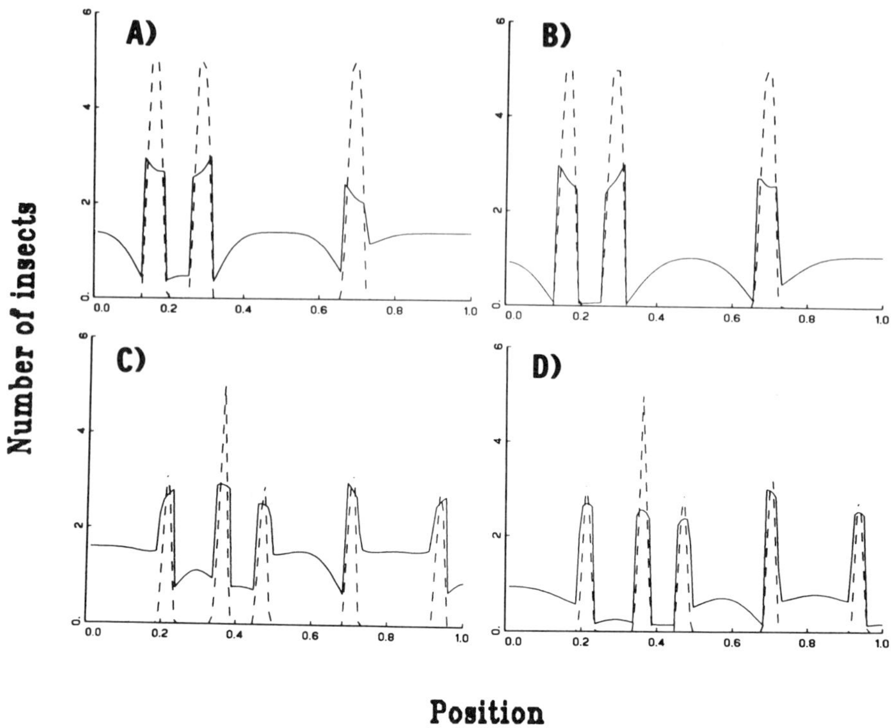

Figure 12.6. Effects on predator aggregation of prey dispersion and different relative importances of random and directed motion. Graphs show spatial distributions of predators (———) and prey (– – –), generated by numerical solutions of equations (12.1a) and (12.1b) at t = one day for two initial prey dispersion patterns and two kinds of predator movement behavior. Vertical axis is number of insects (× 2,000 for the prey curves), and horizontal axis is position along a 1 m row of plants. (a) Few uniformly large prey peaks; random motion high relative to directed motion. (b) Few uniformly large prey peaks; random motion low relative to directed motion. (c) Several smaller prey peaks; random motion high relative to directed motion. (d) Several smaller prey peaks; random motion low relative to directed motion. Prey peaks are randomly distributed along the spatial axis; sizes of prey peaks in (c) and (d) were generated from a negative binomial distribution. Initial predator dispersions were uniform across the spatial axis. For simplicity, the parameters governing between-field predator movement, predator births and deaths, local prey movement, and prey births and deaths have been set to zero. Functional response parameters and predator foraging velocity are as in Kareiva and Odell (1987).

$D(V)$. Note that the effects of strong predator aggregation are more pronounced when the prey are found in large isolated clumps than when the prey peaks are smaller and closer together; in other words, the importance of strong predator aggregation is greatest when clusters of prey are large and far apart. Clearly, modest changes in predator movement can yield major changes in the degree of aggregation, and the effect of these changes depends in part on the nature of the prey dispersion. Furthermore, in terms of predator efficacy in a biocontrol setting, when prey are patchily distributed, strongly aggregating predators will eat more prey than predators in which random motion is more important. The results in Figure 12.6 also illustrate the possibility that the evolutionary advantages of a particular tendency for random relative to directed motion might depend on the expected dispersion of prey. For instance, if prey are usually found in large but widely scattered outbreaks, so that most of the environment is devoid of prey, there may be a premium on strong directed motion (so that predators are not likely to wander out of a prey patch after finding one). The major point is that we might expect ladybug species to be adapted to different prey dispersions through their local movement behavior.

Since we have some information on variation in hunting behavior between species and between individuals of the same species, we can use equation (12.1a) to investigate the consequences of that variation. Both the diffusion coefficient and the aggregation coefficient are proportional to the square of the average velocity of searching predators (Kareiva and Odell, 1987). Using this relationship, we can calculate how many aphids a ladybug is likely to eat if all other parameters are held constant while velocity is varied over the range suggested by our data. Presumably, a rapidly searching predator will encounter and eat more prey than a more sluggish predator, with the actual number eaten depending on both the relative values of $D(V)$ and $A(V)$ and on the response of the predator's feeding rate to prey density. The result also depends, of course, on the exact prey dispersion; for simplicity, we have selected just one prey dispersion to investigate. Figure 12.7 shows the results of a series of numerical studies in which we have calculated the number of aphids that a ladybug is expected to eat in a one-day period for a variety of velocities, ranging from 10% to 150% of the average velocity of *C. septempunctata* foraging on goldenrod. All of these values of predator foraging velocity are well within the range that we have observed for individual *C. septempunctata* foraging on pea plants. The results of this simulation dramatize how, by moderate changes in its velocity, a pred-

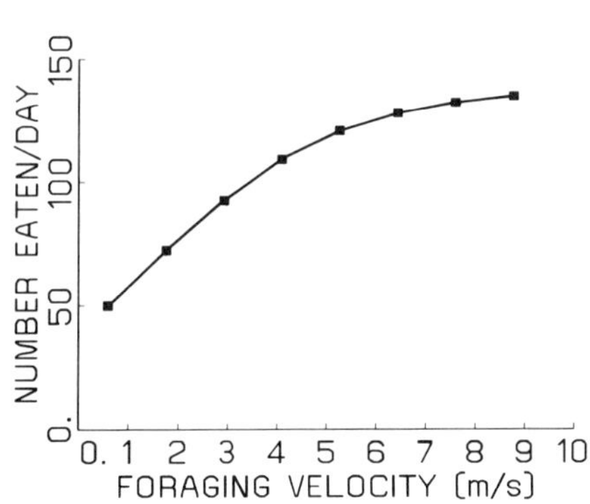

Figure 12.7. Effect of predator foraging velocity on number of prey eaten per predator per day. Vertical axis is number of aphids eaten per ladybug per day; horizontal axis is predator foraging velocity, in m day^{-1}. Results were obtained from numerous numerical solutions of equations (12.1a) and (12.1b). Initial prey dispersion pattern consisted of a single peak of 200 aphids in the center of a 1 m row of plants; initial predator dispersion was uniform across the spatial axis. Functional response parameters were chosen to give a maximum feeding rate of 200 aphids day^{-1} and a feeding rate of 50 aphids day^{-1} at a prey density of 100. Predator movement parameters were as in Figure 12.6 (b) and (d); all other parameters were as in Figure 12.6.

ator can greatly improve its feeding rate. It is clear from these results that foraging velocity should be viewed by biocontrol researchers as an important variable contributing to predator efficacy.

CONCLUSIONS

If finding prey is indeed a major problem for predators, then the effects illustrated in Figures 12.6 and 12.7 should play a major role in the dynamics of predator populations. In particular, predator-turning behavior and foraging velocity must interact strongly with functional-response parameters to determine the success or failure of introduced biological control agents. Additionally, these theoretical results suggest evolutionary pressures that might shape an organism's foraging behavior. We do not mean to imply that aggregation behavior is the "be all and end all" of predator success. But we do think that such behavior deserves more attention, both from the applied perspective and from the evolutionary perspective. There is a clear opportunity for predator species to adapt to habitat structure, to plant morphology, and to prey dispersion through their aggregative foraging behavior.

Acknowledgments

M. Andersen was supported by NSF grant BSR-8501031 while doing this research, and P. M. Kareiva was supported by NSF grant BSR-8517183. Discussions with Garrett Odell were particularly helpful in formulating the model analysis.

REFERENCES

Angelet, G. W., J. M. Tropp, and A. N. Eggert. 1979. *Coccinella septempunctata* in the United States: Recolonizations and notes on its ecology. *Environ. Entomol.* 8:896–901.

Baker, H. G., and G. L. Stebbins. 1965. *The Genetics of Colonizing Species*. New York: Academic Press.

Beddington, J. R., C. A. Free, and J. H. Lawton. 1978. Characteristics of successful natural enemies in models of biological control of insect pests. *Nature* 273:513–519.

Belcher, D., and R. Thurston. 1982. Inhibition of movement of larvae of the convergent lady beetle by leaf trichomes of tobacco. *Environ. Entomol.* 11:91–94.

Comis, D., and M. Heppner. 1986. Battle plans for national assault on aphids. *Agric. Res. (Washington, D.C.)* June/July 10–12.

Crawley, M. 1986. The population biology of invaders. *Philos. Trans. R. Soc. Lond. Ser. Biol. Sci.* 314:711–731.

Curio, E. 1976. *The Ethology of Predation*. New York: Springer-Verlag.

Hall, R., and L. Ehler. 1979. Rate of establishment of natural enemies in classical biological control. *Bull. Entomol. Soc. Am.* 25:280–282.

Hassell, M. 1978. *The Dynamics of Arthropod Predator–Prey Systems*. Princeton, N.J.: Princeton Univ. Press.

———. 1982. Patterns of parasitism by insect parasitoids in patchy environments. *Ecol. Entomol.* 7:365–377.

Hassell, M., and R. M. May. 1974. Aggregation of predators and insect parasites and its effect on stability. *J. Anim. Ecol.* 43:567–594.

———. 1985. From individual behaviour to population dynamics. Pages 3–23 in Sibly, R. M., and R. H. Smith (eds.), *Behavioural Ecology.* Oxford, England: Blackwell.

Hastings, A. 1977. Spatial heterogeneity and the stability of predator–prey systems. *Theor. Pop. Biol.* 12:37–48.

Heads, P., and J. H. Lawton. 1983. Studies of the natural enemy complex of the holly leaf miner: The effects of scale on the detection of aggregative responses and the implications for biological control. *Oikos* 40:267–276.

Huffaker, C., and R. Rabb (eds.). 1984. *Ecological Entomology.* New York: Wiley-Interscience.

Kareiva, P. 1985. Patchiness, dispersal, and species interactions:. Consequences for communities of herbivorous insects. Pages 192–206 in Case, T., and J. Diamond (eds.), *Community Ecology*. New York: Harper & Row.

———. 1987. Habitat fragmentation and the stability of predator–prey interactions. *Nature* 326:388–391.

Kareiva, P., and P. Doak. 1990. Unpublished manuscript.

Kareiva, P., and G. Odell. 1987. Swarms of predators exhibit "preytaxis" if individual predators use area-restricted search. *Am. Nat.* 130:233–270.

Luck, R. F., B. Shepard, and P. Kenmore. 1988. Experimental methods for evaluating arthropod natural enemies. *Annu. Rev. Entomol.* 33:367–391.

Moulton, S., and S. Pimm. 1986. Species introductions to Hawaii. Pages 231–249 in Mooney, H., and J. Drake (eds.), *Ecology of Biological Invasions of North America and Hawaii*. New York: Springer-Verlag.

Murdoch, W., et al. 1985. Biological control in theory and practice. *Am. Nat.* 125:344–366.

Obrycki, J., and M. Tauber. 1984. Natural enemy activity of glandular-pubescent potato plants. *Environ. Entomol.* 13:679–683.

Obrycki, J., et al. 1982. Establishment of a European lady beetle in New York State. *N.Y. Food and Life Sci. Bull.* 94:1–3.

Quilici, S., and G. Iperti. 1986. The influence of host plant on the searching ability of first instar larvae of *Propylea quatordecimpunctata*. Pages 99–106 in Hodek, I. (ed.), *Ecology of Aphidophaga*. The Hague, The Netherlands: Dr. W. Junk.

Rabb, R., and J. Bradley. 1968. The influence of host plants on parasitism of eggs of the tobacco hornworm. *J. Econ. Entomol.* 61:1249–1250.

Renshaw, E. 1982. The development of a spatial predator–prey process on interconnected sites. *J. Theor. Biol.* 94:355–365.

Rotheray, G. 1986. Effects of plant topography and larval morphology on the searching efficiency of aphidophagous Syrphidae. Pages 107–111 in Hodek, I. (ed.), *Ecology of Aphidophaga*. The Hague, The Netherlands: Dr. W. Junk.

Shah, M. 1972. The influence of plant surfaces on the searching behavior of coccinellid larvae. *Entomol. Exp. Appl.* 31:377–380.

Shands, W., and G. Simpson. 1972. Insect predators for controlling aphids on potatoes. 2. In small plots with two kinds of barriers, in small fields, or in large cages. *J. Econ. Entomol.* 65:514–518.

Shands, W., G. Simpson, and C. Gordon. 1972. Insect predators for controlling aphids. *J. Econ. Entomol.* 65:810–817.

Sundby, R. 1966. A comparative study of the efficiency of three predatory insects. *Entomophaga* 11:395–404.

Tamaki, G., and R. Weeks. 1972. Efficiency of three predators used alone or in combination against three insect prey species in a greenhouse study. *Environ. Entomol.* 1:258–263.

Taylor, L. R. 1984. Assessing and interpreting the spatial distributions of insect populations. *Annu. Rev. Entomol.* 29:321–357.

Trumble, J., et al. 1983. Temporal variation in the spatial dispersion patterns of aphids infesting strawberries. *Environ. Entomol.* 12:595–598.

Turchin, P., and P. Kareiva. 1989. Aggregation in *Aphis varians*: An effective strategy for reducing predation risk. *Ecology* 70:1008–1016.

Turnbull, A., and D. Chant. 1961. The practice and theory of biological control of insects in Canada. *Can. J. Zool.* 39:697–753.

Vandermeer, J. H. 1973. On the regional stabilization of locally unstable predator–prey relationships. *J. Theor. Biol.* 41:161–170.

13

THE EVOLUTION OF HABITAT PREFERENCE: AVOIDANCE AND ADAPTATION

Mark D. Rausher

Duke University

INTRODUCTION

One of the most conspicuous types of variation among species of herbivorous insects occurs in the number of different types of host plants that are used. Some insects are extreme specialists, using only one type of host at a given locality (Singer, 1971; Fox and Morrow, 1981). Others are extreme generalists and may typically use plant species from more than 10 families (Scriber and Feeny, 1979; Strong et al., 1984). Still other species exhibit degrees of host specialization between these too extremes.

Evolutionary biologists have employed both optimality models and explicitly genetic models in attempts to understand the reasons for such extensive variation in breadth of host range (Levins and MacArthur, 1969; Jaenike, 1978; Courtney, 1982, 1983; Futuyma, 1983; Rausher, 1984a; Iwasa et al., 1984; Rausher and Englander, 1987). These modeling efforts have led to predictions regarding the conditions under which evolution of a specialized host range will occur. Factors that are predicted to favor a broad host range include large differences among hosts in suitability, large temporal variation in the suitability of individual hosts, general rarity of host plants, and independent regulation of subpopulations on different hosts.

One major limitation of all of these models is that the suitability of a particular type of host plant (or, equivalently, the fitness of the insect on a particular host) is considered a fixed characteristic. In particular, no allowance is made for genetic variation in physiological characteristics of insects that might influence relative fitness on different hosts. Yet such variation often exists in populations of herbivorous insects (Gould, 1979; Via, 1984a; Hare and Kennedy, 1986). When it does, behavior and physiology evolve jointly, with the state of one character influencing the selection pressures experienced by the other. It is not clear that under such circumstances the factors that favor monophagy or polyphagy will be the same as when only behavior can evolve.

Few attempts have been made to model the joint evolution of preference and performance (viability and fecundity) traits. Gould (1984) analyzed a two-locus genetic model numerically in order to examine the joint evolution of avoidance of, and resistance to, pesticides and found that behavioral and physiological monophagy on the one hand and behavioral and physiological polyphagy on the other hand often represent alternative stable states for a population. He showed clearly that the state to which a population evolves can be influenced by initial gene frequencies and that the existence of the specialist state as an equilibrium depends on the existence of a cost associated with physiological adaptation. Castillo-Chavez et al. (1988) derived stability criteria for the equilibria in Gould's model and came to similar conclusions. Other researchers have argued verbally that whether monophagy or polyphagy evolves should depend on the relative magnitude of search costs associated with specialization and costs incurred by generalists in maintaining physiological adaptations to several hosts simultaneously (Dethier, 1954; Feeny, 1975; Rausher, 1983; Futuyma, 1983). Implicit in these arguments is the assumption that an increase in search costs favors polyphagy, while an increase in costs of physiological adaptation favors monophagy. These assumptions, as well as the effects of relative host abundance, mode of population regulation, and intensity of selection (as determined by differences in suitability among hosts) have not been examined quantitatively, however.

Here I present a preliminary analysis of a two-locus model of the joint evolution of preference and performance. Although the model is presented in terms of an insect herbivore and its host plants, the model is general enough to apply to the evolution of habitat breadth in any organism in which habitat selection is under maternal control. The main objective of this analysis is to determine under what conditions monophagy or polyphagy are expected to evolve, given a particular set of ecologically realistic starting conditions (see below).

THE MODEL

The situation modeled here is one in which a herbivorous insect uses two different species of host plants. Viability and fecundity of individuals developing on one host (Host I) are high, while viability and fecundity of individuals developing on the other host (Host II) are low. This situation could arise in any of the following ways.

1. *Invasion of the herbivore's range by a novel plant species*. Under this scenario, the insect is initially monophagous on Host I; Host II does not occur within the region occupied by the insect. Host II produces all of the chemical and physical cues that elicit oviposition by the insect but also contains a toxin or other property, not present in Host I, that reduces viability or fecundity. Invasion by Host II of the region occupied by the insect would then lead to the initial situation described above. An example of such a scenario is provided by the invasion of alpine meadows in the

Rocky Mountains within historical times by the cruciferous herb, *Thlaspi arvense*. This plant is toxic to larvae of *Pontia occidentalis* (Reakirt) and *Artogeia napi* (Linnaeus), though both butterflies oviposit readily on it (Chew, 1975, 1977).

2. *Expansion of the herbivore's range into a region with a novel host*. As in the previous scenario, this one begins with the insect initially being monophagous on Host I. Host II, which has the same properties as described previously, occurs only outside the current region occupied by the insect, while Host I occurs both within that region and in areas in which Host II is found. Expansion of the insects' range to encompass areas with both hosts would then produce the initial situation described above.

3. *Evolution of resistance by one host plant*. In this case, the insect population initially is physiologically adapted to two host species and uses each in proportion to their relative abundances in the environment. Evolution of a toxin or other resistance character then occurs in Host II, reducing the viability or fecundity of the insect on that host. Since the resistance character need not have any direct effect on insect behavior (Bernays and Graham, 1988), the insect would still use both hosts, even though the suitability of Host II is much lower than that of Host I.

The model analyzed here postulates that there is variation at each of two loci in the insect population. One locus, designated here as the "A" locus, influences the oviposition behavior of females. Individuals that are AA at this locus oviposit on the two hosts in proportion to the relative abundances of those hosts in the environment. If Host I constitutes a fraction c of the total number of Host I and Host II individuals, then AA individuals place a proportion c of their eggs on Host I and a proportion $(1 - c)$ on Host II. Individuals of genotype AA are thus generalists. By contrast, individuals that are of genotype aa are assumed to be specialists and place all of their eggs on Host I. Heterozygotes, Aa, are assumed to exhibit no dominance in preference. The proportion of eggs they place on Host I is thus $(1 + c)/2$.

The second locus, designated locus "B," determines the viability of individuals that develop on particular hosts. Allele B is assumed to be the wild-type allele. Consequently, for BB individuals, viability is high on Host I but low on Host II, with the difference in viability on the two hosts designated by the selection coefficient s (Table 13.1). Allele b, by contrast, is assumed to improve viability on Host II. In particular, I assume that individuals of genotype bb have equal viabilities on the two hosts. To allow for the possibility that increased viability on Host II can be achieved

Table 13.1. Viabilities of Genotypes at the B Locus on Hosts I and II

	Viability on	
Genotype	Host I	Host II
BB	1	$1 - s$
Bb	$1 - t$	$1 - t - 0.5s$
bb	$1 - 2t$	$1 - 2t$

only at the expense of a partial reduction in viability on Host I, the fitness of bb individuals on both hosts is taken to be $1 - 2t$. The quantity $2t$ is thus the magnitude of the loss of physiological adaptation to a resistant host (Rausher, 1984b; Gould, 1984). The viability of the heterozygote, Bb, is taken to be intermediate on both hosts (Table 13.1).

The loci A and B are assumed to be unlinked. It is also assumed that the genes at these loci have no direct pleiotropic effects. In particular, it is assumed that genotype at the A locus does not influence viability on either host, while genotype at the B locus does not influence the proportions of eggs laid on the two hosts. There may, however, be an indirect effect of genotype at the A locus on fecundity if degree of behavioral specialization influences the number of eggs that a female can lay in her lifetime. If, for example, a female's fecundity is limited by the time available for oviposition (Courtney, 1986; Rausher, 1985), then a polyphagous insect will have a higher fecundity than a monophagous insect, because it can find and oviposit on more acceptable plants per unit time. This effect is essentially a search cost associated with behavioral specialization. The cost of searching is designated by the parameter K, which can range in value from 0 (no cost) to 1 (full cost). The manner in which this cost is incorporated into the model is described in Appendix 13.1.

The basic life cycle of the insect is as follows: mated females place offspring into one or both habitats in proportions governed by their genotype at the A locus. The offspring develop entirely in their natal habitat, with a probability of survival determined by an individual's genotype at the B locus. When development is completed, surviving individuals mate randomly with respect to habitat of origin, and the cycle then begins again.

Two modes of population regulation are considered here. In the "soft selection" mode (Wallace, 1968), the subpopulations on each host are regulated independently. In particular, each host plant species is assumed here to contribute a fixed proportion of individuals to the mating pool. Such a situation would arise if there were a fixed carrying capacity for each habitat. For convenience, that proportion is taken to be equal to the fraction that that host constitutes of the total number of hosts in the habitat. In the "hard selection" mode, by contrast, regulatory factors act on the population as a whole (e.g., after individuals leave their natal habitat). The fractional contribution of each habitat to the mating pool thus is proportional to the number of eggs placed in that habitat and inversely proportional to the average viability in the habitat.

These assumptions dictate the recursion equations (given in Appendix 13.1) that describe the changes in gene frequencies that are expected to occur at the A and B loci. Using these equations, I ask what happens to these frequencies when alleles a and b are initially rare. Such initial frequencies correspond to a situation in which the insect population is initially behaviorally generalized but physiologically specialized on Host I (all individuals AABB) and has introduced into it by mutation alleles for avoidance of Host II (allele a) and for physiological adaptation to Host II (allele b).

NUMERICAL ANALYSES

Comparison of Hard and Soft Selection

Two sets of numerical analyses were performed, one assuming soft selection and the other assuming hard selection as the mode of population regulation. Initial gene frequencies were 0.005 for alleles a and b. A total of 1,470 runs were performed for each set, with each run corresponding to a different combination of the model's parameter values (Table 13.2). These values were chosen to span the set of all possible combinations of the parameters.

A comparison of these two sets of analyses indicates that mode of population regulation has a striking influence on whether specialization evolves. Under hard selection, the two most common results are the evolution of the generalist genotype (AAbb) and evolution of the specialist genotype (aaBB), with the former outcome slightly more common (Table 13.3). These two outcomes occur in 80% of the runs. In approximately 10% of the runs, no evolution occurs; the population remains behaviorally polyphagous but physiologically specialized. Finally, in another 10% of the runs, an equilibrium is reached in which one or the other locus is polymorphic while the other is fixed. These cases, in which the equilibrium is neutrally stable to perturbation, correspond to situations in which there is either no search cost ($K = 0$) or in which there is no cost of physiological adaptation ($t = 0$). It is clear from these analyses that, under hard selection, either physiological and behavioral specialization or physiological and behavioral generalization can evolve, depending on the values of the parameters.

By contrast, the evolution of specialization does not appear possible under soft selection. In this set of trials, the most common outcome was the evolution of physiological and behavioral polyphagy (fixation of the AAbb genotype), which occurred in approximately 70% of the runs. Evolution of complete specialization (fixation of aaBB) did not ever occur. No evolution away from the initial behavioral polyphagy and physiological specialization occurred in approximately 23% of the runs, while attainment of a polymorphic equilibrium either at the B locus or at both loci occurred in approximately 7% of the runs. It should be noted that, when the equilibrium

Table 13.2. Parameter Values of Model used in Analysis Comparing Evolution Under Hard and Soft Selection[a]

Parameter	Values Used
s	0.05, 0.10, 0.30, 0.50, 0.75, 0.99
$2t$	0, $0.05s$, $0.1s$, $0.3s$, $0.5s$, $0.75s$, $0.9s$
K	0, 0.05, 0.1, 0.3, 0.5, 0.75, 1
c	0.1, 0.3, 0.5, 0.7, 0.9

[a]The outcome of selection was determined for all combinations of indicated parameters. Parameters are explained in the text.

Table 13.3. Comparison of Evolution Under Hard and Soft Selection (Entries indicate number of trials leading to a particular type of equilibrium.)

	Selection Regime	
Equilibrium Type	Hard	Soft
AAbb fixed (behavioral and physiological generalist)	670	1,034
aaBB fixed (behavioral and physiological specialist)	501	0
AABB fixed (no evolutionary change)	200[a]	334
bb fixed, A locus polymorphic	84[b]	0
AA fixed, B locus polymorphic	0	33
A and B loci polymorphic	0	69
Indeterminate	15	0
Total	1,470	1,470

[a]Includes 58 cases in which allele B is nearly fixed and is neutrally stable to invasion by b allele because $t = 0$.
[b]Cases in which allele a is present at appreciable frequencies, corresponding to $K = 0$. The polymorphism is neutrally stable. Random perturbations will eventually carry these cases to fixation of aaBB (see text).

reached involved a polymorphism at the A locus, the frequency of the a allele was seldom high, which means that, in these cases, the population remained essentially behaviorally generalized. Because only under hard selection does the evolution of specialization appear possible, I restrict further analysis of the model to consideration of evolution under hard selection.

Effects of Initial Gene Frequencies

Gould (1984) found that the outcome of selection in his two-locus model of the evolution of preference and performance could depend on the initial frequencies of the a and b alleles. A small shift in initial frequencies could cause the modeled populations to shift from evolution of monophagy to evolution of polyphagy. This dependency on initial frequencies is also a feature of the model analyzed here. The reason for this dependency is that, under many combinations of the parameters, both the AAbb (generalist) and aaBB (specialist) equilibria exist and are locally stable. Consequently, each governs an area of attraction in gene frequency space. Whether monophagy or polyphagy evolves thus will depend upon which region of attraction initial gene frequencies lie in. An example is shown in Figure 13.1, which is a plot of gene frequency trajectories for a specific set of parameter values (see caption for Fig. 13.1). The dashed line is the boundary separating the regions of attraction for specialization ($p_a = 1, p_b = 0$) and generalization ($p_a = 0, p_b = 1$).

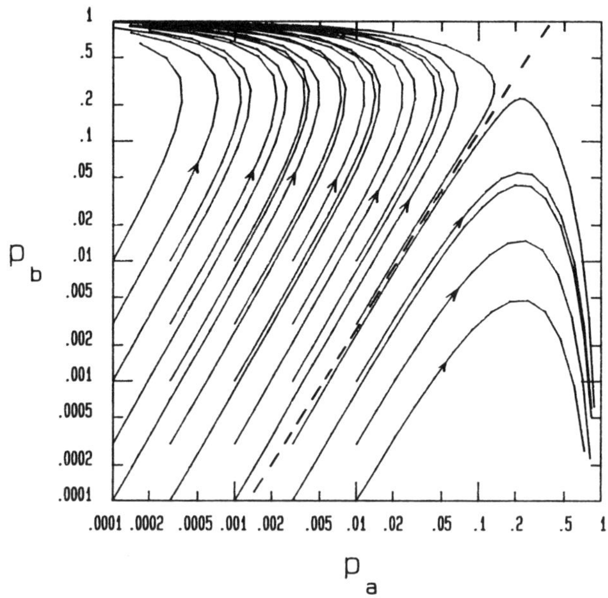

Figure 13.1. Plot of gene frequency trajectories for a particular combination of model parameters. All trajectories converge to either the generalist (AAbb fixed) or specialist (aaBB fixed) equilibrium. A broken line is the attraction boundary (see text). p_a is the frequency of a allele. p_b is the frequency of b allele. The values for the parameters in this example are $c = 0.5$, $s = 0.5$, $t = 0.2$, $K = 0.2$, and $P_4 = 0.75$.

When both specialist and generalist equilibria are locally stable, the position of the boundary between the regions of attraction (hereafter called "attraction boundary") will determine the likelihood of the evolution of specialization. If the attraction boundary is close to the line $p_a = 0$ (i.e., near the x axis of Fig. 13.1), then the region of attraction of the generalist equilibrium is large, that of the specialist equilibrium is small, and most combinations of initial gene frequencies will lead to the evolution of polyphagy. By contrast, if the attraction boundary is close to the line $p_b = 0$ (i.e., the y axis of Figure 13.1), then the region of attraction of the specialist equilibrium will be large and most combinations of initial gene frequencies will lead to specialization.

These considerations suggest that one effect of altering parameters in the model may be to shift the attraction boundary and thus alter the likelihood that a given initial combination of gene frequencies will lead to specialization. In the next section, I will examine this suggestion.

Effects of Changing Model Parameters

To determine the position of the attraction boundary for specialist and generalist equilibria, another set of numerical analyses was performed. For each of the 216 combinations of parameter values listed in Table 13.4, a number of runs were per-

Table 13.4. Parameter Values Used to Analyze Effects of Parameter Change on Position of Attraction Boundary[a]

Parameter	Values Used
s	0.05, 0.30, 0.50, 0.75, 0.95
$2t$	0.0045, 0.025, 0.045, 0.25, 0.45, 0.70
K	0.1, 0.5, 0.9
c	0.1, 0.5, 0.9

[a]All possible combinations of parameters were examined. Combinations with $2t > s$ were not analyzed.

formed. Each run began at a different initial frequency of the a allele, while the frequency of the b allele was held fixed at 0.001. By plotting the trajectories of these runs as in Fig. 13.1, it was possible to determine the approximate position of the attraction boundary when both specialist and generalist equilibria existed. These runs also revealed when the initial state of the population (AABB fixed) was evolutionarily stable, as well as when only the generalist equilibrium or only the specialist equilibrium was stable.

Effects of Search Costs

Increasing the cost of searching, K, decreases the likelihood that specialization will evolve. This effect occurs in two ways. First, increasing K tends to shift the attraction boundary toward the specialist equilibrium (Fig. 13.2A), thus decreasing the likelihood that an arbitrary initial combination of gene frequencies will lead to specialization. This effect is pronounced for some combinations of values of the parameters s, t, and c, whereas, for other combinations, the effect is minor, except for changes in K near its extremes (i.e., near $K = 1$ or $K = 0$). Second, in situations in which the specialist equilibrium is globally stable for low values of K, increasing K may cause that equilibrium to become unstable, while the AABB (physiological specialist, behavioral generalist) equilibrium becomes globally stable. Evolution of specialization is thus prevented, because no evolution occurs.

Effects of Cost of Physiological Adaptation

Increasing the cost of physiological adaptation, t, decreases the likelihood that generalization will evolve, as expected. It does so in some situations by shifting the attraction boundary toward the generalist equilibrium (Fig. 13.2B), thus increasing the likelihood that specialization will evolve. Additionally, for combinations of c, s, and K for which the specialist equilibrium is always unstable, increasing t can destabilize the generalist equilibrium and make the AABB equilibrium become globally stable, thus preventing any evolutionary change.

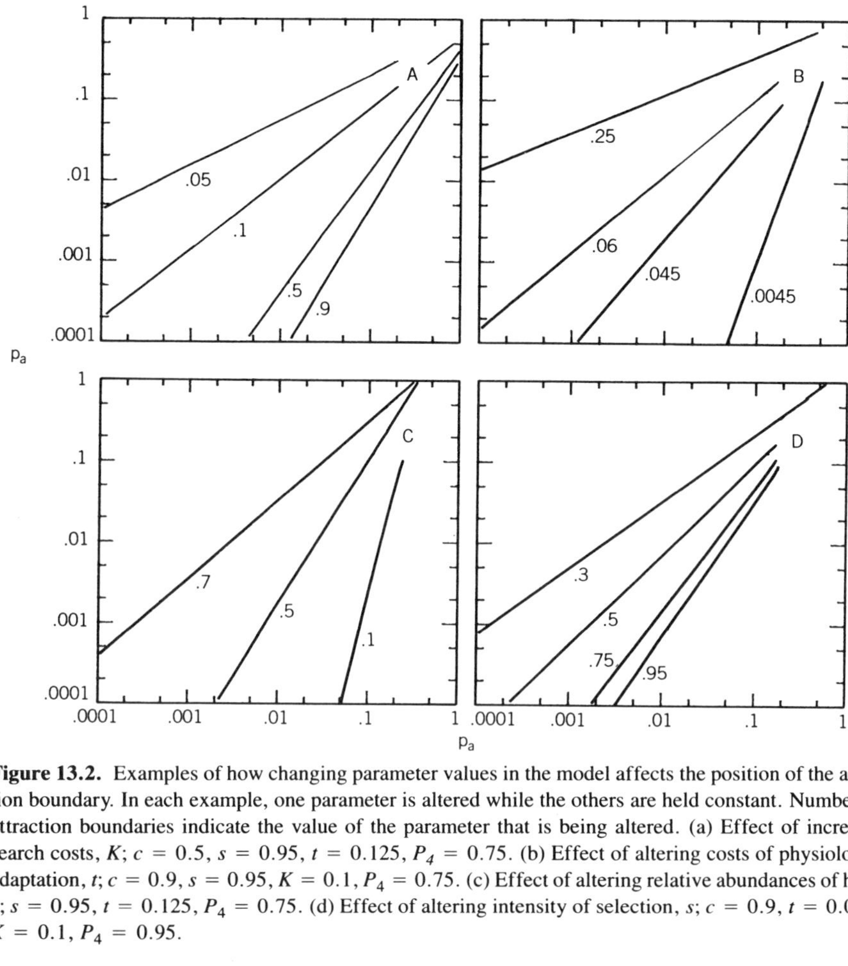

Figure 13.2. Examples of how changing parameter values in the model affects the position of the attraction boundary. In each example, one parameter is altered while the others are held constant. Numbers by attraction boundaries indicate the value of the parameter that is being altered. (a) Effect of increasing search costs, K; $c = 0.5$, $s = 0.95$, $t = 0.125$, $P_4 = 0.75$. (b) Effect of altering costs of physiological adaptation, t; $c = 0.9$, $s = 0.95$, $K = 0.1$, $P_4 = 0.75$. (c) Effect of altering relative abundances of hosts, c; $s = 0.95$, $t = 0.125$, $P_4 = 0.75$. (d) Effect of altering intensity of selection, s; $c = 0.9$, $t = 0.0125$, $K = 0.1$, $P_4 = 0.95$.

Effects of Relative Abundances of Hosts

In the model presented here, the parameter c indicates the fraction that Host I constitutes of the total number of Hosts I and II. There seems to be no intuitive reason for expecting alteration of c to favor the evolution of either polyphagy or monophagy. Increasing c decreases the proportion of "bad" hosts oviposited on by mothers of Bb and BB individuals and thus ameliorates selection against the B allele. Increasing c also decreases the proportion of "bad" hosts oviposited on by Aa and AA individuals and thus ameliorates selection against the A allele. There is no obvious a priori reason to believe that the magnitude of this amelioration will be greater for one locus than for the other and hence no obvious reason to expect changes in c to favor evolution at one locus over the other.

The numerical analyses indicate, however, that increasing c may often increase the likelihood that specialization will evolve. One reason for this is that, for many combinations of *s*, *t*, and *K*, increasing *c* causes the attraction boundary to shift toward the generalist (AAbb) equilibrium (Fig. 13.2C). Another reason is that, for other combinations of these parameters, increasing *c* destabilizes an AABB equilibrium that is globally stable for low values of *c* and converts an unstable aaBB equilibrium into one that is globally stable.

Effects of Intensity of Selection

The effects of altering the intensity of selection, *s*, are not as straightforward as effects associated with altering the other parameters of the model. The numerical analyses revealed that, for a few combinations of *c*, *t*, and *K*, increasing the intensity of selection significantly moves the attraction boundary toward the specialist (aaBB) equilibrium (Fig. 13.2D), thus making the evolution of polyphagy more likely. For a couple of combinations of these parameters, by contrast, increasing *s* has the opposite effect, though the effect on the position of the attraction boundary is rather weak, except as *s* approaches 1.0. For still other combinations, increasing *s* destabilizes a globally stable AABB equilibrium. Depending on the particular values of *c*, *t*, and *K*, this destabilization is accompanied by either the specialist equilibrium or the generalist equilibrium, or both, becoming stable. Thus, decreasing viability on Host II (i.e., increasing *s*) can favor monophagy or polyphagy, depending on environmental conditions and the magnitudes of costs of searching and physiological adaptation.

ANALYTICAL RESULTS

Criteria for Stability of the Equilibria

The numerical analyses described in the preceding section indicate that the only important equilibria for the hard-selection model are (1) AABB fixed, which represents no evolutionary change from the initial condition; (2) AAbb fixed, which corresponds to behavioral and physiological generalization (polyphagy); and (3) aaBB fixed, which corresponds to behavioral and physiological specialization (monophagy). Occasionally, equilibria occur involving polymorphisms at the B locus with AA fixed, with equilibria with the A locus polymorphic occurring even less frequently. However, these minor equilibria are normally neutrally stable (as revealed by numerical perturbation) and occur only when search costs or costs of physiological adaptation are nonexistent. The effect of perturbations of these equilibria is eventually to move the population to the specialist (aaBB) equilibrium if search costs are nonexistent or to the generalist (AAbb) equilibrium if costs of physiological adaptation are nonexistent. Consequently, a virtually complete picture of the stability properties of the model can be obtained by examining the stability of the AABB, AAbb, and aaBB equilibria.

Criteria for the stability of these equilibria (Table 13.5) were determined by standard linear stability analysis (see Appendix 13.2). These criteria do not assume

Table 13.5. Stability Criteria for the Generalist (AAbb Fixed), Specialist (aaBB Fixed), and Initial (AABB Fixed) Equilibria. (No assumptions about dominance at the A locus have been made.)

AABB Fixed Equilibrium

1a. $c > 1 - 2t/s$

1b. $K > s/[s + (1 - s)/cP_4$

1c.[a] $$\frac{0.25c\,[P_4 + K(1 - P_4)]\,[(1 - t - 0.5s) + 0.5sP_4]}{P_4\,[c + K(1 - c)]\,[(1 - s) + sc]} + \frac{0.25[(1 - t - 0.5s) + 0.5sc]}{[(1 - s) + sc]} < 1$$

AAbb Fixed Equilibrium

2a. $c < 1 - 2t/s$

2b.[b] $$\frac{0.25\,[1 + K(1 - P_4)/P_4]\,[(1 - t - 0.5s) + 0.5sP_4]}{[1 + K(1 - c)/c]\,[1 - 2t]} + \frac{0.25[(1 - t - 0.5s) + 0.5sc]}{[1 - 2t]} < 1$$

aaBB Fixed Equilibrium

3a. $K < sP_4/[1 - s(1 - P_4)]$

3b.[c] $$K < \frac{[4(1 + t) + s(1 - P_4)]\,P_4}{[2(1 - t) - s(1 - P_4)]\,[1 - P_4]}$$

3c. $t > 0$

[a]Numerical analyses suggest that this criterion fails only when both criteria 1a. and 1b. fail.
[b]This criterion is always satisfied if $t < 0.4286$.
[c]This criterion is satisfied if $P_4 > 0.33$; therefore, it is always satisfied if there is little dominance at the A locus or if $c >$ 0.33.

anything about dominance at the A locus, although they do assume no dominance at the B locus. Several immediate conclusions may be drawn from these criteria:

1. If there is no cost of physiological adaptation to Host II ($t = 0$), the specialist (aaBB) equilibrium is never stable. Therefore, in order for specialization to evolve, it is necessary for a cost of adaptation to exist, although this cost need not be large.
2. If fecundity is absolutely limited by available search time ($K = 1$), then specialization cannot evolve. This conclusion follows from the observation that $1 > [0.5s(1 + c)]/[1 - s + 0.5s(1 + c)]$ always, which violates stability criterion 3a. Consequently, if search costs are too high, specialization cannot evolve, regardless of the magnitude of costs of adaptation.
3. The initial (AABB) and generalist (AAbb) equilibria cannot simultaneously be stable. This situation is prevented, because c cannot simultaneously satisfy stability criteria 1a and 2a.
4. The initial (AABB) and specialist (aaBB) equilibria cannot simultaneously be unstable. Because $c < 1$, either criterion 1b or 3a must hold. It is possible for both to hold simultaneously, in which case, both equilibria may be stable simultaneously.
5. It can be shown that criterion 3b is important only if $P_4 > 1/3$, which will occur only if c is small and allele A exhibits a fair amount of dominance. If there is no dominance at the A locus, then P_4 is minimally $1/2$ and criterion 3b is always satisfied. The stability of the specialist equilibrium is then governed only by criteria 3a and 3c.
6. Criterion 2b is always satisfied as long as $2t > 0.9372$. Consequently, except when costs of physiological adaptation to Host II are very large, stability of the generalist equilibrium is determined solely by criterion 2a. Recent empirical evidence suggests that costs of adaptation are not large in natural populations of herbivorous insects (see "Discussion and Conclusions") and that, therefore, criterion 2b is seldom important.

Effects of Model Parameters on Stability

Given the stability criteria in Table 13.5, determining the effects on stability of altering the parameters is relatively straightforward. One simply determines whether increasing a given parameter increases or decreases the quantities on the right- or left-hand side of the inequality sign in a given criterion and, thus, whether it makes satisfying that criterion more or less difficult. For example, consider criterion 1a. Increasing the parameter s increases the quantity on the right-hand side of the inequality. For a given set of values of c and t, then, this increase makes it less likely that the criterion will be satisfied. Increasing s thus tends to decrease the likelihood that the initial (AABB) equilibrium will be stable. By contrast, increasing the value of c, keeping t and s constant, increases the likelihood that the criterion will be satisfied and thus tends to increase the likelihood that the equilibrium will be stable.

Similar considerations involving the other criteria and parameters lead to the effects of parameter alteration summarized in Table 13.6. Several inferences may be drawn from these effects. In doing so, it must be kept in mind that criteria 2b and 3b are important under very restricted conditions (see above). Consequently, except as noted, the effects of parameter change on these criteria are ignored:

1. Increasing the search costs (K) decreases the likelihood that the specialist equilibrium is stable, increases the probability that the initial equilibrium is stable, and has no effect on the likelihood that the generalist equilibrium is stable. Increasing search costs thus tends to prevent the evolution of specialization.
2. Increasing the costs of physiological adaptation (t) decreases the probability that the generalist equilibrium is stable, increases the likelihood that the initial equilibrium is stable, and has no effect on the stability of the specialist equilibrium. Increasing t thus tends to favor the evolution of specialization.
3. Increasing the relative abundance of Host I (c) decreases the probability that the generalist equilibrium is stable, has no effect on the stability of the specialist equilibrium, and may increase or decrease the probability that the initial equilibrium is stable. In other words, increasing the relative abundance of Host I tends to favor specialization by destabilizing the generalist equilibrium.
4. Increasing the intensity of selection (s) tends to destabilize the initial equilibrium but tends to increase the likelihood of stability for both the generalist and specialist equilibria. Unfortunately, this result sheds little light on whether increasing s would tend to favor the evolution of specialization or generalization.

These conclusions are in keeping with those issuing from the numerical analyses. In addition, examination of the stability criteria indicates that increasing the dominance of the a allele may stabilize or destabilize all of the equilibria. It might thus be

Table 13.6. Effects of Increases in Values of the Model's Parameters on Likelihood That Equilibria Will be Stable[a]

Equilibrium	Stability Criterion	s	t	c	K	P_4
AABB fixed	1a.	−	+	+	N.A.	N.A.
	1b.	−	N.A.	−	+	−
	1c.	−	+	+ or −	+	+ or −
AAbb fixed	2a.	+	−	−	N.A.	N.A.
	2b.	−	−	−	+	+ or −
aaBB fixed	3a.	+	N.A.	−	−	+
	3b.	+	+	N.A.	−	+ or −

[a]The effect determined from each stability criterion (see Table 13.5) is listed separately. A "+" indicates that an increase in the indicated parameter increases the likelihood that the equilibrium will be stable. A "−" indicates that an increase in the parameter decreases that likelihood. "N.A." indicates that there is no effect.

expected that dominance at the A locus affects the probability that specialization will evolve, but the exact nature of that effect is likely to be complicated and dependent on the values of the parameters of the model.

DISCUSSION AND CONCLUSIONS

Specialization and Mode of Population Regulation

Perhaps the most striking feature of the model analyzed here is the effect of mode of population regulation on the likelihood that behavioral and physiological specialization will evolve. Such specialization is approximately as likely as the evolution of physiological and behavioral generalization under hard selection but appears to be essentially impossible under soft selection. Single-locus soft-selection models of the evolution of host preference (Rausher, 1984a; Rausher and Englander, 1987) suggest the reason for this difference. Under soft selection, overall survivorship on a given host is density dependent. The evolution of behavioral specialization tends to be prevented, because it leads to overcrowding on one host (Host I) and undercrowding on the other host (Host II). Under such conditions, the fitness of behavioral generalists tends to be greater than that of behavioral specialists, because the generalists place a greater proportion of their offspring on an uncrowded host.

This type of selection pressure apparently also acts to prevent behavioral specialization in the two-locus model analyzed here. Since behavioral specialization is prevented, all females will place some offspring on the initially less-suitable host (Host II). Consequently, there will be selection tending to favor improvement of the fitness of individuals that develop on Host II. This type of selection will favor the evolution of physiological generalization as long as the reduction in fitness for individuals developing on Host I is not too severe, that is, as long as the cost of adaptation is not too high. Otherwise, the evolution of physiological generalization will be prevented, and the population will remain physiologically specialized though behaviorally generalized.

Recent surveys of the diet breadth of phytophagous insects (Futuyma and Gould, 1979; Strong et al., 1984; Bernays, 1984) indicate that specialist species far outnumber generalist species (Bernays and Graham, 1988). When interpreted in light of the model presented here, this pattern suggests that regulation of density in herbivorous insects does not normally occur independently on different host plants. Unfortunately, most studies that have attempted to examine factors that regulate insect populations (see Strong et al., 1984, for a compilation of such studies) have not explicitly attempted to determine at what spatial scale density regulation occurs. There is, therefore, little empirical evidence that can be used to examine the validity of this suggestion.

Specialization and Relative Host Abundances

One unexpected result is that, as the proportion of poor hosts (Host II) decreases, evolution of avoidance tends to be favored over evolution of physiological adaptation. It is not obvious why this effect exists. One possibility is that, as the proportion of good hosts decreases, the magnitude of search costs increases, because there are relatively fewer acceptable hosts available to specialists. However, this cannot be the entire explanation, since increasing the proportion of good hosts tends to favor avoidance even when search costs are assumed to be zero ($K = 0$).

Regardless of the causes of this effect, it may have important practical implications for the deployment of pesticides and resistant crop cultivars. One way to minimize the likelihood that pest species would develop resistance to such environmental toxins would be to use them sparingly at first. In the case of pesticides, this would mean that, in a given region, only a small percentage of potentially treatable areas should initially be treated. In the case of resistant crop varieties, this would mean intercropping of resistant and susceptible varieties, with the resistant varieties at low frequencies. This type of initial release would tend to favor the evolution of avoidance. If complete avoidance results within a reasonable amount of time, the use of the pesticide or resistant variety could then be greatly increased. The insect would then not be expected to develop resistance readily, because it would be trapped in the stable equilibrium state of being both behaviorally and physiologically specialized. To ensure the stability of that state, some percentage of potentially treatable areas would have to be left unsprayed (or planted with susceptible cultivars).

Gould (1984) analyzed a model in which level of avoidance was genetically fixed but could be altered by changing environmental factors, such as the repellancy of an insecticide or the fraction of a crop to which a repellent insecticide was applied. He found that increasing the level of avoidance significantly reduced the rate at which physiological adaptation to the insecticide evolved. The model described here extends this result to show that the evolution of such adaptation can actually be prevented completely if behavior causing complete avoidance is permitted to evolve. Whether such avoidance can be brought about within a reasonable amount of time by the appropriate deployment strategy is, of course, a question that must be settled empirically.

Specialization and Costs of Adaptation

Several authors have suggested that specialization by herbivorous insects often occurs because of the existence of costs associated with physiological generalization (see "Introduction" on page 259). According to this argument, generalists maintain a variety of characters (e.g., detoxification mechanisms), each of which adapts an individual for using a particular host species. Because an individual feeds on only one plant species at a given time, most of these adaptations are unused. They therefore confer no benefit on the individual. Maintenance of these unused adaptations is presumed to be costly, however, since it requires energy and/or materials to be diverted from other functions that would enhance fitness. Compared to a generalist feeding on a particular plant species, the fitness of a physiological specialist on that species should be higher, because it does not incur the costs associated with maintaining a battery of unused adaptations. This higher fitness is presumed to favor the evolution of physiological specialization. Once such specialization has been achieved, selection for avoidance of plants to which the insect is not physiologically adapted would follow, yielding behavioral specialization.

Not surprisingly, analysis of the model presented here substantiates this argument. Increasing the cost of physiological adaptation to Host II increases the likelihood that both physiological and behavioral specialization will evolve from an initial state of physiological specialization and behavioral generalization.

Perhaps even more interesting, however, is the result that specialization cannot evolve if there is no cost of physiological adaptation. This prediction is seemingly at odds with two empirical observations. The first is that most herbivorous insects are specialists (Bernays and Graham, 1988). The second is that several recent attempts to detect costs of adaptation generally seem to indicate that such costs are uncommon in herbivorous insects (Gould, 1979; Rausher, 1984b; Via, 1984b; Weber, 1985; Hare and Kennedy, 1986; Futuyma and Philippi, 1987). These two observations would seem to imply that costs of adaptation are not often responsible for specialization in plant-feeding insects (Futuyma and Philippi, 1987; Bernays and Graham, 1988).

Rausher (1988) gives several reasons why this inference may be premature, including the following: (1) most studies that have attempted to detect costs have examined only a few fitness components; (2) genetic differences in general vigor may be expressed under unnatural laboratory conditions, and these differences may mask any negative genetic correlations in performance across hosts that would be indicative of the presence of costs; and (3) selection has eliminated costs that once existed.

Another possible explanation for the apparent rarity of costs of physiological adaptation to new hosts is that such costs commonly exist but attempts to detect them have not been sufficiently sensitive. It seems likely, given the small number of families employed in most studies, that costs on the order of, say, 10% (i.e., a unit gain in fitness on Host II causes a loss in fitness of 0.1 units on Host I) could not be distinguished statistically from the complete absence of costs. If costs of such magnitude can prevent the evolution of polyphagy and hence favor monophagy, then current evidence cannot enlighten us as to whether such costs commonly are responsible for specialization.

The stability properties of the model analyzed here suggest that small costs can, under some circumstances, prevent the evolution of physiological and behavioral generalization. The generalist (AAbb) equilibrium is unstable essentially whenever $2t/s > 1 - c$ (criterion 2a, Table 13.5). The cost of adaptation, measured as the loss in fitness on Host I divided by the gain in fitness on Host II when substituting an a allele for an A allele is $2t/(s - 2t)$. It is easily shown (see Appendix 13.3), that the stability criterion implies

$$2t/(s - 2t) > (1 - c)/c$$

Consequently, the generalist equilibrium will be unstable whenever the cost of adaptation exceeds the right-hand side of the preceding equation. For $c = 0.9$, this means that a cost of 11% will prevent the evolution of polyphagy. In general, small costs will tend to favor specialization whenever Host I is more common than Host II, but the minimum cost necessary increases rapidly as c decreases. For example, a cost of more than 50% (corresponding to a t/s ratio of 0.167) is needed to destabilize the generalist equilibrium whenever $c = 0.67$, that is, whenever Host I is more than twice as abundant as Host II.

Most attempts to measure costs of adaptation probably would be able to detect costs of this magnitude. Failure to do so (ignoring other possible reasons for failure, listed earlier) suggests that any costs that do exist would be able to prevent the evolution of polyphagy only if good hosts are commonly much more abundant than bad hosts. Moreover, even when costs of adaptation prevent the attainment of the generalist equilibrium, they will favor specialization (as opposed to no evolution away from the initial state of behavioral generalization but physiological specialization) only when the specialist equilibrium is stable, that is, when search costs are small or nonexistent. What evidence exists suggests that search costs may be large in some species and absent in others (Courtney, 1982, 1986) but provides little indication of the relative frequency of search costs of different magnitudes. To the extent that search costs exist, however, they restrict further the conditions under which small costs of adaptation can foster the evolution of specialization. One may thus conclude from current evidence and the predictions of the model presented here that search costs may be responsible for the evolution of specialization in some species of herbivorous insects (those that find themselves in situations in which good hosts are much more common than bad hosts) but that monophagy in most species must be caused by some other factor or factors.

Specialization as an Evolutionary Cul-de-sac

In closing, I will outline one possible factor favoring diet specialization at the level of plant families. The analysis presented here suggests that behavioral and physiological specialization may be an evolutionary cul-de-sac from which escape is difficult. Consider a species that is behaviorally and physiologically generalized on two different host species. Suppose further that some ecological factor (e.g., parasitoid load) makes survivorship on one host (Host I) much greater than on the second host (Host II). Finally, suppose that this ecological factor is not one whose adverse effects can be easily ameliorated by a simple genetic change brought about by mutation.

Under these circumstances, one would often expect the insect to evolve behavioral avoidance of the poor host (i.e. specialization; see Fig. 13.3), as predicted by a variety of models (see "Introduction" on page 000). Gilbert (1979) termed this type of specialization "ecological monophagy" (see also Bernays and Graham, 1988). Once behavioral specialization occurs, one would expect a gradual loss of the ability to detoxify secondary chemicals present in Host II but not in Host I. This loss would occur both because of the accumulation of nonfunctional, though neutral, mutations affecting detoxification mechanisms and because selection for improved adaptation to Host I would lead to a decline in fitness on Host II, as long as costs of physiological adaptation are not strictly zero.

Once ability to process the foliage of Host II physiologically is completely (or nearly completely) lost, the population will have reached a state equivalent to the specialist equilibrium in the model. But this equilibrium is stable as long as there is any cost associated with physiological adaptation. (Search costs may be ignored,

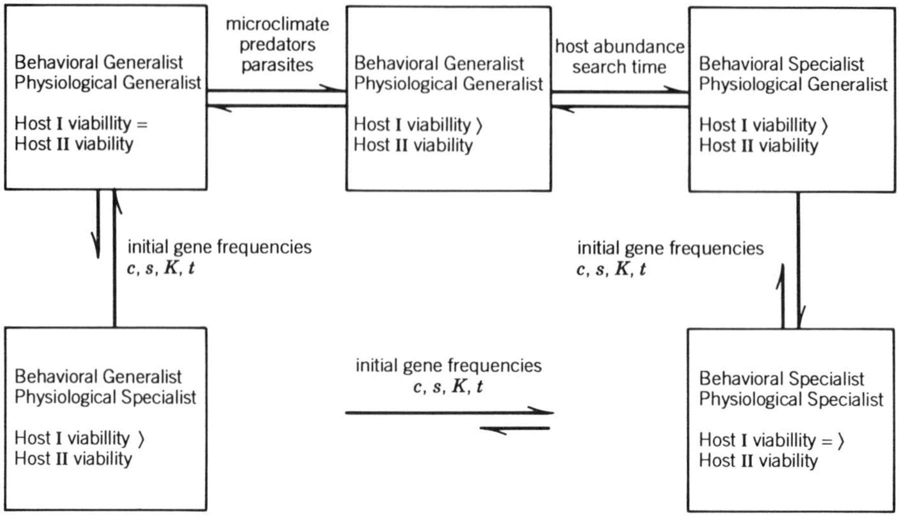

Figure 13.3. Transitions between states of physiological and behavioral specialization. Each box in the diagram represents such a state. Length of arrows between boxes indicate the likelihood that a population will evolve from one state to another. See text for further details.

because ecological monophagy would not have been achieved in the first place if they were prohibitively high.) Consequently, it will be very difficult for the population to expand its host range to use Host II again (or to use any other plant species that is toxic or of low quality). In essence, broadening of the host range will tend to be prohibited, because the population cannot simultaneously put together genotypes that detoxify the secondary compounds of Host II and that modify behavior to permit the insect to use that host.

The model analyzed here also suggests that the generalist equilibrium is often locally stable. This might be taken to imply that physiological and behavioral generalization is also an evolutionary cul-de-sac. However, as outlined earlier, populations may escape from this state through the intermediate state of being physiologically generalized but behaviorally specialized. By contrast, escape from complete specialization is not possible through this route for the reasons outlined earlier. Nor is it likely that a population will be able to escape complete specialization by becoming behaviorally generalized though physiologically specialized. Since the population lacks the ability to detoxify secondary compounds of Host II, there will be no advantage to incorporating that host into the diet, even if search costs increase (e.g., because the relative abundance of Host I decreases; stability criterion 3a is virtually always satisfied if $s = 1$, which will be true if ability to detoxify the secondary compounds of Host II is lacking). Consequently, evolution away from complete generalization would seem much more feasible than evolution away from complete specialization. Such a bias, operating over long periods of time, would result in there being more specialist than generalist species.

This argument is not likely to apply when Host I and Host II are chemically similar, as is often the case, for example, when the plants belong to the same plant family. Under such circumstances, cross-adaptation between the secondary compounds in the two hosts (Gould, 1983) is likely. Loss of contact with Host II will then probably not lead to a drastic reduction in ability to process the foliage of that host. If search costs increased sufficiently because of alterations in the abundances of the two hosts, the specialist equilibrium could become destabilized, leading to the evolution of behavioral generalization and physiological specialization. The evolution of complete generalization might then proceed from this state.

The argument is likely to apply primarily when Host I and Host II are from different families, since families tend to differ markedly in secondary chemical content. Cross-adaptation between the secondary compounds of the two hosts is then less probable, so that loss of contact with Host II is likely to lead to a great reduction in ability to use that host's foliage. The properties of the model presented here thus suggests a possible mechanism underlying what seems to be a general evolutionary trend in herbivorous insects from host generalization to host specialization (Hering, 1951; Price, 1983; Berenbaum, 1981, 1983).

Acknowledgments

Support for the work reported here was provided by a Duke University Research Council grant and by NSF grants BSR 8406780 and BSR 8507359.

REFERENCES

Berenbaum, M. R. 1981. Patterns of furanocoumarin distribution in insect herbivory in the Umbelliferae: Plant chemistry and community structure. *Ecology* 62:1254–1266.

———. 1983. Coumarins and caterpillars: A case for coevolution. *Evolution* 37:163–169.

Bernays, E. A. 1984. Arthropods for weed control in IPM systems. Pages 373–388 in Hoy M. A., and D. Herzog (eds.), *Biological Control in IPM Systems*, New York: Academic Press.

Bernays, E. A., and M. Graham. 1988. On the evolution of host specificity in phytophagous arthropods. *Ecology* 69:886–892.

Castillo-Chavez, C., S. A. Levin, and F. Gould. 1988. Physiological and behavioral adaptation of insects to varying environments: A mathematical model. *Evolution* 42:986–994.

Chew, F. S. 1975. Coevolution of Pierid butterflies and their cruciferous food plants. I. The relative quality of available resources. *Oecologia* 20:117–128.

———. 1977. Coevolution of Pierid butterflies and their cruciferous food plants. II. The distribution of eggs on potential food plants. *Evolution* 31:568–579.

Courtney, S. P. 1982. Coevolution of Pierid butterflies and their cruciferous foodplants. IV. Hostplant apparency and *Anthocharis cardamines* oviposition. *Oecologia* 52:258–265.

———. 1983. Models of hostplant location by butterflies: The effect of search images and search efficiency. *Oecologia* 59:317–321.

———. 1986. The ecology of Pierid butterflies: Dynamics and interactions. *Adv. Ecol. Res.* 15:51–131.

Dethier, V. G. 1954. Evolution of feeding preferences in phytophagous insects. *Evolution* 8:33–54.

Feeny, P. 1975. Biochemical coevolution between plants and their insect herbivores. Pages 3–19 in L. E. Gilbert and P. H. Raven (eds.), *Coevolution of Animals and Plants*. Austin, Tex.: Univ. of Texas Press.

Fox, L. R., and P. A. Morrow. 1981. Specialization: Species property or local phenomenon. *Science* 221:887–893.

Futuyma, D. J. 1979. Associations of plants and insects in a deciduous forest. *Ecol. Monogr.* 49:33–50.

———. 1983. Selective factors in the evolution of host choice by phytophagous insects. Pages 227–244 in Ahmad, S. (ed.), *Herbivorous Insects: Host-Seeking Behavior and Mechanisms*. New York: Academic Press.

Futuyma, D. J., and Gould, F. 1979. Associations of plants and insects in a deciduous forest. *Ecol. Monogr.* 49:33–50.

Gilbert, L. E. 1979. Development of theory in the analysis of insect-plant interactions. Pages 117–154 in Horn, D. R. Mitchell, and G. Stairs (eds.), *Analysis of Ecological Systems*. Columbus, Ohio: Ohio State Univ. Press.

Gould, F. 1979. Rapid host range evolution in a population of the phytophagous mite *Tetranychus urticae* Koch. *Evolution* 33:241–250.

———. 1983. Genetics of plant-herbivore systems: Interactions between applied and basic study. Pages 599–653 in Denno, R. F., and M. S. McClure (eds.), *Variable Plants and Herbivores in Natural and Managed Systems*. New York: Academic Press.

———. 1984. Role of behavior in the evolution of insect adaptation to insecticides and resistant host plants. *Bull. Entomol. Soc. Am.* 30:34–41.

Hare, J. D., and G. C. Kennedy. 1986. Genetic variation in plant-insect associations: Survival of *Leptinotarsa decimlineata* populations on *Solanum carolinense*. *Evolution* 40:1031–1043.

Hering, E. M. 1951. *Biology of the Leaf Miners*. The Hague, The Netherlands: Dr. W. Junk.

Iwasa, Y., Y. Suzuki, and H. Matsuda. 1984. Theory of oviposition strategy of parasitoids. I. Effect of mortality and limited egg number. *Theor. Pop. Biol.* 26:205–227.

Jaenike, J. 1978. On optimal oviposition behavior in phytophagous insects. *Theor. Pop. Biol.* 14:350–356.

Levins, R., and R. MacArthur. 1969. An hypothesis to explain the incidence of monophagy. *Ecology* 50:910–911.

Price, P. W. 1983. Hypotheses on organization and evolution in herbivorous insect communities. Pages 559–596 in Denno, R. F., and M. S. McClure (eds.), *Variable Plants and Herbivores in Natural and Managed Systems*. New York: Academic Press.

Rausher, M. D. 1983. Ecology of host-selection behavior in phytophagous insects. Pages 223–257 in Denno, R. F., and M. S. McClure (eds.), *Variable Plants and Herbivores in Natural and Managed Systems*. New York: Academic Press.

———. 1984a. The evolution of habitat selection in subdivided populations. *Evolution* 38:596–608.

———. 1984b. Trade-offs in performance on different hosts: Evidence from within- and between-site variation in the beetle *Deloyala guttata*. *Evolution* 38:582–595.

———. 1985. Variability for host preference in insect populations: Mechanistic and evolutionary models. *J. Insect. Physiol.* 31:873–889.

———. 1988. Is coevolution dead? *Ecology* 69:898–901.

Rausher, M. D., and R. Englander. 1987. The evolution of habitat preference. II. Evolutionary genetic stability under soft selection. *Theor. Pop. Biol.* 31:116–139.

Scriber, J. M., and P. Feeny. 1979. The growth of herbivorous caterpillars in relation to degree of feeding specialization and to growth form of their foodplants. (Lepidoptera: Papilionidae and Bombycoidea). *Ecology* 60:829–850.

Singer, M. C. 1971. Evolution of food-plant preference in the butterfly, *Euphydryas editha*. *Evolution* 25:383–389.

Strong, D. R., J. H. Lawton, and R. Southwood. 1984. *Insects on Plants: Community Patterns and Mechanisms*. Cambridge, Mass.: Harvard Univ. Press.

Via, S. 1984a. The quantitative genetics of polyphagy in an insect herbivore. I. Genotype-environment interaction in larval performance on different host plant species. *Evolution* 38:881–895.

———. 1984b. The quantitative genetics of polyphagy in an insect herbivore. II. Genetic correlations in larval performance within and among host plants. *Evolution* 38:896–905.

Wallace, B. 1968. *Topics in Population Genetics*. New York: W. W. Norton.

Weber, G. 1985. Genetic variability in host plant adaptation of the green peach aphid, *Myzus persicae*. *Entomol. Exp. Appl.* 38:49–56.

APPENDIX 13.1

In this appendix, I describe the recursion equations for the hard- and soft-selection models and describe the meaning of the various parameters in the equations. Consider first the hard-selection model.

Let $G_1, G_2, \ldots, G_9$ correspond to the frequencies of genotypes AABB, AABb, AAbb, AaBB, AaBb, Aabb, aaBB, aaBb, and aabb, respectively, where the A locus governs oviposition preference and the B locus governs viability on a particular host (see text). It is assumed that loci A and B are unlinked. Then the following quantities can be defined:

$$\begin{aligned} a_{\mathrm{AB}} &= G_1 + 0.5G_2 + 0.5G_4 + 0.25G_5 \\ a_{\mathrm{Ab}} &= G_3 + 0.5G_2 + 0.5G_6 + 0.25G_5 \\ a_{\mathrm{aB}} &= G_7 + 0.5G_4 + 0.5G_8 + 0.25G_5 \\ a_{\mathrm{ab}} &= G_9 + 0.5G_6 + 0.5G_8 + 0.25G_5 \end{aligned} \tag{13A.1}$$

These values are the frequencies of the four different chromosome types in the population at mating. Also, define d_{ij}, the number of chromosomes of type ij (i = A or a, j = B or b) that are contributed by females to offspring that are placed on Host I by

$$\begin{aligned} d_{\mathrm{AB}} &= P_1F_1G_1 + 0.5P_2F_2G_2 + 0.5P_4F_4G_4 + 0.25P_5F_5G_5 \\ d_{\mathrm{Ab}} &= P_3F_3G_3 + 0.5P_2F_2G_2 + 0.5P_6F_6G_6 + 0.25P_5F_5G_5 \\ d_{\mathrm{aB}} &= P_7F_7G_7 + 0.5P_4F_4G_4 + 0.5P_8F_8G_8 + 0.25P_5F_5G_5 \\ d_{\mathrm{ab}} &= P_9F_9G_9 + 0.5P_6F_6G_6 + 0.5P_8F_8G_8 + 0.25P_5F_5G_5 \end{aligned} \tag{13A.2}$$

where P_i is the proportion of eggs placed on Host I by genotype i and F_i is the fecundity of genotype i. Finally, define h_{ij}, the number of chromosomes of type ij that are contributed by females to offspring that are placed on Host II by

$$\begin{aligned}
h_{AB} &= (1-P_1)F_1G_1+0.5(1-P_2)F_2G_2+0.5(1-P_4)F_4G_4+0.25(1-P_5)F_5G_5\\
h_{Ab} &= (1-P_3)F_3G_3+0.5(1-P_2)F_2G_2+0.5(1-P_6)F_6G_6+0.25(1-P_5)F_5G_5\\
h_{aB} &= (1-P_7)F_7G_7+0.5(1-P_4)F_4G_4+0.5(1-P_8)F_8G_8+0.25(1-P_5)F_5G_5\\
h_{ab} &= (1-P_9)F_9G_9+0.5(1-P_6)F_6G_6+0.5(1-P_8)F_8G_8+0.25(1-P_5)F_5G_5
\end{aligned} \quad (13A.3)$$

It is assumed that behavioral generalist (genotype AA) oviposits on host plants in proportion to their abundance in the habitat, so that $P_1 = P_2 = P_3 = c$. Genotype aa is a behavioral specialist, so $P_7 = P_8 = P_9 = 1$.

Fecundities are modeled after Rausher (1985). As one extreme situation, it is assumed that fecundity is limited, at least in part, by the amount of time available for search. In effect, this means that eggs are not limited and total fecundity is proportional to the number of acceptable plants that are discovered in a fixed amount of time. Behavioral specialists (aa) are assumed to oviposit on some constant fraction k of Host I individuals discovered and are assigned a relative fecundity of 1. Behavioral generalists (AA) are assumed to oviposit on the same constant fraction k of both Host I and Host II individuals discovered. If Host II is twice as abundant as Host I, then $c = 1/3$, and the generalist genotype will discover and oviposit on three times as many plants as the specialist genotype. Hence, the fitness of the generalist genotype is 3. In general, the fitness of the generalist genotype is $1/c$. Individuals of genotype Aa are viewed as reducing the fraction of Host II individuals discovered that are acceptable. If, for example, such individuals lay on only $k/2$ of the Host II plants alighted on and Host II is twice as abundant as Host I, then they oviposit on equal numbers of the two hosts. The proportion, P, that Host I constitutes of the total hosts oviposited on is then $1/2$, and the fecundity of such individuals is $2 = 1/P$. In general, then, fecundity is equal to $1/P$.

The difference between the relative fecundity of 1 for aa individuals and $1/P$ for AA and Aa individuals is the maximum cost of searching, that is,

$$K_{\max} = (1/P) - 1$$

At the opposite extreme, there may be no search costs. In this case, $F_1 = F_2 = \ldots = F_9 = 1$. To allow for cases intermediate between these two extremes, we define the parameter K as indicating the relative cost of searching in specialists ($0 < K < 1$). For a given value of K, the fecundity of a genotype that lays a fraction P of its eggs on Host I is then defined as

$$F = 1 + K[(1/P) - 1]$$

When $K = 0$, then F $= 1$ for all genotypes (i.e., for all P). When $K = 1$, then $F = 1/P$, as described earlier for the full-cost case.

Under hard selection, the recursion equations are

$$
\begin{aligned}
TG_1' &= W_{\mathrm{BB,I}}a_{\mathrm{AB}}d_{\mathrm{AB}} && + W_{\mathrm{BB,II}}a_{\mathrm{AB}}h_{\mathrm{AB}} \\
TG_2' &= W_{\mathrm{Bb,I}}(a_{\mathrm{AB}}d_{\mathrm{Ab}} + a_{\mathrm{Ab}}d_{\mathrm{AB}}) && + W_{\mathrm{Bb,II}}(a_{AB}h_{\mathrm{Ab}} + a_{\mathrm{Ab}}h_{\mathrm{AB}}) \\
TG_3' &= W_{\mathrm{bb,I}}a_{\mathrm{Ab}}d_{\mathrm{Ab}} && + W_{\mathrm{BB,II}}a_{\mathrm{Ab}}h_{\mathrm{Ab}} \\
TG_4' &= W_{\mathrm{BB,I}}(a_{AB}d_{\mathrm{aB}} + a_{\mathrm{aB}}d_{\mathrm{AB}}) && + W_{\mathrm{BB,II}}(a_{AB}h_{\mathrm{aB}} + a_{\mathrm{aB}}h_{\mathrm{AB}}) \\
TG_5' &= W_{\mathrm{Bb,I}}(a_{AB}d_{\mathrm{ab}} + a_{\mathrm{AB}}d_{\mathrm{aB}} + a_{\mathrm{ab}}d_{\mathrm{AB}} + a_{\mathrm{aB}}d_{\mathrm{Ab}}) \\
&\quad + W_{\mathrm{Bb,II}}(a_{AB}h_{\mathrm{ab}} + a_{\mathrm{Ab}}h_{\mathrm{aB}} + a_{\mathrm{ab}}h_{\mathrm{AB}} + a_{\mathrm{aB}}h_{\mathrm{Ab}}) \qquad (13A.4) \\
TG_6' &= W_{\mathrm{bb,I}}(a_{Ab}d_{\mathrm{ab}} + a_{\mathrm{ab}}d_{\mathrm{Ab}}) && + W_{\mathrm{Bb,II}}(a_{Ab}h_{\mathrm{ab}} + a_{\mathrm{ab}}h_{\mathrm{Ab}}) \\
TG_7' &= W_{\mathrm{BB,I}}a_{\mathrm{aB}}d_{\mathrm{aB}} && + W_{\mathrm{BB,II}}a_{\mathrm{aB}}h_{\mathrm{aB}} \\
TG_8' &= W_{\mathrm{Bb,I}}(a_{aB}d_{\mathrm{ab}} + a_{\mathrm{ab}}d_{\mathrm{aB}}) && + W_{\mathrm{Bb,II}}(a_{aB}h_{\mathrm{ab}} + a_{\mathrm{ab}}h_{\mathrm{aB}}) \\
TG_9' &= W_{\mathrm{bb,I}}a_{\mathrm{ab}}d_{\mathrm{ab}} && + W_{\mathrm{BB,II}}a_{\mathrm{ab}}h_{\mathrm{ab}}
\end{aligned}
$$

where T is a normalizing constant equal to the sum of all the right-hand sides, and $W_{i,j}$ is the viability of genotype i (i = BB, Bb, or bb) on Host j (j = I or II). Viabilities are given in Table 13.1

Under soft selection, the recursion equations are

$$
\begin{aligned}
G_1' &= (c/T_{\mathrm{I}})\,(W_{\mathrm{BB,I}}a_{\mathrm{AB}}d_{\mathrm{AB}}) && + [(1-c)/T_{\mathrm{II}}]\,(W_{\mathrm{BB,II}}a_{\mathrm{AB}}h_{\mathrm{AB}}) \\
G_2' &= (c/T_{\mathrm{I}})\,[W_{\mathrm{Bb,I}}(a_{\mathrm{AB}}d_{\mathrm{Ab}} + a_{\mathrm{Ab}}d_{\mathrm{AB}}] && + [(1-c)/T_{\mathrm{II}}]\,[W_{\mathrm{Bb,II}}(a_{AB}h_{\mathrm{Ab}} + a_{\mathrm{Ab}}h_{\mathrm{AB}})] \\
G_3' &= (c/T_{\mathrm{I}})\,(W_{\mathrm{bb,I}}a_{\mathrm{Ab}}d_{\mathrm{Ab}}) && + [(1-c)/T_{\mathrm{II}}]\,(W_{\mathrm{BB,II}}a_{\mathrm{Ab}}h_{\mathrm{Ab}}) \\
G_4' &= (c/T_{\mathrm{I}})\,[W_{\mathrm{BB,I}}(a_{AB}d_{\mathrm{aB}} + a_{\mathrm{aB}}d_{\mathrm{AB}})] && + [(1-c)/T_{\mathrm{II}}]\,[W_{\mathrm{BB,II}}(a_{AB}h_{\mathrm{aB}} + a_{\mathrm{aB}}h_{\mathrm{AB}})] \\
G_5' &= (c/T_{\mathrm{I}})\,[W_{\mathrm{Bb,I}}(a_{AB}d_{\mathrm{ab}} + a_{\mathrm{Ab}}d_{\mathrm{aB}} + a_{\mathrm{ab}}d_{\mathrm{AB}} + a_{\mathrm{aB}}d_{\mathrm{Ab}})] \\
&\quad + [(1-c)/T_{\mathrm{II}}]\,[W_{\mathrm{Bb,II}}(a_{AB}h_{\mathrm{ab}} + a_{\mathrm{Ab}}h_{\mathrm{aB}} + a_{\mathrm{ab}}h_{\mathrm{AB}} + a_{\mathrm{aB}}h_{\mathrm{Ab}})] \\
G_6' &= (c/T_{\mathrm{I}})\,[W_{\mathrm{bb,I}}(a_{Ab}d_{\mathrm{ab}} + a_{\mathrm{ab}}d_{\mathrm{Ab}})] && + [(1-c)/T_{\mathrm{II}}]\,[W_{\mathrm{Bb,II}}(a_{Ab}h_{\mathrm{ab}} + a_{\mathrm{ab}}h_{\mathrm{Ab}})] \\
G_7' &= (c/T_{\mathrm{I}})\,(W_{\mathrm{BB,I}}a_{\mathrm{aB}}d_{\mathrm{aB}}) && + [(1-c)/T_{\mathrm{II}}]\,(W_{\mathrm{BB,II}}a_{\mathrm{aB}}h_{\mathrm{aB}}) \\
G_8' &= (c/T_{\mathrm{I}})\,[W_{\mathrm{Bb,I}}(a_{aB}d_{\mathrm{ab}} + a_{\mathrm{ab}}d_{\mathrm{aB}})] && + [(1-c)/T_{\mathrm{II}}]\,[W_{\mathrm{Bb,II}}(a_{aB}h_{\mathrm{ab}} + a_{\mathrm{ab}}h_{\mathrm{aB}})] \\
G_9' &= (c/T_{\mathrm{I}})\,(W_{\mathrm{bb,I}}a_{\mathrm{ab}}d_{\mathrm{ab}}) && + [(1-c)/T_{\mathrm{II}}]\,(W_{\mathrm{BB,II}}a_{\mathrm{ab}}h_{\mathrm{ab}})
\end{aligned}
$$

where T_{I} is the sum of the terms multiplied by c/T_{I} and T_{II} is the sum of the terms multiplied by $(1 - c)/T_{\mathrm{II}}$.

APPENDIX 13.2

In this appendix, I illustrate how the stability properties listed in Table 13.5 were calculated. For illustration, I examine criterion 3a.

To begin, the hard-selection recursion equations (13A.4) are rewritten as eight independent equations, in which G_7 is replaced with $1 - (G_1 + G_2 + G_3 + G_4 + G_5 + G_6 + G_8 + G_9)$ and the equation for G_7' prime is eliminated. The Jacobian, J, of this set of eight equations has the form

$$J = \begin{matrix} 0 & 0 & 0 & 0 & 0 & 0 & 0 & 0 \\ 0 & 0 & 0 & 0 & 0 & 0 & 0 & 0 \\ 0 & 0 & 0 & 0 & 0 & 0 & 0 & 0 \\ a & b & 0 & c & d & 0 & 0 & 0 \\ 0 & e & f & 0 & g & h & 0 & 0 \\ 0 & 0 & 0 & 0 & 0 & 0 & 0 & 0 \\ 0 & 0 & 0 & 0 & i & j & k & l \\ 0 & 0 & 0 & 0 & 0 & 0 & 0 & 0 \end{matrix}$$

where $a, b, \ldots, l$ are positive constants and

$$c=0.5\ [W_{BB,I}P_7F_7+W_{BB,II}(1-P_7)F_7]+0.5[W_{BB,I}P_4F_4+W_{BB,II}(1-P_4)F_4] \quad (13A.5a)$$
$$k=0.5\ [W_{Bb,I}P_7F_7+W_{Bb,II}(1-P_7)F_7]\ +0.5[W_{Bb,I}P_8F_8+W_{Bb,II}(1-P_8)F_8] \quad (13A.5b)$$
$$g=0.25[W_{Bb,I}P_7F_7+W_{Bb,II}(1-P_7)F_7]\ +0.25[W_{Bb,I}P_5F_5+W_{Bb,II}(1-P_5)F_5] \quad (13A.5c)$$

The specialist (aaBB) equilibrium will be stable if, and only if, all of the eigenvalues of J are less than 1 (all are positive, because J is positive). Solving the determinantal equation

$$0 = det\ (J - \mathrm{I})$$

gives the following equation

$$0 = (-\lambda)^5\ (c - \lambda)\ (k - \lambda)\ (g - \lambda),$$

which indicates that there are five zero eigenvalues, as well as the eigenvalues $\lambda = c$, $\lambda = k$, and $\lambda = g$. Consequently, the conditions for stability of the specialist equilibrium are

$$c < 1$$
$$k < 1$$

and (13A.6)

$$g < 1$$

Consider the criterion $c < 1$. From equation (13A.5a), this is equivalent to

$$0.5[W_{BB,I}P_7F_7 + W_{BB,II}(1 - P_7)F_7] + 0.5[W_{BB,I}P_4F_4 + W_{BB,II}(1 - P_4)F_4] < 1$$

But note that $F_7 = P_7 = W_{BB,I}$, so that this reduces to

$$0.5 + 0.5[F_4P_4 + W_{BB,II}F_4(1 - P_4)] - 1$$

From Table 13.1, $W_{BB,II} = 1 - s$, which yields

$$0.5 + 0.5[F_4P_4 + (1 - s)F_4(1 - P_4)] < 1$$

This rearranges to

$$F_4 < 1/[P_4 + (1 - s)(1 - P_4)]$$

For arbitrary cost of searching, K, $F_4 = 1 + K[(1/P_4) - 1]$ (see Appendix 13.1), which yields

$$1 + K[(1/P_4) - 1] < 1/[P_4 + (1 - s)(1 - P_4)]$$

or, upon further rearrangement,

$$K < sP_4/[1 - s(1 - P_4)]$$

as stability criterion 3a. When P for Aa heterozygote individuals shows no dominance, then $P_4 = (1 + c)/2$, and this criterio P_4 becomes

$$K < s/[s + 2(1 - s)/(1 + c)]$$

The remaining criteria in Table 13.5 are derived in similar fashion.

APPENDIX 13.3

Here I show that the stability criterion $c > 1 - 2t/s$ implies that the cost of physiological adaptation is greater than $(1 - c)/c$. Rearranging the stability criterion gives

$$2t/s > 1 - c$$

Dividing each side of this equation into 1 yields

$$s/2t < 1/(1 - c)$$

Subtracting 1 from each side of this equation and combining terms gives

$$(s - 2t)/2t < c/(1 - c)$$

Dividing each side of this equation into 1 then yields

$$2t/(s - 2t) > (1 - c)/c$$

which is the desired result.

SECTION IV

THE PATTERNS OF VARIATION: CASE STUDIES

14

DIVERSITY AND VARIATION OF EUROPEAN CORN BORER POPULATIONS

William B. Showers

U.S. Department of Agriculture, Agricultural Research Service, Corn Insects Research Laboratory

INTRODUCTION

Immediately upon discovery of the European corn borer, *Ostrinia nubilalis* (Hübner) (Lepidoptera: Pyralidae), in North America, populations of this species were demonstrating diversity of voltinism, and it is probable that at least two voltine types of European corn borer populations were introduced. In 1914, Vinal (1917) determined the presence of a bivoltine form of this species in eastern Massachusetts, and, shortly thereafter, Felt (1919) reported that the population near Lake Erie and in western New York State was univoltine.

In addition to the two introduced voltine types, European corn borer populations also came to North America with two distinct sex pheromone communication systems: a *Z* (*cis*) type, using 98:2 mix of (*Z*)- and (*E*)-11-tetradecenyl acetates and an *E* (*trans*) type using a 1:99 blend of (*Z*)- and (*E*)-11-tetradecenyl acetates (Glover et al., 1987).

The European corn borer is widely distributed throughout central and southern Europe, Siberia, northern India, and western China (Caffrey and Worthley, 1927). This species has long been recognized in the Old World as a serious pest of corn, hops, millet, and hemp. There are no records before 1835 of the species being an important economic threat to crops (Caffrey and Worthley, 1927).

Vinal (1917) reported that market sweet-corn growers in the Boston, Massachusetts, area were aware of injury from this species as early as 1913. During 1919, a second infestation was discovered, again on sweet corn, in eastern New York State. A third relatively large area surrounding Lake Erie was determined also to be infested between 1919 and 1921. This area comprised western New York, northwestern Pennsylvania, northern Ohio, and eastern Michigan in the United States and southern Ontario in Canada. All of these finds were on field corn (Caffey and Worthley, 1927). The manner in which the European corn borer gained entrance to North America is not known. It was believed by early researchers, however, that broomcorn, imported for manufacturing purposes from Hungary and Italy between 1909 and 1914 was the probable carrier (Smith, 1920; McLaine, 1922). Figure 14.1 shows the movement of European corn borer across the United States.

The female moth usually deposits a cream-colored mass of eggs (about 20–25 eggs per mass) on the underside of the leaf of a host plant. Eggs hatch in four days, and neonates immediately feed on various external parts of the plant. Larvae undergo four (sometimes five) molts and, usually after the third molt, bore into the stalk or stem of the host plant. Larval feeding destroys phloem and xylem and characteristically produces large cavities within the stalk or stem (Showers et al., 1983).

DIVERSITY WITHIN THE SPECIES

Voltinism

The cause of voltine differences among populations of European corn borer is genetic timing of larval diapause. Beck (1982) defines *diapause* as a genetically determined state of suppressed development that might be induced by environmental factors.

Beck and Apple (1961) reported that photoperiod is the primary factor and temperature the secondary factor inducing diapause in most populations of the European corn borer in North America. They presented evidence that this was not true, how-

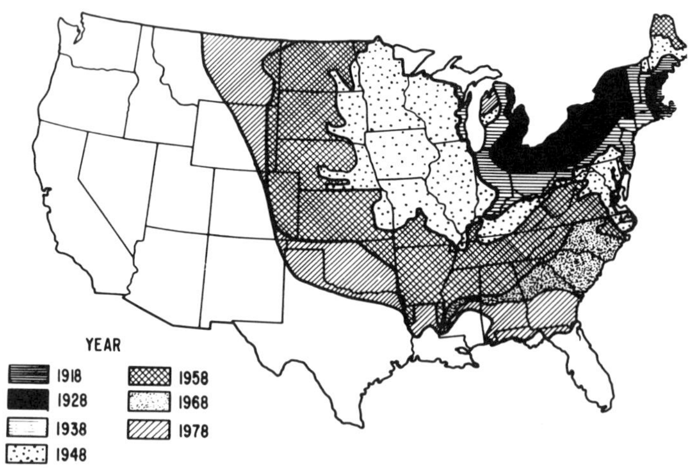

Figure 14.1. Movement of European corn borer across the United States, shown in 10-year periods (Showers, 1979)

ever, for the Massachusetts population, where temperature was as important as photoperiod for inducing, as well as inhibiting, diapause.

European corn borers enter diapause during late summer or early autumn and overwinter as fifth-stage larvae in the stalks of host plants (Caffrey and Worthley, 1927). Overwintering larvae of multivoltine populations will break diapause and develop to adulthood earlier than will larvae of an univoltine population (McLeod, 1976).

Showers (1979) compiled data on corn borer development and calculated the average number of degree days necessary to achieve 50% pupation for three populations of European corn borer: one from the area of the original bivoltine population (Vinal, 1917), the second from an area believed to be inhabited by an univoltine population (Felt, 1919), and the third from an area farther west that had been invaded during the 1940s (Fig. 14.1). The results showed a distinct difference in the numbers of degree days necessary for development of European corn borer populations from Concord, Massachusetts (bivoltine) and those from Toledo, Ohio (univoltine). The reduction in number of degree days over years for the Toledo population suggests that, after 1943, requirements for this population had become similar to the requirements for the bivoltine Muscatine, Iowa, population. The seemingly sudden predominance of a bivoltine population of European corn borer in Ohio could have resulted from a genetic shift of the natural population or an invasion by a two-generation population.

Showers (1979) also gathered records of the number of days at 28°C for 50% pupation of several populations during a period of rapid expansion (1947–1950). The mean number of days to 50% pupation for the invading population in Minnesota (Fig. 14.1) was consistent with the number of days to 50% pupation for the univoltine population of western New York. The invading populations of Illinois, Iowa, and Kentucky, however, required a number of days to 50% pupation consistent with the number of days necessary for the post-1943 bivoltine population of Ohio. Of interest was the relationship between the number of days necessary for 50% pupation of the Massachusetts population (specimens collected near the original identification site) and similar requirements for 50% pupation of populations collected 800–1100 kilometers (km) farther south in eastern Virginia and eastern North Carolina. These results suggested the presence of three voltine ecotypes (uni-, bi-, and multivoltine). An ecotype equals groups of populations with most individual insects having similar environmental tolerances (Showers et al., 1975).

When the area of distribution of a species extends continuously over a wide range of latitudes, diapause induction, duration of dormancy and the interactions between day length and temperature are particularly variable and form a north–south cline. As a result, critical photoperiod and intensity diminish toward the lower latitudes (Showers et al., 1975; Lees, 1980). These phenomena have been accepted as evidence that geographic populations are closely adapted in their phenology to their own part of the geographic range of the species. Theoretically, the insect would receive a false indication of season in other parts of the range (Lees, 1980). Therefore, environmental conditions will affect the rate of dispersal and adaption to the north and south. The route of least resistance for invading populations of European corn borer during the 1940s was west toward the vast Corn Belt of the United States (15,405,600 hectares (ha) of field corn; Bergman et al., 1985). The effects of day length on development, for any given day, would be similar whether a population was in Ohio or 1,920 km west in Nebraska. Summer thermoperiod (degree days/day) would increase as the invading species moved westward (Showers et al., 1975). During the tenure of the univoltine population in Ohio, dispersal averaged about 19 km per year; after the appearance of the bivoltine population, dispersal westward averaged 114 km per year (Palmer et al., 1985).

As expected (Arbuthnot, 1944; Lees, 1980), dispersal northward and southward was slower. In a laboratory study, Beck and Apple (1961) determined that a southern Minnesota population of European corn borer required significantly more degree days for development than the old established population of Massachusetts, although, in their natural habitats, the two populations would be subjected to somewhat similar long day lengths and cool temperatures. They determined that the Massachusetts population (42°30′N latitude) required developmental degree days consistent with those necessary for a southern population from Missouri (36°50′N latitude).

Sparks et al. (1966a, 1966b) studied the diapause of three populations of European corn borer—Minnesota, Iowa, and Missouri—both in the laboratory and under field cages. On the basis of their results, they hypothesized that, in a northern population (Minnesota), genes favoring a single generation would be increased through selection by the environment. Likewise, in a southern population (Missouri), although these genes would remain, they would be diluted each year, because the larvae that entered summer diapause would be exposed to hazards during the long warm season. Chiang et al. (1968) field-tested these three populations at three locations (Waseca, Minnesota; Ankeny, Iowa; and Portageville, Missouri) and determined that the southern population (Missouri) required significantly fewer degree days for development than the northern population (Minnesota). They also concluded that the southern population was more responsive to changes in photoperiod.

Showers et al. (1972) studied the diapause response of populations of European corn borer resulting from reciprocal matings of east coast (Maryland), southern fringe (Alabama), and northern fringe (Minnesota) areas of the species distribution in the United States. These field studies were conducted at Morris, Minnesota (45° 35′N latitude) and Ankeny, Iowa (41°43′N latitude). The authors indicate that the triggering mechanisms for diapause or pupation became operational at photophases between 14 hours 41 minutes to 14 hours 12 minutes at Morris, Minnesota, and 14 hours 22 minutes to 13 hours 55 minutes at Ankeny, Iowa. They found that the percentage diapause of the hybrid populations that included a Minnesota parent was intermediate to the percentage diapause of the parental populations and that diapause and/or development to adulthood were male sex-linked. This phenomenon was most readily seen between the Minnesota and Alabama males at 41°43′N (Fig. 14.2). These results indicate that the introduction of a multivoltine population into an univoltine area could shift the local population toward multivoltinism. This would happen, however, only if the introduced population was synchronous with the local population.

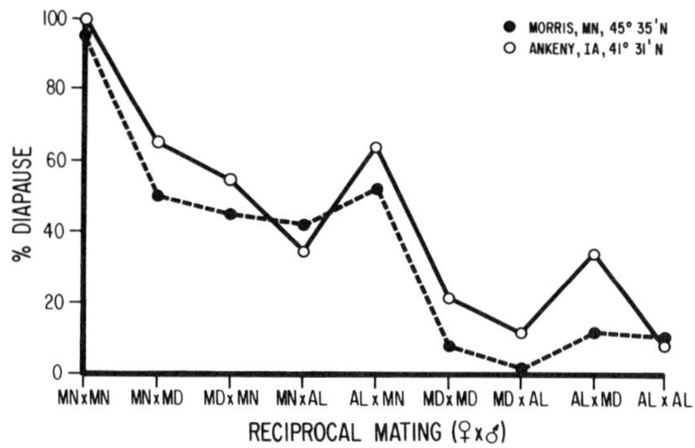

Figure 14.2. Percentage diapause of European corn borer progenies from Alabama, Maryland, and Minnesota parents and their reciprocals (adapted from Showers et al., 1972)

On the basis of the time necessary for European corn borer larvae to break diapause in the spring and a difference in the critical photoperiod (length of scotophase) that induces diapause, McLeod (1976) determined that there are two distinct populations in southwestern Ontario. The univoltine population, however, is just 40 km north of the bivoltine population. These two adaptations (i.e., time to break diapause and critical photoperiod) seemingly assure that most European corn borers in the northern part of the corn-growing area have only a single generation per year. Wressell (1971), however, has noted a gradual shift of the bivoltine population to the northwest in Ontario. The two limiting factors in this spread would seem to be the length of the growing season (about 2,300 degree days before a killing frost) and a photoperiodic threshold that allows pupation during the first generation (McLeod, 1976). Crosses between these two populations (McLeod, 1978) demonstrated that growth rate was female sex-linked, whereas diapause incidence was male sex-linked.

Later, near St. Remi, Quebec, Canada, an area long known to have univoltine European corn borer populations (Hudon, 1959; Showers et al., 1975), McLeod et al. (1979) collected larvae from cornfields infested during the normal time (early July) and from cornfields infested during August. They compared the progeny of the two populations and found that the August population had early emergence in the spring with less intense diapause and that members of the population required a lower critical photoperiod (i.e., comes later in the summer) for diapause induction. McLeod et al. (1979) also presented evidence, which will be discussed later in this chapter, that the bivoltine population was not a result of a genetic shift of the natural (univoltine) population.

Armed with the results of Showers et al. (1972) and McLeod (1978), Reed et al. (1981) crossed both a low-diapausing laboratory population and a low-diapausing field population (multivoltine) of European corn borer with a naturally occurring high-diapausing (univoltine) population of European corn borer. The authors calculated expected diapause frequencies, sex ratios of diapausing progeny, and sex ratios of nondiapausing progeny for sex-linked inheritance by a single-gene locus lacking dominance. When the male parent was univoltine, the progeny had a greater frequency of diapause and a greater proportion of females in the diapausing portion of the populations than when the male parent was from a low-diapause population (multivoltine). All progeny tended to respond, (i.e., go into diapause or continue development through to adulthood) in the same manner as the male parent. In the order Lepidoptera, the male is the homogametic sex; therefore, these results are consistent with sex-linked inheritance.

An extensive study was conducted at seven field locations, north to south, consisting of nine populations from across the distribution range of European corn borer in North America (Showers et al., 1975). The authors determined that, on the basis of diapause response, the species is composed of three ecotypes: northern (univoltine), central (bivoltine), and southern (three or more generations per year). They concluded that there is a propensity for diapause in the north (northern ecotype) but a tendency to delay diapause in the south, and that this is manifested by the central and southern ecotypes as a gradation in diapause response. Palmer et al. (1985) suggested that the European corn borer is typically heterovoltine (partial generations) in North America, especially between areas separating zones of homovoltine (complete generations). Showers et al. (1975) presented evidence that the northern ecotype, even though it does not have an obligatory diapause (Beck and Hanec, 1960; Reed et al., 1981), will diapause during the summer at any latitude of the United States, whereas the central and southern ecotypes will be confronted with certain latitudes that necessitate the diapause response to be greater than it would be either north or south of these latitudes (31°, 36°, or 41°N) (Fig. 14.3). Later, Showers (1979) developed the concept of barrier latitudes. The environmental conditions (photoperiod and temperature) of these latitudes might block the southern expansion of the European corn borer. Of nearly equal importance, however, is whether the population is synchronous with the cropping strategies of the invaded area (Sparks and Showers, 1975). An important factor, particularly in the northern areas invaded by the European corn borer, is the availability of acceptable host types, that is, patchiness or sparseness of corn and other hosts (Palmer et al., 1985).

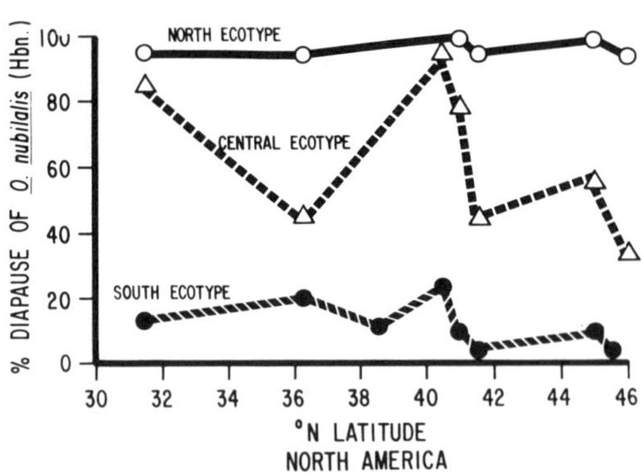

Figure 14.3. Percentage diapause of field-tested populations representing three ecotypes of the European corn borer at several latitudes in the United States (adapted from Showers et al., 1975)

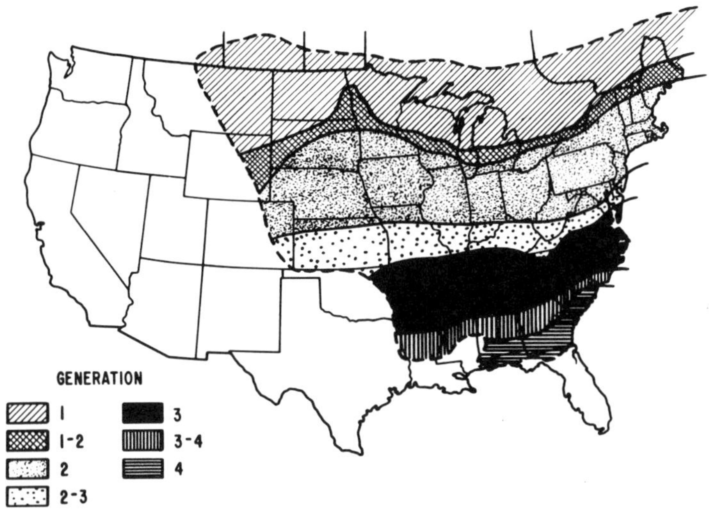

Figure 14.4. Approximate areas of current generation zones of the European corn borer in North America. The broken line represents the outer limit of distribution (Palmer et al., 1985)

Approximate areas of voltine zones for the European corn borer in North America have been developed by Palmer et al. (1985). Voltinism adaptation has often produced heterovoltinism in transition zones (Fig. 14.4). Note the protrusion of one of these transitional zones into the Red River Valley of eastern North Dakota and northwestern Minnesota. This phenomenon will be addressed later in this chapter.

Sex Pheromone Blends

In Iowa, Sparks (1963) studied the mating behavior of European corn borer moths in large outdoor cages and also in cages within a programmable bioclimatic chamber. He applied an ether extract from tips of abdomens of mating European corn borer females to artificial female moths and, using infrared equipment, filmed the behavior of male moths with the artificial moths, He concluded that the extract contained a sex attractant. The presence of a chemical entity taken from female moths was confirmed by Klun (1968). Then, Klun and Brindley (1970) identified *Z*-11-tetradecenyl acetate as a potent male sex stimulant that has the same gas–liquid chromatographic and biological properties as the natural pheromone of the European corn borer female.

Klun and Robinson (1971) field-tested 14 grams (g) of *Z*-11-tetradecenyl acetate per sticky trap and determined that the synthetic compound was as competitive as four virgin females per sticky trap for capturing feral European corn borer males. They also found that *E*-11-tetradecenyl acetate inhibited the male's ability to find the baited trap. These results were similar to those found earlier in New York by Roelofs and Comeau (1971). But they used the redbanded leafroller, *Argyrotaenia velutinana* (Walker), as the test insect. Then Roelofs and Tette (1970) determined that the obliquebanded leafroller, *Choristoneura rosaceana* (Harris), also used *Z*-11-tetradecenyl acetate for sexual communication.

After field-testing various mixtures of *Z*- and *E*-11-tetradecenyl acetates, Klun and Robinson (1972) suggested that many species of Lepidoptera could use the identical chemical structure but respond only to specific concentrations and/or blends.

Determination of the sex pheromone communication system of the European corn borer was further complicated, however, when evidence was presented by Roelofs et al. (1972) that the majority of trapped males in a population of European corn borers near Rochester, New York, were attracted to *E*-11-tetradecenyl acetate, but few were attracted to *Z*-11-tetradecenyl acetate. These divergent findings (Klun and Robinson, 1971; Roelofs et al., 1972) became more understandable, however, after Klun et al. (1973) determined that about 4% *E* isomer with *Z* isomer enhanced the capture of European corn borer males in Iowa and after Kochansky et al. (1975) found that 3–4% *Z* with 96–97% *E* increased the capture of European corn borer males in central New York. Recently, Glover et al. (1987) determined that a 98:2 mix of *Z*- and *E*- or a 1:99 mix of *Z*- and *E*- more appropriately defined the sex pheromone communication system of these European corn borer populations.

Showers et al. (1974) reported that, in Iowa, few European corn borer males were attracted to New York females. *Z*-emitting and responding European corn borer populations are widespread in North America (Showers et al., 1974; Kochansky et al., 1975; Struble et al., 1987), whereas the *E*-emitting and responding populations may be restricted to Quebec (McLeod et al., 1979) and the northeastern United States. Cardé et al. (1975) determined that, in Pennsylvania, European corn borer males were attracted to both 97:3 and 2:98 blends of *Z*- and *E*-11-tetradecenyl acetates. The authors suggested that two distinct populations might coexist.

A classic study was conducted by Klun and Cooperators (1975), during which geometric formulations of 1:0, 97:3, 1:1, 3:97, and 0:1 *Z*:*E* 11-tetradecenyl acetates were placed in sticky-coated traps throughout Europe and eastern North America. Attraction of European corn borer males to the 97:3 blend proved widespread in both Europe and North America. At many of the test sites, however, there were small percentages of males attracted to the formulations in which *E* was predominant. Large numbers of males were attracted to the 3:97 blend, however, only at Bologna, Italy, and Geneva, New York. These results suggest strongly that the *Z* and *E* populations of North America were separately introduced. Heterogeneity of male response also suggests that, in some areas, the two pheromonal types occur sympatrically (Cardé et al., 1975; Klun and Cooperators, 1975; McLeod et al., 1979; Roelofs et al., 1985; DuRant et al., 1986; Klun and Huettel, 1987).

Analyses of the *Z*:*E* ratios of individual females produced from crosses of individual female–male pairs of *Z* or *E* populations and their reciprocals were conducted by Klun and Maini (1979). F_1 female hybrids had an isomeric complement that approximated 35:65 *Z*:*E*. Because the isomer ratio of these F_1 females was biased in favor of a larger proportion of *E* isomer, the authors concluded that incomplete dominance was present and that the genes controlling the sex pheromone secretion are not sex-linked. The frequency of pheromonal types in the F_2 population indicated a ratio of *E* parent to heterozygote to *Z* parent of 16:49:16. This ratio is in agreement with the

Mendelian ratio for breeding within a group heterozygous for one pair of alleles exhibiting incomplete dominance.

Recently Klun and Huettel (1987) conducted studies that indicated assortative mating of *Z, ZE*, and *E* phenotypes. The *Z* pheromone production allele frequency was about four times greater than that of the *E* allele. Their results suggest that moths inheriting *E* alleles are subjected to some undefined selective disadvantage. Progeny tests with males captured in traps baited with females of the three types produced evidence that sex pheromone male response and female production functions may be regulated by separate genetic loci. The authors further suggest that the loci are not always in complement. A given male might carry an allele coding for production of one pheromonal isomer but be genetically predisposed to respond to another.

Using similar populations of *Z, ZE*, and *E* phenotypes, Roelofs et al. (1987) looked for inheritance patterns for sex pheromone (gas chromotographic analyses of pheromone glands) in females, pheromone detection on male antennal olfactory receptor cells (single sensillum recordings), and male pheromone behavioral responses (flight tunnel analyses). Data collected on progeny from reciprocal crosses, subsequent F_2, and maternal and paternal backcrosses produced evidence similar to the results of Klun and Huettel (1987) that the production of female pheromone blend is usually controlled by a single autosomal factor and that pheromone-responding olfactory cells are controlled by another autosomal factor. Results of flight tunnel analyses, however, were interpreted by Roelofs et al. (1987) as follows: the *Z*-paternal-backcross males' response profiles were the same as those of *Z*-parent males, and the *E*-paternal-backcross males' response profiles were the same as those of E-parent males. The maternal backcross response profiles were similar to the expected, half-intermediate and half-paternal male types. These results strongly indicate that male behavioral response to sex pheromone is controlled by a male sex-linked gene.

Sympatry of Voltinism and Sex Pheromone Blends

The European corn borer population of Quebec, Canada, has exhibited univoltinism for many years (Hudon, 1959). During the late 1970s, substantial data were gathered by McLeod et al. (1979) that showed that a bivoltine population was present in the vicinity of St. Rémi, Quebec. This information suggested that either the natural population had shifted genetically and could now produce two generations per year or that there had been a recent invasion into Quebec from New York to the south. McLeod et al. (1979) tested the response of European corn borer males to two isomeric mixtures of 11-tetradecenyl acetate by placing baited traps at three locations. They also developed response curves for each isomeric mixture by using electroantennograms. Only at St. Rémi was there a clear indication that there was a population that responded to 97:3, *E*:*Z*-11-tetradecenyl acetates, similar to the bivoltine population of central New York. Electroantennograms of sample insects from July infestations were similar to *Z*-responding males, whereas electroantennograms of sample insects from August infestations, although significantly distinct, were closer to the responses elicited from antennae of sample insects from Geneva, New York. These results indicate that the bivoltine population found near St. Rémi originated in New York and only recently invaded Quebec.

The areas of central and western New York have been known to harbor the univoltine population of European corn borer since around 1916 (Felt, 1919). During the 1960s, McEwen et al. (1968) reported that a bivoltine population was seemingly predominant in these areas. The authors and others assumed that there was a recent voltinism shift from one to two generations. F. F. Dicke (personal communication), a long-time researcher (1927 to the present) of European corn borer ecology and plant–insect interactions, has been aware of a bivoltine population in western New York since the late 1930s. It was not predominant, however, until the approximate time reported by McEwen et al. (1968).

Recently, Eckenrode et al. (1983) monitored seasonal flights of European corn borer using blacklight traps at 28 locations in central and western New York. Calendar date of peak catch and degree day accumulations indicated that both univoltine and bivoltine populations were present at 16 of the trapping sites.

Sex pheromone blend analyses of glands from individual European corn borer females collected at several of these sites were conducted by Roelofs et al. (1985). They determined that, similar to the Iowa bivoltine population (Klun and Robinson, 1971), a bivoltine population in western New York uses *Z*-11-tetradecenyl acetate as its primary pheromone component. There are at least two sites in central New York, however, where a bivoltine population using *E*-11-tetradecenyl acetate and a univoltine population using *Z*-11-tetradecenyl acetate occur sympatrically, whereas, at another site, just the univoltine *Z*-type population is present. These results support the existence of at least three European corn borer populations: bivoltine *E*, bivoltine *Z*, and univoltine *Z*.

Although there might be some interbreeding of univoltine and bivoltine populations, the results of studies conducted by McEwen et al. (1968), Eckenrode et al. (1983), and Roelofs et al. (1985) suggest that there has been little genetic shift of voltinism or predominant sex pheromone blends. Seemingly, each population has maintained its integrity.

VOLTINISM—GENETIC SHIFT OR DISPERSION?

During the 1920s to early 1970s, the populations of European corn borer in the northern areas of the midwestern United States and the southern areas of Canada (excluding southern Ontario) appeared to be univoltine (Hudon, 1959; Chiang, 1961; Showers et al., 1975; Showers, 1981). As early as 1950, however, a few investigators believed that there was a suggestion of bivoltinism in the Dakotas (Munro and Colberg, 1950) and in Manitoba, Canada (Smith, 1950). Recently, evidence has been presented (Fig. 14.4) to indicate a transition from univoltine to bivoltine populations in the Red River Valley of eastern North Dakota and northwestern Minnesota (Palmer et al., 1985).

Quantitative population records for the European corn borer were kept in Minnesota from 1945 to 1970 and in neighboring South Dakota from 1948 until 1970

(Chiang, 1972). The mean number of larvae per 100 random corn plants was recorded each autumn from several fields in each county of each state except in the northern counties of Minnesota and the western counties of South Dakota, where corn acreage was too small to justify a survey. An univoltine population of the species entered Minnesota's southeastern corner (from Wisconsin) during 1943 (Chiang, 1961), and fronts of distribution moved westward and northward, entering the southeastern corner of North Dakota during 1950. Chiang (1972) discovered a large population buildup or population swell just behind the front of distribution. On the basis of these population swells during the 26-year period, Chiang (1972) hypothesized that a new population swell occurred only when the invading population was significantly different from the existing population. These results might suggest that a bivoltine population could have moved up through southern Wisconsin and entered Minnesota as early as 1952 and that another, possibly more substantial, bivoltine population entered Minnesota on southerly winds from Iowa (Chiang et al., 1965).

Recently, a bivoltine population has been found in Quebec, Canada (McLeod et al., 1979) and possibly in Manitoba and Saskatchewan, Canada (Lilly and Harper, 1982). A study was conducted in Alberta, Canada, between 1983 and 1985 to determine the nature of phenological differences between two populations—one on a plain, the other in a valley—of European corn borer (Lee and Spence, 1986). The authors concluded that differences in pupal weight, moth emergence, time of peak flight, and number of nights per flight period were because of environmental variation and not because of genetic differences. The valley environment was more conducive to bivoltinism. When the plain population was reared in the valley or when the valley population was reared on the plain, there were nonsignificant phenological differences.

The results presented by Eckenrode et al. (1983) and Roelofs et al. (1985) that European corn borer populations can maintain their genetic integrity while in sympatry, plus the information presented by Chiang (1972), McLeod (1976), Palmer et al. (1985), and Lee and Spence (1986) that bivoltine populations are present in historically univoltine northern areas of mid-North America, lend substantial credence to the suggestion by Chiang et al. (1965) that bivoltine populations are moving northward on southerly winds.

In this Chapter, I present research to determine whether this species, long considered a nonmigrating insect, is capable of long-range movement. Pioneer International, Inc., one of the world's largest seed corn producers, decided to plant a seed corn nursery at Johnstown, North Dakota. One of the objectives to be accomplished at this nursery was the development of northern corn hybrids with some degree of resistance to European corn borer attack. The insects to be used in these trials were to be collected near, and progeny reared at, Pioneer's insect-rearing laboratory, Johnston, Iowa. Egg masses from this bivoltine population would be shipped to Johnstown, North Dakota, where certain corn inbreds would receive large numbers of freshly hatched larvae per plant (Guthrie et al., 1960).

With Pioneer's full cooperation, a study was initiated in 1984 to observe the potential movement of a bivoltine population within an univoltine habitat. In addition, a study to quantify aspects of flight behavior using mark-release and recapture of European corn borer adults was carried out in central Iowa.

Methods

To gather baseline information, during 1984, field populations of European corn borer were sampled in the Red River Valley of northwestern Minnesota and northeastern North Dakota, 97°30′W longitude and 48°15′N latitude. This area has long been known to maintain an univoltine population of European corn borer (Showers et al., 1972; Showers et al., 1975).

During June 1985, plant breeders with Pioneer infested approximately 30,000 corn plants in a 1.25-ha block with 50 neonate bivoltine European corn borers per plant. In early September, we collected a minimum of 50 fifth-stage European corn borer larvae 0.2 km from the June-infested corn inbreds. We then collected similar numbers of European corn borer larvae from cornfields concentrically at 8, 16, and 32 km from the infested nursery.

These diapausing larvae, collected during late summer 1984 and 1985, were individually isolated, and the numbers from each sample site maintained as a separate population. Diapause was broken (24-h photophase, 28°C) during the springs of 1985 and 1986. All insects that pupated and eclosed within each population were mated to obtain progeny. Experiments designed as randomized complete blocks with six replicates (trays) in 1985 and 10 replicates (trays) in 1986 were conducted in programmable CE-2 Percival^R Growth Chambers under 14 hours 55 minutes photophase: 9 hours 05 minutes scotophase with a thermoperiod that dropped to 22.2°C at dawn (0445 CST) and rose to 27.8°C at 1600 CST. These conditions were chosen to force most individuals of an univoltine population to diapause but to allow most individuals of a bivoltine population to continue to adulthood (Showers, 1970).

To complement the study of the North Dakota material, a quantification of adult European corn borer movement was initiated during the summer of 1986. Bivoltine European corn borers were reared on meridic diet (Lewis and Lynch, 1970) that contained lipid soluble dyes, Calco Red^R or Sudan Blue^R (Ostle et al., 1983).

There were two release sets (July 5–July 19 and July 25–August 9) with five releases (replicates) per release set. Two modified (with a floor) Texas cone traps (Hartstack et al., 1979) were used to monitor adult emergence. These capture data were extrapolated to the numbers of moths released from four adjacent cages. Recapture data were obtained from 400 Scentry^R wing traps with adhesiveness reinforced with Tack-Trap^R uniformly coated on the interior of the bottom panel. Each trap was charged with 40 micrograms (ug) of 97:3 *Z*:*E*-11-tetradecenyl acetates impregnated in a rubber septum. These sex pheromone traps were placed 2 meters (m) above the soil surface in concentric lines of 200 m (traps 100 m apart), 800 m (traps 200 m apart), 3,200 m (traps 270 m apart), and 6,400 m (traps 300 m apart) from a central release site on the Iowa State University Research Farm, Ankeny, Iowa. With this trap arrangement, distance and direction of male moth movement could be determined. An anemometer (Climatronics^R, Mark III) stationed at the release site recorded wind speed and direction nightly. Traps were observed the day after the night of release and two to three days thereafter. All traps received fresh sex pheromone and new bottom panels after four nights.

Results

At the initiation of the 1985 diapause experiment in the growth chamber, there were 102 individuals of the F_2 from each 1984 North Dakota site and the central Iowa population. Survivorship ranged from 88 to 96 individuals per population. The numbers of adults, indicating bivoltinism per population ranged from 3 of 88 to 64 of 94. The numbers of adults, percentages of nondiapause, and arcsin transformations are presented in Table 14.1.

All of the populations surrounding the potential location of the corn inbred nursery near Johnstown, North Dakota, produced a diapause response consistent with that of an univoltine European corn borer. Under the regimen of the experiment, 68% of the individuals of the known bivoltine population (central Iowa) pupated and became adults. At the initiation of the diapause experiment in 1986, there were 100 individuals of the F_2 from each 1985 North Dakota site and the central Iowa population. Survivorship ranged from 85 to 94 individuals per population. The numbers of adults illustrating bivoltinism ranged from 5 of 88 to 57 of 94. The distance and direction from the corn inbred nursery infested with bivoltine European corn borer, the numbers of adults, percentage nondiapause, and arcsin transformations are presented in Table 14.2.

The F_2 progeny of European corn borers collected September 11, 1985, 0.2 km

Table 14.1. Ability of Several Populations of European Corn Borer to Develop to Adulthood, Simulated July 12 in Iowa, 1985[a]

Distance (km) from Johnstown, N.D., Populations Collected	Direction	Adults/ Insect Forms	Percentage Nondiapause	Arcsin[b]
0.2	. . .	5/90	5.6	13.5[b]
8.0	north	5/96	5.2	13.1[b]
16.0	north	6/88	6.8	15.1[b]
8.0	east	10/91	11.0	19.4[b]
16.0	east	6/87	6.9	15.2[b]
8.0	south	4/89	4.5	12.2[b]
16.0	south	3/88	3.4	10.6[b]
8.0	west	4/89	4.5	12.2[b]
16.0	west	7/95	7.4	15.8[b]
Iowa population	. . .	64/94	68.1	55.6[a]

[a]Progeny of specimens collected in 1984 reared under diapausing conditions for univoltine population; 14 hours 55 minutes photophase: 9 hours 05 minutes scotophase and 22.2°–27.8°C.
[b]Percentages were arcsin transformed and subjected to student-Newman-Keuls'; different letters designate significance at $p < 0.05$.

Table 14.2. Influence of Bivoltine Population of European Corn Borer in an Univoltine Area, 1986[a]

Distance (km) from Johnstown, N.D., Populations Collected	Direction	Adults/ Insect Forms	Percentage Nondiapause	Arcsin[b]
0.2	. . .	56/89	62.9	52.6[a]
8.0	north	40/94	42.6	41.0[a]
16.0	north	47/85	55.3	47.9[a]
32.0	north	36/85	42.4	40.4[a]
8.0	east	5/88	5.7	14.3[b]
16.0	east	6/86	7.0	15.5[b]
53.0	east	12/96	12.5	20.7[b]
8.0	south	37/92	40.2	39.2[a]
16.0	south	9/94	9.6	18.5[b]
8.0	west	15/93	16.1	23.6[b]
16.0	west	7/92	7.6	16.5[b]
Iowa population	. . .	57/94	60.6	51.4[a]

[a] F_2 progeny of specimens collected in 1985 reared under diapausing conditions for univoltine population; 14 hours 55 minutes photophase: 9 hours 05 minutes scotophase and 22.2°–27.8°C.
[b] Percentages were arcsin transformed and subjected to student-Newman-Keuls'; different letters designate significance at $p < 0.05$.

from the infested corn inbred block, produced numbers of adults substantially in agreement with a bivoltine population. This result strongly suggests that the collected borers were the offspring of the Iowa population introduced as neonates into the corn inbred nursery during June. The influence of the bivoltine introduction is discernible 32 km north of the infested nursery (Table 14.2). There is some indication that the offspring of the bivoltine introduction expanded southward for 8 km; however, the progeny of population collections from 8 and 16 km east or west produced adults consistent with an univoltine population of European corn borer (53 km east, Table 14.2).

It takes about 450 accumulated centigrade degree days (Showers et al., 1983) from egg hatch to egg deposition in a bivoltine population of European corn borer. During 1985, this phenology would have occurred in the Johnstown, North Dakota, area by the first week in August (infestation with neonates June 20–21). Nightly wind direction for the period from July 28 to August 13 was obtained from the National Weather Service, Grand Forks, North Dakota. There were some northerly winds during the period, but the prevailing winds were from the south (Fig. 14.5). These results (Table 14.2 and Fig. 14.5) indicate that the bivoltine European corn borer introduced near Johnstown, North Dakota, was capable of moving at least 32 km north on the prevailing winds. Eggs were deposited, and elements of a second generation had achieved fifth instar and entered diapause by September 11–12, 1985.

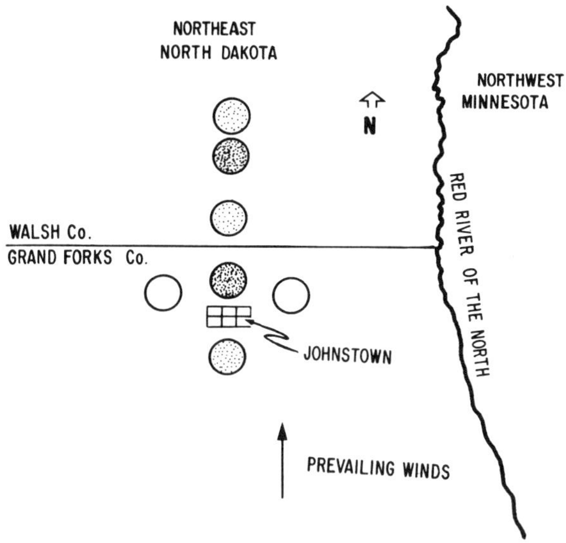

Figure 14.5. Relationship of southerly winds and movement of a bivoltine population of European corn borer from an infested inbred corn nursery. Stippling indicates bivoltine influence. Northeastern North Dakota, 1985

In central Iowa during the summer 1986, two sets of five releases per set of European corn borer adults marked with lipid soluble dyes were conducted. The estimated average numbers of emerged adults, morphologically normal (capable of flying), and available males for recapture in traps baited with 97:3 *Z*:*E*-11-tetradecenyl acetates are presented in Table 14.3. During the five releases of July 5–19, more males were available for recapture; however, for illustrative purposes, the two release sets were combined, and data statistically analyzed with 10 replications (releases). An average of 154 males were recaptured per release. This number is 1.08% of males available (14,312) for recapture per release. The data were partitioned over distance recaptured from release site per night and night of recapture after release (Fig. 14.6).

Table 14.3. Estimated Average Numbers of Emerged Adults, Normal Adults, and Males Available for Recapture in Traps Baited with Synthetic Sex Pheromone, Central Iowa, 1986

	Release Sets[a]		
Average of Five Releases	July 5–19	July 25–Aug. 7	Overall Average
Number emerged	61,970	45,050	53,510
Number normal	30,985	22,493	26,739
Available males	15,493	13,131	14,312

[a]Five releases, of four night intervals each, per release set.

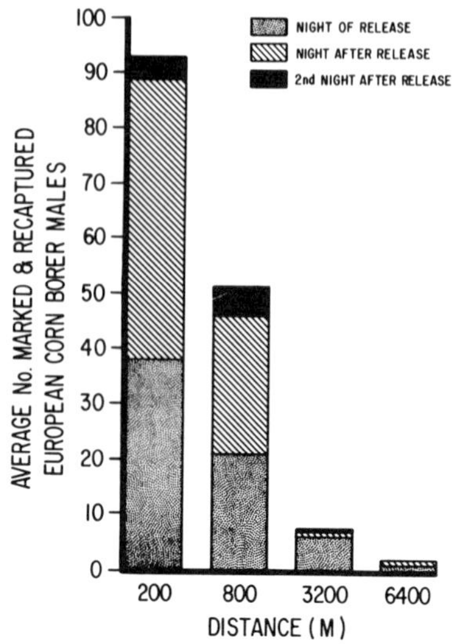

Figure 14.6. Average recapture of marked European corn borer adult males at various distances (m) and nights, central Iowa, 1986

The 200-m traps had the largest recapture, with movement occurring in all directions. Recapture of male moths at 800 m was approximately half of the recapture at 200 m. Although statistically nonsignificant at 800 m, the average recapture of male moths was greatest (29 versus 22) within the northern or northwestern traps. The winds were variable during the nights of the 10 releases, but they were predominantly from the south or southeast at 5–13 kmph. The recapture of male moths at 800 m might have been influenced by these winds. Conversely, all males recaptured at 3,200 m (3.2 km) or 6,400 m (6.4 km) were in traps placed along the north or northwest of these specific trap lines. These results agree with the directional influence found in North Dakota (Table 14.2).

At 200 m, an average of 41% (38 males) were recaptured during the night of release. Approximately 55% (51 males) were recaptured during the night after release. The percentage of nightly recapture at 800 m was similar; 41% (21 males) were recaptured during the night of release, and 49% (25 males) were recaptured during the night after release (Fig. 14.6). Essentially, these results indicate random movement, with possibly some nonrandom influence from winds at 800 m. Eighty percent of the males recaptured at 3,200 m, however, were consistently recaptured during the night of release (6 males/release). Just 13% (1 male) was recaptured during the night after release (Fig. 14.6). I suggest that these night-of-release captures at 3,200 m represent the initiation of long-range movement. By the time the portion of the population that was involved with long-range movement reached 6,400 m, most available males would have been flying at altitudes above the sex pheromone (2 m above the surface) so that the odor would not have been perceivable.

DISCUSSION AND CONCLUSIONS

Reed et al. (1981) determined that progeny of an univoltine European corn borer male had a greater frequency of diapause and a greater proportion of females in the diapausing portions of the populations than when the male was from a low-diapause population. These results support earlier reports (Showers et al., 1972; McLeod, 1978) that diapause is genetically linked to the male and, depending on scotophase and temperature, that the univoltine male or the multivoltine male could express dominance (Showers, 1981). If, however, in a given geographic area of sympatry, the univoltine and bivoltine populations were asynchronous, interbreeding would be minimal (Eckenrode et al., 1983; Roelofs et al., 1985).

The criteria necessary for asynchrony would be (1) time to emergence in the spring and (2) the critical photoperiodic threshold for diapause induction (McLeod, 1976; McLeod et al., 1979). Do the voltine ecotypes presented by Showers et al. (1975) fit these criteria? Are there really significant differences between ecotypes, or is there a gradation of responses, depending on latitude? A gradation would indicate plasticity within the species rather than true ecotypic response.

Showers et al. (1975) demonstrated that bivoltine (central ecotype) or multivoltine (southern ecotype) populations of European corn borer, if placed at northern latitudes, would develop as fast as, or faster than, if these populations were at more southern latitudes (Fig. 14.3). These results suggest that the bivoltine (central ecotype) and multivoltine (southern ecotype) populations exhibit a gradation of diapause response. Diapause of the northern ecotype (univoltine) (Fig. 14.3), however, fits the criteria of McLeod (1976) and McLeod et al. (1979).

Other important factors in determining the ecotype that will be found in any area are the length of the growing season, the position of the critical photoperiod within the season, and the type and abundance of plant host (Sparks and Showers, 1975; McLeod, 1976; Palmer et al., 1985).

Chiang et al. (1965) presented evidence that European corn borer adults captured in blacklight traps in southern Minnesota might have come north on southerly winds out of the bivoltine area of northern Iowa. Munro and Colberg (1950) suggested that bivoltine populations might be present in the Dakotas, and Smith (1950) indicated some evidence of this phenomenon in Manitoba, Canada. These populations did not sustain themselves, however, and, through the years, univoltine populations remained predominant (Palmer et al., 1985; Hudon and LeRoux, 1986).

The monitoring studies conducted in the Red River Valley of northeastern North Dakota and the mark-release-recapture study in central Iowa (Table 14.2; Fig. 14.5; Fig. 14.6) indicate that a bivoltine population of European corn borer will move on southerly winds and that a portion of that population will move beyond 3.2 km during the night of release. Bivoltine invasions of northern areas inhabited by the univoltine population of European corn borer may have been occurring for years (Munro and Colberg, 1950; Smith, 1950), but only recently has establishment been demonstrable (McLeod, 1976; McLeod et al., 1979) (Fig. 14.4). Heterovoltinism in transition zones (Palmer et al., 1985) might not be an adaptation of voltinism but, rather, a sympatry of voltinism.

In areas where the univoltine and bivoltine populations are sympatric, there seems to be very little interbreeding between the two (McLeod et al., 1979; Eckenrode et al., 1983). This may also be true where sympatry of sex pheromone types occur (Cardé et al., 1975; Roelofs et al., 1985). Asynchrony of voltine populations could occur because of length of the growing season and abundance of host type. If the season is long, some univoltine individuals might proceed to adulthood, mate, and deposit eggs that produce larvae killed by frost. Therefore, the next year, there would be fewer univoltine moths in the emerging population. A long season, with adequate host plants, however, would be optimal for a bivoltine population, and large numbers would emerge the next spring. During a shorter season, most members of an univoltine population would enter diapause and emerge the next year. Many second-generation individuals of a bivoltine population, however, would be killed by frost.

From historical data, these phenomena are difficult to ascertain. Usually an average time to emergence is obtained. Quite often, these averages, calculated over years, will show a replacement of the univoltine European corn borer with a bivoltine European corn borer (Showers, 1979). This replacement is suggestive of interbreeding of the two populations, but this probably is not true (McLeod et al., 1979; Eckenrode et al., 1983).

Although the European corn borer is polyphagous and has been reported on nearly every robust herbaceous wild and cultivated plant with stems large enough for the larvae to enter (Hudon and LeRoux, 1986), it is primarily a pest of corn. In the northern areas of the United States and the southern areas of Canada (excluding southern Ontario) where, until recently, corn has not been in abundance, the insect has been found in wild grasses (Dicke, 1932) and in a number of broad-leaved weeds, as well as potato, *Solanum tuberosum* L., bean, *Vicia faba* L., and eggplant, *S. melongena* L. (Hudon and LeRoux, 1986). Many of these hosts, along with sweet corn, would allow development of a single generation of European corn borer whether the parent population was univoltine or bivoltine but would not allow the development of a second generation within the shortened growing season of these northern latitudes. For years, hybrid field corn was not overly suited for these northern areas and was sparse. Also, much of it was cut early for silage. Again, a potential second generation of European corn borer would have been eliminated.

During the past 10 years, there have been a number of corn hybrids developed for the Canadian prairie provinces and the northern United States. Thus, there has been a dramatic increase in corn plantings in North Dakota (Fig. 14.7). Along the Red River Valley of northeastern North Dakota, land in hybrid field corn during 1975 was less than 1,000 ha; in 1985, 13,000 ha. Farther south (District 6-EC, Fig. 14.7), plantings increased from 5,000 ha in 1975 to 48,000 ha in 1985. With hybrid corn expansion in this area came the establishment of the bivoltine European corn borer (Fig. 14.4).

Based on the evidence presented in Tables 14.1, 14.2, and Figures 14.4 through 14.7, and the results of Chiang (1972), McLeod (1976), Palmer et al. (1985), and Lee and Spence (1986), it is my theory that the bivoltine European corn borer has been invading the historically univoltine areas of Minnesota, North Dakota, and the

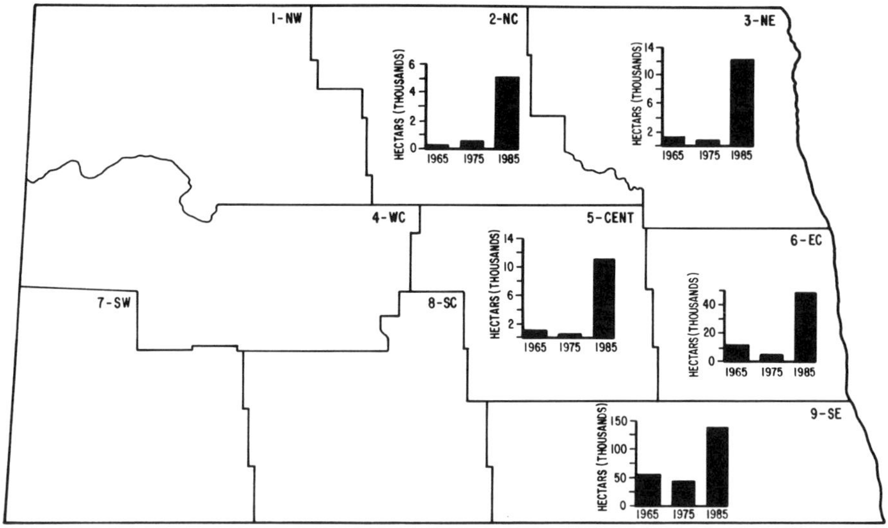

HECTARS OF GRAIN CORN (THOUSANDS)

Figure 14.7. Hectares of hybrid grain corn planted in central and eastern North Dakota, 1965, 1975, and 1985

Canadian prairie provinces for years (Munro and Colberg, 1950; Smith, 1950; Chiang et al., 1965). But, until recently, the bivoltine population has not been able to become established because of the sparseness of corn and the prevailing cropping practices. The habitats introduced and maintained by the agricultural communities favored an univoltine European corn borer. Now, however, with large plantings of hybrid field corn (Fig. 14.7) that remain succulent through August, the advantage has shifted toward the invading bivoltine European corn borer.

The valleys (Lee and Spence, 1986) with warmer temperatures and, thus, succulent corn lasting later into the season will, in my opinion, serve as reservoirs for the wind-borne spread of the bivoltine population of the European corn borer. I suggest that the univoltine population of European corn borer will, because of asynchrony with the bivoltine population, remain a distinct (McLeod, 1976; McLeod et al., 1979; Eckenrode et al., 1983) but nearly unnoticed entity. Continued monitoring of European corn borer populations of northern tier central United States and prairie provinces of Canada will provide further verification of this theory, and resulting information can be used to explain dispersal and adaptation of the European corn borer throughout North America. These studies will add significantly to the knowledge on the evolution of immigrant insect populations.

Acknowledgments

The assistance of the following members of my research project is sincerely appreciated and acknowledged: W. H. Hendrix, III, M. E. Derrick, D. Gunnarson, (all of the Department of Entomology, Iowa State University), and J. Andrew, U.S. Department of Agriculture, Agri-

cultural Research Service. The research on European corn borer adult movement in North Dakota could not have been carried out without the cooperation and assistance of M. Weiss, North Dakota State University, and R. Luedtke, Pioneer International, Inc. For critical review of the manuscript, I gratefully acknowledge H. C. Chiang, C. J. Eckenrode, D. G. R. McLeod, and W. L. Roelofs. This is Journal Paper No. J-12905 of the Iowa Agriculture and Home Economics Experiment Station, Ames, Iowa, Project No. 2749.

REFERENCES

Arbuthnot, K. D. 1944. Strains of the European corn borer in the United States. *U.S. Dep. Agric. Tech. Bull.* 869.

Beck, S. D. 1982. *Insect Photoperiodism,* 2nd ed. New York: Academic Press.

Beck, S. D., and J. W. Apple. 1961. Effects of temperature and photoperiod on voltinism of geographical populations of the European corn borer, *Pyrausta nubilalis. J. Econ. Entomol.* 54:550–558.

Beck, S. D., and W. Hanec. 1960. Diapause in the European corn borer, *Pyrausta nubilalis* (Hbn.). *J. Insect Physiol.* 4:304–318.

Bergman, P. W., R. A. Davis, S. N. Fertig, F. Kuchler, R. McDowell, C. Osteen, A. L. Padula, K. L. Smith, and R. F. Torla. 1985. Pesticide assessment of field corn and soybeans: Corn Belt States. National Agricultural Pesticide Impact Assessment Program (NAPIAP) Ext. Serv. U.S. Dep. Agric., Washington D.C.

Caffrey, D. J., and L. H. Worthley. 1927. A progress report on the investigations of the European corn borer. *U.S. Dep. Agric. Bull.* 1476.

Cardé, R. T., J. Kochansky, J. F. Stimmel, A. G. Wheeler, Jr., and W. L. Roelofs. 1975. Sex pheromone of the European corn borer (*Ostrinia nubilalis*): *cis*- and *trans*-responding males in Pennsylvania. *Environ. Entomol.* 4:413–414.

Chiang, H. C. 1961. Fringe populations of the European corn borer, *Pyrausta nubilalis*: Their characteristics and problems. *Ann. Entomol. Soc. Am.* 54:378–387.

———. 1972. Disperson of the European corn borer (Lepidoptera: Pyralidae) in Minnesota and South Dakota, 1945 to 1970. *Environ. Entomol.* 1:157–161.

Chiang, H. C., V. Sisson, and M. A. Ewert. 1965. Northerly movement of corn borer moth, *Ostrinia nubilalis,* in southern Minnesota. *Proc. Minn. Acad. Sci.* 33:17–19.

Chiang, H. C., A. J. Keaster, and G. L. Reed. 1968. Differences in ecological responses of three biotypes of *Ostrinia nubilalis* from the north central United States. *Ann. Entomol. Soc. Am.* 61:140–146.

Dicke, F. F. 1932. Studies on the host plants of the European corn borer, *Pyrausta nubilalis* Hübner, in southeastern Michigan. *J. Econ. Entomol.* 25:868–878.

———. 1989. Personal communication.

DuRant, J. A., D. G. Manley, and R. T. Cardé. 1986. Monitoring of the European corn borer (Lepidoptera: Pyralidae) in South Carolina using pheromone traps. *J. Econ. Entomol.* 79:1539–1543.

Eckenrode, C. J., P. S. Robbins, and J. T. Andaloro. 1983. Variations in flight patterns of European corn borer (Lepidoptera: Pyralidae) in New York. *Environ. Entomol.* 12:393–396.

Felt, E. P. 1919. The European corn borer. *Cornell Ext. Bull.* 31:35–48.

Glover, T. J., X. H. Tang, and W. L. Roelofs. 1987. Sex pheromone blend discrimination by male moths from *E* and *Z* strains of European corn borer. *J. Chem. Ecol.* 13:143–151.

Guthrie, W. D., F. F. Dicke, and C. R. Neiswander. 1960. Leaf and sheath feeding resistance to the European corn borer in eight inbred lines of dent corn. *Ohio Agric. Res. Dev. Cent. Res. Bull.* 860.

Hartstack, A. W., J. A. Witz, and D. R. Buck. 1979. Mothtraps for the tobacco budworm. *J. Econ. Entomol.* 72:519–522.

Hudon, M. 1959. Some aspects of seasonal development of the European corn borer, *Ostrinia nubilalis* (Hbn.) (Lepidoptera: Pyralidae) in southwestern Ontario. *Que. Soc. Prot. Plants Rep.* 41:121–128.

Hudon, M., and E. J. LeRoux. 1986. Biology and population dynamics of the European corn borer (*Ostrinia nubilalis*) with special reference to sweet corn in Quebec. I. Systematics, morphology, geographical distribution, host range, economic importance. *Phytoprotection* 67:39–54.

Klun, J. A. 1968. Isolation of a sex pheromone of the European corn borer. J. Econ. Entomol. 61:484–487.

Klun, J. A., and T. A. Brindley. 1970. *Cis*-11-tetradecenyl acetate, a sex stimulant of the European corn borer. *J. Econ. Entomol.* 63:779–780.

Klun, J. A., and M. D. Huettel. 1987. Genetic regulation of sex pheromone production and response: Interaction of sympatric phermonal races of the European corn borer, *Ostrinia nubilalis* (Lepidoptera: Pyralidae). *J. Chem. Ecol.* 14:2047–2061.

Klun, J. A., and S. Maini. 1979. Genetic basis of an insect chemical communication system: The European corn borer. *Environ. Entomol.* 8:423–426.

Klun, J. A., and J. F. Robinson. 1971. European corn borer moth: Sex attractant and sex attraction inhibitors. *Ann. Entomol. Soc. Am.* 64:1083–1086.

———. 1972. Olfactory discrimination in the European corn borer and several pheromonally analogous moths. *Ann. Entomol. Soc. Am.* 65:1337–1340.

Klun, J. A., O. L. Chapman, K. C. Mattes, P. W. Wojtkowski, M. Beroza, and P. E. Sonnet. 1973. Insect sex pheromones: Minor amount of opposite geometrical isomer critical to attraction. *Science* 181:661–663.

Klun, J. A., and Cooperators. 1975. Insect sex pheromones: Interspecific pheromonal variability of *Ostrinia nubilalis* in North America and Europe. *Environ. Entomol.* 4:891–894.

Kochansky, J., R. T. Cardé, J. Liebherr, and W. L. Roelofs. 1975. Sex pheromone of the European corn borer, *Ostrinia nubilalis* (Lepidoptera: Pyralidae) in New York. *J. Chem. Ecol.* 1:225–231.

Lee, D. A., and J. R. Spence. 1986. Phenology of the European corn borer (*Ostrinia nubilalis*) in Alberta. *Can. Entomol.* 118:943–956.

Lees, A. D. 1980. Insect diapause—stability and change. *Nature (Lond.)* 286:446.

Lewis, L. C., and R. E. Lynch. 1970. Treatment of *Ostrinia nubilalis* larvae with Fumidil B to control infections caused by *Perezia pyraustae. J. Inveretebr. Pathol.* 15:43–48.

Lilly, C. E., and A. M. Harper. 1982. Status of the European corn borer in Alberta. Pages 12–13 in Sear, L. J. L., K. K. Krogman, and T. G. Atkinsons (eds.), *Research Highlights—1981*. Lethbridge, Canada: *Agric. Can. Res. Stn.*

McEwen, F. L., J. A. Adams, A. C. Davis, and H. B. Rinick, Jr. 1968. Corn borer in western New York. *N.Y. Food Life Sci. Q.* 1:15–16.

McLaine, L. S. 1922. The spread of European corn borer in southern Ontario. *J. Econ. Entomol.* 15:227–231.

McLeod, D. G. R. 1976. Geographical variation of diapause termination in the European corn borer, *Ostrinia nubilalis* (Lepidoptera: Pyralidae) in southwestern Ontario. *Can. Entomol.* 108:1403–1407.

———. 1978. Genetics of diapause induction and termination in the European corn borer, *Ostrinia nubilalis* (Lepidoptera: Pyralidae), in southwestern Ontario. *Can. Entomol.* 110:1351–1353.

McLeod, D. G. R., C. Ritchot, and T. Nagei. 1979. Occurrence of a two-generation strain of the European corn borer, *Ostrinia nubilalis* (Lepidoptera: Pyralidae), in Quebec. *Can. Entomol.* 111:233–236.

Munro, J. A., and W. J. Colberg. 1950. European corn borer occurrences in North Dakota. *Entomol. Soc. Ont. Annu. Rep.* 80:20–21.

Ostle, K. R., G. L. Hein, L. G. Higley, L. V. Kaster, and W. B. Showers. 1983. European corn borer (Lepidoptera: Pyralidae) development, larval survival, and adult vigor on meridic diets containing marker dyes. *J. Econ. Entomol.* 77:118–120.

Palmer, D. F., T. C. Schenk, and H. C. Chiang. 1985. Dispersal and voltinism adaption of the European corn borer in North America, 1917–1977. *Minn. Agric. Exp. Stn. Bull.* AD-SB-2716.

Reed, G. L., W. D. Guthrie, W. B. Showers, B. D. Barry, and D. F. Cox. 1981. Sex-linked inheritance of diapause in the European corn borer: Its significance to diapause physiology and environmental response of the insect. *Ann. Entomol. Soc. Am.* 74:1–8.

Roelofs, W. L., and A. Comeau. 1971. Sex attractants in Lepidoptera. Pages 91–114 in Sitahori, A. (ed.), *Chemical Releases in Insects, III.* New York: Gordon and Breach.

Roelofs, W. L., and J. Tette. 1970. Sex pheromone of the obliquebanded leafroller moth. *Nature (Lond.)* 226:1172.

Roelofs, W. L., R. T. Cardé, R. J. Bartelt, and P. G. Tierney. 1972. Sex attractant trapping of the European corn borer in New York. *Environ. Entomol.* 1:606–608.

Roelofs, W. L., J. W. Du, X. H. Tang, P. S. Robbins, and C. J. Eckenrode. 1985. Three European corn borer populations in New York based on sex pheromones and voltinism. *J. Chem. Ecol.* 11:829–836.

Roelofs, W. L., T. Glover, X. H. Tang, I. Sreng, P. Robbins, C. Eckenrode, C. Lüfstedt, B. S. Hansson, and B. O. Bengtsson. 1987. Sex pheromone production and perception in European corn borer moths is determined by both autosomal and sex-linked genes. *Proc. Natl. Acad. Sci. (USA)* 84:7585–7589.

Showers, W. B. 1970. Ecological effects on F_1 progenies from reciprocal matings of three biotypes of the European corn borer, *Ostrinia nubilalis* (Hübner). Ph.D. Dissertation. Iowa State Univ. (Libr. Cong. Card No. Mic. 71–14262). 246 pages.

———.1979. Effect of diapause on the migration of the European corn borer into the southeastern United States. Pages 420–430 in Rabb, R. L., and G. G. Kennedy (eds.), *Movement of Highly Mobile Insects: Concepts and Methodology in Research.* Raleigh, N.C.: North Carolina State Univ. Press.

———. 1981. Geographic variation of the diapause response in the European corn borer. Pages 97–111 in Denno, R. F., and H. Dingle (eds.), *Insect Life History Patterns, Habitat and Geographic Variation.* New York: Springer-Verlag.

Showers, W. B., T. A. Brindley, and G. L. Reed. 1972. Survival and diapause characteristics of hybrids of three geographical races of the European corn borer. *Ann. Entomol. Soc. Am.* 65:450–457.

Showers, W. B., G. L. Reed, and H. Oloumi-Sadeghi. 1974. European corn borer: Attraction of males to synthetic lure and to females of different strains. *Environ. Entomol.* 3:51–58.

Showers, W. B., H. C. Chiang, A. J. Keaster, R. E. Hill, G. L. Reed, A. N. Sparks, and G. J. Musick. 1975. Ecotypes of the European corn borer in North America. *Environ. Entomol.* 4:753–760.

Showers, W. B., J. F. Witkowski, C. E. Mason, F. L. Poston, S. M. Welch, A. J. Keaster, W. D. Guthrie, and H. C. Chiang. 1983. Management of the European corn borer. North Cent. Reg. Publ. 22. Ames, Iowa: Iowa State Univ.

Smith, C. A. S. 1950. The first records of European corn borer in western Canada. *Entomol. Soc. Ont. Annu. Rep.* 80:18–19.

Smith, H. E. 1920. Broom corn, the probable host in which *Pyrausta nubilalis* Hübn. reached America. *J. Econ. Entomol.* 13:425–430.

Sparks, A. N. 1963. The use of Infra-red photography to study mating of the European corn borer. *Proc. North Cent. Branch Entomol. Soc. Am.* 18:96.

Sparks, A. N., and W. B. Showers. 1975. Current status of European corn borer in south Georgia. *J. Ga. Entomol. Soc.* 10:347–352.

Sparks, A. N., T. A. Brindley, and N. D. Penny. 1966a. Laboratory and field studies of F_1 progenies from reciprocal matings of biotypes of the European corn borer. *J. Econ. Entomol.* 59:915–921.

Sparks, A. N., H. C. Chiang, A. J. Keaster, M. L. Fairchild, and T. A. Brindley. 1966b. Field studies of European corn borer biotypes in the midwest. *J. Econ. Entomol.* 59:922–928.

Struble, D. L., J. R. Byers, D. G. R. McLeod, and G. L. Ayre. 1987. Sex pheromone components of an Alberta population of European corn borer, *Ostrinia nubilalis* (Hbn.) (Lepidoptera: Pyralidae). *Can. Entomol.* 119:291–299.

Vinal, S. C. 1917. The European corn borer, a recently established pest in Massachusetts. *Mass. Agric. Exp. Stn. Bull.* 178:147–152.

Wressell, H. B. 1971. A comparison of European corn borer, *Ostrinia nubilalis* (Hübner) moth flight to light traps in southwestern Ontario on the relationship to borer infestation. *Proc. Entomol. Soc. Ont.* 102:63–71.

15

GEOGRAPHIC AND GENETIC VARIATION AMONG ALFALFA WEEVIL STRAINS

Ting H. Hsiao

Utah State University

INTRODUCTION

Exotic insect pests are ideal for studying habitat-related variations in insects. First, these pests are abundant and easy to locate, because each of them is usually associated with a specific crop. Second, they are widely distributed through the direct or indirect action of humans. Because survey records are available that pinpoint exactly when the pest colonized a given area, spatial and temporal changes in the population since its establishment can be determined. Studying habitat-related variations in cosmopolitan pests that have been introduced into different geographic regions could provide information on the pest's ecological and genetic diversities. However, the short evolutionary time available after colonization of a new habitat means that only minute changes may be detectable unless more than one race of a pest species has been introduced.

The alfalfa weevil, *Hypera postica* (Gyllenhal), complex in the United States is an excellent subject for studying genetic variability in an introduced pest. It has been a major pest of alfalfa in North America since it was discovered at the beginning of this century. There have been three documented introductions of the pest, which has spread rapidly to all 48 contiguous states of the United States, as well as several regions in Canada and Mexico.

The alfalfa weevil has mainly one generation per year and overwinters or aestivates as adults. The species is oligophagous. Although it has been reported to feed on a number of legumes, its host range is restricted to various native and cultivated *Medicago* species and, to some extent, *Melilotus, Trifolium,* and *Vicia* (Essig and Michelbacher, 1933). Both larvae and adults feed on the foliage. The female typically lays from several hundred to over 3,000 eggs in the alfalfa stem in early spring

(Coles and Day, 1977). Larvae hatch in late spring and early summer, pass through four instars, and then spin cocoons and pupate inside. Adults, after a short feeding period, enter aestival or overwintering diapause. Because of its high reproductive potential, the weevil can severely damage hay crops.

Its economic importance has resulted in extensive research on the alfalfa weevil; over 2,000 articles have been written about the species worldwide (Wood et al., 1978). This discussion focuses on geographic and genetic diversity among weevil strains in the United States, and attempts to clarify the taxonomic status of these introduced weevils. The relevance of genetic studies in integrated pest management of the alfalfa weevil is also discussed.

TAXONOMIC STATUS, GEOGRAPHIC DISTRIBUTION, AND SPREAD IN NORTH AMERICA

The alfalfa weevil is a cosmopolitan pest. It is native to Asia, Europe, and North Africa (see Commonwealth Distribution Maps of Pests No. 456). Its distributional range in Asia extends to Xinjiang, China.

These weevils were introduced into the United States on three separate occasions. The first introduction was discovered in the Salt Lake Valley of Utah in 1904 (Titus, 1910). The weevil is believed to have come from Western Europe near the borders of France, Italy, and Switzerland (Davis, 1967). It spread quickly to all adjacent states and was found in California in 1923. By early 1930, it was found in the northern and central parts of California (Essig and Michelbacher, 1933). This introduction, which is now referred to as the "western weevil," was confined to 12 western states for nearly 50 years. Since 1954, the western weevil was found in Alberta, Canada (Hobbs et al., 1959).

In 1939, a second introduction was discovered in Yuma, Arizona (Wehrle, 1940). This introduction, named the "Egyptian alfalfa weevil," *H. brunneipennis* Boheman, spread to southern and central California between 1950 and 1965 (van den Bosch and Marble, 1971), and more recently to New Mexico and northern Mexico. It is confined to the hot desert climate. The Egyptian alfalfa weevil has been traced to the Nile Valley and the Mediterranean region (Michelbacher and Leighly, 1940; Wehrle, 1940; Davis, 1967).

The third introduction, discovered in Maryland in 1952, was named the "eastern strain" of *H. postica* to distinguish it from the western strain (Poos and Bissell, 1953). The precise origin of the eastern weevil is not known, but it probably came from an area of Europe different from that of the western weevil. The eastern weevil appears to be the most vigorous and adaptable of the three introductions. It has spread to all eastern and central states, as well as to Ontario and Quebec, Canada (Mailloux and Pilon, 1975).

Since about 1970, the eastern weevil has invaded the habitats of the western weevil in central Nebraska and, later, in Kansas and Colorado. Overlapping zones between the western and the Egyptian weevils have recently been detected in four western states, including Arizona, New Mexico, Nevada, and Utah (Hsiao, 1990). At present, alfalfa weevils are found in all 48 contiguous states. The following discussion refers to all three weevil introductions in North America as strains rather than as two separate species. Evidence to support this contention will be discussed.

Taxonomically, there is considerable confusion concerning the species name of the alfalfa weevil. *Phytonomus variabilis* Herbst, *P. posticus* Gyllenhal, and *H. variabilis* (Herbst) have been used by European workers for many years. Titus (1911) was first to adopt the present name *Hypera postica* (Gyllenhal). Barber (1924) and Warner (1962) supported its use in North American literature. Both *Hypera* and *Phytonomus* are large genera, each with over 100 described species (Csiki, 1934). In continental Europe and the Middle East, where the alfalfa weevil is thought to be indigenous, there are several closely related species or strains (Kippenberg, 1983).

Zaslavskiy (1961) reviewed several *Phytonomus* species in the Soviet Union. He indicated that *P. variabilis, P. transsylvanicus* Petri, and *P. murina* F. are closely related species that are difficult to distinguish morphologically. Franz (1949) compared male genitalia of *P. transsylvanicus* and *P. variabilis* from different parts of Europe and concluded that they were races of the same species. *P. murina* (*P. fuscocinerus* Marshall), which is somewhat larger in size than *H. postica* and feeds on the same food plants, probably belongs to the same species complex. The species name *H. brunneipennis* Boheman first appeared in 1834 (Csiki, 1934). This species is probably widely distributed in North Africa (Michelbacher and Leighly, 1940). Because its geographic distribution overlaps that of *H. postica* and because of uncertainty concerning its taxonomic identity, the species name *H. brunneipennis* was seldom used in European literature (Csiki, 1934; Wood et al., 1978). Weevils discovered in Yuma, Arizona, in 1939 were identified by L. L. Buchanan as *H. brunneipennis* (Wehrle, 1940). The common name, Egyptian alfalfa weevil, was adopted because of its supposed Middle East origin and to distinguish it from the alfalfa weevil, *H. postica*. In the current literature, *H. brunneipennis* refers exclusively to the Egyptian alfalfa weevils found in California and Arizona.

Morphological comparisons have not identified any diagnostic feature distinctive of the three weevil introductions (Michelbacher and Leighly, 1940; Pienkowski et al., 1969). In field populations, the Egyptian weevil is somewhat larger than the western weevil; the eastern weevil is intermediate in size. Until recently, identification of weevil species and strains in North America was generally based on the geographic origin of the specimens. The discovery of mixing zones between weevil strains invalidates this practice. A method of identification based on difference in mandible structures between weevil strains has been proposed recently, but its diagnostic value in field samples has not been tested (Bland, 1984). Clearly, biosystematic studies are needed to clarify the existing taxonomic confusion.

BIOLOGICAL AND ECOLOGICAL DIFFERENCES AMONG WEEVIL STRAINS

Despite the lack of morphological distinctions, several biological and ecological differences between weevil strains have been reported (Davis, 1967; Armbrust and Gyrisco, 1982; van den Bosch et al., 1982). For example, field observations showed that the western weevil prefers to cocoon in the ground litter, but the Egyptian and eastern weevils prefer the aerial parts of plants. During hibernation, the Egyptian weevil aggregates under tree bark or beneath boards, whereas the western and eastern weevils scatter in or outside fields. Weevil strains also differ in their response to the hymenopterous parasite, *Bathyplectes curculionis* (Thomson). Eggs of this parasite are encapsulated strongly by larvae of eastern and Egyptian weevils, but only weakly by the western weevil (van den Bosch and Dietrick, 1959; Salt and van den Bosch, 1967; Puttler, 1967). For this reason, the parasite is more effective on the western weevil than on the Egyptian or eastern weevil. Climatic adaptation is another obvious difference. The Egyptian weevil thrives in hot climates, whereas the western weevil appears to do well in cold climates. The weevils also differ in their seasonal phenology and diapause characteristics. When reared under a nondiapause photoperiod regime (eight- or nine-hour photophase), the western weevil had an extended preoviposition period that was significantly longer than in the eastern and Egyptian strains (Rosenthal and Koehler, 1968; Schroder and Steinhauer, 1976a). The rate of larval and pupal development of the eastern weevil was also shorter than that of the western and European weevils (Schroder and Steinhauer, 1976a). In warmer climates, the Egyptian weevil and the eastern weevil resume feeding and oviposition in the fall and begin a new generation during the winter and early spring months. These weevils aestivate during the late spring and summer months. The western weevil typically reproduces in the spring and produces one generation during the late spring and early summer months. It enters adult diapause in late summer and overwinters until the following spring. In several areas, some weevils, regardless of strains, produce a distinct second generation (White et al., 1969; Loan et al., 1983; Hsiao, 1990). Such a bivoltine life cycle has also been observed in Europe and Asia (Saringer and Deseo, 1966; Hsiao, 1990). Evidently, climatic conditions affect the life history strategy of the species. Thus, life history traits, as discussed earlier, are useful for describing weevil adaptation to varied environmental conditions but are not reliable criteria for differentiating between weevil strains.

REPRODUCTIVE COMPATIBILITY BETWEEN WEEVIL STRAINS

Reproductive compatibility is used as an indicator of relatedness between insect species and populations. Blickenstaff (1965) first reported partial reproductive incompatibility between alfalfa weevil strains. Eggs of the cross between Maryland (eastern) female and Utah (western) male are infertile, whereas the reciprocal cross is fertile, but a distorted sex ratio results in predominantly female progeny. This phenomenon has been confirmed independently in several other studies (White et al., 1972) (Table 15.1). Schroder and Steinhauer (1976b) reported similar results with the Maryland weevil and weevils from France and Germany. However, reciprocal crosses between the two European weevils and the Utah weevil or within the European weevils were fully compatible. Klostermeyer (1978), and Klostermeyer and Manglitz (1979) found incompatibilities within field populations in Nebraska. Recently, hybridization experiments between all three weevil strains confirmed that the cross between eastern female and western male was infertile (Hsiao and Hsiao,

Table 15.1. Cross-Compatibility Between Alfalfa Weevil Strains

Cross (Female x Male)	Progeny Viable	Sex Ratio of Progeny	Reference
Eastern (MD) x Eastern (NE)	Yes	ND	4, 5
Eastern (MD) x Western (UT)	No	NA	1, 2, 4
Eastern (MD) x Western (NE)	No	NA	4, 5
Eastern (CO) x Western (UT)	No	NA	7
Eastern (NE) x Western (UT)	No	NA	4, 5
Eastern (MD) x Egyptian (CA)	Yes	Normal	2
Eastern (MD) x France	No	NA	6
Eastern (MD) x Germany	No	NA	6
Western (NE) x Western (UT)	Yes	ND	4, 5
Western (UT) x Eastern (MD)	Yes	Distorted	1, 2, 4
Western (UT) x Eastern (CO)	Yes	Distorted	7
Western (UT) x Egyptian (CA)	Yes	Distorted	2
Western (UT) x France	Yes	Normal	6
Western (UT) x Germany	Yes	Normal	6
Western (UT) x Netherlands	Yes	Normal	3
Egyptian (CA) x Western (UT)	No	NA	2
Egyptian (CA) x Eastern (MD)	Yes	Normal	2
France x Eastern (MD)	Yes	Distorted	6
France x Western (UT)	Yes	Normal	6
Germany x Eastern (MD)	Yes	Distorted	6
Germany x Western (UT)	Yes	Normal	6
Netherlands x Western (UT)	Yes	Normal	3

Abbreviations: CA = California, Co = Colorado, MD = Maryland, NE = Nebraska, UT = Utah, NA = not applicable, ND = not determined.

Sources: (1) Blickenstaff 1965; (2) Hsiao and Hsiao 1985a; (3) Hsiao, unpublished data; (4) Klostermeyer 1978; (5) Klostermeyer and Manglitz, 1979; (6) Schroder & Steinhauer 1976b; (7) White et al. 1972.

1985a). Similarly, infertility was observed in the cross between Egyptian female and western male. The reciprocal cross was, however, fertile. The most significant finding was that reciprocal crosses between eastern and Egyptian weevils are fully reproductively compatible (Table 15.1). Clearly, cross-mating data do not support the Egyptian weevil as a separate species from the eastern weevil. However, partial incompatibility indicates a distinct biological difference between the western and the other two weevil strains.

CYTOGENETIC AND ALLOZYME VARIATIONS AMONG WEEVIL STRAINS

Genetic studies have been conducted to clarify the taxonomic and biological relationships between weevil strains. Cytogenetic analyses were carried out initially. A preliminary study by Klostermeyer (1978) detected no differences in the chromosome number and karyotypes in the three weevil strains. A more extensive study by Hsiao and Hsiao (1984) compared the chromosome number and karyotypes of eight populations, including three of the eastern strain, two each of the western strain and the Egyptian weevil, and one from The Netherlands. All weevil populations have the same 22 chromosomes, 2n = 20 + XYp, which is identical to that reported for five other *Hypera* species (Takenouchi, 1972; Klostermeyer, 1978). Chromosomal karyotypes of all populations had identical patterns and arrangements. Furthermore, the C-banding pattern of their chromosomes revealed no distinct variations (Hsiao and Hsiao, 1984). Thus, the cytogenetic evidence does not support the existence of separate species among weevil populations.

Gel electrophoresis of enzymes or proteins is a useful technique for detecting inter- and intraspecific variations among insects (Berlocher, 1984). Sell et al. (1978) examined 5 putative gene loci from 4 enzymes by starch gel electrophoresis in order to assess genetic variations between eastern and western weevils. There was considerable variability of these enzymes between the two weevil strains and among geographic populations. Hsiao and Stutz (1985) used polyacrylamide gel electrophoresis to examine 22 putative gene loci from 12 enzymes of six weevil populations, two each from eastern (Beltsville, Maryland; Washington County, Illinois), western (Logan, and St. George, Utah), and Egyptian (Yuma, Arizona; Westmorland, California) weevils. The mean heterozygosity of these populations was 0.231 ± 0.005 (S.E.), and the average proportion of polymorphic loci was 0.536 ± 0.008. The allozyme data were used to calculate Nei's genetic identity and distance for the three weevil strains, which were 0.968 ± 0.007 and 0.033 ± 0.007, respectively. The fixation index (F_{st}) of these populations was 0.024, indicating that only 2.4% of the measured variation was due to genetic differences between populations. The other 97.6% of the variation was due to within-population variation. Compared with data available from *Drosophila* and other insects (Ayala, 1975; Jacobson and Hsiao, 1983), these values suggest only intraspecific differences between the six alfalfa weevil populations studied.

An UPGMA (unweighted pair-group arithmetic average) dendrogram of Nei's genetic identity (I) for the six weevil populations was prepared from the data of Hsiao and Stutz (1985) (Fig. 15.1). The I values between populations of each strain were closely clustered. Also, the values between Egyptian and eastern weevil populations were much closer (a mean of 0.995) than between the western and the other two populations (0.946). Thus, the Egyptian and eastern weevils are genetically more similar to each other than to the western weevil. However, the overall high-genetic identity values of the three weevil strains imply that they are genetically very similar and have not differentiated sufficiently to be recognized as separate species or even subspecies.

Additional evidence of genetic uniformity of the North American weevils was the scarcity of a vestigial-wing mutation in the field populations. Hsiao and Hsiao (1985b) showed that the mutation is due to a dominant autosomal gene; normal-wing individuals carry the recessive trait. In a field survey of over 100 localities in North America, the mutant was found only in an eastern weevil population from Maryland and two Egyptian weevil populations from California. It was not found in any west-

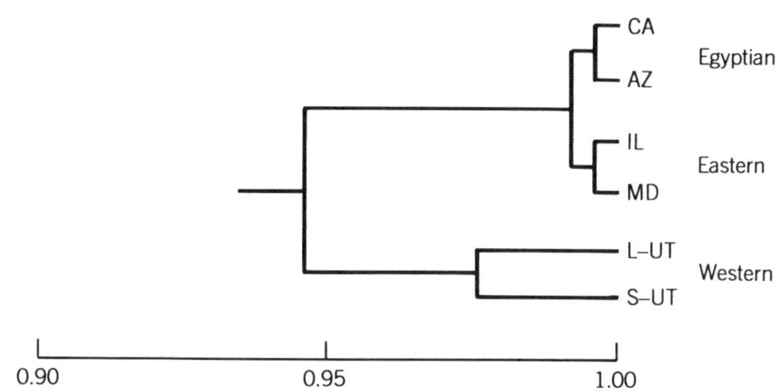

Figure 15.1. UPGMA dendrogram of Nei's genetic identity (I) of six alfalfa weevil populations. Data from Hsiao and Stutz (1985). *Abbreviations:* AZ = Yuma, Arizona; CA = Westmorland, California; IL = Washington County, Illinois; MD = Beltsville, Maryland; L-UT = Logan, Utah; S-UT = St. George, Utah.

ern weevils. In contrast, vestigial-wing mutants were widespread in several European weevil populations that were examined (Hsiao, 1990), which suggests that very few weevils of the three strains successfully colonized North America. The founder stocks of these strains probably consist of a relatively limited genetic load of the parent species. This view is supported by the similarity of genetic identity values between populations of each weevil strain, as revealed by allozyme analysis (Fig. 15.1).

RICKETTSIA AS THE CAUSE OF CYTOPLASMIC INCOMPATIBILITY

As noted earlier, the cytogenetic and allozyme data did not indicate that there was sufficient genetic differentiation between alfalfa weevil strains in the United States to justify their taxonomic status as separate species or subspecies. However, evidence of partial reproductive incompatibility between weevil strains seemingly contradicted such a contention. The central issue lies in what causes reproductive incompatibility. Klostermeyer (1978) restored cross-incompatibility between eastern and western weevils by treatment with tetracycline, which suggests that microorganisms may affect incompatibility. Hsiao and Hsiao (1985c) found, by light and electron microscopic examinations, a symbiotic microorganism in the western weevil. The microorganism possessed a trilaminar membrane. Its size (about 0.5–1 μm in diameter) was between that of ordinary bacteria and viruses, and was typical of rickettsialike microorganisms. This rickettsia, tentatively identified as *Wolbachia postica,* was not found in the eastern and Egyptian weevils. The microorganism is transmitted maternally and is, therefore, present in the hybrids of the compatible cross between females of the western weevil and males of the eastern or the Egyptian weevil (Hsiao and Hsiao, 1985a). Figure 15.2 explains the role of rickettsia in cross-incompatibility between weevil strains. Egyptian and eastern weevils are fully compatible, because they do not harbor rickettsia. The crosses between western females

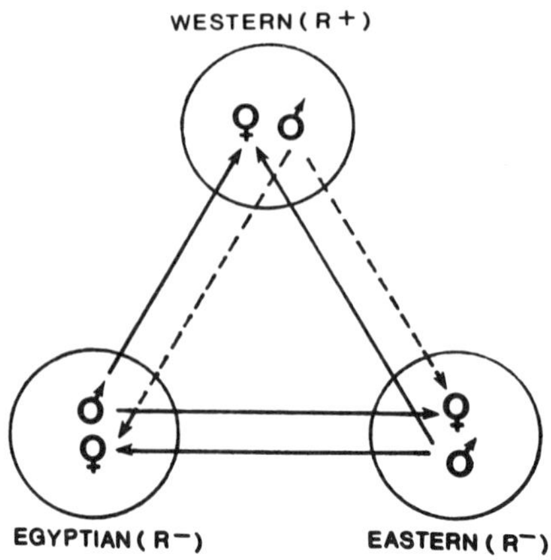

Figure 15.2. Rickettsia relationships to cross-compatibility between alfalfa weevil strains. R+ = rickettsia-positive, R− = rickettsia-negative.

and males of Egyptian or eastern weevils produce viable progeny but have a high proportion of females. The progeny carry rickettsia from the female parent. The only infertile crosses are between western males and females of Egyptian or eastern weevils. Apparently, rickettsia are the barrier to egg fertility.

Yen and Barr (1971, 1974) were the first to discover the presence of a rickettsialike microorganism, *Wolbachia pipientis* Hertig, as the cause of cross-incompatibility within subspecies of *Culex pipiens* L. Some 20 strains of rickettsia have now been suggested as being involved in cross-incompatibility (Subrarao, 1982). Other mosquitoes, in which evidence for cytoplasmic incompatibility has been found, include *Aedes scutellaris* (Smith-White and Woodhill, 1954) and *A. polynesiensis* (Wright and Wang, 1980; Wright and Barr, 1981). Cytoplasmic incompatibility due to *Wolbachia* has also been reported for the almond moth, *Ephestia cautella* (Brower, 1976; Kellen et al., 1981) and the raisin moth, *Cadra figulilella* (Kellen et al., 1981). The alfalfa weevil is apparently the first Coleoptera in which a *Wolbachia* is found to be responsible for cytoplasmic incompatibility. A *Wolbachia* like microorganism may also be involved in the unidirectional incompatibility of *Drosophila simulans* (Hoffmann and Turelli, 1988). A mycoplasmalike microorganism has been implicated in the cytoplasmic incompatibility of *D. paulistorum* (Ehrman and Kernaghan, 1971). Possible involvements of microorganisms in cytoplasmic incompatibility have been reported for the European cherry fruit fly, *Rhagoletis cerasi* (Boller et al., 1976), the rice planthopper, *Laodelphax striatellus* (Noda, 1984), and the flour beetle, *Tribolium confusum* (Wade and Stevens, 1985).

The precise mechanism of rickettsia-induced incompatibility in insects is not known. Barr (1982) hypothesized that, when a mosquito sperm develops in the presence of rickettsia, it takes on rickettsia matter, which disguises the sperm. When this sperm enters an uninfected oocyte, the oocyte fails to recognize the sperm and treats it as a foreign body. Fertilization is disrupted or terminated. If the oocyte is infected with the same kind of rickettsia, it recognizes the sperm and fertilization proceeds normally. Since the rickettsia of the alfalfa weevil causes the same type of reproductive incompatibility as that found in mosquitoes, a mechanism similar to that just described above may be assumed.

Our recent experiments showed that rickettsia-free western weevils obtained by treatment with tetracycline were completely reproductively compatible with eastern or Egyptian weevils (Hsiao, 1990). Incompatibility within the western weevils was induced when rickettsia-free females crossed with males harboring rickettsia. The restoration of incompatibility proves that all weevil strains are potentially reproductively compatible. The rickettsia is a cytoplasmic factor that does not influence the genetic constitution of the weevil. Therefore, its role as a reproductive barrier does not contradict with cytogenetic and allozyme findings mentioned in the previous section.

STATUS OF POPULATION OVERLAPPING AND FIELD HYBRIDIZATION BETWEEN WEEVIL STRAINS

The spread of the weevil in North America is well documented because of its economic importance. By 1930, the western weevil was recorded in eight states from the West Coast to western Nebraska. Eastward movement of the weevil then appeared to be slow for the next 50 years. The eastern weevil spread to all eastern and central states within 20 years. By early 1970, the eastern weevil had reached the area inhabited by the western weevil. An obvious difference between eastern and western weevils has been the relative rate of spread from the initial areas of infestation (Blickenstaff, 1965; Klostermeyer and Manglitz, 1979; Manglitz et al., 1981). The initial spread of the western weevil was erratic but was more or less equal in all directions. The estimated rate of spread over a 60-year period was 16–32 km per year. Under certain favorable conditions, it apparently could spread as far as 80–100 km in a season (Michelbacher and Leighly, 1940). By comparison, the spread of the eastern weevil is much more rapid, estimated at 80–120 km per year.

Manglitz (1974) was the first to report the presence of both eastern and western weevils in Nebraska. He found that the eastern weevil had reached Dawson County by 1974. Klostermeyer and Manglitz (1979) showed that these weevils overlapped in a 240-km zone in central Nebraska by 1977. The eastern weevil apparently has not spread west of the Rocky Mountains.

Klostermeyer and Manglitz (1979) originally demonstrated mixing zones using single-pair hybridization to determine cross-incompatibility. Such a procedure is very tedious and time consuming. The discovery of rickettsia in the western weevil made it possible to use the presence or absence of the microorganism as a convenient diagnostic trait to detect mixing zones between weevil strains. Our laboratory used this technique to examine over 100 alfalfa weevil populations from 10 western states and mapped out the distribution and overlapping zones of all three weevil strains (Fig. 15.3). Gravid females from each location were dissected, and ovarian smears were individually prepared to determine if rickettsia were present (Hsiao and Hsiao, 1985c). Weevils that are rickettsia-positive belong to the western strain or its hybrid (dark circle). In areas west of the Rocky Mountains, the absence of rickettsia indicates that the weevils are of the Egyptian strain (open circle). A mixing zone is evident where both rickettsia-negative and rickettsia-positive weevils were found in the same field (partially shaded circle). The same system applies to the eastern weevil in areas east of the Rocky Mountains.

In western states, the region occupied by the western weevil remains nearly the same since early 1930 (Essig and Michelbacher, 1933). The Egyptian weevil has, however, expanded its range to Arizona, New Mexico, southern Nevada, and the central and southern regions of California. It was surprising to find pockets of the western weevil in the coastal areas of Santa Cruz and Monterey (Fig. 15.3). These weevils appear to be remnants of the original western weevil that have inhabited central California since 1933 (Michelbacher and Leighly, 1940). Their presence indicates that the western weevil has not been completely displaced by the Egyptian weevil in the region. Two overlapping zones between the Egyptian and western weevils have been detected for the first time in southeastern Utah, bordering Nevada and Arizona, and in the four-corners area of Utah, Colorado, Arizona, and New Mexico.

All areas east of the distribution map (Fig. 15.3) are presently occupied by the eastern weevil, which has continued to disperse westward. The mixing zone between eastern and western weevils now covers central Nebraska and Kansas to eastern Colorado. Data from rickettsia examination verify the findings of single-pair matings by Klostermeyer and Manglitz (1979), who originally described a mixing zone between eastern and western weevils in central Nebraska. Interestingly, the eastern weevil is

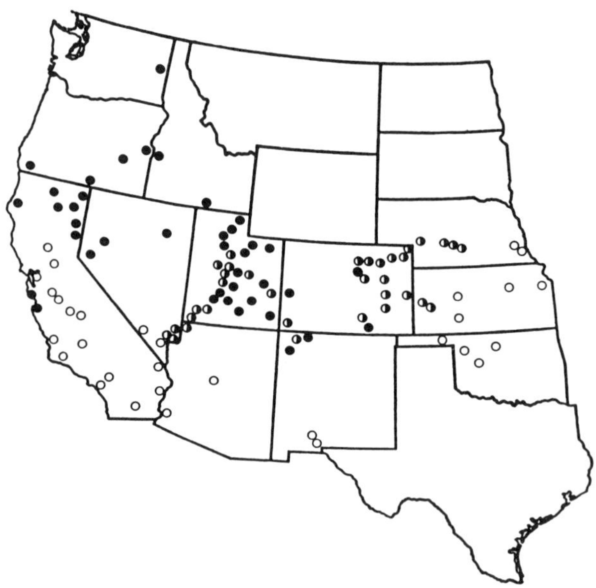

Figure 15.3. Current distribution of alfalfa weevil strains and their overlapping zones. Open circle, rickettsia-negative population; solid circle, rickettsia-positive population; partial shaded circle, overlapping population.

spreading westward more rapidly than the western weevil. Manglitz et al. (1981) estimated that the eastern weevil moves westward in excess of 80 km per year, whereas the western weevil moves eastward at about 16 km per year. The estimated width of the overlapping zone in central Nebraska was 240 km in 1977 (Klostermeyer and Manglitz, 1979); the zone has now expanded to over 480 km. The Rocky Mountains appear to be a barrier to the advancement of the eastern weevil. The distribution map shows that the western weevil occupies the colder region of the Northwest, whereas the Egyptian weevil is found in warmer climates.

The presence of both rickettsia-positive and rickettsia-negative weevils in the mixing zone raises the possibility of cross-mating. Mating behavior of the eastern weevil has been investigated extensively (LeCato and Pienkowski, 1970a, 1970b, 1972). Both sexes are polygamous. In the laboratory, fertility of once-mated females was slightly reduced because of a shortage of sperm during the latter part of the oviposition period. Plural matings are considered to be common among field weevils. In the laboratory, interstrain matings between eastern and western weevils occur freely without behavioral barriers (Blickenstaff, 1969; Armbrust et al., 1970; White et al., 1972; Klostermeyer and Manglitz, 1979). There were also no behavioral barriers to matings between western and Egyptian weevils (Hsiao and Hsiao, 1985a). However, in laboratory mating studies that included both weevil strains in a free-choice situation, the eastern male was more aggressive and mated more frequently with females of both strains than the western males (Armbrust et al., 1970). The same study also showed that the eastern and western males preferred females of like strain. Blickenstaff (1969), however, found no consistent preferential matings within or between strains. As just explained, viable progeny are produced from the cross between rickettsia-positive females and rickettsia-negative males; the sex ratio of this cross is distorted. Furthermore, the cross-mating yields only rickettsia-positive hybrids. In the reciprocal cross, mating apparently proceeds normally but the sperms are incapable of fertilizing the eggs. Thus, when interstrain mating is prevalent, the rickettsia-negative strain would be at a disadvantage and probably would be gradually eliminated. This situation has not yet been observed in the previously mentioned overlapping zones. Instead, several other relationships have been observed. In central California, where the western weevil was first reported in 1933 and subsequently infested six counties by 1940 (Michelbacher and Leighly, 1940), the invasion of the Egyptian weevil from the south has completely displaced the western weevil, especially in the inland valleys. The western weevil has been detected in coastal areas of central California. Results of our survey (Fig. 15.3) are in agreement with the previous description of the spread of Egyptian weevil in California (van den Bosch and Marble 1971). However, we have not been able to find an overlapping zone between the two weevil strains. Manglitz and his colleagues studied the overlapping zone between eastern and western weevils in central Nebraska. They did not obtain any evidence suggesting field hybridization from comparison of sex ratio data or hybrid vigor (Klostermeyer and Manglitz, 1979).

Our laboratory is currently investigating the overlapping zones between western and Egyptian weevils in southern Utah. Preliminary data show that hybridization occurs in some individuals of the population. This is based on the findings of the absence or low egg fertility among field-collected females that have been mated (sperms in spermatheca) but that are rickettsia-negative. These females represent the Egyptian weevils that have mated only with western males or with males of both strains. Female alfalfa weevils copulated repeatedly during an oviposition period of several months in the field. Hence, an incompatible mating could be followed by a compatible intrastrain mating, and partial fertility could be restored. Under these circumstances, hybridization would have little effect on the coexistence of the two strains. Nevertheless, a unidirectional gene flow toward the rickettsia-positive western weevils is expected. Under field conditions, hybridization can be influenced by behavioral and ecological factors, such as mating preference and temporal isolation due to phenological difference or habitat preference. Information available thus far suggests that field hybridization would have a minor impact on the weevil population in the overlapping zone, because the reduction in egg fertility can be compensated by the high reproductive potential of the species.

CONCLUSION

The alfalfa weevil complex in North America provides an example of how geographic and genetic diversity of an introduced pest may affect pest management. The origins of these three weevil strains have been traced to various regions of Europe and the Middle East. The strains are now well adapted to their new habitats. It appears that relatively few individuals of each strain initially colonized North America, as indicated by the absence or scarcity of the vestigial-wing mutation (see earlier discussion). Such a founder effect implies that the gene pool of each strain is probably far more limited than the indigenous populations of the species. However, there are considerable differences in ecological, physiological, and behavioral traits between weevil strains. Such variations affected the suitability of the three weevil strains as hosts for the hymenopterous parasite, *B curculionis* (Maund and Hsiao, 1991). It is well documented that the eastern and Egyptian weevils are more effective in encapsulating the parasite eggs and, consequently, are less suitable hosts for the parasite than the western weevil. Behavioral differences between the eastern and western weevils in response to parasitism hindered the establishment of *Tetrastichus incertus* (Ratzeburg) in the western weevil (Volker and Simpson, 1975). Most recently, we discovered considerable differences in the suitability of the three weevil strains as hosts for the adult parasite, *Microctonus aethiopoides* Loan. This parasite has been established in the eastern United States since 1970 and is considered to be the most effective natural enemy of the eastern weevil in the northeastern states during the past decade (Day, 1981). Results of our laboratory studies indicate that parasite development is unsatisfactory in the western weevil (Hsiao, 1986), due to the presence of rickettsia in the western weevil. Removal of the rickettsia restored parasite development in the western weevil (Hsiao, 1990). The previous examples emphasize the importance of matching biotypes of the alfalfa weevils to its parasites in biological control programs.

Insects of agricultural crops introduced into new habitats often become serious pests for a variety of reasons, the most obvious of which is the lack of natural enemies to regulate populations. This is undoubtedly an important factor, as is demonstrated by the successful biological control of some exotic pests following the introduction of native natural enemies (van den Bosch et al., 1982). However, it is not always apparent why biological controls are not effective in other instances. In the past, biological control programs involving exotic pests have stressed the importance of obtaining exotic natural enemies from a climatic region similar to that in which the pest is found. This practice would ensure ecological synchrony between the pest and its parasite. The importance of matching biotypes of the pest and its parasites to ensure physiological synchrony has been overlooked. The past attempted colonization of several natural enemies introduced for the biological control of the alfalfa weevil probably failed due to a lack of physiological synchrony. Thus, the genetic variability of pest biotypes should be evaluated as a part of a comprehensive biological control program. Many of the methods developed from genetic studies of the alfalfa weevil complex are indispensable tools for identifying pest complexes and for monitoring population dynamics and pesticide resistance. Genetic information of insect pests is clearly vital in formulating a sound pest management strategy.

Acknowledgments

This work is partially supported by USDA Western Region IPM Special Grants Program No. 85-CRSR–2–2693. I would like to thank two anonymous reviewers for helpful comments on the manuscript. Journal Paper No. 3536 of Utah Agricultural Experiment Station, Logan, Utah.

REFERENCES

Armbrust, E. J., and G. G. Gyrisco. 1982. Forage crops in insect pest management. Pages 443–463 in Metcalf, R. L., and W. H. Luckmann (eds.), *Introduction to Insect Pest Management*, 2nd ed. New York: John Wiley & Sons.

Armbrust, E. J., C. E. White, and S. J. Roberts. 1970. Mating preference of eastern and western United States strains of the alfalfa weevil. *J. Econ. Entomol.* 63:674–675.

Ayala, F. J. 1975. Genetic differentiation during the speciation process. *Evol. Biol.* 8:1–75.

Barber, H. S. 1924. The generic names of the clover and alfalfa weevils *Hypera* and *Phytonomus*. *Proc. Entomol. Soc. Wash.* 26:216.

Barr, A. R. 1982. Symbiont control of reproduction in *Culex pipiens*. Pages 153–158 in Steiner, W. W. M., W. J. Tabachnick, K. S. Rai, and S. Narang (eds.), *Recent Developments in the Genetics of Insect Disease Vectors*. Champaign, Ill.: Stipes.

Berlocher, S. H. 1984. Insect molecular systematics. *Annu. Rev. Entomol.* 29:403–433.

Bland, R. G. 1984. Mouthpart sensilla and mandibles of adult alfalfa weevil *Hypera* and the Egyptian alfalfa weevil *H. brunneipennis* (Coleoptera: Curculionidae). *Ann. Entomol. Soc. Am.* 77:720–724.

Blickenstaff, C. C. 1965. Partial intersterility of eastern and western U. S. strains of the alfalfa weevil. *Ann. Entomol. Soc. Am.* 58:523–526.

———. 1969. Mating competition between eastern and western strains of the alfalfa weevil, *Hypera postica. Ann. Entomol. Soc. Am.* 62:956–958.

Boller, E. F., K. Rus, V. Vallo, and G. L. Bush. 1976. Incompatible races of European cherry fruit fly, *Rhagoletis cerasi* (Diptera: Tephritidae), their origin and potential use in biological control. *Entomol. Exp. Appl.* 20:237–247.

Brower, J. H. 1976. Cytoplasmic incompatibility: Occurrence in a stored-product pest, *Ephestia cautella. Ann. Entomol. Soc. Am.* 69:1011–1015.

Coles, L. W. and W. H. Day. 1977. The fecundity of *Hypera postica* from three locations in the eastern United States. *Environ. Entomol.* 6:211–212.

Csiki, E. 1934. Coleoptera, Curculionidae, Subfamily Hyperinae. Pages 1–66 in Schenkling, S. (ed.), *Coleopterorum Catalogus*. Berlin: Dr. W. Junk.

Davis, D. W. 1967. How different are the eastern and western forms of the alfalfa weevil? *Proc. Utah Acad. Sci. Arts Lett.* 44:353–357.

Day, W. H. 1981. Biological control of the alfalfa weevil in the northeastern United States. Pages 361–374 in Papavizas, G. C. (ed.), *Biological Control in Crop Protection*. Beltsville *Symp. Agric. Res.* 5. London: Allanheld, Osmun & Co. Publishers.

Ehrman, L., and R. P. Kernaghan. 1971. Microorganismal basis of the infectious hybrid sterility in *Drosophila paulistorum. J. Hered.* 62:66–71.

Essig, E.O., and A. E. Michelbacher. 1933. The alfalfa weevil. *Calif. Agric. Exp. Stn. Bull.* 567.

Franz, H. 1949. Zur Kenntnis der Rassenbildung bei Kafern der Ostalpinen Fauna. *Zentralbl. Gesamte Entomol.* 3:3–23.

Hobbs, G. A., W. O. Numni, and J. F. Virostek. 1959. History of the alfalfa weevil, *Hypera postica* (Gyll.) (Coleoptera: Curculionidae), in Alberta. *Can. Entomol.* 59:562–565.

Hoffmann, A. A., and M. Turelli. 1988. Unidirectional incompatibility in *Drosophila simulans:* Inheritance, geographic variation and fitness effects. *Genetics* 119:435–444.

Hsiao, T. H. 1986. Alfalfa weevil biotype problem in relation to biological control. *Report 30th North American Alfalfa Improvement Conference.* Abstract, p. 34.

———. 1990. Unpublished data.

Hsiao, C., and T. H. Hsiao. 1984. Cytogenetic studies of alfalfa weevil (*Hypera postica*) strains (Coleoptera: Curculionidae). *Can. J. Genet. Cytol.* 26:348–353.

———. 1985c. Rickettsia as the cause of cytoplasmic incompatibility in the alfalfa weevil, *Hypera postica* (Gyllenhal). *J. Invertebr. Pathol.* 45:244–246.

Hsiao, T. H., and C. Hsiao. 1985a. Hybridization and cytoplasmic incompatibility among alfalfa weevil strains. *Entomol. Exp. Appl.* 37:155–159.

———. 1985b. Inheritance of a vestigial-wing mutation in the alfalfa weevil, *Hypera postica* (Gyllenhal). *Entomol. Exp. Appl.* 37:229–233.

Hsiao, T. H., and J. M. Stutz. 1985. Discrimination of alfalfa weevil strains by allozyme analysis. *Entomol. Exp. Appl.* 37:113–121.

Jacobson, J. W., and T. H. Hsiao. 1983. Isozyme variation between geographic populations of the Colorado potato beetle, *Leptinotarsa decemlineata* (Coleoptera: Chrysomelidae). *Ann. Entomol. Soc. Am.* 76:162–166.

Kellen, W. R., D. F. Hoffmann, and R. A. Kwock. 1981. *Wolbachia* sp. (Richettsiales: Rickettsiaceae) a symbiont of the almond moth, *Ephestia cautella*: Ultrastructure and influence on host fertility. *J. Invertebr. Pathol.* 37:273–283.

Kippenberg, H. 1983. Unterfamilie Hylobiinae. Pages 121–154 in Freude, H., K. W. Harde, and G. A. Lolse (eds.), *Die Kafer Mitteleuropas II*. Krefeld, Germany: Goecke & Evers Verlag.

Klostermeyer, L. E. 1978. Eastern and western alfalfa weevils, *Hypera postica* (Gyllenhal): Distribution in Nebraska, chromosome comparisons, and effects of antibiotics on cross-incompatibility. Ph.D. dissertation, Univ. of Nebraska, Lincoln. 73 pages.

Klostermeyer, L. E., and G. R. Manglitz. 1979. Distribution of eastern and western alfalfa weevil in Nebraska determined by cross-matings. (Coleoptera: Curculionidae). *J. Kansas Entomol. Soc.* 52:209–214.

LeCato, G. L., and R. L. Pienkowski. 1970a. Frequency and duration of mounting and copulation by the alfalfa weevil in the laboratory. *Ann. Entomol. Soc. Am.* 63:1548–1552.

———. 1970b. Sexual responsiveness of the male alfalfa weevil, *Hypera postica*, as affected by prior contact with other alfalfa weevils. *Entomol. Exp. Appl.* 13:462–466.

———. 1972. Fecundity, egg fertility, duration of oviposition, and longevity of alfalfa weevils from eight mating and storage conditions. *Ann. Entomol. Soc. Am.* 65:319–323.

Loan, C. C., J. C. Guppy, F. Meloche, and D. G. Harcourt. 1983. Development of a second generation of the alfalfa weevil, *Hypera postica* (Coleoptera: Curculionidae), in Eastern Ontario. *Proc. Entomol. Soc. Ont.* 114:61–67.

Mailloux, G., and J. G. Pilon. 1975. Morphologie, biologie et ecologie des populations de *Hypera postica* (Gyllenhal) (Coleoptera: Curculionidae) au Quebec. *Mem. Entomol. Soc. Quebec* No. 4, 151 pages.

Manglitz, G. R., L. E. Klostermeyer, and D. L. Keith. 1981. Comparisons of eastern and western strains of the alfalfa weevil in Nebraska. *J. Econ Entomol.* 74:581–588.

Maund, C. M., and T. H. Hsiao. 1991. Differential encapsulation of two *Bathyplectes* parasitoids among alfalfa weevil strains, *Hypera postica* (Gyllenhal). *Can. Entomol.* 123:197–203.

Manglitz, G. R., L. E. Klostermeyer, and D. L. Keith. 1981. Comparisons of eastern and western strains of the alfalfa weevil in Nebraska. *J. Econ Entomol.* 74:581–588.

Michelbacher, A. E., and J. Leighly. 1940. The apparent climatic limitations of the alfalfa weevil in California. *Hilgardia* 13:101–139.

Noda, H. 1984. Cytoplasmic incompatibility in a rice planthopper. *J. Hered.* 75:345–348.

Pienkowski, R. L., F. K. Hsieh, and G. L. LeCato. 1969. Sexual dimorphism and morphometric differences in the eastern, western, and Egyptian alfalfa weevil. *Ann. Entomol. Soc. Am.* 62:1268–1269.

Poos, F. W., and T. L. Bissell. 1953. The alfalfa weevil in Maryland. *J. Econ. Entomol.* 46:178–179.

Puttler, B. 1967. Interrelationship of *Hypera postica* (Coleoptera: Curculionidae) and *Bathyplectes curculionis* (Hymenoptera: Ichneumonidae) in the eastern United States with particular reference to encapsulation of the parasite eggs by the weevil larvae. *Ann. Entomol. Soc. Am.* 60: 1031–1038.

Rosenthal, S. S., and C. S. Koehler. 1968. Photoperiod in relation to diapause in *Hypera postica* from California. *Ann. Entomol. Soc. Am.* 61:531–534.

Salt, G., and R. van den Bosch. 1967. The defense reactions of three species of *Hypera* (Coleoptera: Curculionidae) to an ichneumon wasp. *J. Invertebr. Pathol.* 9:164–177.

Saringer, G., and K. V. Deseo. 1966. Effect of photoperiod and temperature on the diapause of the alfalfa weevil (*Hypera variabilis* Hrbst.) *Acta Phytopathol. Acad. Sci. Hung.* 1:353–363.

Schroder, R. F. W., and A. L. Steinhauer. 1976a. Effects of photoperiod and temperature regimens on the biology of European and United States alfalfa weevil populations. *Ann. Entomol. Soc. Am.* 69:701–706.

———. 1976b. Studies of cross-mating among strains of the alfalfa weevil from the United States and Western Europe. *Proc. Entomol. Soc. Wash.* 78:1–5.

Sell, D. K., E. J. Armbrust, and G. S. Whitt. 1978. Genetic differences between eastern and western populations of the alfalfa weevil. *J. Hered.* 69:37–50.

Smith-White, S., and A. R. Woodhill. 1954. The nature and significance of non-reciprocal fertility in *Aedes scutellaris* and other mosquitoes. *Proc. Linn. Soc. N. S. W.* 79:163–176.

Subrarao, S. K. 1982. Cytoplasmic incompatibility in mosquitoes. Pages 313–342 in Steiner, W. W. M, W. J. Tabachnick, K. S. Rai, and S. Narang (eds.). *Recent Developments in the Genetics of Insect Disease Vectors.* Champaign, Ill.: Stipes.

Takenouchi, Y. 1972. Chromosome study on four species of the genus *Hypera* (Curculionidae: Coleoptera). *J. Hokkaido Univ. Educ. Sect. IIB* 23:1–6.

Titus, E. G. 1910. The alfalfa weevil. *Utah Agric. Exp. Stn. Bull.* No. 110.

———. 1911. The genera *Hypera* and *Phytonomus* (Coleoptera, family Curculionidae) in America, north of Mexico. *Ann. Entomol. Soc. Am.* 4:383–494.

van den Bosch, R., and E. J. Dietrick. 1959. The interrelationships of *Hypera brunneipennis* (Coleoptera: Curculionidae) and *Bathyplectes curculionis* (Hymenoptera: Ichneumonidae) in southern California. *Ann. Entomol. Soc. Am.* 52:609–616.

van den Bosch, R., and V. L. Marble. 1971. The Egyptian alfalfa weevil, the threat to California alfalfa. *Calif. Agric.* 25(5):3–4.

van den Bosch, R., P. S. Messenger, and A. P. Gutierrez. 1982. *An Introduction to Biological Control.* New York: Plenum.

Volker, K. C., and R. G. Simpson. 1975. Behavior of alfalfa weevil larvae affecting the establishment of *Tetrastichus incertus* in Colorado. *Environ. Entomol.* 4:742–744.

Wade, M., and L. Stevens. 1985. Microorganism mediated reproductive isolation in flour beetle (Genus *Tribolium*). *Science* 227:527–528.

Warner, R. E. 1962. The scientific name of the alfalfa weevil (Coleoptera: Curculionidae). *Coleopt. Bull.* 16:107–108.

Wehrle, L. P. 1940. The discovery of an alfalfa weevil (*Hypera brunneipennis* Boheman) in Arizona. *J. Econ. Entomol.* 33:119–121.

White, C. E., E. J. Armbrust, and J. Ashley. 1972. Cross-mating studies of eastern and western strains of alfalfa weevil. *J. Econ. Entomol.* 65:85–89.

White, C. E., E. J. Armbrust, J. R. DeWitt, and S. J. Roberts. 1969. Evidence of a second generation of the alfalfa weevil in southern Illinois. *J. Econ. Entomol.* 62:509–510.

Wood, K. A., E. J. Armbrust, D. P. Bartell, and B. J. Irwin. 1978. The literature of arthropods associated with alfalfa. V. A bibliography of the alfalfa weevil, *Hypera postica* (Gyllenhal), and the Egyptian alfalfa weevil, *Hypera brunneipennis* (Boheman) (Coleoptera: Curculionidae). *Ill. Agric. Exp. Stn. Spec. Publ.* 54.

Wright, J. D., and B. Wang. 1980. Observations on Wolbachiae in mosquitoes. *J. Invertebr. Pathol.* 35:200–208.

Wright, J. D., and A. R. Barr. 1981. *Wolbachia* and the normal and incompatible eggs of *Aedes polynesiensis* (Diptera: Culicidae). *J. Invertebr. Pathol.* 38:409–418.

Yen, J. H., and A. R. Barr. 1971. New hypothesis of the cause of cytoplasmic incompatibility in *Culex pipiens* L. *Nature* 232:657–658.

———. 1974. Incompatibility in *Culex pipiens*. Pages 97–118 in Pal, R., and M. J. Whitten (eds.), *The Use of Genetics in Insect Control*. Amsterdam: Elsevier/North Holland Publ. Co.

Zaslavskiy, V. A. 1961. Review of the species of the genus *Phytonomus* Schonh. (Coleoptera: Curculionidae) in the fauna of the USSR. *Entomol. Rev.* 40:342–348.

16

VARIABLE HOST QUALITY AND EVOLUTION IN THE MEXICAN BEAN BEETLE

Margaret E. Saks

California Institute of Technology

INTRODUCTION

Many studies of plant–insect interactions address the factors that restrict particular insect groups to feeding on particular plant species. These studies often indicate that insect diets are determined by the qualitative features (often, allelochemicals) of available plant species (Feeny, 1975, 1976; Futuyma, 1976; Rhoades and Cates, 1976; Rosenthal and Janzen, 1979; Berenbaum, 1983). A less frequently addressed problem, but nevertheless a problem of substantial importance, is the influence of intraspecific variation in plant characteristics on the evolution of the life-history traits of associated insect pests (Denno and McClure, 1983). Whereas the presence or absence of plant features is likely to be an important determinant of the evolution of insect host utilization patterns, quantitative variation in plant features may well influence the evolution of insect life-history traits once an insect and plant group become associated.

Leaf nutritional quality is one plant feature that could influence the evolution of insect life-history traits. In examinations of the nutritional components of insect diets, Fraenkel (1969) and Beck and Reese (1976) concluded that similarities among plants in their nutritional quality and among insects in their dietary needs were too great to influence the host utilization patterns of insects. Nevertheless, McNeill and Southwood (1978) and Mattson (1980) suggested that quantitative differences in foliar nitrogen concentration might influence the evolution of insect life-history traits. These workers reached this conclusion based on a substantial body of literature (as reviewed in Scriber and Slansky, 1981; Slansky and Rodriguez, 1987) that shows that nitrogen is limiting to the growth of phytophagous insects. Therefore, the types and sources of intraspecific variation in resource nutritional quality, and their influence on the evolution of life-history traits, are of substantial interest.

The association of the Mexican bean beetle, *Epilachna varivestis* Mulsant (Coleoptera: Coccinellidae), with its host plant, the common bean, *Phaseolus vulgaris* (Leguminosae), provides a useful model system for examining the general problem of the influence of the environment on the population dynamics, population structure, and evolution of phytophagous insects. The nutritional quality of *P. vulgaris* foliage was chosen as the environmental parameter of interest, since this chemical aspect of plants has received little attention with respect to the above problems. The first goal of the research was to elucidate the sources and levels of variation in foliar nutritional quality. This information was then used to examine the influence of variation in host nutritional quality on the flight behavior of beetles from North Carolina and on the evolution of life-history traits and their physiological determinants in native (Mexican) and introduced (United States) populations of *E. varivestis*.

SOURCES OF VARIATION IN FOLIAR PROTEIN

Studies of the sources and levels of variation in the nutritional quality of *P. vulgaris* leaves involved determinations of the protein concentration of leaves collected from U.S. and Mexican cultivars that differed in growth habit. The Coomassie dye-binding assays of Bradford (1976) and Peterson (1983) were used to measure protein. Protein, rather than nitrogen, was assayed because the Leguminosae contain nitrogen-bearing compounds (ureides; Thomas and Schrader, 1981) of no known nutritive value to insects without hemolymph symbionts (Cochran, 1975; Wilson and Stinner, 1984).

There are two likely sources of variation in foliar nutritional quality. The first source includes abiotic factors, such as soil nitrogen availability, which could directly influence foliar nitrogen concentration (Runge, 1983). The second source is plant genotype. This source was of particular interest for two related reasons. *P. vulgaris* was domesticated approximately 7,000 years ago (Burkart and Brücher, 1953; Kaplan, 1965; Gentry, 1969) and, since the early 1900s, has undergone intense artificial selection for growth habit. The determinate (bush) cultivars that now predominate in the United States have only a few nodes, have short internode distances, flower synchronously, and mature in approximately 65 days (Ojehomon, 1966). In contrast, the ancestral indeterminate (vining) cultivars that still predominate in southern Mexico characteristically produce leaves at many nodes, have long internode distances, flower asynchronously, and usually take 100 days or more to mature (Evans and Davis, 1978). Because growth habit, as well as other plant traits (Hilu, 1983; White, 1984; Gottlieb, 1984; Schlichting, 1986), are genetically determined, and because pleiotropic effects among traits are quite common (Falconer, 1981), foliar nitrogen could be indirectly influenced by a breeding program whose main goal is selection for plant characteristics that influence yield.

The influence of plant genotype (i.e., growth habit) on variation in foliar protein was examined in two ways. The first set of analyses focused on the influence of plant phenological stage on the protein in terminal leaves collected from plants of both growth habits. The analyses for determinate plants revealed that leaves from nonre-

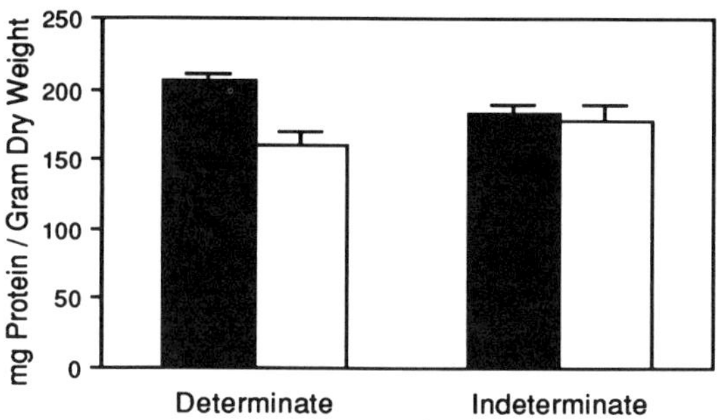

Figure 16.1. The protein concentration of terminal leaves (mean ± SEM) collected from prereproductive (solid bars) and reproductive (setting seed; open bars) determinate and indeterminate cultivars of *Phaseolus vulgaris*.

productive plants had about 30% more protein than those from plants that were setting seed ($t_{[16]} = 4.48$, $p < 0.001$). In contrast, there was no effect of plant reproductive stage on the protein concentration of the terminal leaves collected from indeterminate plants ($t_{[13]} = 0.40$, *ns*; Fig. 16.1). Thus, whereas plant phenological stage is a source of temporal variation in foliar protein for determinate cultivars (presumably due to the translocation of nitrogen from leaves to developing seeds (Thimann, 1980; Stoddard and Thomas, 1982), the relationship between reproduction and foliar protein is more complex for indeterminate cultivars.

The second set of analyses examined the influence of reproduction on the distribution of protein within the canopies of plants of each growth habit. Leaves were collected from basal and terminal nodes of reproductive plants. Because flower phenology is asynchronous in indeterminate cultivars, leaves collected from basal nodes were associated with developing seeds, whereas those collected from terminal nodes were associated with vegetative growth. The terminal leaves of indeterminate plants had 40% higher protein concentrations than did the basal leaves ($F_{[1,24]} = 59.24$, $p < 0.001$; Fig. 16.2). In contrast, there was no effect of leaf position for the determinate plants ($F_{[1,24]} = 0.07$, *ns*; Fig. 16.2).

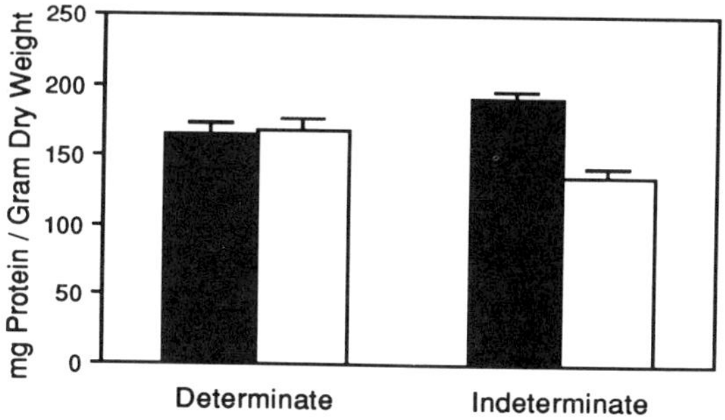

Figure 16.2. The protein concentration (mean ± SEM) of leaves collected from terminal nodes (solid bars) and from basal nodes (open bars) of determinate and indeterminate *Phaseolus vulgaris* cultivars that were setting seed.

Both sets of analyses indicate that plant phenology is a source of variation in foliar protein for determinate as well as indeterminate cultivars. They also indicate that growth habit provides an additional source of variation. For determinate cultivars, temporal variation in foliar protein is strongly associated with plant phenological stage, and there is little spatial variation within the plant canopy once reproduction begins. For indeterminate cultivars, spatial variation in foliar protein within the plant canopy is associated with the onset of reproduction, but temporal variation is dependent on the reproductive state of the particular node with which a leaf is associated. Both sets of results are consistent with the hypothesis that, in addition to influencing plant morphology and flower phenology, selection for plant growth habit, presumably as a consequence of pleiotropy, has affected the availability of protein in *P. vulgaris* leaves.

The effects of the observed patterns of foliar protein variation on the evolution of *E. varivestis* life-history traits are mediated by the insect's behavior and developmental plan as well as by local cultivation practices. Determinate *P. vulgaris* cultivars predominate in the United States. Their short maturation times enable farmers to plant sequentially throughout the growing season. Thus, plants in one field can be setting seed and will therefore be low in foliar protein, while plants in a neighboring plot can be prereproductive and thus will have relatively high levels of protein. Moreover, the relationship between *E. varivestis* longevity (hatch to death) and the maturation times of determinate cultivars is such that foliar protein will be changing, even within a given field, during the lifetime of an individual *E. varivestis*. Indeterminate cultivars are favored in southern Mexico. Due to their relatively long maturation times and to the time needed for drying seed in the field, only one crop per year

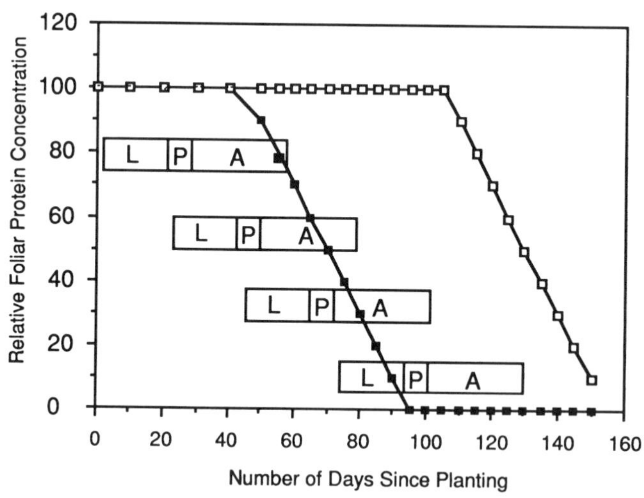

Figure 16.3. A schematic representation of the relationship between the developmental cycle of *Epilachna varivestis* and temporal changes in the protein concentration of the terminal leaves of determinate (closed squares) and indeterminate (open squares) cultivars of *Phaseolus vulgaris*. Boxes represent the length of the larval (L), pupal (P), and adult (A) stages. Values indicated on the y-axis are purely for illustrative purposes.

is generally planted. Thus little field to field variation in foliar protein is expected. Moreover, the terminal foliage of these cultivars is protein-rich even when basal leaves at reproductive nodes are senescing. Because *E. varivestis* is negatively geotactic, individuals in southern Mexico are able to find a protein-rich resource throughout their lifetimes. As a consequence of the differences in the cultivars favored in the United States and southern Mexico, and the behavioral characteristics and longevity of *E. varivestis* relative to the maturation times of locally available hosts, an individual in the United States has a higher probability of feeding on a nutritionally variable resource within its lifetime than does an individual in southern Mexico (Fig. 16.3).

NUTRITIONAL VARIATION AND FLIGHT BEHAVIOR

The analyses of temporal variation in the concentration of protein in the foliage of determinate *P. vulgaris* (above) showed that the nutritional complexity of a habitat can be high, even for plants such as *P. vulgaris* that are commonly cultivated in monocultures. Moreover, these studies showed that foliage quality can change within the lifetime of individual beetles in the United States. Although plant diversity is commonly emphasized in studies of insect movement probabilities (Root, 1973; Risch, 1981; Kareiva, 1983), food quality is also known to influence insect flight behavior (Rankin and Riddiford, 1977; Rankin and Rankin, 1980). Therefore, tethered flight assays (modifications of Dingle, 1965; Rankin and Rankin, 1980) were used to examine whether host nutritional quality might affect the distribution and abundance of *E. varivestis* by affecting the flight behavior of adults. The specific hypothesis was that insect movement probabilities (flight times) could be influenced by temporal variation in the nutritional quality of a plant that is commmonly cultivated in monocultures.

Beetles collected in North Carolina were used in these experiments. They were fed either protein-rich foliage from prebloom (young) *P. vulgaris* or protein-poor foliage from mature (senescent) *P. vulgaris* (cv. Long Tendergreen), according to the regimen indicated in Figure 16.4a. Individuals either fed on the same quality host throughout their juvenile and adult stages (i.e., were host constant) or fed on one type of host during their juvenile stage and on the alternate host during their adult stage (i.e., were host switched). The flight behavior of both males and females was assayed every other day from day 5 to day 15 posteclosion [preliminary tests and Blau and Stinner (1983) showed that younger adults did not fly more than a minute or so]. The median flight time (i.e., over the six flight tests) for each individual was used in analyses that examined the effects of the treatments on beetle flight. The data were analyzed using survival statistics (SAS Institute, 1985), since flight data have statistical properties similar to those generally associated with survival data.

The experimental design (Fig. 16.4a) resulted in four treatment groups with sample sizes that ranged from 26 to 32 individuals for each sex. The treatment groups will be referred to by indicating the juvenile and then the adult host (i.e., Y-S signifies that animals were fed young foliage in the juvenile stage and senescent foliage in the adult stage). Using this design, it was possible to examine flight behavior, not

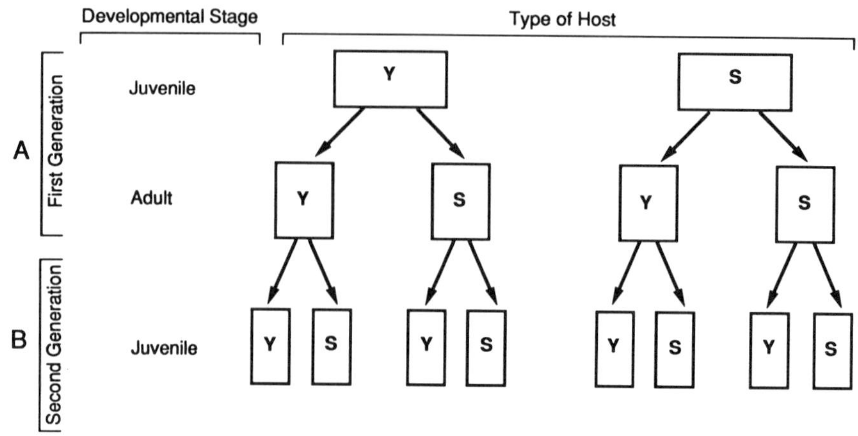

Figure 16.4. Schematic representation of the experimental design used to examine the influence of temporal change in foliar nutritional quality on the flight behavior, reproduction, and progeny pupal weight of *Epilachna varivestis*. Animals were fed either young (Y) or senescent (S) foliage from a determinate *Phaseolus vulgaris* cultivar. (a) The design used for experiments in which adult-expressed traits were examined (i.e., flight behavior and rate of reproduction). (b) The design used for examining the effects of parental-host history on the hatch weight and pupal weight of progeny.

only with respect to different types of hosts, but also with respect to effects due to their utilization during particular stages of beetle development.

The effect of the juvenile host was examined by comparing animals that had been fed the same adult host but different juvenile hosts (i.e., first analysis, Y-Y versus S-Y, Fig. 16.5a and 16.5c; second analysis, S-S versus Y-S, Fig. 16.5b and 16.5d). These analyses revealed that females that were fed young foliage in the juvenile stage consistently tended to fly longer than those that were fed senescent foliage in the juvenile stage, both when they were fed young foliage in the adult stage ($p < 0.15$; quantiles (min): Y-Y 25% = 8.0, 50% = 17.8, 75% = 31.5; S-Y 25% = 4.2, 50% = 10.5, 75% = 24.0) and when they were fed senescent foliage in the adult stage ($p < 0.15$; quantiles (min): Y-S 25% = 12.0, 50% = 20.5, 75% = 45.0; S-S 25% = 6.0, 50% = 16.5, 75% = 29.5). Parallel analyses for males did not reveal even a slight effect of the juvenile host on their flight times (Fig. 16.5).

The effect of the adult host on flight was examined by comparing the flight times of animals in the Y-Y and Y-S treatments (Fig. 16.5a and 16.5d) and by comparing the flight times of animals in the S-S and S-Y treatments (Fig. 16.5b and 16.5c). Although the effects of the adult host were not significant for females, there was a tendency for females to fly longer when fed senescent than when fed young foliage in the adult stage, both when the juvenile host was young foliage and when it was senescent foliage. In contrast, the adult host had a large effect on male flight. Males that were fed young foliage in the adult stage flew significantly longer than those that were fed senescent foliage when the juvenile host was young foliage ($p < 0.01$; quantiles (min): Y-Y 25% = 9.0, 50% = 14.8, 75% = 23.5; Y-S 25% = 4.5, 50% = 7.0, 75% = 13.0) as well as when it was senescent foliage ($p < 0.04$; quantiles

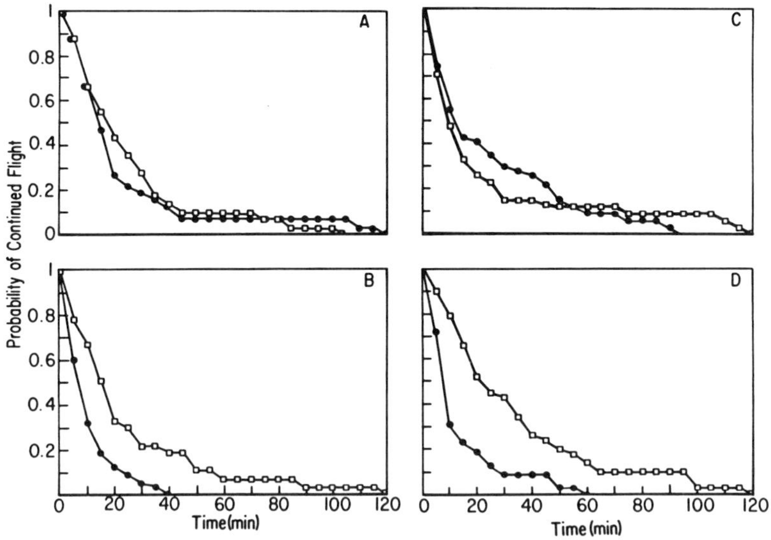

Figure 16.5. The probability that a male (closed circle) or a female (open square) will continue to fly if it has already flown for a given length of time. The values for each host treatment (juvenile followed by adult host) are presented in separate panels: (a) Y-Y; (b) S-S; (c) S-Y; (d) Y-S. (Note that the labels for panels b and d were inadvertently swapped in Saks et al., 1988).

(min): S-Y 25% = 5.5, 50% = 13.0, 75% = 44.0; S-S 25% = 0.5, 50% = 7.5, 75% = 14.0).

The relationship between age at first oviposition and age at peak flight was also examined for females in each treatment group. These analyses showed that, although treatments influenced female age at first oviposition (mean age ranged from 8 to 11 days across treatments), females made their longest flight prior to oviposition, at approximately 7 days posteclosion, regardless of treatment. These results superficially suggest that *E. varivestis* displays an oogenesis flight syndrome (Johnson, 1969). However, the analyses did not reveal the significant decrease in the length of the post- compared with pre-oviposition flights that characterizes this syndrome; rather, females reached the characteristic flight time for each particular treatment on about day 7 and continued to fly for approximately the same length of time at older ages, even after they began to oviposit (Saks et al., 1988). Thus, it seems that the distribution of eggs between the natal patch and patches to which females might disperse depends on (1) whether host nutritional quality is sufficient to enable females to reproduce prior to day 7 posteclosion and (2) the influence of host nutritional quality on flight tendency (but not the age of peak flight).

In general, the analyses showed that host quality affected the flight times of both male and female *E. varivestis*. However, there was an important difference between the sexes in the nature of this effect. Male flight was only affected by the type of host consumed in the adult stage, whereas female flight tended to be influenced by both the juvenile and the adult host. More important, for females, the influence of a particular host on flight behavior depended on the time in development that the host was consumed. Females fed young foliage during juvenile development flew longer than those fed senescent foliage. However, females fed young foliage during adult development flew less than those fed senescent foliage. Comparisons of the median flight times of females in the S-Y and the Y-S treatments (Fig. 16.5c and 16.5d, respectively) showed that those in the Y-S treatment flew significantly longer than those in the S-Y treatment ($p < 0.02$). This difference between Y-S and S-Y females clearly indicates that females do not fly in response to the average nutritional quality of their hosts; rather, the effects of young and senescent foliage on the flight times of adult females were a function of the time in development at which each host was consumed.

Females flew significantly longer than males when they were fed senescent foliage in the adult stage (S-S, $p < 0.01$; Y-S, $p < 0.0002$; Fig. 16.5b and 16.5d). However, the flight times of males and females did not differ when they were fed young foliage in the adult stage (Fig. 16.5a and 16.5c). The sexual dimorphism in flight seems to be due to the previously mentioned differences between males and females in both the type of host that elicited flight and the stage in development at which flight behavior was determined.

NUTRITIONAL QUALITY AND BEETLE MOVEMENT IN THE FIELD

The results of the tethered flight assays suggested that the different flight behaviors of males and females when fed senescent foliage in the adult stage might lead to locally biased sex ratios. This general idea was tested in the field in upstate New York. Two neighboring bean fields were located. The first field had been infested with *E. varivestis* soon after planting. At the time of sampling, the plants had set seed and eclosing adults were feeding on senescent foliage. It is referred to here as the "natal field." The second field was planted later in the season, and surveys of insects and of damage to foliage showed that it had been only recently occupied by *E. varivestis*.

Beetles were collected in the natal field and placed in one of two age classes. Those in the young adult class ($n = 80$) were either reared from pupae collected in the field or had soft or very light-colored elytra (adults with light-colored elytra are generally less than five days old; Saks, 1990). Beetles do not fly more than a minute or so when they are younger than five days old (Blau and Stinner, 1983; Saks et al. 1988). Therefore, these collections of new adults could be used to examine local sex ratio prior to flight. Individuals were designated as mature ($n = 182$) if their cuticles were hardened and tanned. Differences in the sex ratio for each age class would indicate differences between sexes in emigration from this natal field.

Figure 16.6 shows that sex ratio was 50:50 for the collections of new adults. In contrast, sex ratio was male biased in the collections of the (mature) adults that were capable of flight. Approximately 70% of the nondispersing beetles were males. Sex ratio was skewed toward females in the neighboring field that had not been initially infested with beetles but that was accumulating dispersing beetles. Thus, differential

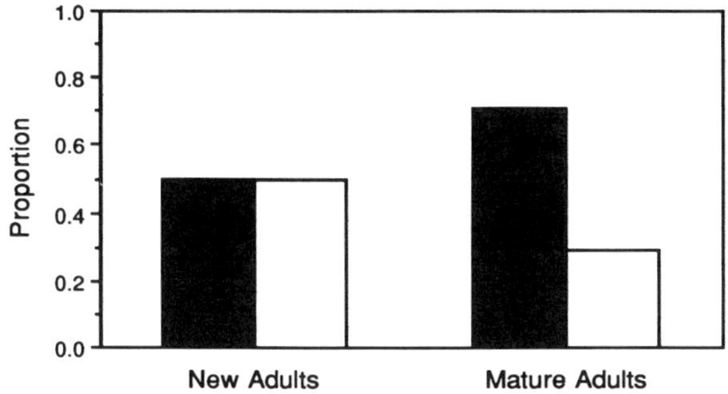

Figure 16.6. The proportion of males (closed bars) and females (open bars) in collections of new and mature adults in a bean field that contained senescent foliage (see text for age–class criteria).

mortality of males and females probably does not account for the biased sex ratio that was observed for mature adults in the natal field.

These field data, as well as the data on tethered flight (mentioned previously), indicate that local sex ratio can be influenced by the different flight behaviors of males and females when they are fed senescent foliage. A locally male-biased sex ratio should increase male–male competition for mates and should thereby intensify sexual selection on traits that increase the probability that a male will father offspring. In fact, U.S. male *E. varivestis* appear to produce a biochemical factor with just such an effect (Saks, 1990).

NUTRITIONAL VARIATION AND PHENOTYPIC EVOLUTION IN U.S. AND MEXICAN BEETLES

The analyses of foliar protein (discussed earlier) showed that protein declines with plant age in the terminal foliage of the determinate *P. vulgaris* cultivars that now predominate in the United States but not in the terminal foliage of the indeterminate cultivars that predominate in southern Mexico. This observation was of particular interest, because *E. varivestis* has only recently (since around 1920) invaded the eastern United States (Howard, 1930; Gordon, 1975). Thus, the general question of whether plant-breeding strategies influence phenotypic evolution in associated insect pests could be examined by comparing the life-history traits of U.S. and Mexican beetles.

A second, more specific, question was also addressed. Theories that relate temporal variation in the environment to evolution generally either delimit the conditions that maintain genetic variation (Kimura, 1954; Dempster, 1955) or examine the influence of environmental heterogeneity on components of fitness and their underlying genetic basis (Levins, 1968; Hedrick et al., 1976; Wiens, 1976; Hedrick, 1986). Although they differ substantially in focus, these approaches similarly assume that the relationship between organism longevity and environmental periodicity is a critical determinant of evolutionary outcomes. They do not, however, explicitly consider the influence of environmental variation during development on life-history trait evolution. Because foliar protein varies within the lifetime of an individual beetle in the United States but not in Mexico, it was possible to examine the effect of within-development variation in resource quality on life-history evolution by comparing traits of beetles from Mexico and the United States.

Beetles from cultures initiated with pupae collected in 1980 in North Carolina (NC) and New York (NY) in the United States and in the state of Oaxaca in southern Mexico (MEX) were standardized on a common diet in the laboratory. The experiments examined the reproductive rate of adults (Fig. 16.4a), the weight of their hatchlings, and the pupal weight of progeny (Fig. 16.4b). The experiments thus spanned two generations.

Reproductive rate was measured for individuals that were fed either protein-rich foliage from prebloom (young) or protein-poor foliage from senescent (i.e., that had set pods) *P. vulgaris* (cv. Long Tendergreen) during larval development and then fed the same or the alternate host during the adult stage. Treatments in which beetles were fed the same quality host in their juvenile and adult stages (i.e., were host constant) mimicked conditions in Mexico. Those in which beetles were fed different quality hosts during their juvenile and adult stage (i.e., were host switched) mimicked conditions in the United States. Each individual's (parent's) pupal weight was measured within 24 hours of pupation, and single-pair matings ($n = 52$) were established within each population and treatment group at adult eclosion. Adults were fed for 18–21 days posteclosion (depending on eclosion date). Reproductive rate was calculated as the number of hatched larvae produced by each pair per day of reproduction.

The weight of hatchlings produced by parents in each population and treatment group was also measured. The hatched larvae produced by each pair were frozen and lyophilized, and the siblings from each clutch were weighed en masse to the nearest 1 μg. In addition, a subset of newly hatched progeny ($n = 390$) was reserved to examine the effects of parental-host history on progeny pupal weight. These hatchlings were fed either young or senescent foliage during juvenile development (Fig. 16.4b) and were weighed within 24 hours of pupation.

The reproductive rate of U.S. and MEX beetles did not differ when animals were host constant. However, the mean reproductive rate of host-switched MEX beetles was less than 50% that of similarly treated U.S. beetles (Fig. 16.7). The significant population-treatment interaction term revealed in an analysis of variance (ANOVA) of the data ($F_{[3,44]} = 4.22, p < 0.01$) strongly suggests that NC and NY beetles have diverged from their Mexican ancestors in response to the cultivation practices that now predominate in the United States.

The data additionally revealed a peculiar relationship between female body size and reproductive rate. Over all populations, the pupal weight of parents that were fed young foliage in their juvenile stage ($\overline{X}$ (mg) $\pm$ SEM: females, $48.1 \pm 0.7, n = 30$; males, $42.5 \pm 0.5, n = 45$) was significantly greater ($p < 0.001$ for both sexes, Student's *t*-test) than that of individuals that were fed senescent foliage (females, $39.7 \pm 0.6, n = 41$; males, $36.0 \pm 0.5, n = 33$). Yet large females did not have consistently higher reproductive rates than smaller females. Product moment corre-

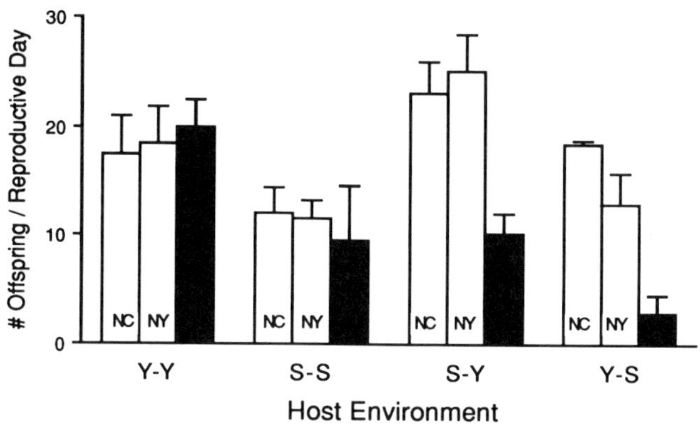

Figure 16.7. The rate of offspring production of NC, NY, and MEX (closed bars) adults in each of four host treatments. The specific host treatments are indicated on the x-axis (juvenile followed by adult host). Data express the mean number of offspring per reproductive day ± SEM.

lations between female weight and rate of offspring production revealed that body size was positively correlated with reproductive rate for host-constant ($r = 0.527$, $n = 28$, $p < 0.01$) but not host-switched ($r = -0.260$, $n = 24$, *ns*) females. In particular, even when MEX females were heavy (i.e., due to feeding on young foliage in the juvenile stage), they had lower reproductive rates than smaller MEX females when both groups were fed senescent foliage in the adult stage (compare Y-S and S-S). Moreover, MEX females that were switched from young to senescent foliage had lower reproductive rates than similarly treated U.S. females (Fig. 16.7). Thus, the fat body reserves reflected by pupal weight did not provide MEX females the reproductive advantage anticipated from studies of other insects (Engelmann, 1970) or that anticipated from the correlation observed (above) for host constant females.

ANOVA for weight of hatchlings did not reveal an effect of treatment for any of the tested populations. Since weight of hatchlings and egg weight are positively correlated in *E. varivestis* (Saks, 1990), it seems that parental-host history has little, if any, effect on the quantity of yolk supplied to eggs. In contrast, there were striking effects of parental-host history on the pupal weight of progeny, regardless of the type of food they consumed during larval development (Fig. 16.8). Moreover, these effects differed for NC, NY, and MEX beetles. The pupal weight of MEX progeny (both sexes) of host-switched parents was significantly lower than that of progeny from host-constant parents [Fig. 16.8, modified *t*-test (Hochberg, 1974), $p < 0.05$]. In contrast, parental-host switching had no effect on NY progeny pupal weight and had a positive effect on progeny pupal weight for NC beetles (Fig. 16.8, modified *t*-test, NY males and females, *ns*; NC males and females, $p < 0.05$). Since the weight of progeny at hatch was unaffected by the host treatment of parents (previous discussion), it seems that the observed second generation effects of parental-host history are either due to differences among the populations in the quality of yolk supplied to eggs or are due to differences in larval growth rate (Saks, 1983).

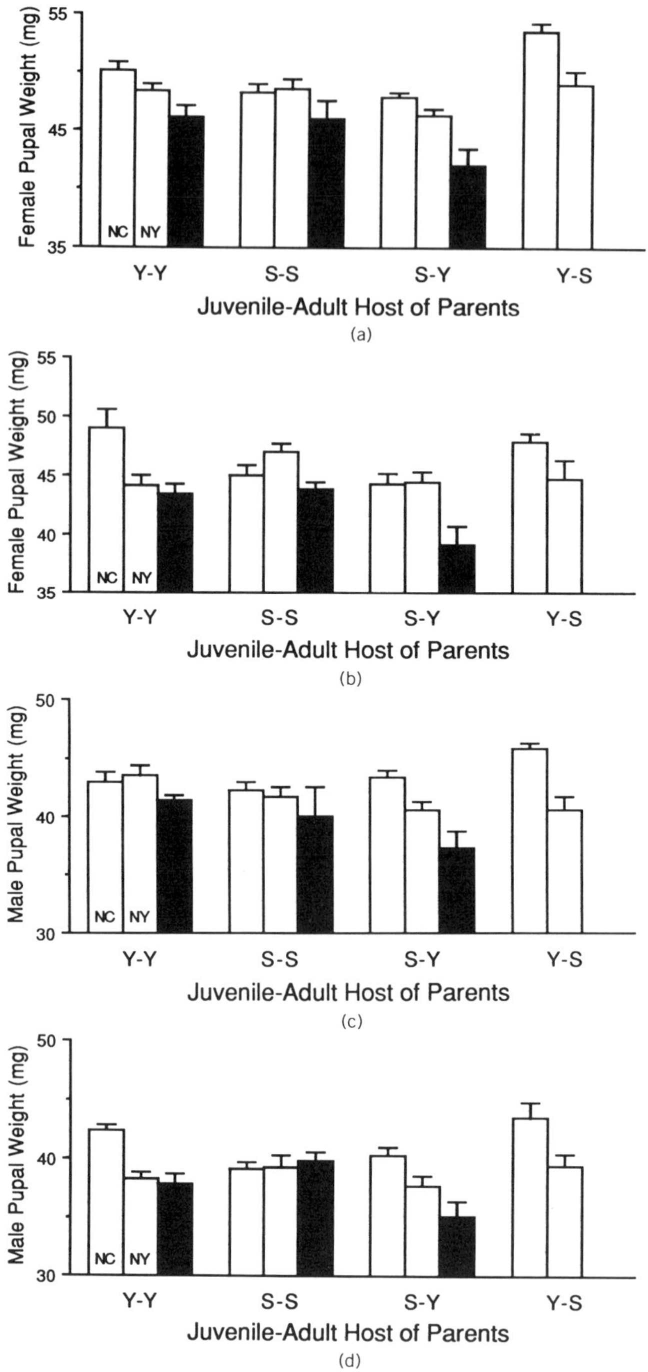

Figure 16.8. The pupal weight of NC (first open bar in each grouping), NY (second open bar in each grouping), and MEX (closed bars) progeny. The x-axis gives the host history of their parents (juvenile followed by adult host). The progeny were either fed young [panels (a) and (c)] or senescent [panels (b) and (d)] bean foliage during larval development. Data express the mean pupal weight (± SEM) of female [panels (a) and (b)] and male [panels (c) and (d)] progeny. Progeny were not available from MEX parents that were given the Y-S treatment, since this treatment severely depressed offspring production (see Fig. 16.7).

The patterns in the progeny pupal weight and parental reproductive rate data were similar. Host switching had a negative effect on both traits of MEX but not U.S. beetles. Since U.S. beetles have a higher probability than MEX beetles of feeding on nutritionally different hosts during their lifetimes, it seems that NC and NY beetles have diverged from their Mexican ancestors in both reproductive rate and some aspect of progeny growth in response to the cultivation practices that predominate in each country.

Several lines of evidence suggest that temporal variation in host nutritional quality, rather than differences in average nutritional quality or a high probability of feeding on protein-poor foliage, accounts for the divergence among populations. U.S. and MEX beetles did not differ in either reproductive rate or progeny pupal weight when they were host constant but fed a double dose of senescent foliage. Yet they did differ when they were fed senescent foliage during only one developmental stage (Figs. 16.7 and 16.8). Thus, beetles are not adapting to senescent foliage per se. In addition, beetles are not "averaging" the environments to which they are exposed. For example, the reproductive rate of MEX beetles that were fed young foliage in the juvenile stage and senescent foliage in the adult stage was significantly lower than that of beetles fed the same hosts but in the opposite order (Fig. 16.7). Thus, the effect of a given type of host on a trait, such as reproductive rate, depends on the other hosts on which an individual feeds and on the time during development at which these are encountered. Finally, the adult host does not seem to take complete precedence over the juvenile host in determining reproductive rate. MEX beetles that were fed young foliage in both the juvenile and the adult stages had higher reproductive rates than those that were fed senescent foliage in the juvenile stage and young foliage in the adult stage (Fig. 16.7). The observations therefore suggest that components of fitness can be affected, not only by the type of host consumed at the time that particular traits are expressed, but also by the host(s) previously consumed by that individual or its parents.

NUTRITIONAL VARIATION AND REPRODUCTIVE PHYSIOLOGY

In organisms with complex life cycles (e.g., holometabolous insects), body size is determined at a discrete time in the life cycle. Moreover, its phenotypic value is strongly influenced by larval nutrition (above and see also Scriber and Slansky, 1981; Slansky and Rodriguez, 1987) as well as by an individual's genotype. The high degree of plasticity in insect body size suggests that individuals attain sizes appropriate to the environments in which they are raised. However, the evolution of a complex life cycle precludes any "backward morphological development" for tracking environmental changes that occur after the adult stage is reached. Although it is possible to envision an adaptive advantage to this type of developmental stasis, the negative association between body size and reproductive rate for host-switched MEX beetles (previously discussed) indicates that the evolution of a complex life cycle may impose a constraint on reproductive fitness when resources utilized during the juvenile and adult stages differ markedly in quality. These observations raised the following problem: given the historical constraints inherent in the evolution of a complex life cycle with distinct developmental stages, what sorts of mechanisms have U.S. beetles evolved that account for their reproductive flexibility in a temporally varying environment?

Juvenile resources are clearly a critical determinant of reproductive success for species that feed only as larvae (e.g., lepidopterans that have atrophied adult mouthparts). The quality of juvenile resources influences the quantity of stored fat body reserves, which, in turn, is positively associated with body size. Consequently, pupal weight is a good predictor of fecundity for such species (Engelmann, 1970).

The relationship between pupal weight and reproduction is not as clear for species, such as *E. varivestis,* that feed in both the juvenile and adult stages. Species with this type of life-history plan can potentially utilize juvenile as well as adult resources for reproduction, and the degree of utilization of each type of resource might be expected to be a species specific characteristic. However, the results described earlier for *E. varivestis* suggest that this is not necessarily the case, because the relationship between body size and reproductive rate differed between U.S. and MEX beetles when they were host switched. Moreover, because large MEX adults reproduced less than similarly sized U.S. adults when they were host switched, it seemed that the populations might differ in their use of juvenile reserves for reproduction when individuals were faced with a nutritionally stressful adult host. A series of experiments was undertaken to test this idea.

Beetle cultures were initiated with pupae collected in 1985 from locations close to those from which beetles were obtained in 1980 for the reproductive rate experiment (discussed earlier) and were standardized in the laboratory prior to the experiments. Experimental animals were reared on bean foliage (cv. Long Tendergreen) throughout development, except when they were starved (see following discussion).

The first experiment examined the use of juvenile reserves for reproduction. In this experiment (Fig. 16.9a), juvenile reserves were traced by feeding late-fourth (last) instar larvae on leaf discs that were painted with protein hydrolysate that was labeled with ^{14}C. The amount of radiolabel ingested by each individual was determined by subtracting the amount of radiolabel remaining on leaf fragments from the amount initially applied. After receiving radiolabel, larvae were reared to pupation on nonradiolabeled foliage. Females were weighed within 24 hours of adult eclosion. The mean weight of the NC and MEX females used in this experiment (and that following) was 45 milligrams (mg). The use of reserves for reproduction was examined for females in two body size categories: those that weighed more than 45 mg and those that weighed 45 mg or less.

Single-pair matings between radiolabeled females and nonradiolabeled males were established on day 6 posteclosion. Eggs were collected from these pairs, and the experiment was terminated by sacrificing females on day 13 posteclosion (n = 61). Control groups of females were also sacrificed on the day of adult eclosion and on day 6 posteclosion. The latter age approximately coincides with the end of the obligate prereproductive period and the beginning of the reproductive period for *E. varivestis*. The amount of ^{14}C in eggs and female bodies was measured. Values reported here for reproductive effort are the amounts of radiolabel in all eggs produced by each female and are standardized to the estimated amount of radiolabel in that female's body at the onset of reproduction.

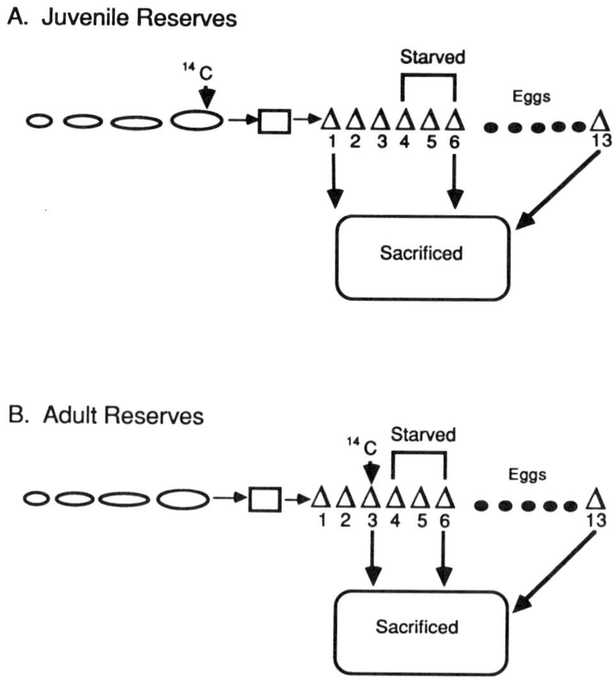

Figure 16.9. Schematic representation of the experimental design used to trace the use of juvenile- and adult-obtained reserves for reproduction by U.S. and Mexican *Epilachna varivestis*. The diagram indicates the time in development at which ^{14}C was administered and the age (days posteclosion) of adult females when sacrificed. Ovals represent the four larval instars, squares represent the pupal stage, and triangles represent the adult stage. (a) The protocol used for tracing juvenile-obtained radiolabel. (b) The protocol used for tracing adult-obtained radiolabel.

There were two treatment groups in the experiment: females that were consistently well fed and females that were well fed as juveniles and then nutritionally stressed as adults. Nutritionally stressed females were starved for 48 hours from days 4 to 6 posteclosion (Fig. 16.9a). The treatments were analogous to the host constant (Y-Y) and the host switched (Y-S) treatments used in the reproductive-rate experiment described earlier.

The second experiment examined the use of adult reserves for reproduction (Fig. 16.9b). Adult reserves were traced by injecting three day-old females with the radiolabeled protein hydrolysate ($n = 58$). Except for this difference in the timing and manner of radiolabel administration, the protocol was the same as that for the juvenile reserve experiment.

The results revealed that females that were well fed as adults allocated a greater proportion of adult than juvenile reserves to reproduction (Fig. 16.10a). However, when females were starved, approximately equal proportions of their juvenile and adult reserve pools were allocated for reproduction (Fig. 16.10b). These observations have two implications. First, they indicate that different physiological mechanisms determine the allocation of each type of reserve to reproduction. They thus suggest that these mechanisms can be independently influenced by selection. Sec-

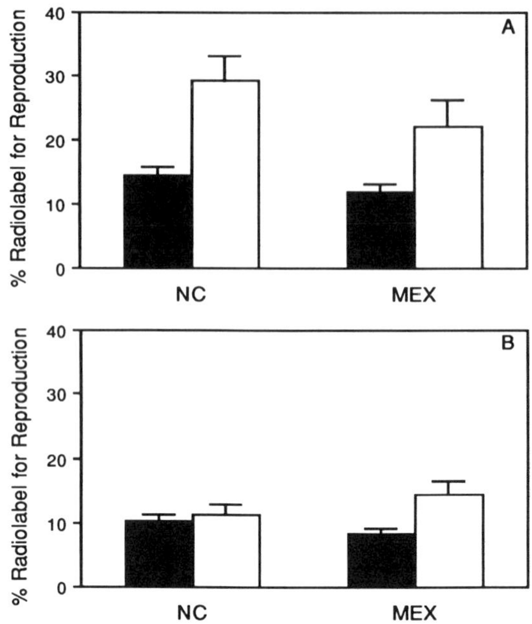

Figure 16.10. Percent (mean ± SEM) of juvenile- (closed bars) and adult-obtained (open bars) radiolabel allocated to reproduction by North Carolina (NC) and Mexican (MEX) *Epilachna varivestis* females. (a) Values for females that were fed. (b) Values for females that were starved.

ond, they indicate that the relative importance of each type of reserve for reproduction depends on the nutritional conditions that females encounter in the adult stage.

Three-way ANOVAs were used to examine the influence of population, nutrition, and body size on the allocation of juvenile- and adult-obtained radiolabel for reproduction. Starvation decreased the proportion drawn from both reserve pools. It decreased the use of juvenile reserves by approximately 5% ($F_{[1,53]} = 12.53$, $p < 0.001$) and the use of adult reserves by approximately 15% ($F_{[1,50]} = 14.97$, $p < 0.0003$). However, the use of each type of reserve differed with respect to the interactions that were detected among factors. For adult reserves, there was a significant interaction between population and nutrition ($F_{[1,50]} = 3.90$, $p < 0.05$) but no interaction between size and the other factors (Fig. 16.11). In contrast, there was a significant size–population interaction ($F_{[1,53]} = 3.87$, $p < 0.05$) and a significant size–population–nutrition interaction ($F_{[1,53]} = 7.49$, $p < 0.01$) in the analysis for juvenile reserves. When females were fed, body size and reproductive effort were positively associated for MEX but not for NC females. Conversely, when females were starved, body size and reproductive effort were negatively associated for MEX females but positively associated for NC females (Fig. 16.12).

In the reproductive rate experiment, large MEX females reproduced less than smaller MEX females when both groups were fed senescent foliage in the adult stage. In addition, MEX females that were fed young and senescent foliage in the juvenile and adult stages, respectively, reproduced less than U.S. females given the

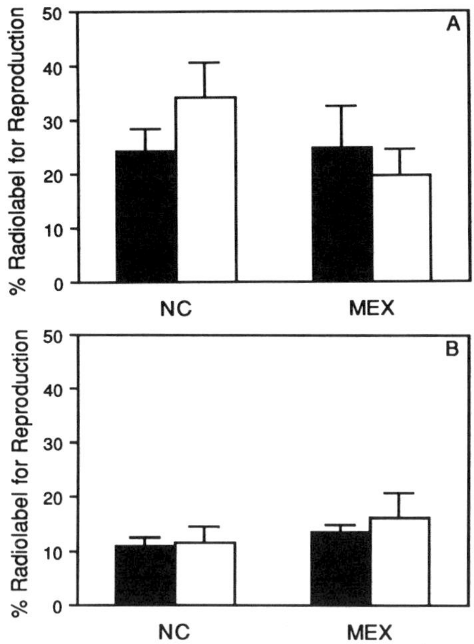

Figure 16.11. Percent (mean ± SEM) of adult-obtained radiolabel that was allocated to reproduction by small (adult eclosion weight was ≤45 mg; solid bars) and large (adult eclosion weight was > 45 mg; open bars) female *Epilachna varivestis* from North Carolina (NC) and Mexico (MEX). (a) Values for females that were fed. (b) Values for females that were starved.

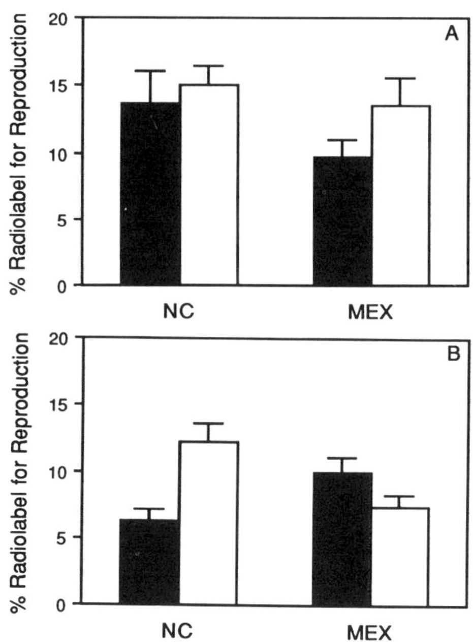

Figure 16.12. Percent (mean ± SEM) of juvenile-obtained radiolabel that was allocated to reproduction by small (adult eclosion weight was ≤45 mg; solid bars) and large (adult eclosion weight was > 45 mg; open bars) female *Epilachna varivestis* from North Carolina (NC) and Mexico (MEX). (a) Values for females that were fed. (b) Values for females that were starved.

same host treatment (see Fig. 16.7). This suggests that the juvenile fat body reserves of host-switched (i.e., Y-S) MEX females were not used in proportion to their quantity but that those of U.S. females were. The results of the radiolabeling experiments also suggest that the large U.S. but not large MEX females are able to use juvenile reserves for reproduction when they are nutritionally stressed in the adult stage (Fig. 16.12). Therefore, the results of these two different types of experiments are consistent with the idea that the higher reproductive rates of U.S. compared to MEX females when resources vary from high to low quality during beetle development is due—at least in part—to changes in the use of juvenile reserves for reproduction.

DISCUSSION AND CONCLUSIONS

Artificial selection for growth habit in the common bean, *Phaseolus vulgaris*, has led to changes in the temporal and spatial availability of protein in its foliage. Since determinate cultivars predominate in the United States, whereas indeterminate cultivars predominate in Mexico, the pests of *P. vulgaris* face quantitatively different resources in each country.

The relationship between the period of environmental fluctuations and the longevity of particular organisms has long been thought to be the critical aspect of temporal variation that influences evolution (Levins, 1968; Hedrick, 1986). However, these studies of *Epilachna varivestis* suggest that variation in resource quality during the development of an organism, rather than the organism's longevity with respect to environmental periodicity, may be a key parameter that influences behavior and evolution. Several characteristics of *E. varivestis*, including (1) flight behavior, (2) local sex ratio, (3) life-history traits, and (4) certain physiological determinants of reproduction, were affected by the time in development at which beetles consumed particular hosts and by the type of host(s) previously consumed by that individual or its parents. Thus, simply knowing that an environment varies or that it has various states may not be sufficient for the development of an understanding of the influence of environmental variation on behavior and evolution; rather, it may be necessary to examine explicitly the relationship between particular environmental states and organism development before formulating predictions about evolution in heterogeneous environments.

Flight behavior is likely to be a key determinant of distribution, abundance, and gene flow in most insect species. For *E. varivestis,* these ecologically and evolutionarily important parameters are additionally mediated by resource variation. The population dynamics of this insect does not necessarily depend solely on the diversity of plants within its habitat but may be influenced also by the nutritional quality of its primary host plant, because this factor has an effect on flight behavior. In addition, because resource quality affects dispersal probabilities, it may also influence levels of genetic variation in natural populations of *E. varivestis*. Finally, differences in the flight behavior of males and females when they are fed senescent foliage seem to cause local sex ratio to be male biased. A possible consequence of this phenomenon is the evolution of biochemical factors in males that increase their chance of fathering offspring (Saks, 1990) when male–male competition for mates is intense. Although males in other insect species produce similar types of factors (Chen, 1984), little is known about the ecological conditions that influence their evolution and maintenance in natural populations. It is possible that the evolution of male factors in *E. varivestis* may be promoted by the differential influence of host nutritional quality on male and female flight behavior.

The parallels between the experiments that examined the influence of temporal variation in host nutritional quality on reproductive rate and on the use of juvenile and adult reserves for reproduction lead to two main conclusions. First, the evolution of a complex life cycle in which phenotypes are fixed in particular developmental stages is not necessarily adaptive when the environment changes from one of relative constancy to one that is temporally varying. Second, the historical constraints inherent in the evolution of any type of complex life cycle may cause selection to operate, not on the developmental plan itself, but on physiological processes that link one developmental stage to another.

In general, the results of these studies of *E. varivestis* showed that temporal variation in host nutritional quality, rather than its average state, can have a strong influence on insect behavior and evolution. Because a temporally varying environment can influence a variety of traits that interact within the same individual, it is important to identify the suite of phenotypic responses that an organism has to a variable environment in order to make clear predictions about its evolution. Flight behavior is likely to affect gene flow among populations and can affect selection coefficients on life-history traits if it influences an individual's exposure to a particular environment. In addition, the phenotypes of nonbehavioral life-history traits (e.g., body size) may affect selection coefficients on traits that are physiologically interrelated. Thus, the suite of phenotypic responses that an organism has when challenged with a novel environment is likely to be an important determinant of the particular evolutionary trajectory that the species will take in the new environment.

Acknowledgments

I am grateful to M. K. Saks, J. A. Sifuentes, and J. L. Carrillo for facilitating my collecting trips in Mexico and to D. J. Futuyma, R. E. Stinner, and M. A. Rankin, who collaborated at various times on aspects of this work. L. Morrow, M. L. Vallone, A. Ramirez, and J. Keever provided indispensable help in the laboratory. F. Gould, J. Jaenike, J. Siebenaller, and A. Zera graciously took time from their own work for discussions with me. The research was supported by a Sigma Xi Grant in Aid of Research and a Postdoctoral Fellowship from the University of Rochester to M. E. Saks, by National Science Foundation Grants BSR-8306000 to D. J. Futuyma, DEB-8214598 to R. E. Stinner, and BSR-8605196 to J. Jaenike, and by a U. S. Department of Agriculture Competitive Grant 84-CRCR-1-1458 to M. A. Rankin.

REFERENCES

Beck, S. D., and J. C. Reese. 1976. Insect–plant interactions: Nutrition and metabolism. *Recent Adv. Phytochem.* 10:41–92.

Berenbaum, M. 1983. Coumarins and caterpillars: A case for coevolution. *Evolution* 37:163–179.

Blau, W. S., and R. E. Stinner. 1983. Temporal flight patterns in the Mexican bean beetle (Coleoptera: Coccinellidae) and their relation to weather. *Environ. Entomol.* 12:1047–1054.

Bradford, M. M. 1976. A rapid and sensitive method for the quantitation of microgram quantities of protein utilizing the principle of protein-dye binding. *Anal. Biochem.* 72:248–254.

Burkart, A., and H. Brücher. 1953. *Phaseolus aborigineus* Burkart, die mutmassliche andine Stammform der Kulturbohne. *Der Züchter* 23:65–72.

Chen, P. S. 1984. The functional morphology and biochemistry of insect male accessory glands and their secretions. *Annu. Rev. Entomol.* 29:233–255.

Cochran, D. G. 1975. Excretion in insects. Pages 177–282 in Candy, D. J., and B. A. Kilby (eds.), *Insect Biochemistry and Function*. New York: Wiley & Sons.

Dempster, E. R. 1955. Maintenance of genetic heterogeneity. *Cold Spring Harbor Symp. Quant. Biol.* 20:25–32.

Denno, R. F., and M. S. McClure (eds.). 1983. *Variable Plants and Herbivores in Natural and Managed Systems*. New York: Academic Press.

Dingle, H. 1965. The relation between age and flight activity in the milkweed bug *Oncopeltus*. *J. Exp. Biol.* 42:269–283.

Engelmann, F. 1970. *The Physiology of Insect Reproduction*. Oxford, United Kingdom: Pergamon.

Evans, A. M., and J. H. C. Davis. 1978. Breeding *Phaseolus* beans as grain legumes for Britian. *Appl. Biol.* 3:1–42.

Falconer, D. S. 1981. *Introduction to Quantitative Genetics*, 2nd ed. New York: Longman.

Feeny, P. 1975. Biochemical coevolution between plants and their herbivores. Pages 3–19 in Gilbert, L. E., and P. R. Raven (eds.), *Coevolution of Animals and Plants*. Austin, Tex.: Univ. of Texas Press.

———. 1976. Plant apparency and chemical defense. *Recent Adv. Phytochem.* 10:1–40.

Fraenkel, G. 1969. Evaluation of our thoughts on secondary plant substances. *Entomol. Exp. Appl.* 12:473–486.

Futuyma, D. J. 1976. Food plant specialization and environmental predictability in Lepidoptera. *Am. Nat.* 110:285–292.

Gentry, H. S. 1969. Origin of the common bean, *Phaseolus vulgaris*. *Econ. Bot.* 23:55–69.

Gordon, R. D. 1975. A revision of the Epilachninae of the Western Hemisphere (Coleoptera: Coccinellidae). *U.S. Dept. Agric. Tech. Bull.* 1493.

Gottlieb, L. D. 1984. Genetics and morphological evolution in plants. *Am. Nat.* 123:681–709.

Hedrick, P. W. 1986. Genetic polymorphism in heterogeneous environments: A decade later. *Annu. Rev. Ecol. Syst.* 17:535–566.

Hedrick, P. W., M. E. Ginevan, and E. P. Ewing. 1976. Genetic polymorphism in heterogeneous environments. *Annu. Rev. Ecol. Syst.* 7:1–32.

Hilu, K. W. 1983. The role of single-gene mutations in the evolution of flowering plants. *Evol. Biol.* 16:97–128.

Hochberg, Y. 1974. Some generalizations of the T method in simultaneous inference. *J. Multiv. Anal.* 4:224–234.

Howard, N. F. 1930. The Mexican bean beetle in the East and its control. *U. S. Dep. Agric. Farmers Bull.* 1624.

Johnson, C. G. 1969. *Migration and Dispersal of Insects by Flight*. London: Methuen.

Kaplan, L. 1965. Archeology and domestication in American *Phaseolus* (beans). *Econ. Bot.* 19:358–368.

Kareiva, P. 1983. Influence of vegetation texture on herbivore populations: Resource concentration and herbivore movement. Pages 259–289 in Denno, R. F., and M. S. McClure (eds.), *Variable Plants and Herbivores in Natural and Managed Systems*. New York: Academic Press.

Kimura, M. 1954. Process leading to quasi-fixation of genes in natural populations due to random fluctuations in selection intensities. *Genetics* 39:280–295.

Levins, R. 1968. *Evolution in Changing Environments*. Princeton, N. J.: Princeton Univ. Press.

McNeill, S., and T. R. E. Southwood. 1978. The role of nitrogen in the development of insect/plant relationships. Pages 77–98 in Harborne, J. B. (ed.), *Biochemical Aspects of Plant and Animal Coevolution*. New York: Academic Press.

Mattson, W. J., Jr. 1980. Herbivory in relation to plant nitrogen content. *Annu. Rev. Ecol. Syst.* 11:119–161.

Ojehomon, O. O. 1966. The development of flower primordia in *Phaseolus vulgaris* (L.) Savi. *Ann. Bot.* 30:487–492.

Peterson, G. L. 1983. Determination of total protein. *Methods Enzymol.* 91:95–110.

Rankin, M. A., and S. Rankin. 1980. Some factors affecting presumed migratory flight activity of the convergent ladybeetle, *Hippodamia convergens* (Coccinellidae: Coleoptera). *Biol. Bull.* 158:356–369.

Rankin, M. A., and L. M. Riddiford. 1977. Hormonal control of migratory flight in *Oncopeltus fasciatus*: The effects of the corpus cardiacum, corpus allatum and starvation on migration and reproduction. *Gen. Comp. Endocrinol.* 33:309–321.

Rhoades, D. F., and R. G. Cates. 1976. A general theory of plant antiherbivore chemistry. *Recent Adv.. Phytochem.* 10:168–213.

Risch, S. J. 1981. Insect herbivore abundance in tropical monocultures and polycultures: An experimental test of two hypotheses. *Ecology* 62:1325–1340.

Root, R. B. 1973. Organization of a plant–arthropod association in simple and diverse habitats: The fauna of collards (*Brassica oleracea* L.). *Ecol. Monogr.* 43:95–124.

Rosenthal, G. A., and D. H. Janzen (eds.). 1979. *Herbivores: Their Interactions with Secondary Plant Metabolites*. New York: Academic Press.

Runge, M. 1983. Physiology and ecology of nitrogen nutrition. Pages 163–200 in Lange, O. L., P. S. Nobel, C. B. Osmond, and H. Ziegler (eds.), *Physiological Plant Ecology III. Responses to the Chemical and Biological Environment*. Berlin: Springer-Verlag.

Saks, M. E. 1983. Evolutionary divergence of native and introduced populations of the Mexican bean beetle, Epilachna varivestis (Coleoptera: Coccinellidae). Ph.D. dissertation, SUNY-Stony Brook.

———. 1990. Unpublished data.

Saks, M. E., M. A. Rankin, and R. E. Stinner. 1988. Sexually differentiated flight responses of the Mexican bean beetle to larval and adult nutrition. *Oecologia* 75:296–302.

SAS Institute Inc. 1985. Lifetest procedure. Pages 529–557 in *User's Guide. Statistics*, SAS Institute Inc., Cary, North Carolina.

Schlichting, C. D. 1986. The evolution of phenotypic plasticity in plants. *Annu. Rev. Ecol. Syst.* 17:667–693.

Scriber, M. J., and F. Slansky, Jr. 1981. The nutritional ecology of immature insects. *Annu. Rev. Entomol.* 26:183–211.

Slansky, F., Jr., and J. G. Rodriguez (eds). 1987. *Nutritional Ecology of Insects, Mites, Spiders and Related Invertebrates*. New York: Wiley & Sons.

Stoddard, J. L., and H. Thomas. 1982. Leaf senescence. Pages 592–636 in Boulter, D., and B. Parthier (eds.), *Nucleic Acids and Proteins in Plants I. Structure, Biochemistry and Physiology of Proteins*. Berlin: Springer-Verlag.

Thimann, K. V. (ed.). 1980. *Senescence in Plants*. Boca Raton, Fla: CRC Press.

Thomas, R. J., and L. E. Schrader. 1981. Ureide metabolism in higher plants. *Phytochemistry* 20:361–371.

White, J. 1984. Plant metamerism. Pages 15–47 in Dirzo, R., and J. Sarukhán (eds.), *Perspectives on Plant Population Ecology*. Sunderland, Mass.: Sinauer.

Wiens, J. A. 1976. Population responses to patchy environments. *Annu. Rev. Ecol. Syst.* 7:81–120.

Wilson, K. G., and R. E. Stinner. 1984. A potential influence of *Rhizobium* activity on the availability of nitrogen to legume herbivores. *Oecologia* 61:337–341.

17

CAUSES OF HOST-ASSOCIATED VARIATION IN INSECT HERBIVORES: AN EXAMPLE FROM FALL ARMYWORM

Dorothy P. Pashley

Louisiana State University

INTRODUCTION

Many polyphagous species are composed of groups of specialists, each of which uses only a portion of the host repertoire of the entire species (Fox and Morrow, 1981; Rausher, 1984; Via, 1984a, 1984b, 1986; Futuyma and Peterson, 1985; Hare and Kennedy, 1986; Futuyma and Philippi, 1987; Pashley, 1988a). When resource usage varies spatially or temporally (populations are allopatric or allochronic, respectively), differences among populations are likely to be maintained by an absence of gene flow. The situation is less clear in sympatry (or synchrony), because populations have the opportunity to interbreed. In these cases, to clarify taxonomic status and evolutionary associations, it is necessary to test experimentally for natural selection on characters that are observed to differ and to quantify reproductive compatibilities between taxa.

A variety of terms has been applied to taxa exhibiting host-associated variation. They have been referred to as strains, biotypes, ecomorphs, ecotypes, or races, and, after careful study, some have been reclassified as cryptic or sibling species (Mayr, 1970; Jaenike, 1981; Diehl and Bush, 1984; Futuyma and Peterson, 1985). The distinction among these categories depends upon the geography and the phenology of populations, which of several factors contribute to differences, and whether or to what degree taxa are interbreeding.

This Chapter focuses on the resolution of factors responsible for host-associated variation in herbivorous insects and discusses terminology currently used to describe such taxa. A sequence of steps for distinguishing various causes is outlined and illustrated with recently gathered data for two newly discovered host strains (a term used here for taxa of unknown status) of a major agricultural pest, the fall armyworm (Lepidoptera: Noctuidae).

RESOLVING CAUSES OF HOST-ASSOCIATED VARIATION

Causes of host-associated differentiation can be systematically addressed by answering a sequence of questions (Fig. 17.1). Initially, an insect must feed on more than one plant species, or, because definitions of polyphagy vary (Scriber, 1973; Slansky, 1976; Futuyma, 1976), more than one plant genus or family. If an insect feeds on more than one host, but, after careful study, no feature (e.g., behaviorial, morphological, biochemical) distinguishes individuals on different hosts, the species is likely to be a generalist containing polyphagous individuals.

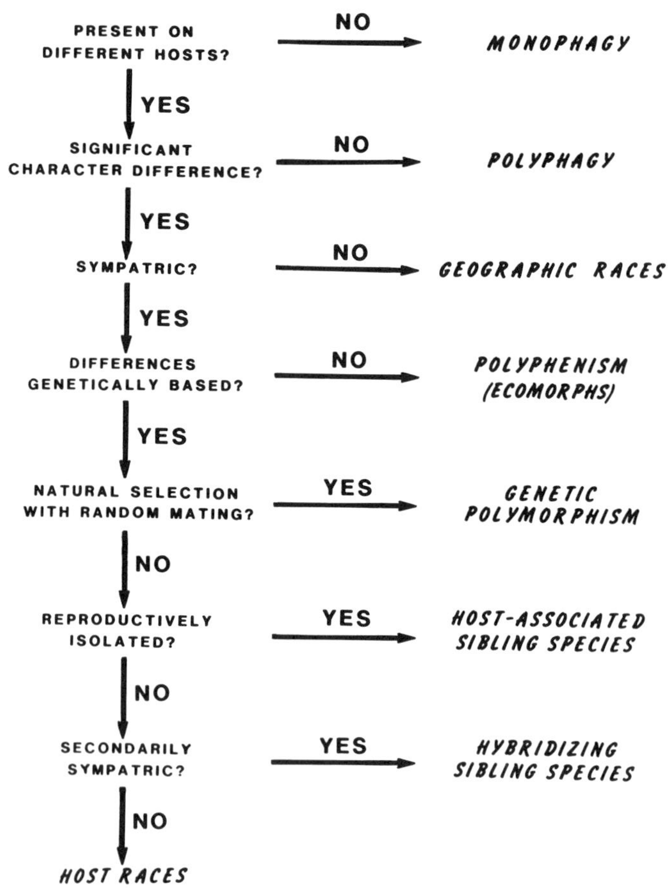

If particular characteristics are associated with host use and ranges are allopatric, strains represent geographic races, whose differences may have been caused by any number of factors having to do with the occupation of different environments or by genetic drift. Geographically disjunct taxa have been referred to as races, ecotypes, semispecies, subspecies, and species (Mayr, 1970; Cracraft, 1983; Diehl and Bush, 1984; McKitrick and Zink, 1988). Reproductive compatibilities are essentially unknown in these situations, because taxa never encounter each other in nature and thus their taxonomic designation is somewhat semantic. This same confusion arises in a less commonly encountered situation, in which taxa are completely allochronic.

If strains are sympatric, a number of experiments can be performed to determine their status. In some cases, variation may be induced entirely by environmental differences associated with hosts. In other words, regardless of an individual's history or genetic background, it acquires a phenotype induced by the host upon which it develops. This is termed "nongenetic polyphenism" or "ecomorphic variation." Alternatively, variation may be genetically based and partially or completely independent of environmental induction. For discrete characters, such as allozymes, for which variation exists within a host strain, crosses suffice to verify Mendelian inheritance. If crosses are not possible, adherence to Hardy–Weinberg expectations suggests that variation is genetically based.

Separating genetic and environmental influences for continuous traits is accomplished by a quantitative genetic approach (Falconer, 1981) involving split-progeny rearings on different hosts within each of numerous family groups. This type of experiment simultaneously addresses several questions. Effects of the host plant on the phenotypic expression of insect traits are readily apparent for both strains, because the same genetic backgrounds are exposed to different hosts. Inherent genetic differences between strains are evident by comparing responses of traits when strains are reared on the same hosts. Finally, partitioning of variance within and among families in each strain in relation to hosts (genotype-by-environment interactions) allows the determination of whether genetic variation exists that might facilitate further divergence within strains, given the proper environmental stimulus.

If strain differences have a genetic basis, there are two possible causes for host-associated variation. Natural selection regimes may differ on each host with respect to characters of interest, or taxa may not be randomly mating, in which case, gene flow between hosts is reduced. In the former case, strains mate randomly each generation and place eggs in each habitat, but selective mortality of certain genotypes during development results in different ultimate genotypic frequencies on each host. These differences are referred to as "genetic polymorphisms." Cross-rearing studies reveal the importance of natural selection, since shifts in genotype frequencies or metric character means should be apparent by comparing survivors within strains on each host.

If natural selection is not solely responsible for maintaining genetic differences, strains may exhibit some degree of assortative mating, remain associated with a particular host type, and lay their eggs in that habitat. Adaptation to the host environment may play a role in divergence between strains, but, in contrast to the previous situation, genetic change is not impeded by the homogenizing effects of gene flow each generation.

The remaining steps in the assignment of sympatric host-associated strains to a taxonomic category are related to the degree to which strains interbreed. If strains are completely reproductively isolated, they represent host-associated sibling species. If they are only partially isolated, they are either host races (Diehl and Bush, 1984) or secondarily sympatric species that are capable of hybridizing to some degree.

Two approaches can be taken to determine if reproductive isolation is complete. Laboratory crosses provide information on postreproductive isolation if viability is reduced in Fl progeny and on prereproductive barriers if behavioral or mechanical differences prohibit mating. When no viable progeny are produced in the laboratory, it can generally be assumed that taxa are reproductively isolated in the field. However, if no pre- or postreproductive barriers are observed, it cannot be assumed that strains mate in nature, because the laboratory environment may modify behavior (Mayr, 1970; Diehl and Bush, 1984).

A second approach involves studies of interbreeding in nature. If feasible, mated pairs may be captured and each member identified to a strain. If pheromones are involved in mate attraction, individuals caged within a trapping device can be used to attract members of the opposite sex that can then be identified to a strain. In this way, cross-attraction between strains can be quantified. Field observations using captive females may verify that responding males are actually capable of coupling. Mated females may be returned to the laboratory to examine mating success and progeny viability.

Separating host races from hybridizing species is difficult because it often requires the use of inferences about the history of the taxa. If host shifts occurred in recent times and events were well documented, it might be possible to categorize the taxa as sympatrically generated host races. This appears to be the case for *Rhagoletis pomonella*, a tephritid fruit fly that has shifted from native hawthorns to introduced fruits (cherry and apple) in the last 150 years (Bush, 1969; Feder et al., 1988; McPheron et al., 1988). Unfortunately, for many species, information on the time and geography of host shifts is not available. In these situations, distinguishing the two categories must be based upon comparisons between observed life-history traits and those characteristics specifically proposed for host races (Bush, 1975a, 1975b; Diehl and Bush, 1984). According to Bush (1975b), species likely to form host races mate on their host plants, remain closely associated physically with the plants throughout their life stages, and exhibit prereproductive as opposed to postreproductive barriers. If these requirements are violated, taxa are probably not host races but are, instead, hybridizing species. Secondarily, sympatric-hybridizing species may be prevalent in agricultural pests whose distributions are tied to human manipulations of plants.

The ease with which one proceeds through this sequence of steps may vary with the species studied or the types of characters examined. If taxa on different hosts have no diagnostic characters (i.e., they exhibit frequency differences), wild individuals may need to be captured and artificially marked and released, increasing the difficulty of addressing certain questions. Many agriculturally important insects are amenable to approaches presented here, because they can easily be reared in a laboratory setting and will readily mate in captivity. The fall armyworm fits into this category, and, in the following section, I will detail research relevant to resolving the taxonomic status of its two host-associated strains.

HOST-ASSOCIATED VARIATION IN FALL ARMYWORM

The fall armyworm, *Spodoptera frugiperda* (J. E. Smith), is primarily a Neotropical insect that invades temperate regions during warmer months and occurs in high densities on various crops (Sparks, 1979). Although it has been reported to feed on species from 23 plant families, its major hosts are taxonomically diverse grasses (Poaceae) that include such important crops as corn, sorghum, rice, and various forage grasses (Luginbill, 1928; Pashley, 1988b). It is not a monophagous species by any definition.

It had been assumed that the fall armyworm is a polyphagous species lacking populational or individual specificity (Luginbill, 1928). This initial assumption was refuted when significant allozymic differences were detected between a population that was collected feeding on rice in Puerto Rico and nine other populations that were collected feeding on corn in continental America (Pashley et al., 1985). Genetic variation was initially thought to represent geographic isolation of Puerto Rico rather than host-associated differentiation. However, a survey of different host plants in Louisiana revealed that the two strains were sympatric and synchronic but tended to use different hosts (Pashley, 1986, and Fig. 17.2). One strain (designated the rice

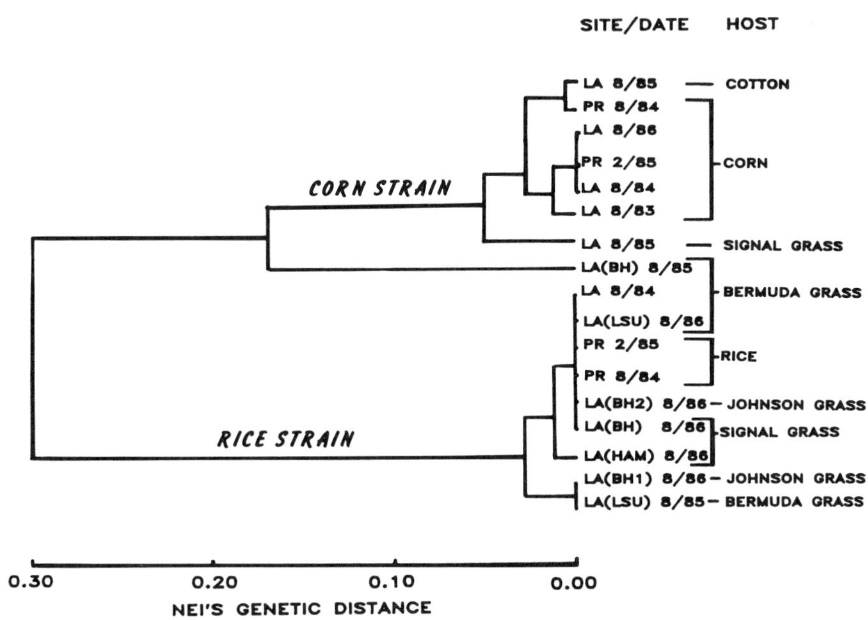

Figure 17.2. Genetic relationships among fall armyworm collected from different hosts plants. Genetic distances are based on three polymorphic loci used to separate strains (esterase, hydroxybutyrate dehydrogenase, and peptidase). PR = Puerto Rico, LA = Louisiana (sites in parentheses are different locations in LA).

strain because it was first discovered on that host) feeds on rice and Bermuda, Johnson, and signal grasses. The other, referred to as the "corn strain," feeds primarily on corn and cotton. It does, however, feed on hosts of the rice strain when they are grown in proximity to corn (as is LA 8/85 signal grass in Fig. 17.2) and occasionally when they are not grown in proximity to corn [LA(BH) 8/85 Bermuda grass in Fig. 17.2]. Host fidelity does not appear complete in either strain.

Five allozyme loci have been determined to differ between the strains (Pashley, 1986), and crossing experiments within strains have verified Mendelian inheritance of all except a few of the families examined at the esterase locus (Pashley, 1990). Problems with esterase relate to difficulties in scoring female genotypes, particularly as they age. Esterase is the most distinctive locus between strains, and, although departures from Mendelian ratios are apparent in a few crosses, results verify the maintenance of strain-associated alleles in parents and progeny.

Cross-rearing studies of each strain reared on its own and the other strain's host did not result in significant shifts in electrophoretic genotypes within strains (Pashley, 1986). Therefore, natural selection was not likely to be solely responsible for maintaining observed differences. I concluded that genetically based host-related differences represented inherent strain differences and not a selectively mediated genetic polymorphism.

Other types of characters have subsequently been determined to differ between the strains. Developmental differences have been detected and analyzed using a quantitative genetic approach (Pashley, 1988a). Although the host plant had a significant effect on the variation at most traits (larval weight, larval duration, pupal weight, and survival), there were significant between-strain differences in rates of development. Other differences have been noted between strains in their response to insecticides (Pashley et al., 1987b) and their development on varieties of Bermuda grass (Pashley et al., 1987a).

The correct choice in the remaining dichotomy between host races and species (Fig. 17.1) has not been resolved completely. Laboratory crossing experiments indicated unidirectional premating isolation (Pashley and Martin, 1987). Females of the corn strain did not mate with males of the rice strain, but females of the rice strain did mate with males of the corn strain. In backcrosses, hybrid males produced by the latter cross mated with both parental females, but hybrid females did not mate with either type of male. This incompatibility is likely to be behavioral and not mechanical, because, in a later experiment (Whitford et al., 1988), mating did occur when corn strain females of a long-established laboratory colony (Perkins, 1979) were placed with rice strain males, and hybrids were viable. Thus, at present, no postreproductive barriers have been detected in either of the reciprocal matings.

Studies of mating behavior in nature have been conducted and suggest differential cross-attraction between strains. In August 1986, virgin females of each strain were placed at several Louisiana sites and captured males were identified to strain using electrophoresis (Pashley et al., 1990). Both types of males exhibited a preference for females of their own strain. Most rice strain males (119/166 = 72%) were attracted to rice strain females, although some (28%) went to corn females. Corn strain males, likewise, preferred corn strain females (12/16 = 75%), but some were attracted to rice strain females (25%).

A collaborative study is currently underway in Georgia (Carpenter and Pashley, 1990), and preliminary results support findings in Louisiana. Although sample sizes are very small, 75% of the rice strain males preferred rice strain females and 86% of the corn strain males preferred corn strain females. Considering the similarity of results from both locations, it is likely that there is a high degree of assortative mating, maintained by either pheromonal or calling behavior differences. Field observations will determine if matings occur when nonstrain males are attracted.

If the fall armyworm strains are determined to be hybridizing to some degree in nature, it is unlikely that they represent host races in the initial stages of speciation. They fit few of the biological criteria established for host races (Bush, 1975b). Although they probably often mate on their larval host plants, adults require nectar and may forage on nonhost plants, where it is possible that some mating occurs. In addition, they are considered long-distance migrants and, thus, spend some proportion of their life away from their host environment (Sparks, 1979). Contributing to taxonomic uncertainty is the observation that mating barriers appear to be prereproductive, suggestive of a recent speciation event.

CONCLUSION

In addition to providing a theoretical framework for taxonomic decision making, the steps presented in the flowchart constitute practical guidelines for understanding biological phenomena responsible for host-associated variation in nature. It is critical that entomologists using "polyphagous" insects to address both applied and basic questions consider the possibility that insects collected from one habitat may not perform in the same manner as insects collected from a different habitat. It is relatively easy to establish whether variation exists by measuring or quantifying a few different types of characters in individuals collected from different hosts. In fall armyworm, nearly every character that has been examined has exhibited a significant strain difference. If experiments described here are not possible, it is essential that a voucher series be established with precise habitat-related data so that future discoveries of variation or sibling species will allow reinterpretation of previous studies.

Acknowledgments

Research on fall armyworm biology was supported by USDA Grant 86-CRCR-1-2027. The manuscript was improved by comments from Bruce McPheron, David Pashley, and Bob Zink.

REFERENCES

Bush, G. L. 1969. Sympatric host race formation and speciation in frugivorous flies of the genus *Rhagoletis* (Diptera: Tephritidae). *Evolution* 23:237–251.

———. 1975a. Modes of animal speciation. *Annu. Rev. Ecol. Syst.* 6:334–364.

———. 1975b. Sympatric speciation in phytophagous parasitic insects. Pages 187–206 in Price, P. W. (ed.), *Evolutionary Strategies of Parasitic Insects and Mites*. New York: Plenum.

Carpenter, J. E., and D. P. Pashley. 1990. Unpublished data.

Cracraft, J. 1983. Species concepts and speciation analysis. *Current Ornithol.* 1:159–187.

Diehl, S. R., and G. L. Bush. 1984. An evolutionary and applied perspective of insect biotypes. *Annu. Rev. Entomol.* 29:471–504.

Falconer, D. S. 1981. *Introduction to Quantitative Genetics.* New York: Longman.

Feder, J. L., C.A. Chilcote, and G.L. Bush. 1988. Genetic differentiation between sympatric host races of *Rhagoletis pomonella. Nature* 336:61–64.

Fox, L. R., and P. A. Morrow. 1981. Specialization: Species property or local phenomenon? *Science* 211:887–893.

Futuyma, D. J. 1976. Food plant specialization and environmental predictability in Lepidoptera. *Am. Nat.* 110:285–292.

Futuyma, D. J., and S. C. Peterson. 1985. Genetic variation in the use of resources by insects. *Annu. Rev. Entomol.* 30:217–238.

Futuyma, D. J., and T. E. Philippi. 1987. Genetic variation and covariation in responses to host plants by *Alsophila pometaria* (Lepidoptera: Geometridae). *Evolution* 41:269–279.

Hare, J. D., and G. G. Kennedy. 1986. Genetic variation in plant–insect associations: Survival of *Leptinotarsa carolinense. Evolution* 40:1031–1043.

Jaenike, J. 1981. Criteria for ascertaining the existence of host races. *Am. Nat.* 117:830–834.

Luginbill, P. 1928. The fall armyworm. *U.S. Dep. Agric. Tech. Bull.* No. 34. 92 pp.

Mayr, E. 1970. *Populations, Species, and Evolution.* Cambridge, Mass.: Harvard Univ. Press.

McKitrick, M. C., and R. M. Zink. 1988. Species concepts in ornithology. *Condor* 90:1–14.

McPheron, B. A., D. C. Smith, and S. H. Berlocher. 1988. Genetic differences between host races of the apple maggot fly. *Nature* 336:64–66.

Pashley, D. P. 1986. Host associated genetic differentiation in fall armyworm: A sibling species complex? *Ann. Entomol. Soc. Am.* 79:898–904.

———. 1988a. Quantitative genetics, development and physiological adaptation in sympatric host strains of fall armyworm. *Evolution* 42:93–102.

———. 1988b. The current status of fall armyworm host strains. *Fla. Entomol.* 71:227–234.

———. 1990. Unpublished data.

Pashley, D. P., A. M. Hammond, and T. N. Hardy. 1990. Unpublished data.

Pashley, D. P., S. J. Johnson, and A. N. Sparks. 1985. Genetic population structure of migratory moths: The fall armyworm (Lepidoptera: Noctuidae). *Ann. Entomol. Soc. Am.* 78:756–762.

Pashley, D. P., and J. A. Martin. 1987. Reproductive incompatibility between host strains of fall armyworm (Lepidoptera: Noctuidae). *Ann. Entomol. Soc. Am.* 80:731–733.

Pashley, D. P., S. S. Quisenberry, and T. Jamjanya. 1987a. Impact of fall armyworm (Lepidoptera: Noctuidae) host strains on the evaluation of Bermuda grass resistance. *J. Econ. Entomol.* 80:1127–1130.

Pashley, D. P., T. C. Sparks, S. S. Quisenberry, T. Jamjanya, and P. Dowd. 1987b. Two fall armyworm strains feed on corn, rice and Bermuda grass. *La. Agric.* 30:8–9.

Perkins, W. D. 1979. Laboratory rearing of the fall armyworm. *Fla. Entomol.* 62:87–91.

Rausher, M. D. 1984. Tradeoffs in performance on different hosts: Evidence from within- and between-site variation in the beetle *Deloyala guttata*. *Evolution* 38:582–595.

Scriber, J. M. 1973. Latitudinal gradients in larval feeding specialization of the world Papilionidae (Lepidoptera). *Psyche* 80:355–373.

Slansky, F., Jr. 1976. Phagism relationships among butterflies. *J. N.Y. Entomol. Soc.* 84:91–105.

Sparks, A. N. 1979. A review of the biology of the fall armyworm. *Fla. Entomol.* 62:82–87.

Via, S. 1984a. The quantitative genetics of polyphagy in an insect herbivore. I. Genotype-environment interaction in larval performance on different host plant species. *Evolution* 38:881–895.

———. 1984b. The quantitative genetics of polyphagy in an insect herbivore. 11. Genetic correlations in larval performance within and among host plants. *Evolution* 38:896–905.

———. 1986. Genetic covariation between oviposition preference and larval performance in an insect herbivore. *Evolution* 40:778–785.

Whitford, F., S. S. Quisenberry, T. J. Riley, and J. W. Lee. 1988. Mating compatibility, ovipositional preference, and larval development of two electrophoretically differentiated fall armyworm colonies. *Fla Entomol.* 71:234–243.

18

ADAPTATIONS OF *DIABROTICA* TO HABITAT MANIPULATIONS

James L. Krysan

U.S. Department of Agriculture, Agricultural Research Service, Yakima Agricultural Research Laboratory

INTRODUCTION

This case study concerns two rootworm pests of corn [the northern, *Diabrotica barberi* Smith and Lawrence (NCR) and the western, *D. virgifera virgifera* LeConte (WCR) (Coleoptera: Chrysomelidae)], species for which habitat modifications by agricultural practices have had a profound influence on the insects' roles as pests. The key cultural practice for controlling these rootworms is crop rotation, a habitat manipulation that is widely recognized as one of the classic examples of cultural control of an insect (Herzog and Funderbunk, 1986). I will outline the history of these insects as pests, discuss their responses to agricultural practices, and examine the entomological interpretation of the responses in terms of the knowledge of rootworm biology.

It is now recognized that the NCR and WCR have important biological differences, causing each species to respond in its own way to spatial and temporal changes due to different cropping practices. However, from the mid-1960s to the mid-1970s, it was widely thought that the western and northern corn rootworms were interbreeding, and, therefore, they were often considered to be a single pest (Chiang, 1973; Luckmann et al., 1974) Fundamental research in the 1970s and 1980s has provided new perspectives from which to view the response of rootworms to crop rotation.

BIOLOGY

The general biology of the species of the *virgifera* group, which includes the NCR and the WCR, has been recently reviewed in detail (Krysan and Smith, 1987); here I will summarize those features pertinent to a discussion of the response of *Diabrotica* to crop rotation. The NCR and the WCR share a number of important traits but, also, differ in biological features especially pertinent to this review. Both species have, at most, one generation per year. The larvae are functionally monophagous and have very low mobility relative to the size of cornfields. The eggs are deposited in the soil in late summer (Hein and Tollefson, 1985) and enter diapause, which accommodates overwintering (Chiang, 1973; Krysan, 1982).

In the NCR, there is great variation in the time required for diapause termination. In most samples, a few eggs hatch within a few weeks without exposure to cold (Patel and Apple, 1967; Chiang et al., 1972; Krysan, 1982; Krysan et al., 1984); most eggs require chill to hatch, and the percent of hatch increases with the duration of chill (Krysan, 1982; Krysan et al., 1984). It was generally thought that the eggs diapaused for one year, but there have been three reports in which the following percentages of NCR eggs diapaused for two years before hatching: 0.3% (Chiang, 1965); 40% (Krysan et al., 1984); and 9–47% for four populations (Krysan et al., 1986). In all three studies, eggs had been placed in the soil out of doors to establish what proportion diapaused for two years before hatching. The contribution of additive genetic variance to this variability in time required for diapause termination is unknown, but evidence to be discussed in a later section indicates that the incidence of one or two years of diapause in a population may depend on the selection pattern imposed by rotation schemes.

The egg of the WCR also diapauses, but the effect of the environment on hatch is different from the NCR. Eggs of the WCR do not require a chill to hatch (Krysan et al., 1977; Krysan, 1978; Krysan, 1982). Furthermore, study of eggs under field conditions showed that exposure to cold had no effect on diapause termination (Krysan et al., 1984). Following oviposition, embryos develop to a specific "diapause stage" (Krysan, 1972), and, after the passage of time, postdiapause development ensues and leads eventually to hatch. The rate of diapause development is highly variable within any population but is clearly independent of temperature (Krysan, 1978, 1982). Among the several studies of diapause and its termination in the WCR, there is no evidence of prolonged diapause (George and Ortman, 1965; Chiang et al., 1972); most important, there was no evidence of prolonged diapause when eggs passed the winter in the field (Krysan et al., 1984). In summary, the NCR can be dormant for one or two years, but the WCR can be dormant for only one year. The difference is obviously important for adaptation to crop rotation.

Adults of the WCR and NCR have different feeding behaviors. Following emergence, WCR adults generally feed first on the succulent reproductive parts of corn, and, when those tissues dry, the WCR tends to feed on corn foliage (Ludwig and Hill, 1975; Hill and Mayo, 1980). The NCR also feeds on reproductive tissues, but, once they are no longer palatable, the NCR tends not to feed on corn leaves (Lance and Fisher, 1987); rather, it turns to succulent floral parts of a variety of plants that commonly occur in or near the corn ecosystem (Forbes, 1896; Webster, 1913; Hill and Mayo, 1980; Cinereski and Chang, 1968). I conducted a feeding choice experiment in which WCR and NCR beetles were given a choice between corn foliage and the artificial diet of Guss et al. (1976). The extent of feeding was assigned to three classes: (1) none; (2) minimal (a single, small, feeding scar); (3) heavy (more than one scar). In this experiment, beetles were held individually in small plastic cages for one week prior to examination of the leaves for feeding damage. For WCR beetles ($n = 48$), 11 ate no foliage, 11 ate minimally, and 26 ate heavily; for the NCR ($n = 56$), 45 did not feed, 9 ate minimally, and 2 ate heavily. This brief study supports many field observations of differences in feeding by NCR and WCR. The WCR is seldom collected outside of cornfields, whereas the NCR is commonly collected on flowering parts of plants growing near cornfields (Ludwig and Hill, 1975; Hill and Mayo, 1980; Branson and Krysan, 1981).

HISTORY OF ROOTWORMS AND CROP ROTATION

Humans have caused the NCR and WCR to be pests in the Corn Belt of the North Central United States by imprudent habitat manipulations. The pestiferous nature of the two species arose in different times and places but was created by the same strategy: planting corn each year on the same ground.

The NCR first became a pest of corn in about 1870 in Missouri and/or Illinois (Forbes, 1882). Webster (1895, 1913) summarized evidence that the species had been collected in some areas where corn was not then present (e.g., upstate New York, Nova Scotia, and New Brunswick). Throughout his career, Webster traced the diffusion of the NCR as a pest of corn from Illinois to Ohio. He concluded that the NCR had been an innocuous resident of native prairie and did not feed on corn until humans began planting corn year after year on the same land. Wheat was the principal crop of Illinois up to about 1862, but, when economics changed, corn became the predominant crop and it was planted without rotation (Webster, 1913). Economics were not the only factor. In one county in Illinois, farmers quit planting wheat because of the chinch bug; instead, they planted corn continuously, and the crop was devastated by the NCR (Forbes 1882). Based on his knowledge of NCR biology, Forbes (1882) proposed crop rotation as the obvious method for control. The standard became a so-called long rotation in which the land was out of corn for two years or more (Bigger, 1932).

It appears that humans "created" this pest by providing an extensive area of habitat suitable for the selection of a corn host race. The pest was then controlled by rotations that prevented the rootworm from tracking the larval food resource. Subsequently, the putative host race persisted on corn, despite widespread rotation.

The second habitat change that produced a pest concerns the WCR and began during the mid-twentieth century. The WCR was first collected in 1865 (LeConte 1868) in western Kansas and was first reported as a pest in Colorado (Gillette, 1912). The WCR was not a consequential pest until corn was planted continuously in a given field (Tate and Bare, 1944, 1946; Bryson et al., 1953). The problem arose in the western portions of Kansas and Nebraska, and in time the range of the species extended eastward. Infestations occurred only in fields planted continuously with corn. Despite educational programs showing that the WCR depends on continuous corn to become a pest, many farmers continued to plant corn every year in the same land, and entomologists began testing soil insecticides for rootworm control (Muma et al., 1949). The availability of fertilizers and effective insecticides, together with certain economic advantages, led to the planting of corn without rotation. Resident populations of the WCR attained pest levels; planting of continuous corn over large contiguous areas allowed the rapid expansion of the range of the WCR. It now occurs in the United States from western Idaho to the eastern seaboard (Chiang, 1973; Krysan and Smith, 1987). Clearly, agriculturists created the habitat that first produced the pest and that then led to the expansion of its range. Incidentally, the rootworms quickly became resistant to the highly effective chlorinated hydrocarbon insecticides (Ball and Weekman, 1962), and it has been suggested that the insecticidally resistant WCR was actually a better competitor and that its rate of range expansion was greater than that of the susceptible strain (Metcalf, 1982).

I conclude that planting of corn without rotation at two different places and at two different times produced pests from two previously innocuous insects. Both insects responded in the same way to this one practice—a response that stemmed from shared life-history traits.

In sharp contrast, the next stage in the response of these two pests to crop rotations involves critical differences between the species and occurred where they share ranges; where corn was planted continously, the WCR was so dominant that the NCR was nearly excluded (Hill, 1967; Ortman and Fitzgerald, 1964; Hill and Mayo, 1980), but, where rotation was practiced, the NCR dominated (Hill and Mayo, 1980).

Throughout the 1970s, corn in some areas was routinely rotated with soybean or a small grain and such acreage increased during the 1970s and 1980s (Krysan et al., 1986). Where corn was rotated there were occasional, typically unpublished, reports of damage to corn by the NCR (Gould, 1971; Anonymous, 1976). More recently, numerous well-documented reports have been received of pest levels of NCR feeding on the roots of corn where corn was rotated annually (Krysan et al., 1986; Ostlie, 1987). The practice of spatial and temporal habitat modification by annual rotation of corn with a nonhost crop has begun to fail as a control practice in some areas. In all cases reported, the NCR, not the WCR, has been the problem.

BIOLOGICAL MECHANISM OF ADAPTATION TO ANNUAL ROTATION

Typically, the rotation of crops occurs within a single farm unit, so the fields are usually adjacent. The spatial and temporal setting provided by annual rotation of corn with an alternate crop is diagrammed in Figure 18.1. Two possible mechanisms will be examined to explain the presence of larval NCR in year 3 in the hypothetical cornfield of Figure 18.1: (1) eggs could have been deposited among the corn plants in year 1 and remained dormant for 2 years; (2) adults that emerged from corn in year 2 could have laid eggs in the noncornfield and the eggs could have hatched the following spring. Either mechanism could account for the presence of rootworm larvae in corn under annual rotation.

Some evidence can be interpreted to favor the theory that eggs are deposited among the noncorn plants in the field in the preceding year. Hill and Mayo (1980)

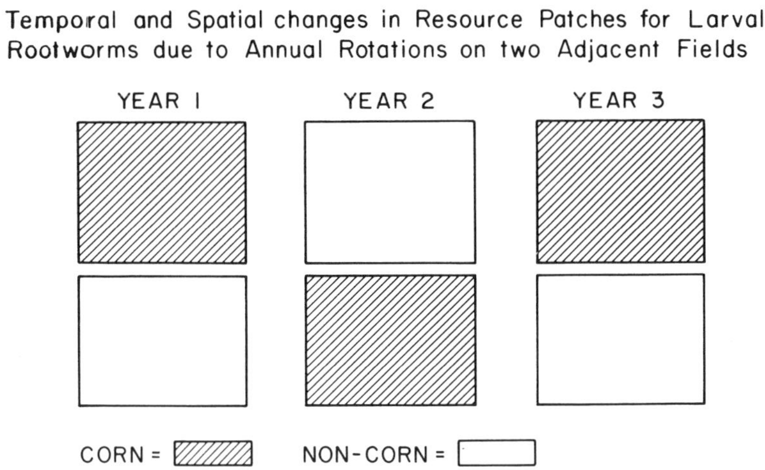

Figure 18.1. Model of temporal and spatial changes in resource patches for larval rootworms due to annual rotations of corn with an alternate crop in two adjacent fields.

found large numbers of NCR, but few WCR, outside of cornfields. Furthermore, in areas where corn was rotated, the highly dispersive NCR prospered, while the nondispersive WCR was much less successful. Therefore, the dispersal behavior of the NCR was correlated with its survival under rotation, and the implicit corollary was generally accepted; the NCR oviposit in the noncornfields into which they disperse (Hill and Mayo, 1980; Branson and Krysan, 1981). However, there is no direct evidence favoring that mechanism. Shaw et al. (1978) found NCR beetles in weedy soybean fields but found no eggs there. Johnson and Turpin (1985) did not detect enhanced oviposition near foxtails, common weedy hosts of adult NCR. In postpollination corn, Cinereski and Chiang (1968) found females whose gut contents proved that they had fed outside that cornfield. Furthermore, Cinereski and Chiang did not find eggs outside cornfields, so they concluded that females left the cornfield to forage for food and then returned to cornfields to oviposit. In a field cage study, when NCR were given a choice between a plot of corn (pollinating or postpollinating) or small grain stubble with flowering weeds, the corn was chosen for oviposition by a significant ratio of nearly 7:1 (Gustin, 1984). This result occurred, despite the fact that the only pollen available for adult feeding was on weeds in the small grain stubble. Finally, in an elegant behavorial study using directional malaise traps, Lance, Elliott, and Hein (1989) found that movements were closely related to sites for feeding and oviposition; when corn silks dried, female NCR tended to move away from corn toward alternate foods. Ovarian stages of females moving toward corn indicated that they were ready to oviposit. From these several studies, I conclude that there is no direct evidence connecting oviposition in noncornfields with the appearance of NCR larvae in corn under rotation.

The alternate hypothesis is that eggs deposited among the corn plants lie dormant through two winters before hatching; evidence summarized next indicates prolonged diapause is the more plausible mechanism. I think prolonged diapause was overlooked as a mechanism because of the widely observed dispersal behavior of the NCR.

Chiang (1965) placed a sample of NCR eggs in the soil out of doors and found that 0.3% of the eggs hatched after two winters; he considered that percentage too low to be of economic significance, and the matter was dropped. Chiang (1965) provided no information on the rotational practices prevailing in the area where the egg samples were obtained. Two decades later, when eggs were obtained from NCR beetles collected in an area where annual rotation prevailed, some 40% were dormant through two winters before hatching (Krysan et al., 1984). There was a bimodal pattern of hatch; some hatched after one year and some after two years. Krysan et al. (1986) found that the proportion of eggs with one versus two years of diapause bore the expected relationship to the cropping pattern in the field of origin; the eggs laid by beetles from three fields with annual rotations included 40–47% with prolonged diapause, while eggs laid by beetles from a field with a long history of continuous corn included 8% eggs with prolonged diapause. Samples of adults collected from one site in two different years (1981 and 1983) yielded essentially the same percentage of eggs with prolonged diapause (Krysan et al., 1986). Beetles from those areas where corn was rotated laid about equal numbers, with one and two years of dormancy, so those dormant for two years would be as likely to cause an economic infestation in the presence of rotation as would those dormant for one year in the absence of rotation.

In 1983, a government program to reduce corn production (the Payment in Kind, or so-called PIK program) produced a fortuitous aberration in the pattern of rotation of corn and soybeans. Those fields provided the "experimental" plots illustrated in Figure 18.2. Without the PIK program, the entire field illustrated would have been planted with corn in 1983; instead, that part of the field in so-called PIK cover was not planted with corn. The entire field was planted to soybean in 1984 and corn in 1985. Rootworm damage to corn in 11 such fields in 1985 was essentially confined to that portion of the field planted to corn in 1983. Similar observations in Minnesota yielded identical results (Krysan et al., 1986). The pattern of damage in 1985 can best be explained by the presence of eggs with prolonged diapause that had been deposited in the field in 1983. If the larval infestation had been due to oviposition in 1984, it should have been across the entire field. Incidentally, field histories indicated no difference in weed distribution between the two halves of the field in 1984.

Results of some earlier studies on the effect of rotation patterns on NCR populations can now be best explained by prolonged diapause. Bigger (1932) reported that, in experimental fields at a single location, annual rotations failed to control NCR, whereas rotations that included corn only every third or fourth year were successful. In several rotation schemes, the crop immediately preceding corn was small grain. Clearly, if dispersal to, and oviposition in, the small grain crop during the preceding year had been the mechanism for infestation of those fields, the number of years a field had been out of corn should have been irrelevant. Also, Hill and Mayo (1980) noted that NCR prevailed where annual rotations were practiced.

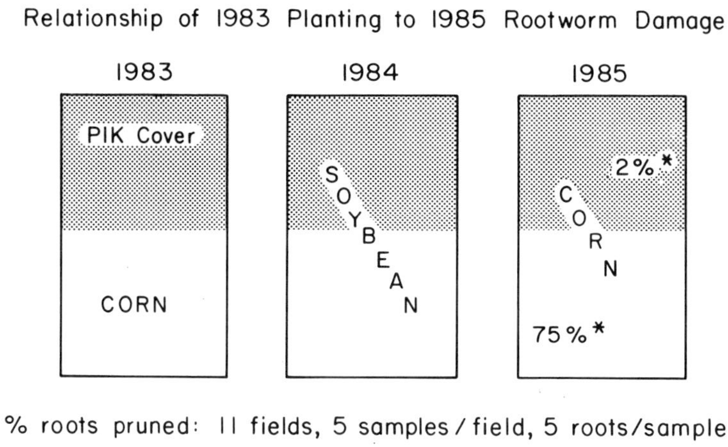

Figure 18.2. Relationship of plantings in 1983 and 1984 to rootworm damage in 1985. An unusual rotation sequence was imposed by the "Payment in Kind" (PIK) program of 1983. The so-called PIK cover crop was not corn, and the area thus planted is shaded for all years. The entire field was planted with soybean in 1984 and corn in 1985 (data from Krysan et al., 1986).

SPECIES DIFFERENCES IN ADAPTATION TO CROPPING SEQUENCES

Since the 1970s entomologists have been able to observe rootworm populations under two distinct and widespread cropping regimes; continuous corn and annual rotations. In some areas where the NCR and WCR occur together, there are striking differences in their survival under the two regimes; rotation favors the NCR, and the planting of corn in the same field each year favors the WCR (Hill and Mayo, 1980). In contrast to the NCR, the WCR is poorly adapted to crop rotation, because it is unable to be dormant for more than one winter (Krysan et al., 1984). Also, it is highly unlikely that the WCR could become adapted to crop rotation by oviposition in the alternate crop, because adults tend to remain within cornfields (Hill and Mayo 1980; Lance, 1990, personal communication).

Continuous corn production satisfies the basic ecological needs of both species, but in that habitat the WCR displaces the NCR (Ortman and Fitzgerald, 1964; Hill and Mayo, 1980). An early hypothesis that the NCR disappeared due to introgression via hybridization in conjunction with dominance of the WCR phenotype (Chiang, 1973; Hintz and George, 1979) has been tested and rejected (Krysan and Guss, 1978; Krysan and Smith, 1987). An alternate hypothesis, that the WCR competitively displaces the NCR (Ortman and Fitzgerald, 1964), has some support. Piedrahita et al. (1985) found in the laboratory that larval WCR displaced NCR from the roots of corn. Naranjo and Sawyer (1987) reported that the NCR has a much lower reproductive potential than the WCR and argued that that could contribute to its observed displacement by the WCR. Further, there is evidence of indirect effects on

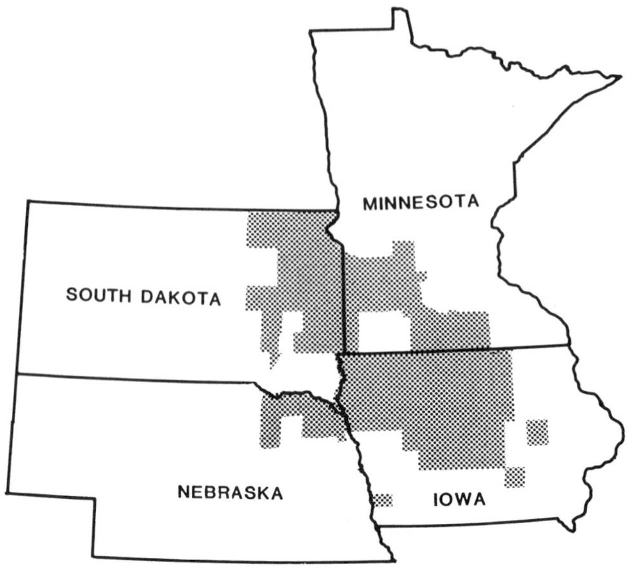

Figure 18.3. Areas (by county and indicated by shading) with corn rootworm damage in fields under annual rotation (after Ostlie, 1987 and Foster, 1987).

reproductive potential, which would make the WCR better adapted to the corn habitat than the NCR. The NCR cannot use mature tissues of corn for reproduction, so, for a consequential part of their reproductive period, many females must forage away from corn to acquire nutrients for reproduction (Naranjo and Sawyer, 1987). Presumably, there is some cost in the dispersal to, and return from, alternate hosts. Thus, the temporal habitat sequence (corn planted yearly) that satisfies the phenological requirements of each species individually results in interspecific competition where cohabitation occurs. Conversely, the annual rotation of corn, while catastrophic for the WCR, favors those NCR that possess the trait for prolonged diapause. In some locales under annual rotation, the NCR survives well enough to be a consequential pest (Foster, 1987).

The geographic extent of the prolonged diapause trait in the NCR is unknown, but, if the damage detected in corn that is rotated is a fair indicator, it occurs in portions of Iowa, Nebraska, South Dakota, and Minnesota, as shown in Figure 18.3. Nothing is known of the genetic or environmental contributions to the expression of extended diapause. The fact that extended diapause was expressed in 40–47% of the eggs from fields with annual rotation but only 9% of the eggs from a field with a history of continuous corn suggests that there may be additive genetic variance for this trait. Direct evidence has not appeared, but the NCR is highly variable geographically in color (Krysan et al., 1983), isozymes (McDonald et al., 1985), and insecticide resistance (Hamilton, 1965, Patel and Apple, 1966; Krysan and Sutter, 1986), so duration of diapause might also vary geographically.

EVOLUTIONARY INTERPRETATION

Recent advances in the biosystematics of *Diabrotica* bear on the history and origin of the NCR as a pest of corn. Webster (1913) had suggested (discussed earlier) that the NCR predated corn in the north central United States and did not feed on corn until humans planted the crop continuously in large areas of the upper Midwest. The theory is supported by recent systematic (Krysan et al., 1983; Krysan and Smith, 1987) phylogenetic (Krysan et al., 1989), behavioral (Yaro et al., 1987), and paleontological (Schwert and Morgan, 1980) studies that place the NCR together with *D. cristata* (Harris) and *D. longicornis* (Say) in a uniquely Nearctic species complex clearly present in the north central and north eastern United States long before native Americans introduced corn to the area. Traits that the NCR shares with *D. cristata* are especially revealing. As adults the two species are unique among *Diabrotica* in that they feed on the flowers of a variety of prairie forbs but tend not to feed on cucurbits (Wiesenborn and Krysan, 1980). Further, they share attraction by certain cyclic propanoids (Yaro et al., 1987). Thus, the precorn habitat of the NCR might have been similar to that of *D. cristata*. Larvae of *D. cristata* feed on roots of big blue stem (*Andropogon gerardi*) (Yaro and Krysan, 1986), a perennial grass. A perennial larval host should provide critical ovipositional cues to a mobile adult whose larval offspring feed the following year. Corn is an annual, but it becomes an ecological perennial when it is planted each year on the same ground. Perhaps the NCR larvae fed on the roots of a perennial grass and humans satisfied that critical life-history need by planting corn continuously. Once a "race" became adapted to corn, it survived at nonpest levels, despite crop rotation during the first half of the twentieth century.

There is no information on the putative original stock from which the NCR corn host race was formed. The complex geographic variation in color (Krysan et al., 1983) and isozymes (McDonald et al., 1985) observed in present-day NCR is clinal in nature and bears no obvious evidence of races; however, all material studied was collected in or near cornfields. No systematic search has been made for a relict race of the NCR. Given the present ubiquity of corn in the range of the NCR, that could be a difficult search.

The entomological community often perceives diapause as a mechanism of adaptation to seasonal cycles in climate, whereas some specialists have emphasized the seasonal availability of food as the primary cyclic event driving the evolution of dormancy mechanisms. The evolution of semivoltinism in the NCR in response to annual rotations of corn is a compelling example of the role of food availability.

FUTURE DIRECTIONS

The mechanism of seasonal adaptation of the NCR to continuous and annual rotations of its host bear ecophysiological and, surely, genetic components that merit dissection by coordinated study. Among the ecophysiological parameters, moisture availability, a key parameter for the WCR (Krysan, 1978; Branson et al., 1982) can probably be ignored for the NCR because of the relatively high level of moisture available in the soils across its range. The effect of temperature on the rate of diapause development should be established. The existing data (Krysan, 1982; Krysan et al., 1984) suggest that the diapause that imposes semivoltinism requires for termination two chill periods separated by a warm interval. A controlled test of this hypothesis would be most informative. Once that information is at hand, additive genetic variance can be definitively measured.

Once the genetic components are defined, studies on short-range geographic variation would be appropriate. There is ample evidence of geographic variation over broad areas for prolonged diapause and other traits (Krysan et al., 1983, 1986; Krysan and Sutter, 1986; McDonald et al., 1985) and some evidence of localized geographic variation (Patel and Apple, 1966). It is not impossible that NCR populations adapt to the cropping practices of individual farmers.

At another level, studies could address the adaptation for, and competition between, the NCR and WCR under the two primary rotational schemes in field cage experiments. The system seems ideal for basic studies of the adaptive significance of host and life-history trait differences as expressed in two such closely related species.

Acknowledgments

I wish to thank G. G. Kennedy, North Carolina State University; Dave Lance, USDA-ARS, Brookings, South Dakota; and Tom Unruh, USDA, ARS, Yakima, Washington for critical review of the manuscript; and Dave Lance for permission to cite from an unpublished manuscript.

REFERENCES

Anonymous. 1976. *U.S. Dep. Agric. Coop. Plant Pest Rep.* 1:690–693.

Ball, H. J., and G. T. Weekman. 1962. Insecticide resistance in the western corn rootworm in Nebraska. *J. Econ. Entomol.* 55:439–441.

Bigger, J. H. 1932. Short rotation fails to prevent attack of *Diabrotica longicornis* Say. *J. Econ. Entomol.* 25:196–199.

Branson, T. F., and J. L. Krysan. 1981. Feeding and oviposition behavior and life cycle strategies of *Diabrotica*: An evolutionary view with implications for pest management. *Environ. Entomol.* 10:826–831.

Branson, T. F., J. Reyes Rueda, and H. Valdes Martinez. 1982. Field biology of Mexican corn rootworm, *Diabrotica virgifera zeae* (Coleoptera: Chrysomelidae), in Central Mexico. *Environ. Entomol.* 11:1078–1083.

Bryson, H. R., D. A. Wilbur, and C. C. Burkhardt. 1953. The western corn rootworm, *Diabrotica virgifera* LeConte in Kansas. *J. Econ. Entomol.* 46:995–999.

Chiang, H. C. 1965. Survival of northern corn rootworm eggs through one and two winters. *J. Econ. Entomol.* 58:470–472.

———. 1973. Bionomics of the northern and western corn rootworms. *Annu. Rev. Entomol.* 18:47–72.

Chiang, H. C., J. A. Mihm, and M. B. Windels. 1972. Temperature effects on hatching of eggs of western and northern corn rootworms. *Proc. North Cent. Branch Entomol. Soc. Am.* 27:127–131.

Cinereski, J. E., and H. C. Chiang. 1968. The pattern of movements of adults of the northern corn rootworm inside and outside of corn fields. *J. Econ. Entomol.* 61:1531–1536.

Forbes, S. A. 1882. The corn root worm. *Annu. Rep. Ill. State Entomol.* 12:10–30.

———. 1896. Insect injuries to the seed and root of Indian corn. *Ill. Agric. Exp. Stn. Bull.* 44:282–296.

Foster, D. E. 1987. Extended diapause of northern corn rootworm. *Crops, Soils and Pests Newsl., Iowa State Univ.* IC-454: p. 207.

George, B. W., and E. E. Ortman. 1965. Rearing the western corn rootworm in the laboratory. *J. Econ. Entomol.* 58:375–377.

Gillette, C. P. 1912. *Diabrotica virgifera* Lec. as a corn root-worm. *J. Econ. Entomol.* 5:364–366.

Gould, G. E. 1971. The corn rootworm problem in Indiana. *Proc. Indiana Acad. Sci.* 80:267–273.

Guss, P. L., T. F. Branson, and J. L. Krysan. 1976. Adaptation of a dry diet for adults of the western corn rootworm. *J. Econ. Entomol.* 69:503–505.

Gustin, R. D. 1984. Effect of crop cover on oviposition of the northern corn rootworm, *Diabrotica longicornis barberi* Smith and Lawrence. *J. Kans. Entomol. Soc.* 57:515–516.

Hamilton, E. W. 1965. Aldrin resistance in corn rootworm beetles. *J. Econ. Entomol* 58:296–300.

Hein, G. L., and J. J. Tollefson. 1985. Seasonal oviposition of northern and western corn rootworms (Coleoptera: Chrysomelidae) in continuous cornfields. *J. Econ. Entomol.* 78:1238–1241.

Herzog, D. C., and J. E. Funderbunk. 1986. Ecological basis for habitat management and pest control. Pages 217–250 in Kogan, M. (ed.), *Ecological Theory and Integrated Pest Management.* New York: John Wiley & Sons.

Hill, R. E. 1967. Changes in distribution and dominance of rootworm species in Nebraska, 1948–1966. *Proc. North Cent. Branch Entomol. Soc. Am.* 22:163–164.

Hill, R. E., and Z. B. Mayo. 1980. Distribution and abundance of corn rootworm species as influenced by topography and crop rotation in eastern Nebraska. *Environ. Entomol.* 9:122–127.

Hintz, A. M., and B. W. George. 1979. Successful laboratory hybridization of *Diabrotica virgifera* (western corn rootworm) and *Diabrotica longicornis* (northern corn rootworm) (Coleoptera: Chrysomelidae). *J. Kans. Entomol. Soc.* 52:324–330.

Johnson, T. B., and F. T. Turpin. 1985. Northern and western corn rootworm (Coleoptera: Chrysomelidae) oviposition in corn as influenced by foxtail populations and tillage systems. *J. Econ. Entomol.* 78:57–60.

Krysan, J. L. 1972. Embryonic stage of *Diabrotica virgifera* (Coleoptera: Chrysomelidae) at diapause. *Ann. Entomol. Soc. Am.* 65:768–769.

———. 1978. Diapause, quiescence, and moisture in the egg of the western corn rootworm, *Diabrotica virgifera. J. Insect Physiol.* 24:535–540.

———. 1982. Diapause in the Nearctic species of the *virgifera* group of *Diabrotica*: Evidence for tropical origin and temperate adaptations. *Ann. Entomol. Soc. Am.* 75:136–142.

Krysan, J. L., and P. L. Guss. 1978. Barriers to hybridization between *Diabrotica virgifera* and *D. longicornis barberi* (Coleoptera: Chrysomelidae). *Ann. Entomol. Soc. Am.* 71:931–934.

Krysan, J. L., and R. F. Smith. 1987. Systematics of the *virgifera* species group of *Diabrotica*. *Entomography* 5:375–484.

Krysan, J. L., and G. R. Sutter. 1986. Aldrin susceptibility as an indicator of geographic variability in the northern corn rootworm, *Diabrotica barberi* (Coleoptera: Chrysomelidae). *Environ. Entomol.* 15:427–430.

Krysan, J. L., T. F. Branson, and G. Diaz Castro. 1977. Diapause in *Diabrotica virgifera* (Coleoptera: Chrysomelidae): A comparison of eggs from temperate and subtropical climates. *Entomol. Exp. Appl.* 22:81–89.

Krysan, J. L., D. E. Foster, T. F. Branson, K. R. Ostlie, and W. S. Cranshaw. 1986. Two years before the hatch: Rootworms adapt to crop rotation. *Bull. Entomol. Soc. Am.* 32:250–253.

Krysan, J. L., J. J. Jackson, and A. C. Lew. 1984. Field termination of egg diapause in *Diabrotica* with new evidence of prolonged diapause in *D. barberi* (Coleoptera: Chrysomelidae). *Environ. Entomol.* 13:1237–1240.

Krysan, J. L., I. C. McDonald, and J. H. Tumlinson. 1989 Phenogram based on allozymes and its relationship to classical biosystematics and pheromone structure among eleven Diabroticites (Coleoptera: Chrycomelidaos). *Ann. Entomol. Soc. Am.* 82:574–581

Krysan, J. L., R. F. Smith, and P. L. Guss. 1983. *Diabrotica barberi* (Coleoptera: Chrysomelidae) elevated to species rank based on behavior, habitat choice, morphometrics, and geographical variation of color. *Ann. Entomol. Soc.* Am. 76:197–204.

Lance, D. R., 1990. Personal communication.

Lance, D. R., and J. R. Fisher. 1987. Food quality of various plant tissues for adults of the northern corn rootworm (Coleoptera: Chrysomelidae). *J. Kans. Entomol. Soc.* 60:462–466.

Lance, D. R., N. C. Elliott, and G. L. Hein. 1989. Flight of *Diabrotica* spp. at the borders of cornfields and its relationship to ovarian stage in *D. barberi*. *Entomol. Exp. Appl.* 50:61–67.

LeConte, J. L. 1868. New Coleoptera collected on the survey for the extension of the Union Pacific Railway, E.D. from Kansas to Fort Craig, New Mexico. *Trans. Am. Entomol. Soc.* 2:49–59.

Luckmann, W. H., H. C. Chiang, E. E. Ortman, and M. P. Nichols. 1974. A bibliography of the northern corn rootworm, *Diabrotica longicornis* (Say), and the western corn rootworm, *Diabrotica virgifera* LeConte (Coleoptera: Chrysomelidae). *Ill. Nat. Hist. Surv. Biol. Notes* No. 90.

Ludwig, K. A., and R. E. Hill. 1975. Comparison of gut contents of adult western and northern corn rootworms in northeast Nebraska. *Environ. Entomol.* 4:435–438.

McDonald, I. C., J. L. Krysan, and O. A. Johnson. 1985. Genetic variation within and among geographic populations of *Diabrotica barberi* (Coleoptera: Chrysomelidae). *Ann. Entomol. Soc. Am.* 78:271–278.

Metcalf, R. L. 1982. Insecticides in pest management. Pages 217–277 in Metcalf, R. L., and W. H. Luckmann (eds.), *Introduction to Insect Pest Management*. 2nd ed. New York: Wiley-Interscience.

Muma, M. S., R. E. Hill, and E. Hixson. 1949. Soil treatments for corn rootworm control. *J. Econ. Entomol.* 42:822–824.

Naranjo, S. E., and A. J. Sawyer. 1987. Reproductive biology and survival of *Diabrotica barberi* (Coleoptera: Chrysomelidae): Effect of temperature, food, and seasonal time of emergence. *Ann. Entomol. Soc. Am.* 80:841–848.

Ortman, E. E., and P. J. Fitzgerald. 1964. Developments in corn rootworm research. *Proc. 19th Ann. Hybrid Corn Ind. Res. Conf.* 19:38–45.

Ostlie, K. R. 1987. Extended diapause: Northern corn rootworm adapts to crop rotation. *Crops and Soils Magazine*, June–July, pp. 23–25.

Patel, K. K., and J. W. Apple. 1966. Chlorinated hydrocarbon resistant northern corn rootworms in Wisconsin. *J. Econ. Entomol.* 59:522–525.

———. 1967. Ecological studies on the eggs of northern corn rootworms. *J. Econ. Entomol.* 60:496–500.

Piedrahita, O., C. R. Ellis, and J. P. Bogart. 1985. Interaction of northern and western corn rootworm larvae (Coleoptera: Chrysomelidae) in a controlled environment. *Environ. Entomol.* 14:138–141.

Schwert, D. P., and A. V. Morgan. 1980. Paleoenvironmental implications of a late glacial insect assemblage from northwestern New York. *Quat. Res.* (N.Y.) 13:93–110.

Shaw, J. T., J. H. Paullus, and W. H. Luckmann. 1978. Corn rootworm oviposition in soybeans. *J. Econ. Entomol.* 71:189–191.

Tate, H. D., and O. S. Bare. 1944. Corn rootworms. *Nebr. Agric. Exp. Stn. 58th Annu. Rep.* 58–59.

———. 1946. Corn rootworms. *Univ. Nebr. Agric. Exp. Stn. Bull.* 381, 12 pp.

Webster, F. M. 1895. On the probable origin, development and diffusion of North American species of the genus *Diabrotica*. *J. N. Y. Entomol. Soc.* 3:158–166.

———. The western corn rootworm. *U.S. Dep. Agric. Bur. Entomol. Bull.* 8:1–8.

Wiesenborn, W. D., and J. L. Krysan. 1980. A survey for *Diabrotica cristata* (Coleoptera: Chrysomelidae) on relict prairies of eastern South Dakota and southwestern Minnesota. *Proc. S. D. Acad. Sci.* 59:130–137.

Yaro, N. D., and J. L. Krysan. 1986. Host relationships of *Diabrotica cristata* (Coleoptera: Chrysomelidae). *Entomol. News* 97:11–16.

Yaro, N. D., J. L. Krysan, and T. L. Ladd, Jr. 1987. *Diabrotica cristata* (Coleoptera: Chrysomelidae): Attraction to eugenol and related compounds compared with *D. barberi* and *D. virgifera*. *Environ. Entomol.* 16:126–128.

19

INFLUENCE OF HABITAT ON THE PEST STATUS AND MANAGEMENT OF *HELIOTHIS** SPECIES ON COTTON IN THE SOUTHERN UNITED STATES

J. R. Bradley, Jr.

North Carolina State University

INTRODUCTION

Heliothis species are pests or potential pests of cotton wherever the crop is grown in the United States. The bollworm, *Heliothis zea* (Boddie), also commonly named corn earworm, tomato fruitworm, and soybean podworm because of its wide host range, was first recognized as a pest of cotton in 1820. It had become a prominent pest of cotton in some areas of the Cotton Belt by 1841 (Quaintance and Brues, 1905). The tobacco budworm, *Heliothis virescens* (F.), is the other *Heliothis* pest of cotton in the United States. However, it more recently attained pest status, since it was first reported as damaging cotton in Louisiana in 1934 (Folsum, 1936) and was not a pest of great concern until the late 1940s following the introduction of organic insecticides (Brazzel et al., 1953). These two *Heliothis* species comprise what is commonly termed the "bollworm complex," since larvae of both species often occur together, feeding symptoms on cotton are inseparable, and the immature stages of the species are indistinguishable with the unaided eye.

The "bollworm complex" requires more expenditures for its control and causes more crop damage, despite control attempts, than any other insect pest or related pest group on cotton. The average annual losses due to crop damage and costs of control were estimated to exceed $1 billion (Knipling and Stadelbacher, 1983). *H. zea* is the more consistent pest over the Cotton Belt, as it attains economically dam-

*Author's note: *Heliothis zea* has been changed to *Helicoverpa zea* since preparation of this manuscript.

aging population levels in most regions annually. In contrast, *H. virescens* occurs more sporadically, but during certain years, or in specific regions of the Cotton Belt (e.g., lower Rio Grande Valley), it can be a devastating pest of cotton. *H. virescens* is more difficult to manage with insecticides, as it is inherently more tolerant and has historically developed insecticide resistant strains much more rapidly than has *H. zea*.

Both species of bollworms are polyphagous and utilize cultivated and wild hosts as food resources; however, *H. zea* has a much broader host range among cultivated plants. Neunzig (1969) reported that only tobacco and cotton were important cultivated hosts of *H. virescens* in North Carolina, while *H. zea* commonly utilized 16 crop hosts. As with earlier studies (Quaintance and Brues, 1905; Isely, 1935; Barber, 1937), Neunzig observed corn to be preferred by the bollworm and tobacco by the tobacco budworm. All studies agree on these host preferences and the concept of a wider host range in *H. zea*.

Local adult movement and oviposition patterns of *Heliothis* species are determined largely by host preferences and availability. Generalized models regarding host preferences may not be representative of specific situations because host plant variety, fitness of host plant, and phenological stage of growth are factors affecting the preference index (Johnson et al., 1975; Adkisson, 1958; Dicke, 1939; Freeman et al., 1967; Isely, 1935; Stadelbacher and Scales, 1973; Thomas and Dunman, 1931; Parsons, 1939, 1940). Subtle differences in any of the these host characteristics may result in wholesale shifts in egg distribution and larval population patterns among various plant communities (wild and cultivated) available to the mobile adult *Heliothis*.

Habitat may be defined as the native environment or region in which an animal or plant lives; agroecosystems may encompass portions of the habitats of many organisms and thus affect their population dynamics. The spatially and temporally arranged host plants constitute the primary components of the *Heliothis* species habitat. Other important components of these habitats are physical features (e.g., soil type), other organisms with which *Heliothis* interact, and the conditioning effects of weather. Modern farmers mold agroecosystems and thereby structure portions of the *Heliothis* habitat to meet crop production goals with very little foresight into the potential for insect management.

This chapter illustrates how habitat has influenced the pest status and management of *Heliothis* species on cotton over the past 100 years or more that *Heliothis* species have been known as pests of cotton. For reasons of clarity, the traditional Cotton Belt extending from North Carolina southwestward through Texas will be the stage. The irrigated desert valley cotton production areas of the Southwest represent relatively recent production systems and are so dissimilar structurally that it is more appropriate to discuss them in detail elsewhere.

EARLY HISTORY OF BOLLWORM PROBLEMS ON COTTON

The southeastern U.S. Coastal Plain extends from North Carolina southwestward through Alabama. Cotton was king throughout this region from the late 1700s until the U.S. Civil War. Following the Civil War, a tide of immigrants began to move westward from the more eastern cotton production states to farming regions west of the Mississippi River. Many were cotton farmers, and wherever they moved they planted cotton to the exclusion of other crops and neglected the crop diversification they had practiced in the East. Plantations and farms of large size were the rule, and the tenant system found its maximum development there. The large areas planted with cotton as compared to other crops and the naturally high soil fertility, which produced rank, succulent cotton plants, were important habitat factors that resulted in more frequent losses to bollworm.

At the turn of the century, Quaintance and Brues (1905) presented an excellent description of the bollworm problem across the U.S. Cotton Belt. At that time, bollworms were particularly destructive from Alabama westward through Texas, but, from Alabama eastward through North Carolina, bollworms were of no economic concern. This difference in bollworm pest status between the two regions was thought to have been due entirely to the influence of habitat on population levels. The particular structural changes in the western cotton agroecosystems that Quaintance and Brues (1905) thought contributed to increased bollworm populations in the latter part of the nineteenth century were the following:

1. Larger farm units composed of the majority of the cultivated acres being planted with cotton in the western areas in contrast to smaller, more diverse farms consisting of multicrop production systems in the east.
2. Less acreage planted to preferred, alternate hosts of *Heliothis* (e.g., peas), which served as trap crops and directly reduced bollworm populations on cotton. The ratio of cotton to peas in North Carolina and South Carolina was 11:1 and 13:1, respectively, in 1899, whereas the cotton to pea ratio for Louisiana and Texas that year was 91:1 and 205:1, respectively.
3. Corn grown west of the Mississippi River was planted earlier and matured earlier, thereby acting as a nursery crop that released large numbers of *H. zea* moths that immigrated to nearby succulent cotton fields to initiate the next generation. On the other hand, the general practice of planting late corn for forage and silage in the East resulted in corn acting as a trap crop that protected cotton from bollworm attack all season long. Farmers in the Carolinas, Georgia, Florida, and Alabama unconsciously managed bollworms on cotton by providing a succession of hosts that were more preferred than cotton throughout the period of time that cotton was susceptible to bollworm attack.
4. The inherent soil fertility in the river valleys of Mississippi, Louisiana, Arkansas, and Texas was conducive to the growth of large, succulent cotton plants that characteristically matured late and supported high numbers of bollworms during late season when no other hosts were available. Conversely, the eastern states were characterized by soils with low fertility that produced a smaller cotton plant that "cut out" early and was a very poor late-season host for bollworms.

The previous examples of how habitat differences affected *Heliothis* population levels and damage potential on cotton provided the basis for insect pest management nearly a century ago. Early entomologists could observe *Heliothis* populations as they were naturally distributed and as they developed on available hosts without interference and the environmental imbalances associated with present-day insecticides. They developed a knowledge base through years of observation and trial and error that provided a clear insight into the ways in which habitat influenced spatial and temporal distribution of *Heliothis* species within the agroecosystem. They observed that insects had definite preferences for certain crop hosts and that certain phenological stages of crops (e.g., silking corn) were particularly preferred. The concepts of host preference and stage specific preference were recognized, though perhaps not understood, by the beginning of the twentieth century. Entomologists during the early part of this century utilized these principles as the basis for managing the bollworm complex on cotton. Their strategy was to make the cotton crop less attractive to insect pests than surrounding host crops and to mature the crop as early as possible, in order to escape late-season ravages. This strategy was effective where it was employed, and losses to bollworms in cotton were minimal, particularly in the diversified agroecosystems of the Southeast. However, in those years when weather prevented early production of the cotton crop, bollworms caused serious damage to the crop in late season, since there were no insecticides to employ in defense of the crop when biological and cultural tactics failed to keep bollworm populations below damaging levels.

Cotton was produced in the southeastern United States and in eastern and southern Texas until the cotton boll weevil (*Anthonomus grandis grandis* Boheman) completed its eastward spread in the 1920s. The components of agroecosystem structure (e.g., small field sizes, crop diversity) that had prevented bollworms from reaching their pest potential in the Southeast favored the boll weevil. Small fields surrounded by ideal overwintering habitat resulted in maximum weevil production, and crop losses were disastrous (Hunter and Coad, 1923; Gaines, 1957). Between World War I and World War II, cotton production dramatically shifted from the rain grown areas of the Southeast to the weevil-free areas of the High Plains of Texas and the irrigated valleys of the Southwest. Only in the mid-South along the lower Mississippi River did cotton retain its previous status. Since many of the soil types of the Southeast Coastal Plain possessed poor structure, and low water-holding capacity, and were highly eroded from generations of continuous farming, cotton production potential was less there than in other parts of the Cotton Belt. To further confound productivity problems, soil heterogeneity, slope, and drainage created field sizes that were not compatable with the mechanization of modernized agriculture. Once the boll weevil became firmly entrenched and the mule was being replaced by the tractor, it was inevitable that much of the area would revert to forests and more diversified agroecosystems. What had formerly been wall-to-wall farming became a mosaic of forests interspersed with a variety of crops, such as corn, tobacco, peanut, sorghum, and soybean, as well as cotton.

For the most part, the bollworm complex was very effectively managed on cotton in the Southeast largely through habitat modifications (i.e., cultural controls) from the time of the boll weevil invasion until the organic insecticide era. The central strategy of entomologists was avoidance of the problem through employing corn and other hosts (e.g., peas) as trap crops, combined with early maturity of cotton (Isely, 1935; Lincoln and Isely, 1947; Thomas and Dunnam, 1931). However, entomologists (Isely, 1935) warned that corn would protect cotton through the trap crop phenomenon only if corn matured late; early maturing corn would act as a nursery crop for late-season infestations on cotton. Lincoln and Isely (1947) discussed five outbreaks of bollworms in Arkansas during the period 1925–1940 when no insecticidal controls were available. Factors correlated to these outbreaks were

1. An acreage of early corn greater than the cotton acreage
2. Very little late-planted corn to act as a late-season trap crop
3. Dry weather, which shortened the silking period of corn and resulted in a shorter than normal trap crop period

Johnson et al. (1975) confirmed the observations of early investigators who reported that the oviposition preference of *H. zea* for corn at silking was 5.5 times greater than that of cotton at flowering. After flowering, the preference decreased with maturity of the plant, regardless of plant species.

The first half of the twentieth century passed with the bollworm complex being considered relatively minor as compared to the boll weevil, largely because of habitat modifications that minimized the frequency and severity of bollworm outbreaks. However, the widespread use of the organic insecticides of the future would dramatically change this relationship.

ORGANIC INSECTICIDE ERA

The new organic insecticides impacted cotton production in a manner unprecedented by any other event in history. For the first time, control of cotton insect pests, including the boll weevil, could be guaranteed. In fact, these new insecticides were so effective that certain insect pests [e.g., cotton leafworm, *Alabama argillacea* (Hübner)] lost their pest status overnight (Newsom and Brazzel, 1968). It appeared that the final solution to insect pest problems in cotton was at hand, and many people thought that these pesticide tools would eradicate all major insect pests. Production stability and higher cotton yields seemed a certainty.

There was wholesale adoption of yield maximization strategies by all segments of the agricultural community with little or no concern for the impacts upon arthropod pests. Habitat manipulations (e.g., cultural controls) that had been employed successfully to lower the mean level of abundance of many arthropod pests, including bollworms, were completely abandoned in quest of increased yields of quality cotton (Bottrell and Adkisson, 1977). Doutt and Smith (1971) described this period when new organic insecticides and other crop production inputs were being used extensively in quest of extremely high yields of quality cotton as the exploitation phase of cotton production.

A protective blanket of effective, inexpensive insecticides allowed cotton breeders to develop high-yielding varieties that would produce fruit as long as the season allowed and would positively respond to irrigation and increased amounts of fertilizers. This surge of new technology produced the desired result; long-season, maximum-input production systems allowed the cotton plant to more completely express genetic yield potential and to produce longer-staple lint that was highly sought by textile mills (Walker and Niles, 1971).

The "honeymoon" period of insect control with the organic insecticides lasted less than a decade, as populations of the boll weevil were found to be resistant to these insecticides in 1955 (Roussel and Clower, 1955). This was quickly followed by the development of chlorinated hydrocarbon resistant strains of tobacco budworm and cotton bollworm (Brazzel, 1963, 1964; Graves et al., 1963, 1967; Graves and Clower, 1975). Within 15 years from the introduction of organic insecticides, resistance affected all major pest species in some or all of their areas of distribution (Newsom and Brazzel, 1968). The immediate response to resistance was to seek a chemical solution to the problem by tank mixing insecticides (e.g., toxaphene-DDT) or by changing to the alternative class of insecticides (e.g., organophosphates), but these proved to be short-term solutions (Canerday, 1974; Harris et al., 1972; Wolfenbarger et al., 1973).

The development of resistant strains of all important arthropod pests was not the only major problem during the early years of the organic insecticide era. The elevation of secondary and potential pests to the status of major importance, the resurgence of treated populations, the adverse effects on nontarget organisms, and the pollution of the environment with toxic, persistent chemicals posed increasingly difficult problems. These pesticide-induced problems have been exhaustively discussed by other authors (van den Bosch et al., 1956; Ripper, 1956; Newsom, 1967; Cope, 1971; Newsom, 1974; Croft and Brown, 1975; Bradley et al., 1987) and will receive no further elaboration here. Nevertheless, the point is made that these problems were created by total reliance on insecticides for insect pest control on cotton for approximately two decades. The proven techniques of cultural control of cotton insects through habitat manipulation had been completely abandoned in favor of production systems that primarily sought to maximize yields without regard to environmental costs. Among the costs of this thrust were the transformation of the cotton bollworm from an occasional to a key pest over sizable portions of the Cotton Belt and the elevatation of the tobacco budworm, an insect not previously considered a noteworthy pest of cotton, to major pest status. *Heliothis* species had become the most destructive pests of cotton during the initial 20 years of the organic pesticide era for the following reasons:

1. Frequent and widespread use of synthetic organic insecticides
   a. Destroyed natural enemies of target and nontarget pests, resulting in resurgence and secondary pest outbreaks
   b. Often delayed cotton fruit set and maturity, thereby increasing late-season susceptibility of the crop to insect pests
2. Changes in cotton production practices that effected a better *Heliothis* habitat included
   a. Improved cotton varieties that required a longer growing season

   b. Increased use of nitrogen fertilizers that prolonged the growing season
   c. The change from cultural weed control to herbicides, which often delayed cotton maturity
   d. Adoption of irrigation, resulting in delayed crop maturity
   e. Increased field sizes and associated reduction in refugia for natural enemies
3. Some changes in alternate host production that decreased their trap crop effects were
   a. The reduction in agroecosystem diversity and shift toward monoculture through reduced planting of alternate crop hosts of *Heliothis* species (e.g., corn, alfalfa, peas)
   b. Improved corn hybrids that could be planted earlier and would mature earlier, with the net result being reduced trap crop effect for corn earworm.
   c. Earlier maturity and more thorough postharvest destruction of tobacco, reducing its value as a trap crop for tobacco budworm

It was painfully apparent to most cotton entomologists by the 1960s that a new strategy had to be employed against cotton insect pests; the problems spawned by total reliance on chemical control for two decades were too serious to ignore any longer. No single tactic (e.g., insecticides) could be expected to provide long-term management of the bollworm complex, particularly under the constraint of habitat manipulations that favored maximum production of insect pests.

INTEGRATED PEST MANAGEMENT ERA

The concept of integrated pest management (IPM) was conceptualized out of the failures of the chemical control era. In IPM, various combinations of methods are utilized in a compatible manner to obtain the best control with the least disruption of the environment (Flint and van den Bosch, 1981). A prerequisite for successful implementation of the IPM concept is more specific knowledge of the system components and a clearer understanding of how these components interact, as well as how they are conditioned by the environment. Often the first line of defense in IPM involves habitat modifications that are designed to lower the mean level of pest abundance, just the opposite of what had occurred during the heyday of the organic insecticide era. This may be achieved for a single-crop host over one or more generations, or it may lead to the overall reduction in the seasonal pest population level. Obviously, lowering the mean level of pest abundance results in thresholds being met less frequently and management costs being reduced.

Texas Production Systems

The state of Texas encompasses a very large, ecologically diverse land mass in which approximately 50% of the U.S. cotton crop is produced. Cotton is grown in seven distinct regions of Texas, each of which represents unique ecological areas that require bollworm management systems tailored to meet local needs. Some regions (e.g., Texas Rolling Plains, Texas Blackland Prairies) have a very low frequency of bollworm outbreaks, because they are dry-land, low-input production systems. In contrast, production systems of the Texas river valleys (e.g., Texas Central River Valleys, Texas Lower Rio Grande Valley) could be generally characterized as high-input yield maximization systems prior to the IPM era. These high-input systems have been subject to intensive insecticide programs for *Heliothis* and have experienced the inherent problem of insecticide resistance.

Texas led the way in the development of IPM systems structured upon insect pest habitat changes achieved through modification of the cotton production systems. In some regions of Texas (e.g., Lower Rio Grande Valley), the levels of insecticide resistance had made cotton production unprofitable by the mid-1960s, so the industry was forced to seek alternative management systems.

The thrust of the Texas program was based upon a system of "short-season cotton production" that resulted in crop maturity in 135–140 days, as compared with the 150- to 160-day long-season concept of the 1950s and 1960s (Walker et al., 1979; Heilman et al., 1977; Namken and Heilman, 1973). The system was modified to meet the specific needs of each of the major cotton production regions in Texas, but the common objective was to escape late-season ravages of bollworms by producing a cotton plant that was less attractive to ovipositing bollworm moths and less favorable for larval survival at key times during the season (Rummel et al., 1986). The concept was made possible through the development of determinate, rapid-fruiting cotton varieties that had acceptable fiber properties (Bird and Percey, 1979). To enhance the benefits from the short-season system, the use of nitrogen fertilizer was minimized, and, in some irrigated crop production areas, water management was a tactic used to avert *Heliothis* damage. Dry-land cotton fields support much lower bollworm populations than irrigated fields; for example, insecticide use for dry-land and irrigated cotton fields in the Lower Rio Grande Valley averaged 3 and 11 applications per season, respectively (Norman et al., 1979). Nevertheless, where irrigation is utilized, the *Heliothis* hazard can be effectively reduced by not irrigating during the 7- to 10-day period prior to peak oviposition (Slosser, 1980). Through this short-season, low-input approach, the cotton field may be transformed into a habitat of minimal acceptance and utilization by *Heliothis* species, whereas the systems that seek maximum yields are characterized by cotton fields that reach their bollworm production potential, unless copious amounts of insecticides are applied to protect them.

The agroecosystem structure in most Texas cotton production regions is simple with respect to *Heliothis* species and host crop diversity, as alternate crop hosts have only minor effects on *Heliothis* populations on cotton. A notable exception to this generalization occurs in the Texas High Plains, where *Heliothis* species were not regarded as important pests until corn became a major crop in the mid-1970s and thus provided an early season host for *H. zea*. Today, the Texas High Plains agroecosystem shares more host crop similarities with the Southeast than the other Texas cotton production regions, since corn acts as a trap crop for *H. zea* during early season and is a nursery crop for later season bollworm problems on cotton. The *H.*

zea infestations in Texas High Plains cotton are frequently of an acute nature, characterized by a sudden onset, a relatively short duration, and a rapid decline. Damaging infestations are typically late season in occurrence, seldom appearing until corn has matured in late July or early August, which is very similar to the pattern of host crop utilization by *H. zea* in North Carolina (Bacheler, 1985). The use and proper management of well-adapted, rapid-fruiting, early-maturing cotton varieties remain the best means of reducing damage during the late-season *Heliothis* attacks. The risk of *Heliothis* damage is less for cotton plants that have achieved good boll set and that have reduced fruiting by early August. Such plants are considered less attractive as oviposition sites, and the survival of newly hatched larvae appears to be greatly reduced (Rummel et al., 1986).

The Texas High Plains provides an excellent example of how recent habitat alterations have changed the pest status and management of *Heliothis* on cotton. Humans created a problem where there had not been one previously through the introduction of a preferred host crop, corn, into the system. To minimize the economic impact of this habitat change, the cotton field habitat had to undergo dramatic alterations associated with the adoption of a short-season production system.

Mississippi River Delta Systems

Cotton production in the Mississippi River Delta region occupies lands adjacent to the Mississippi River and its major feeder streams from southern Missouri through portions of Tennessee, Arkansas, Mississippi, and Louisiana. The delta region is the relatively flat land associated with the basin of the Mississippi River. Most of the area is not truly a geographic delta, but the label has been traditionally used to describe the large, contiguous region of row crop production contained within the river's floodplain. This fertile floodplain is very conducive to large-scale mechanized agriculture, and, as a result, row crops are produced on about 80% of the land area. Cotton, soybean, rice, and wheat are the major crops produced in the region and as a group comprise more than 90% of the total crop acreage (Luttrell et al., 1986).

The adaptability of insects to habitat alterations incurred through changes in host availability may be demonstrated by reviewing the changing pattern of corn production and its effect on *H. zea* in the delta over the past half-century. According to Isley (1935) and Lincoln (1972), the seasonal abundance of *H. zea* is strongly influenced by the availability and sequence of favorable hosts. Isley (1935) considered a one-crop system as being unfavorable to *H. zea* and that a combination of legumes for the first generation and corn for the second generation was a necessary combination for developing *H. zea* outbreaks in cotton. Lincoln et al. (1967) reported that corn acreage had progressed steadily downward in the delta since the mid-1940s. The drastic reduction in acreage planted with corn, which had long been regarded as the key host for successful buildup of *H. zea* populations, did not result in a decrease in the economic importance of *H. zea* as predicted. Instead, *H. zea* simply adapted to this habitat change and began utilizing cotton and soybean as major host crops for the second, as well as subsequent, generations. However, it is likely that, in absence of broad-spectrum insecticides, applied to cotton for control of plant bugs and boll weevil, indigenous predators and parasites would have much more effectively lowered *Heliothis* populations where they occurred on one crop (e.g., cotton) for several successive generations.

Wild hosts play a particularly important role in the population dynamics of *Heliothis* species in the delta. A few key species of wild hosts, primarily crimson clover and wild geranium, totally support the first generation of cotton bollworm and tobacco budworm throughout the region since crop hosts are unavailable during early spring, when the first generation develops in the area (Table 19.1). Entomologists in Mississippi and Arkansas have documented the dependence of the first larval generation of both *Heliothis* pest species on these wild host plants for survival and have related their development on wild hosts to subsequent generations on crop hosts (Mueller and Phillips, 1983; Mueller et al., 1984; Snow et al., 1966; Stadelbacher, 1979; Stadelbacher, 1981). The area serving as habitat for the early-season wild hosts in the delta has been estimated at 3.5% of the total rural area (Stadelbacher, 1982). Stadelbacher, (1982) and Knipling and Stadelbacher (1983) proposed that the most economical and ecologically sound approach for management of *Heliothis* in the delta would be to implement methods for the management of the early-season hosts and/or first larval generations while they were at the lowest numbers and were restricted to a relatively small area. A properly conducted, area-wide management system imposed on the non-crop habitat might lower the general equilibrium position of *Heliothis* species to the extent that control of subsequent generations on crop hosts would be unwarranted. Obviously, the success of this approach is dependent upon the extent of long-range migration of *Heliothis* moths from outside the delta (Mueller et al., 1984).

Table 19.1. Generalized Host Patterns of *Heliothis* Species in North Carolina and the Mississippi Delta

	Major Host Plants	
Generation	North Carolina	Mississippi Delta
F_1	Corn Tobacco Wild hosts	Wild hosts
F_2	Corn Tobacco	Cotton Soybean
F_3	Cotton Soybean Peanut Vegetables Tobacco	Cotton Soybean
F_4	Cotton Soybean Vegetables Tobacco Wild hosts	Cotton Soybean Wild hosts

Management of *Heliothis* species on cotton in the Mississippi Delta is made more difficult by the occurrence of multiple generations on the crop. Certainly, this lack of host diversity increases the probability of insecticide resistance occurring, particularly in tobacco budworm, which has a more limited host range than the cotton bollworm. In fact, strains of tobacco budworm from the Mississippi Delta have now been reported as resistant to the pyrethroids (Luttrell et al., 1987). Plapp and Campanhola (1986) also reported the development of resistance to the pyrethroids in tobacco budworms from the Brazos River Valley of Texas, which has a very similar *Heliothis* habitat to that of the Mississippi Delta. In both areas, tobacco budworm populations are almost exclusively associated with cotton over multiple generations.

The early-maturing, determinant-type cotton cultivars used successfully in Texas to control *Heliothis* species are not adapted to the Mississippi Delta. However, agronomic practices and management decisions related to tillage, cultivar selection, planting date, planting design, fertilization, application of plant growth regulators, irrigation, and crop residue destruction are habitat manipulations that have been demonstrated to be extremely important to *Heliothis* management in the delta (Luttrell et al., 1986). When these management tactics are properly applied, *Heliothis* species populations require minimum amounts of insecticide for their control in the Mississippi Delta.

Southeastern Coastal Plain Systems

In order to appreciate the influence of habitat on the pest status and management of *Heliothis* species on cotton in the Southeast, it is necessary to appreciate the great diversity of crop hosts that are produced within the region in contrast to the restrictive crop hosts available to *Heliothis* in other regions (e.g., Mississippi Delta). The Southeast may be characterized as a multiple-crop host habitat, whereas the Mississippi Delta is a two-crop system with respect to *Heliothis* (Table 19.2). Cotton is a minor component of the *Heliothis* crop host habitat in the Southeast, whereas it is a major component of the Mississippi Delta crop host habitat, through which passes a

Table 19.2. Hectareages for the Major Crop Hosts of *Heliothis* Species in the Southeast and Mississippi Delta Regions of the United States

	Crop (× 1,000)				
State	Corn	Cotton	Soybean	Tobacco	Peanut
North Carolina	720	39	740	109	63
South Carolina	196	42	680	19	6
Georgia	432	70	840	15	257
Alabama	180	124	560	0	88
Mississippi	48	418	1,320	0	0
Arkansas	22	188	1,620	0	0
Louisiana	38	260	1,000	0	0

Source: USDA Agricultural Statistics Board, National Agricultural Statistics Service, 1984.

significant proportion of the *Heliothis* biomass, particularly tobacco budworm. In contrast, corn is the single most important crop host of *H. zea* throughout the Southeast, whereas it is no longer an important factor in the population dynamics of *H. zea* in the Mississippi Delta.

The specific role that corn plays in the population dynamics of H. zea varies within the Southeast. For example, in North Carolina, where the preferred crop host, corn, stays green until mid-July, problems with *H. zea* do not usually occur on cotton until late July or early August. Conversely, in Georgia, which also has a large corn problem in cotton. In both areas, corn serves as a trap crop for first-generation *H. zea* moths, but is also a nursery crop for the development of a very large second-generation (in corn ears). This massive influx of second generation moths from corn into cotton and other crop hosts simply occurs earlier in Georgia.
into cotton and other crop hosts simply occurs earlier in Georgia.

Even within a particular state, crop host habitat variations have been observed to affect *Heliothis* populations on cotton. For example, corn is produced in all areas of North Carolina where cotton is grown, but the ratios of these host crops vary greatly. Relatively local differences in crop mix have been observed to affect population levels of *H. zea*; larval populations on cotton are usually higher in regions of the state where a higher proportion of crop habitat is comprised of corn.

The intercrop relationship between corn and cotton with respect to *H. zea* is a dynamic process that has been greatly influenced by the production and management practices used in field corn. Earlier in this chapter, the decline of corn as a major component of the *H. zea* habitat in the Mississippi Delta and the resultant impact on *H. zea* was documented. Because relative attractiveness varies with phenological state of all hosts, agricultural practices that alter the temporal relationship of crop growth stages may turn a trap crop into a nursery crop. Through the 1950s' full season, open-pollinated corn varieties acted as a trap crop for *H. zea* through early August, resulting in only minor bollworm problems in North Carolina cotton fields. The shift to early-season corn hybrids in the 1960s is undoubtedly responsible in large part for the relatively recent increase in the severity of *H. zea* problems in cotton; earlier maturity in corn reduced its trap crop effectiveness to a point that it no longer provided season-long protection for cotton. From a management perspective, the presence of corn in the system (as it is currently produced) constitutes a two-edged sword: although it protects cotton from bollworms until late July, it serves as a nursery crop from which moths originate and invade cotton and other crops.

In contrast to other states in the Southeast, North Carolina typically has only minor problems with *H. virescens* on cotton. The general opinion is that tobacco serves as the preferred host for *H. virescens* and that, in agroecosystems where tobacco is produced in large acreage, *H. virescens* concentrates its ovipositional activity on tobacco, thus lessening the risk to cotton. Unlike corn, tobacco acts as a trap crop throughout the summer and does not exhibit the nursery crop phenomenon. Tobacco remains attractive to *H. virescens* moths following harvest of the mature leaves, and insect management practices (cultural and chemicals) aimed at protecting tobacco severely limit production of adults that could migrate to cotton. Apparently, the tobacco to cotton ratio has an important effect upon this relationship, because

H. virescens' problems occur on cotton in both South Carolina and Georgia, despite the occurrence of tobacco in both states (Table 19.2).

Habitat structure may affect not only size but also quality of a pest population. The potential for the development of resistance in *Heliothis* species to the pyrethroids is a case in point. Several investigators (Comins, 1977; Georghiou and Taylor, 1977a; Taylor and Georghiou, 1979; Tabashnik and Croft, 1982) have shown that immigration of susceptible individuals into a deme under selective pressure can substantially reduce the development of resistance. The strength of this effect depends, in part, on the ratio of susceptible individuals to resistant individuals (Taylor and Georghiou, 1979). All else being equal, a higher ratio of susceptible to resistant individuals slows the development of resistance. This ratio may be affected by the proportion of a population in untreated hosts that act as a refugium (Georghiou and Taylor, 1977b).

Throughout the Southeast, corn acts as a refugium in which *H. zea* develop under very little insecticide selection. In North Carolina, the majority of the *H. zea* population is subject to insecticide pressure in only one generation out of four each year (Kennedy and Margolies, 1985), and that pressure is far from uniform, as only limited acreages of certain crop hosts may be treated. The patchiness and diversity of habitat within the Southeast has likewise retarded the selection for resistance to pyrethroids in *H. virescens*, particularly in areas where tobacco is produced, since pyrethroids are not labeled for tobacco (Riley, 1988). While the larger farm units and less host crop heterogeneity of the Mississippi Delta and Texas are more amenable to modern farming techniques, these systems are more likely to experience the disastrous effects of insecticide resistance in *Heliothis* species than the diverse systems of the Southeast.

Management of *Heliothis* species on cotton in the Southeast is based upon crop production tactics designed for early crop maturity. Such a system allows the cotton crop to escape much of the late-season damage potential of *Heliothis* species and significantly narrows the window of crop susceptibility.

CONCLUSIONS

The case has been presented to support the hypothesis that habitat is a major factor affecting pest status and management of *Heliothis* species on cotton. The problem originated early in the nineteenth century in response to farming activities that caused habitat alterations favorable to increased abundance of *Heliothis* species on all crop hosts. The pest status of bollworms continuously rose with the evolution of agroecosystems as subsistence farming yielded to mechanization, organic pesticides, fertilization, and irrigation. The capacity for rapid evolution in *Heliothis* species in response to new physico-chemical environments and locations requiring new host relationships is apparent from the success these species have enjoyed wherever farmers have planted cotton, despite the organic chemicals and other tactics they have employed to protect the crop. Successful management has been achieved through IPM systems founded upon habitat manipulations that lower the mean level of *Heliothis* species abundance or that result in asynchronism of the susceptible "boll-set" stage of the cotton plant and peak *Heliothis* populations.

Acknowledgment

This is Paper No. 11460 of the Journal Series of the North Carolina Agricultural Research Service, Raleigh, N.C. 27695-7643.

REFERENCES

Adkisson, P. C. 1958. The influence of fertilizer application on populations of *Heliothis zea* (Boddie) and certain insect predators. *J. Econ. Entomol.* 51:757–759.

Bacheler, J. S. 1985. Pyrethroid use patterns in the U.S., as influenced by *Heliothis* complex species composition and population dynamics. Pages 120–122 in *Proc. Beltwide Cotton Prod. Res. Conf.*, New Orleans, La. Natl. Cotton Council, Memphis, Tenn.

Barber, G. W. 1937. Seasonal availability of food plants of two species of *Heliothis* in eastern Georgia. *J. Econ. Entomol*. 30:150–158.

Bird, L. S., and R. E. Percy. 1979. Advancements in developing multi-adversity (diseases, insects and stress) resistant glandless, and glanded, okra leaf, frego bract and glabrous cottons. Pages 230–232 in *Proc. Beltwide Cotton Prod. Res. Conf.*, Phoenix, Ariz. Natl. Cotton Council, Memphis, Tenn.

Bottrell, D. G., and P. L. Adkisson. 1977. Cotton insect pest management. *Annu. Rev. Entomol.* 22:451–481.

Bradley, J. R., Jr., M. C. Laster, H. N. Pitre, S. H. Roach, and J. W. Smith. 1987. Interactions of insecticides and arthropod natural enemies of *Heliothis* spp. Pages 7–15 in Theory and tactics of *Heliothis* population management: II-Insecticidal and insect growth regulator control. *South Coop. Ser. Bull.* 329.

Brazzel, J. R. 1963. Resistance to DDT in *Heliothis virescens*. *J. Econ. Entomol.* 56:571–574.

———. 1964. DDT resistance in *Heliothis zea*. *J. Econ. Entomol.* 57:455–457.

Brazzel, J. R., C. Lincoln, L. D. Newsom, F. J. Williams, J. S. Roussel, and G. Barnes 1953 Bollworm and tobacco budworm as cotton pests in Louisiana and Arkansas. *La. Agric. Exp. Stn. Tech. Bull.* 482. 47 pp.

Canerday, T. D. 1974. Response of bollworm and tobacco budworm in Georgia to methyl parathion. *J. Econ. Entomol.* 67:299.

Comins, H. N. 1977. The development of insecticide resistance in the presence of immigration. *J. Theor. Biol.* 64:177–197.

Cope, O. B. 1971. Interactions between pesticides and wildlife. *Annu. Rev. Entomol.* 16:325–364.

Croft, B. A., and A. W. A. Brown. 1975. Responses of arthropod natural enemies to insecticides. *Annu. Rev. Entomol.* 20:285–335.

Dicke, F. F. 1939. Seasonal abundance of the corn earworm. *J. Agric. Res.* 59:237–258.

Doutt, R. L., and R. F. Smith. 1971. The pesticide syndrome. Pages 3–15 in Huffaker, C. B. (ed.), *Biological Control*. New York: Plenum.

Flint, M. C., and R. van den Bosch. 1981. *Introduction to Integrated Pest Management*. New York: Plenum.

Folsum, J. W. 1936. Notes on little-known cotton insects. *J. Econ. Entomol.* 29:282–285.

Freeman, M. E., D. M. Daugherty, and R. D. Jackson. 1967. Damage to soybeans by *Heliothis zea*. *Proc. North Cent. Branch Entomol. Soc. Am.* 22:13–14.

Gaines, J. C. 1957. Cotton insects and their control in the United States. *Annu. Rev. Entomol.* 2:319–338.

Georghiou, G. P., and C. E. Taylor. 1977a. Genetic and biological influences in the evolution of insecticide resistance. *Environ. Entomol.* 70:319–323.

———. 1977b. Operational influences in the evolution of insecticide resistance. *Environ. Entomol.* 70:653–658.

Graves, J. B., and D. F. Clower. 1975. Status of resistance in *Heliothis* species. Pages 142–145 in *Proc. Beltwide Cotton Prod. Res. Conf.*, New Orleans, La. Natl. Cotton Council, Memphis, Tenn.

Graves, J. B., D. F. Clower, and J. R. Bradley. 1967. Resistance of the tobacco budworm to several insecticides in Louisiana. *J. Econ. Entomol.* 60:887–888.

Graves, J. B., J. S. Roussel, and J. R. Phillips. 1963. Resistance to some chlorinated hydrocarbon insecticides in the bollworm, *Heliothis zea*. *J. Econ. Entomol.* 56:442–444.

Harris, F. A., J. B. Graves, S. J. Nemec, S. B. Vinson, and D. A. Wolfenbarger. 1972. Insecticide resistance. Pages 17–27 in Distribution, abundance, and control of *Heliothis* species in cotton and other host plants. *South. Coop. Ser. Bull.* 196.

Heilman, M. D., M. J. Lukefahr, L. N. Namken, and J. W. Norman. 1977. Field evaluation of a short-season production system in the Lower Rio Grande Valley of Texas. Pages 80–83 in *Proc. Beltwide Cotton Prod. Res. Conf.*, Atlanta, Ga. Natl. Cotton Council, Memphis, Tenn.

Hunter, W. D., and B. R. Coad. 1923. The boll weevil problem. *U.S. Dept. Agric. Farmers Bull.* 30 pages.

Isely, D. 1935. Relation of hosts to abundance of cotton bollworm. *Arkansas Agric Exp. Stn. Bull.* 320. 29 pp.

Johnson, M. W., R. E. Stinner, and R. L. Rabb. 1975. Ovipositional response of *Heliothis zea* (Boddiel) to its major hosts in North Carolina. *Environ. Entomol.* 4:291–297.

Kennedy, G. G., and D. C. Margolies. 1985. Mobile arthropod pests: Management in diversified agroecosystems. *Bull. Entomol. Soc. Am.* 31:21–27.

Knipling, E. F., and E. A. Stadelbacher. 1983. The rationale for areawide management of *Heliothis* (Lepidoptera: Noctuidae) populations. *Bull. Entomol. Soc. Am.* 29:29–37

Lincoln, C. 1972. Seasonal occurrence. Pages 2–7 in *Distribution, Abundance, and Control of Heliotha's Species in Cotton and other Host Plants. South Coop. Ser. Bull.* 169.

Lincoln, C., and D. Isely. 1947. Corn as a trap crop for the cotton bollworm. *J. Econ. Entomol.* 40:437–438.

Lincoln, C., J. R. Phillips, W.H. Whitcomb, G.C. Dowell, W. P. Boyer, K. O. Bell, Jr., G. L. Dean, E. J. Matthews, J. B. Graves, L. D. Newsom, D. F. Clower, J.R. Bradley, Jr., and J. L. Bagent. 1967. The bollworm–tobacco budworm problem in Arkansas and Louisiana. *Arkansas Agric. Exp. Stn. Bull.* 720. 67 pages.

Luttrell, R. G., J. R. Phillips, and T. R. Pfimmer. 1986. Cultural control in mid-South U.S. cropping systems. Pages 28–37 in Theory and tactics of *Heliothis* population management: I-Cultural and biological control.

Luttrell, R. G., R. T. Roush, A. Ali, J. S. Mink, M. R. Reid, and G. L. Snodgress. 1987. Pyrethroid resistance in field populations of *Heliothis virescens* (Lepidoptera: Noctuidae) in Mississippi in 1986. *J. Econ. Entomol.* 80:985–989.

Mueller, T. F., and J. R. Phillips. 1983. Population dynamics of *Heliothis* spp. in spring weed hosts in southeastern Arkansas: Survivorship and stage-specific parasitism. *Environ. Entomol.* 12:1846–1850.

Mueller, T. F., V. E. Harris, and J. R. Phillips. 1984. Theory of *Heliothis* (Lepidoptera: Noctuidae) management through reduction of the first spring generation: A critique. *Environ. Entomol.* 13:625–634.

Namken, L. N., and M. D. Heilman. 1973. Determinate cotton cultivars for more efficient cotton production on medium-textured soil in the Lower Rio Grande Valley of Texas. *Agron. J.* 65:953–956.

Neunzig, H. H. 1969. The biology of the tobacco budworm and the corn earworm in North Carolina with particular reference to tobacco as a host. *N.C. Agric. Exp. Stn. Tech. Bull.* 196. 76 pp.

Newsom, L. D. 1967. Consequences of insecticide use on non-target organisms. *Annu. Rev. Entomol.* 12:257–286.

———. 1974. Predator–insecticide relationships. *Entomophaga Mem. Hors. Ser.* 7:13–23

Newsom, L. D., and J. R. Brazzel. 1968. Pests and their control. Pages 367–405 in Elliot, F. C., M. Hanover, and W.K. Porter, Jr. (ed.), *Advances In Production and Utilization of Quality Cotton: Principles and Practices.* Ames, Iowa: Iowa State Univ. Press.

Norman, J. W., J. L. Henson, D. A. Wolfenbarger, J. A. Harding, E. V. Gage, R. D. Parker, and G. Collins. 1979. Effects of water management practices on economics of insect control on cotton, Lower Rio Grande Valley, Texas. *J. Econ. Entomol.* 72:367–370.

Parsons, F. S. 1939. Investigations on the cotton bollworm, *Heliothis armigera*, Hubn. (*obsoleta*, Fabr.). Part I. The annual march of bollworm incidence and related factors. *Bull. Entomol. Res.* 30:321–337.

———. 1940. Investigations on the cotton bollworm, *Heliothis armigera*, Hubn. (*obsoleta*, Fabr.). Part III. Relationships between oviposition and the flowering curves of food-plants. *Bull. Entomol. Res.* 31:147–177.

Plapp, F. W., and C. Campanhola. 1986. Synergism of pyrethroids by chlordimeform against susceptible and resistant *Helliothis*. Pages 167–169 in *Proc. Beltwide Cotton Prod. Res. Cont.*, Las Vegas, Nev., Natl. Cotton Council, Memphis, Tenn.

Quaintance, A. L., and C. T. Brues. 1905. The cotton bollworm. *U.S. Dep. Agric. Bur. Entomol. Bull.* 50. 155 pp.

Riley, S.L. 1988. An overview of the status of pyrethroid resistance in the U.S. during 1987. Pages 228–236 in *Proc. Beltwide Cotton Prod. Res. Conf.*, New Orleans, La. Natl. Cotton Council, Memphis, Tenn.

Ripper, W. E. 1956. Effect of pesticides on balance of arthropod populations. *Annu. Rev. Entomol.* 1:403–438.

Roussel, J. S., and D. F. Clower. 1955. Resistance to the chlorinated hydrocarbon insecticides in the boll weevil (*Anthonomis grandis* Boh.). *La. Agric. Exp. Stn. Circ.* 41.

Rummel, D. R., J. F. Leser, J. E. Slosser, G. J. Puterka, C. W. Neel, J. K. Walker, J. H. Benedict, M. D. Heilman, L. N. Namken, J. W. Norman, and J. H. Young 1986. Cultural control of *Heliothis* spp. in southwestern U.S. cropping systems. Pages 38–53 in Theory and tactics of *Heliothis* population management: I-Cultural and biological control.

Slosser, J. E. 1980. Irrigation timing for bollworm management in cotton. *J. Econ. Entomol.* 78:356–359.

Snow, J. W., J. H. Hamm, and J. R. Brazzel. 1966. *Geranium carolinianum*, as an early host for *Heliothis zea* and *H. virescens* (Lepidoptera: Noctuidae) in the southeastern United States, with notes on associated parasites. *Ann. Entomol. Soc. Am.* 59:506–509.

Stadelbacher, E. A. 1979. *Geranium dissectium*: An unreported host of the tobacco budworm and the bollworm and its role in their seasonal and long term population dynamics in the delta of Mississippi. *Environ. Entomol.* 8:1153–1156.

———. 1981. Role of early-season wild and naturalized host plants in the buildup of the F_1 generation of *Heliothis zea* and *H. virescens* in the delta of Mississippi. *Environ. Entomol.* 10:766–770.

———. 1982. An overview and simulation of tactics for management of *Heliothis* spp. on early-season wild host plants. Pages 209–212 in *Proc. Beltwide Cotton Prod. Res. Conf.*, Las Vegas, Nev. Natl. Cotton Council, Memphis, Tenn.

Stadelbacher, E. A., and A. L. Scales. 1973. Technique for determining oviposition preference of the bollworm and tobacco budworm for varieties and experimental stocks of cotton. *J. Econ. Entomol.* 66:418–421.

Tabashnik, B. E., and B. A. Croft. 1982. Managing pesticide resistance in crop-arthropod complexes: Interactions between biological and operational factors. *Environ. Entomol.* 11:1137–1144.

Taylor, C. E., and G. P. Georghiou. 1979. Suppression of insecticide resistance by alteration of dominance and migration. *J. Econ. Entomol.* 72:105–109.

Thomas, F. L., and E. W. Dunnam. 1931. Factors influencing infestation in cotton by *Heliothis obsoleta* Fab. *J. Econ. Entomol.* 24:815–821.

van den Bosch, R., H. T. Reynolds, and E. J. Dietrick. 1956. Toxicity of widely used insecticides to beneficial insects in California cotton and alfalfa fields. *J. Econ. Entomol.* 49:359–363.

Walker, J. K., R. F. Frisbie, and G. A. Niles. 1979. *Heliothis* species in short-season cottons in Texas. *South. Coop. Ser. Bull.* 231:31–43.

Walker, J. K., Jr., and G. A. Niles. 1971. Population dynamics of the boll weevil on modified cotton types: Implications for pest management. *Tex. Agric. Exp. Stn. Bull.* 1109. 14 pp.

Wolfenbarger, D. A., M. J. Lukefahr, and H. M. Graham. 1973. LD_{50} values of methyl parathion and endrin to tobacco budworms and bollworms collected in the Americas and hypothesis on the spread of resistance in these lepidopterans to these insecticides. *J. Econ. Entomol.* 66:211–216.

20

THE ROLE OF DEMIC ADAPTATION IN COLONIZATION AND SPREAD OF SCALE INSECT POPULATIONS

Lawrence M. Hanks

University of California at Riverside

Robert F. Denno

University of Maryland

INTRODUCTION

Dispersal allows an herbivorous insect to escape deterioration of resources but, at the same time, may increase the risk of mortality (Gadgil, 1971; Hamilton and May, 1977; Denno 1983). Colonization of new host plant individuals is essential for the survival of herbivores and the persistence and spread of their populations. Within a habitat, variation among plants in their suitability as hosts has a profound impact on herbivore dispersal and host colonization. Nonhost plants stand as obstacles to dispersing insects, and those insects colonizing these plants will suffer very low fitness (Rosenthal and Janzen, 1979; Kareiva, 1983). Variability among *conspecific* host plants within habitats further contributes to this patchiness in the distribution of available resources (Edmunds and Alstad, 1978; Whitham, 1983). These sources of host variability impose selective pressures that may encourage adaptive divergence of herbivore subpopulations (Templeton, 1981).

The potential for genetic differentiation of subpopulations depends on the relative contribution of contrasting forces: Gene flow between subpopulations renders them genetically homogeneous, while mutation, genetic drift, and natural selection are diversifying influences (Futuyma, 1986). Species having poor dispersal capabilities may be distributed as reproductively isolated local populations that represent genetically distinct "demes" (Selander, 1970).

Some ecological characteristics of herbivores promote the development of genetic polymorphism among demes in host utilization in response to selection imposed by host plants (Mitter and Futuyma, 1983; Futuyma and Peterson, 1985): (1) each individual herbivore experiences only one host patch (the environment is coarse grained); (2) herbivore populations in different patches are regulated independently of one another; (3) initial heritable variation exists within populations such that some herbivore genotypes experience substantially higher fitness in different host patches; (4) there is a substantial selective advantage for the different genotypes in their respective patches; (5) there is little movement between hosts, or, alternatively, patches are extensive relative to the mobility of the herbivore species (there is limited gene flow); (6) individuals mate with others from the same patch type (assortative mating); (7) females select for oviposition patches of the same type as their natal patch.

Scale insects are ideal for studying the impact of host plant variability on herbivore populations because of their weak dispersal capabilities and sedentary nature. These traits are also responsible for their frequent accidental introduction into new geographic areas by transportation of infested plant materials (DeBach et al., 1971; Beardsley and Gonzalez, 1975). Introduced scale species characteristically have "general-purpose" genotypes, resistant to the genetic changes that accompany adaptation to local environments (Baker, 1965). In fact, most introduced scale pests are polyphagous (e.g., see Beardsley and Gonzalez, 1975), and this alone may preadapt them for colonization and exploitation of novel host species (Mitter and Futuyma, 1983).

Edmunds and Alstad (1978) concluded that, because scale insects are sedentary, subpopulations on individual trees are, to some extent, reproductively isolated. Chemical variability among hosts that imposes divergent selective pressures on these subpopulations can therefore produce genetic differentiation of scale demes. Edmunds and Alstad proposed that demic adaptation increasing scale fitness on one tree in turn reduced fitness on other trees, ultimately hindering population spread between trees. This trade-off provided an explanation for the common occurrence of infested trees neighboring scale-free trees. In fact, contagious distribution is characteristic of many scale species (Miller and Kosztarab, 1979); demic adaptation of scale populations to variable host tree individuals may be responsible for this phenomenon.

Our objective in this chapter is to examine the potential for the development of demic adaptation in scale insect populations, the empirical evidence for demic adaptation, and its impact on host colonization and population spread. The questions we specifically address are the following:

1. Does scale insect natural history promote reproductive isolation of populations?
2. Do case studies of scale species provide evidence that demic adaptation does occur?
3. What significance does demic adaptation have in the colonization and spread of scale insects?

4. What factors prevent or limit the extent of demic adaptation in scale populations?
5. What other environmental factors influence the distribution and spread of scale populations, and which of these might cause their contagious distribution?

PERTINENT NATURAL HISTORY OF SCALE INSECTS

The Coccoidea, scale insects in general, include 20 families (Miller and Kosztarab, 1979), including the ground pearls, ensign scales, lac scales, soft scales, gall-like scales, pit scales, cochineal insects, armored scales, mealybugs, and eriococcids. The species of this superfamily represent a great diversity in form and natural history. Extant coccoid families illustrate a clear evolutionary sequence in adult female morphology: primitive scale families bear similarities to the aphids, while others show a progression of decreasing leg size and reduced mobility, culminating in the highly advanced armored scales (Diaspididae) with adult females that are entirely without appendages (Miller and Kosztarab, 1979).

All adult female scales are apterous. Though females of many families have legs and are fully capable of movement, they nevertheless are quite sedentary and, on perennial hosts, are likely to remain on a single host for their entire lives (Goux, 1944; Miller and Kosztarab, 1979; D. R. Miller, personal communication). Thus, oviposition usually occurs on the host plant of the mother. This outcome is analogous to that predicted by the "Hopkins host selection principle," which states that female insects tend to oviposit on their larval host species (see Smith and Cornell, 1979). This behavior appears to be uncommon in phytophagous insects (Futuyma and Peterson, 1985).

Long-distance dispersal of scale insects, an important form of gene flow between populations, is primarily limited to a short wandering phase in the first instar nymph or "crawler". After egg hatch, crawlers move to a new site, where they insert their stylets and begin to feed on plant sap. Crawlers may be incapable of walking distances greater than 1 meter (Melis, 1943). They commonly settle to feed within one to four hours after hatching and travel only a short distance from the mother (reviewed in Hanks and Denno, in prep.).

Although walking may be important in the location of better-quality feeding sites within a single perennial host, rarely will it result in *inter*plant movement. Great numbers of crawlers settle on their natal branches rather than abandon their host (Nielsen and Johnson, 1973; Edmunds and Alstad, 1978), although crawlers of some species do show behaviors that may improve their chances of aerial dispersal (Greathead, 1972; Moran et al., 1982; Washburn and Frankie, 1985). Bliss et al. (1935) exposed crawlers of camphor scale, *Pseudaonidia duplex* Cockerell, to 12.5 mph winds and found that 50% settled without dispersing from the host. Stephens and Aylor (1978) reported a complete absence of aerial dispersal in crawlers of red pine scale, *Matsucoccus resinosae* Bean and Godwin, exposed to wind speeds of 5 miles per second. These crawlers displayed considerable resistance to being blown from the host. Crawlers of white peach scale, *Pseudaulacaspis pentagona* Targioni Tozzetti, firmly grasped the bark when exposed to an airstream, only leaving the branch when the force tore them free (Hanks and Denno, in prep).

Aerial dispersal of crawlers is critical for colonization of new hosts, but it is usually effective over only relatively short distances (Beardsley and Gonzalez, 1975; Rabkin and Lejeune, 1954). Crawlers typically have a flattened body form and two filamentous caudal setae that improve their aerial buoyancy, enabling them to remain aloft in even very slight breezes (Beardsley and Gonzalez, 1975; Stephens and Aylor, 1978; Wainhouse, 1980; Moran et al., 1982). During aerial dispersal, crawlers behave as inert particles, moving primarily downwind (Quayle, 1916; Brown, 1958; Cornwall, 1960; McClure, 1977; Wainhouse, 1980; Moran et al., 1982; Augustin, 1986). Numbers of trapped crawlers typically peak at a distance of less than 15 m from the source plant, dropping off dramatically before they are 75 m away (McClure, 1977; Stephens and Aylor, 1978; Wainhouse, 1980; Moran et al., 1982; Augustin, 1986). Great numbers of dispersing crawlers fall to the ground (McClure, 1977, 1979; Wainhouse, 1980; Stephens and Aylor, 1978), where they die due to their limited ability to move over soil (Beardsley and Gonzalez, 1975).

Small energy reserves and susceptibility to temperature and humidity extremes limit the time that crawlers can remain viable during aerial transport (Beardsley and Gonzalez, 1975; Alstad and Edmunds, 1983). The probability that an individual crawler will reach a new host is very small. Burns and Donley (1970) estimated that only 8% of aerially dispersing crawlers of tulip tree scale, *Toumeyella liriodendri* (Gmelin), were successful in colonizing neighboring trees. A poor ability to colonize new hosts limits the effectiveness of *Dactylopius austrinus* De Lotto as a biological control agent for *Opuntia* cactus (Moran et al., 1982). Wainhouse (1979) observed that colonization of new hosts by crawlers was very inefficient in the felted beech scale, *Cryptococcus fagisuga* Lindinger.

Gene flow between scale populations may also occur in the form of aerial dispersal of adult males. Male scales are very small and may be apterous or winged. They lack functional mouthparts and are usually active for only a few hours (Tashiro and Beavers, 1968; Beardsley and Gonzalez, 1975; Howell and Williams, 1976; Miller and Kosztarab, 1979). Although winged males have very weak flight capabilities and cannot fly upwind in even slight breezes, they may be blown far downwind from their source (Rice and Moreno, 1970; Rotundo et al. 1980). Males often do not show an inclination to take flight but mate with neighboring females immediately after emergence (Lellakova-Duskova, 1963; Doane, 1966; Alstad et al., 1980; Rotundo et al., 1980; Tashiro and Moffitt, 1968; Hanks and Denno, in prep.).

In summary, behavioral and demographic attributes of the dispersing forms of scale insects restrict gene flow between populations on different perennial host plants. Adult females may be completely sessile, or, if they are capable of walking, are likely to remain on their original host plant for oviposition. Crawlers tend to remain on the natal host, and aerial dispersal between hosts is inefficient and marked by high levels of mortality. Males have a very short adult life span, do not move easily between host plants, and often show a tendency to mate with local females.

EMPIRICAL EVIDENCE FOR DEMIC ADAPTATION IN SCALE INSECTS

A few early researchers have recognized that adaptation of scale populations to their current hosts might inhibit the colonization of new hosts. Goux (1944) proposed that low mobility leads to isolation of scale colonies and the development of biological races that differ in host species preference. He suggested that crawlers of polyphagous species were limited in their host range by a preference for the host species on which their parents fed.

Habib (1957) studied *Parthenolecanium corni* Bouche, a polyphagous soft-scale species that shows marked variation in morphology on different host species. He found survivorship to be higher in many cases when transfers of scales were made to the same species as the original host. Furthermore, survivorship of the progeny produced in these transfers was often higher in transfers to the original host species than in transfers to other host species. He concluded that populations of the scale could adapt to their host plant species.

These early works set the stage for the demic adaptation hypothesis: herbivore populations can adapt to the characteristics of host individuals. Edmunds and Alstad (1978) based their hypothesis on their study of black pineleaf scale, *Nuculaspis californica* Coleman (Diaspididae), on ponderosa pine, *Pinus ponderosa* (Lawson). Pine branchlets infested with fecund scales were transferred from infested trees in one area to scale-free trees in a different area, as well as back to each original infested host. Survivorship of transferred scales was significantly higher on the original hosts, suggesting that demic adaptation of scale populations to individual trees had occurred. The experimental design of Edmunds and Alstad (1978), however, did not provide a conclusive test for demic adaptation. Transfer results may have been confounded by habitat effects, since scale-free trees all occurred in a noninfested area far removed from the infested stand (Edmunds and Alstad, 1978; see also Unruh and Luck, 1987). Inhibition of parasites is known to cause outbreaks of this scale (Edmunds, 1973); therefore, differences in levels of parasitism between infested and scale-free stands was likely to have been responsible for these patterns. Furthermore, transfers were not made between infested trees, so it is possible that survivorship was higher on infested trees than on scale-free trees, merely because of differences in general susceptibility to scale attack in trees growing in the infested stand.

Demic adaptation studies that followed (discussed later) have controlled for the possibility of habitat effects and host suitability differences by using trees that all initially supported scale populations. In these studies, scales were transferred as eggs or crawlers from each infested tree to other infested trees (here referred to as "*inter*tree" transfers), as well as transferred back to the original tree ("*intra*tree" transfers). Each host was, therefore, initially proven to be susceptible to scale attack, thus eliminating the possibility that survivorship would be higher on the original host due to general quality alone. Under the hypothesis of demic adaptation, it would be predicted that scales transferred between and within infested trees should show highest survivorship on their original host.

Rice (1983) also tested for demic adaptation of black pineleaf scale populations, but on sugar pine, *Pinus lambertiana* Douglas. In concordance with the demic adaptation hypothesis, there was a tendency for scale survivorship to be greater in intratree transfers than in intertree transfers, but these differences were not statistically significant. This experimental design was complicated by the use of clones of trees as receptors and donors rather than the individual trees themselves (Rice, 1983).

Unruh and Luck (1987), working with pinyon needle scale, *Matsucoccus acalyptus* Herbert (Margarodidae), on pinyon pine, *Pinus monophylla* (Torrey and Frémont), found no indication of higher survivorship in intratree transfers. In a separate study, Cobb and Whitham (in prep.) have also found no evidence for demic adaptation of this scale on pinyon pines.

In a study of the parthenogenetic felted beech scale, *Cryptococcus fagisuga* (Cryptococcidae), on beech, *Fagus sylvatica* L., Wainhouse and Howell (1983) reported equal survivorship when scales were transferred to trees related by clone line to their original host and to unrelated trees in a seed orchard habitat. However, transfers between five infested forest trees (that occurred in four different forested areas) showed higher numbers of surviving adult females in intratree transfers for three of five trees, with a significant overall ANOVA. The remaining two trees showed very low survivorship of scales, regardless of their source.

Hughes (1960) performed transfers of white peach scale, *Pseudaulacaspis pentagona* (Diaspididae), among different host species and reported that, in some cases, survivorship was slightly higher in transfers made to plants of the same species as the original host. This suggests that local scale populations were partially adapted to their host species. Hanks and Denno (in prep.) found no significant differences in survivorship of white peach scale between intra- and intertree transfers for eight sets of transfers with white mulberry, *Morus alba* L., the host. Transfers were later carried out between infested mulberries that were much older, each of which was isolated from other scale-bearing trees by at least 250 m. This design was intended to improve the opportunity for adaptation to hosts, since a greater time period was available for adaptation and between-tree gene flow was minimized. Rather than transferring scales between many trees, pairs of trees were used to improve statistical sensitivity by increasing sample size.

Results of this study showed no significant difference in settling success between intratree transfers and intertree transfers for one pair of mulberry trees (Fig. 20.1a). Survivorship, however, was higher for both sexes in intratree cages, that is, on the original host trees (Fig. 20.1b). Transfers were also made between populations of white peach scale on trees of different species. Settling success was higher in the intratree cages for both sexes in transfers from a population on a mulberry tree and one on a *Catalpa speciosa* Warder to other mulberry and catalpa trees (Fig. 20.2a). There were no significant differences in survivorship (Fig 20.2b). Males appeared to settle more readily on their original host in transfers between a different mulberry population and one on walnut, *Juglans nigra* L., an uncommon host for this scale (Fig. 20.3a) The survivorship data shows a similar relationship, although these differences were not significant (Fig. 20.3b)

In summary, there exists some empirical evidence in support of the demic adaptation hypothesis in the studies using felted beech scale (Wainhouse and Howell, 1983) and white peach scale (Hanks and Denno, in prep.). However, because of methodological problems (Edmunds and Alstad, 1978) or statistically insignificant results (Rice, 1983), it is not certain that demic adaptation occurs in populations of black pineleaf scale. Additionally, two studies of pinyon needle scale have illustrated a clear lack of adaptation to host tree individuals in this species (Unruh and Luck, 1987; Cobb and Whitham, in prep.).

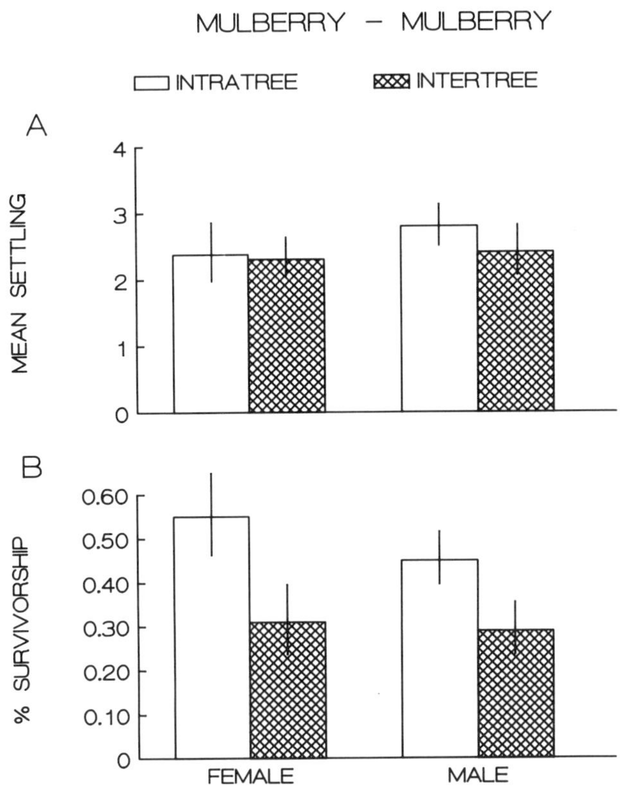

Figure 20.1. (a) Mean settling success and standard error bars for *Pseudaulacaspis pentagona* in intra-tree and intertree transfers using two *Morus alba* hosts. ANOVA was performed on mean values for cages within tree branches. There were no significant differences between treatments for either sex (df = 22). (b) Percent survivorship of *P. pentagona*. Survivorship was higher in the intratree transfers for both sexes, but differences were only significant for females (Females: F = 3.01, $p < .05$, df = 19; Males: df = 21).

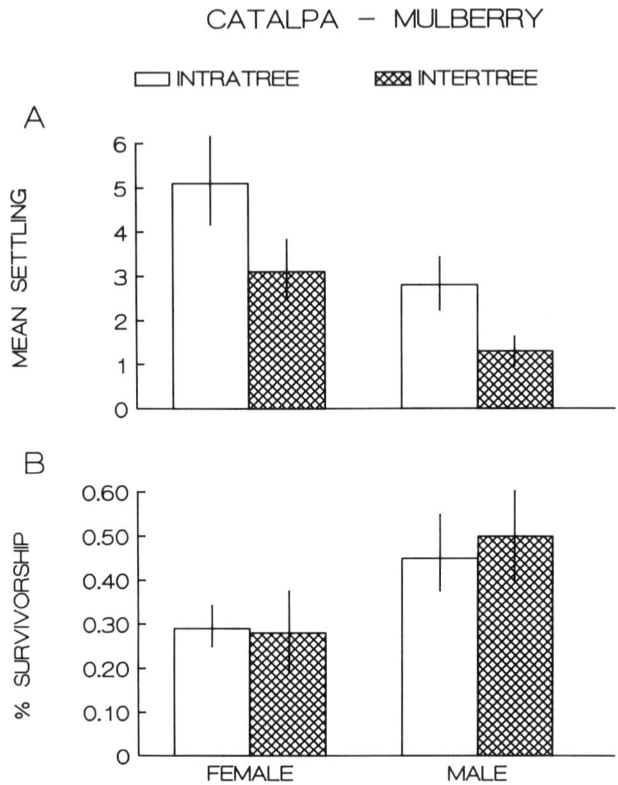

Figure 20.2. (a) Mean settling success and standard error bars for *Pseudaulacaspis pentagona* in transfers from infested *Morus alba* and *Catalpa speciosa* hosts to the original catalpa host, one different catalpa tree, and two other mulberry trees. All trees were infested with scale. ANOVA was performed on mean values for cages within tree branches. Settling success was significantly higher in the intratree transfers for both sexes (Females: $F = 9.94$, $p < .05$, $df = 35$; Males: $F = 4.27$, $p < .01$, $df = 35$). (b) Percent survivorship of *P. pentagona*. No significant differences occur in either sex ($df = 27$).

DISCUSSION

When the dispersal stage of a species is brief and associated with high levels of mortality, the risk of dispersal may outweigh its benefits (Ward, 1987). Scale crawlers presumably have no control over the direction of their aerial dispersal and have a small probability of reaching a new host. Perhaps due to selection arising from the risk of dispersal, crawlers often do not display a readiness to leave a perennial host. There are other possible selective advantages in not dispersing: for example, because female scales are sedentary and oviposition is likely to take place on the maternal host, scales crawlers hatch on the host plant to which they are likely to be best adapted.

For most herbivorous insects, polymorphism in host plant utilization requires the evolution of a correlation between behaviors that limit gene flow among patches (such as selection of the appropriate patch for feeding, mating, and oviposition) and

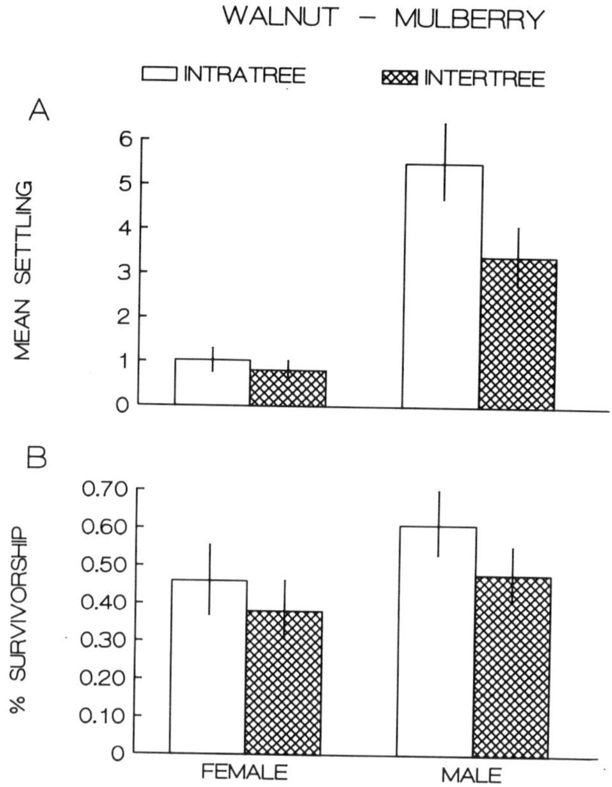

Figure 20.3 (a) Mean settling success and standard error bars for *Pseudaulacaspis pentagona* in intratree and intertree transfers using *Morus alba* and *Juglans nigra* trees. ANOVA was performed on mean values for cages within tree branches. Settling success was significantly higher in the intratree transfers for males, but there was no significant difference for females (Females: df = 35; Males: F = 5.5 , $p < .05$, df = 35). (b) Percent survivorship of *P. pentagona*. Survivorship was higher in the within transfers for both sexes, but these differences were not significant (Females: df = 24; Males: df = 33).

fitness in those patches (Futuyma and Peterson, 1985; Futuyma and Mayer, 1980; Via and Lande, 1985). Pleiotropism or linkage disequilibrium of loci governing these traits may provide this correlation, but their improbability reduces the chances of maintaining polymorphism by this means (Mitter and Futuyma, 1983; Futuyma and Peterson, 1985). This genetic correlation requirement may be eliminated in the scale insects, because gene flow between populations is naturally limited by males mating with local females and by crawlers settling on their natal host. Oviposition on the maternal host not only reduces gene flow but also, as mentioned earlier, places progeny on the most suitable host. These traits, therefore, improve the likelihood that host polymorphism will be maintained within scale insect populations.

Factors Limiting Demic Adaptation

The evidence for demic adaptation of scale populations to host tree phenotype is not overwhelming. In the felted beech scale and white peach scale systems, demic adaptation was indicated, although the level of adaptation (as measured by scale performance in reciprocal transfers between hosts) was not very dramatic. An explanation for the limitation of demic adaptation and its occurrence in some scale species but not in others lies in the characteristics of scale and host species.

If disruptive selective forces imposed by hosts are weak, they may not promote genetic differentiation of scale subpopulations. Differences among hosts, at least in phenotype, have been indicated in all four scale species by persistent, contagious distributions, with infested and scale-free trees standing in proximity (Edmunds, 1973; Edmunds and Alstad, 1978; Wainhouse and Howell, 1982, 1983; Wainhouse et al., 1988; Unruh and Luck, 1987; Hanks and Denno, in prep.). This suggests that differences between hosts are important in influencing scale colonization and so might represent important selective forces acting on scale populations.

Within-habitat genetic variability among conspecific host trees (Wainhouse, 1980; Rice, 1983; Unruh and Luck, 1987; Hanks and Denno, in prep.) might have directly imposed divergent selective pressures that dictate these scale distributions. A considerable amount of chemical variability has been found within pinyon pine populations (Unruh and Luck, 1987). Chemical variability among ponderosa pine trees is also extensive (Edmunds and Alstad, 1981). However, by positioning their feeding stylets away from terpene-containing resin canals in pine needles (Edmunds and Alstad, 1981), black pineleaf scales may not experience the chemical variability that was hypothesized as governing their distribution.

The white peach scale data (Hanks and Denno, in prep) suggest that selective pressures may strongly encourage differentiation when populations occur on different host species rather than on conspecific hosts. A similar effect is found in aphids, in which adaptation to host genotype is more common where populations feed on different species of hosts rather than cultivars of one host species (Futuyma and Peterson, 1985).

For polymorphism in host utilization to arise, there must also be initial genetic variability within the herbivore population upon which selection can act. In black pineleaf scale (Edmunds and Alstad, 1978), felted beech scale (Wainhouse and Howell, 1983), and white peach scale (Hanks and Denno, in prep.), there was considerable variation expressed in the performance of scales from different populations in transfers to the same hosts. These differences were sometimes due to scales from certain populations being more vigorous colonizers on all hosts. It is possible that these differences between scale populations were not due to variability in scale genotype but to a phenotype that is a reflection of factors such as the quality of the original host (Flanders, 1970; Futuyma and Peterson, 1985). Another possibility is that genetic drift in reproductively isolated populations was responsible for these differences in scale vigor. Habitat effects and genetic drift may have been most important in influencing scale phenotype in the forest beech trees of Wainhouse and Howell (1983) and urban trees chosen by Hanks and Denno (in prep.), since these trees were isolated from other infested hosts and gene flow was at a minimum. Unruh and Luck (1987) suggest that such nongenetic or nonadaptive genetic factors are responsible for survivorship differences between pinyon needle scale populations transferred between trees on different mountains.

The time period available for adaptation and the generation time of scale insects will influence the extent of adaptive divergence of subpopulations. For example,

McClure (1983) found adaptation of *Fiorinia externa* Ferris and *Nuculaspis tsugae* (Marlatt) populations to individual hemlock trees in only a 12-year period and has suggested (McClure, personal communication) that the greater differentiation in *N. tsugae* populations was due to its being bivoltine while *F. externa* is univoltine. The orchard beech trees used by Wainhouse and Howell (1983) were 35 years of age, providing a maximum of 35 generations for adaptation of the univoltine felted beech scale. Trees of 15 years of age were studied by Hanks and Denno (in prep.), but, because white peach scale is trivoltine in the Maryland area, there was a maximum of 45 generations available for adaptation. Pinyon needle scale is univoltine, and Unruh and Luck (1987) used pines ranging from 40 to 100 years of age. The ages of the sugar pines of Rice (1983) were not reported; however, black pineleaf scale is univoltine. It is clear that host age and scale generation time do not explain the exclusive occurrence of demic adaptation in felted beech scale and white peach scale populations.

Gene flow between populations on different hosts may limit demic adaptation. In all four scale species, crawlers showed limited dispersal behavior, with a tendency to remain on the natal tree (Alstad and Edmunds, 1983; Wainhouse and Howell, 1983; Unruh and Luck, 1987; Hanks and Denno, in prep.). Males of the three sexual species (*N. californica, M. acalyptus*, and *P. pentagona*) showed little dispersal behavior or commonly mated with local females (McCambridge and Pierce, 1964; Alstad et al., 1980; Unruh and Luck, 1987; Hanks and Denno, in prep.). Demic adaptation seems most likely to occur in the felted beech scale, *C. fagisuga*, since parthenogenetic species are considered more likely to develop host utilization polymorphisms (Futuyma and Peterson, 1985). Gene flow by male dispersal is, of course, nonexistent, and movement of crawlers between trees is very inefficient (Wainhouse, 1980). Furthermore, the persistence of coadapted gene complexes in the absence of recombination (Mitter and Futuyma, 1983; Felsenstein, 1988) may further encourage adaptation of felted beech scale populations to host individuals.

The distribution and abundance of host plants strongly influences the potential for gene flow between herbivore populations. Most studies concerned with the effects of vegetational distribution pattern on herbivore populations have dealt with insects that are capable of directed flight (see Kareiva, 1983, for review). The impact of host distribution is likely to be greater for weakly dispersing herbivores, like scale insects, that experience environmental variation as very coarse grained (Levins, 1968). Where hosts are interspersed with nonhost species, gene flow will decline as a consequence of lower colonization efficiency in the aerially dispersing crawlers and males. It is likely that other tree species in the forest habitat may have insulated felted beech scale populations from gene flow that would undermine demic adaptation. Transfers of this scale within an orchard revealed no indication of demic adaptation, but transfers between these forest trees revealed that some populations were adapted to their host trees. Similarly, white peach scale populations were not adapted to hosts that neighbored other infested trees but showed some degree of adaptation to hosts that stood in isolated settings. Hosts of black pineleaf scale and pinyon leaf scale, on the other hand, occur in more pure stands. In fact, Rice (1983) used scale donor trees that occurred together in an infested arboretum. Even though distances between these coniferous hosts might have exceeded the usual aerial dispersal distance of the scales (Unruh and Luck, 1987), dispersal between these hosts would nevertheless have been facilitated by trees occurring in pure stands. Gene flow between neighboring trees may therefore have been responsible for the apparent lack of population differentiation in black pineleaf scale and pinyon needle scale.

Some of these demic adaptation studies were carried out in areas where scale populations occurred in outbreak densities due to ineffective control by predators and parasites (Rice, 1983; Wainhouse and Howell, 1983; Hanks and Denno, in prep.). Pinyon needle scale can occur in high densities as well (McCambridge and Pierce, 1964; Unruh, personal communication), but it is notable in lacking any parasites or specialist predators (Unruh and Luck, 1987). Scale insects have most likely diversified under more natural conditions, where population densities are much lower. Low-density populations may be more conducive to demic adaptation due to greater levels of inbreeding and reduced gene flow (Templeton, 1980). Thus, the lack of evidence for a great degree of adaptation in scale insects might be due to the density of the populations studied. Future demic adaptation studies might best be conducted using populations of lower density.

Polymorphisms, Trade-offs, and Polyphagy

Evidence for variation in preference and utilization of host plants within species of herbivorous insects is abundant, but there are few studies documenting this variation as inherited and persistent (Mitter and Futuyma, 1983; Futuyma and Peterson, 1985). The demic adaptation hypothesis of Edmunds and Alstad (1978) depends on heritable variation in host utilization among demes but further assumes negative genetic correlations between fitness on different trees: adaptation that improves relative fitness on one host is presumed to be accompanied by a reduction in relative fitness on other hosts. This represents a trade-off for the scale species, since adaptation that improves performance on one host individual will ultimately restrict colonization between hosts. Experimental tests for this type of trade-off, using various insect herbivores presented with different host species, have reported, at most, weak, negative genetic correlations (Gould, 1979; Rausher, 1984; Via, 1984; Futuyma and Philippi, 1987; but see Rausher, 1988). If increased fitness through adaptation to one host individual is not accompanied by lowered fitness on other hosts, demic adaptation can not be tested in the manner just discussed. In this event, performance of scales on one tree may improve through adaptation but with no loss in fitness for ancestral hosts. Consequently, performance on novel trees may be high due to preadaptation (Edmunds and Alstad, 1978; Gould, 1979). In the absence of this genetic trade-off, the fitness of scale individuals could improve through adaptation without limiting population spread to new hosts. By this means, a polyphagous scale may adapt to a novel host species but still maintain a generalist capacity. White peach scale is extremely polyphagous (Davidson et al., 1983), but it shows some ability to adapt to different host individuals and species (Hanks and Denno, in prep.). This ability to colonize and to adapt to novel hosts may explain why this pest has a nearly worldwide distribution (Davidson et al., 1983).

Host polymorphism that operates within habitats has been discovered primarily in herbivore species with narrow host ranges (Futuyma and Peterson, 1985). These more specialized species, such as felted beech scale, may adapt to individual host phenotypes while maintaining high fitness on other host phenotypes and so be generalized in the utilization of their host species (Mitter and Futuyma, 1983). Where gene flow undermines demic adaptation, as may be the case when hosts grow in proximity to one another, selection may favor a single "compromise" genotype (Mitter and Futuyma, 1983). This may explain the uniform performance of black pineleaf scale and pinyon needle scale on different host individuals (Rice, 1983; Unruh and Luck, 1987).

A high degree of demic adaptation that restricts herbivores to single hosts might further constitute overspecialization (Kuris and Norton, 1985), since the inevitable death of the host will result in the complete extinction of the adapted genotype. The absence of data showing strong negative correlations in scale populations may suggest a role for demic adaptation in "fine tuning" populations to their immediate environment without a loss in their ability to colonize new hosts.

The Contagious Distribution of Scale Insects

Where demic adaptation of scale populations was detected (Wainhouse and Howell, 1983; Hanks and Denno, in prep.), differences in fitness between adapted versus nonadapted scales on individual hosts did not appear great enough to suggest that scales could not colonize and survive on novel hosts. Furthermore, demic adaptation was only detected in populations that occurred on hosts that did not neighbor other infested hosts. It is therefore unlikely that demic adaptation contributes significantly to the characteristic contagious distribution of scale insects within habitats. Other factors may better account for these patterns.

First, variability among hosts in genotypic resistance to scale attack will lead to a contagious distribution, even in the absence of demic adaptation of scales (Flanders, 1970). Wainhouse and Deeble (1980), for example, found that general resistance in one beech clone line explained much of the patchy distribution of felted beech scale in an orchard setting. Wainhouse et al. (1988) reviewed the evidence that beech tree resistance may reflect histological characteristics of the bark that prevent feeding and that the scale may be incapable of adapting to this form of resistance.

Second, variability in host phenotype can strongly affect scale distribution (Flanders, 1970). Mattson et al. (1988) concluded that herbivores that are more intimately associated with their host (such as borers, gall formers, leaf miners, and scale insects) often show greater tree-to-tree differences in population density than do foliage feeding species, and suggested that this is primarily due to environment-related variation in nutrients and defensive compounds. Reduced host quality due to water stress (Hanks and Denno, in prep.), edaphic factors (Flanders, 1970; McClure, 1977, 1986), and tree age (Edmunds and Alstad, 1978; Wainhouse and Howell, 1983; Unruh and Luck, 1987) may limit the spread of scale populations.

Finally, differential mortality inflicted by scale enemies on different hosts may further influence scale distribution (McClure, 1979; Hanks and Denno, in prep.).

These environmental factors may interact in ways that complicate the distribution of scale populations. For example, predators exclude white peach scale populations from the highest quality trees growing in natural settings, while host water stress in urban settings further restricts the spectrum of available hosts. These selective forces acting together confine large-scale populations to a narrow window of low-quality hosts in disturbed areas (Hanks and Denno, in prep.).

Demic Adaptation and the Introduction of Scale Pests

The introduction of scales into new regions provides a means of examining scale populations under very different environmental regimes. McClure (1983, 1986) studied the ecology of two hemlock scale species, both in their native Japan and in the United States. Japanese scale populations were low due to close synchrony between scale and parasitoid populations, but, where trees stood outside their natural range, scale populations were greater due to a change in host plant physiology. In the United States, fecundity was greater on Japanese hemlocks than on North American species, and population densities were high due to asynchrony between scale and parasitoid populations. On both continents, however, scales showed significantly higher fecundity when transferred to the host species of the mother, regardless of the host's country of origin. Furthermore, this adaptation to the host species occurred over a relatively short period of time. It is notable that the extent of adaptation was probably not great enough to prevent spread of populations between host species. This improved performance on maternal hosts is good evidence for the role of demic adaptation in short-term fine tuning of populations—improving utilization of different host species in very different environments.

SUMMARY AND CONCLUSION

We conclude the following:

1. Scale insect natural history does promote reproductive isolation of subpopulations on perennial hosts.
2. Case studies of scale species do provide evidence that demic adaptation occurs, but they indicate that it is not a general phenomenon.
3. The influence of demic adaptation on the colonization and spread of scale populations within habitats is insignificant.
4. Gene flow appears to be the most important factor undermining adaptation of scale subpopulations to individual host plants: the distribution of hosts within habitats may be very important in determining the impact of this gene flow.
5. Any of a number of environmental factors may limit scale populations and work in concert to produce the contagious distribution characteristic of many scale species.

We suggest that the sedentary nature and weak dispersal abilities of scale insects, in conjunction with variation in host suitability, limit the colonization of new hosts more than does specialization of scales to host genotype.

Acknowledgments

This is Scientific Article Number A-4866, Contribution Number 7897 of the Maryland Agricultural Experiment Station Department of Entomology. Funding for the white peach scale research was provided in part by the Gahan Scholarship of the Department of Entomology, University of Maryland. Thanks go to B. Benrey, J. A. Davidson, M. S. McClure, C. Mitter, K. L. Olmstead, G. K. Roderick, T. R. Unruh, and to two anonymous reviewers for their comments on an early draft.

REFERENCES

Alstad, D. N., G. F. Edmunds, Jr. 1983. Adaptation, host specificity and gene flow in the black pineleaf scale. Pages 413–426 in Denno, R. F., and M. S. McClure (eds.), *Variable Plants and Herbivores in Natural and Managed Systems*. New York: Academic Press.

Alstad, D. N., G. F. Edmunds, Jr., and S. C. Johnson. 1980. Host adaptation, sex ratio, and flight activity in male black pineleaf scale. *Ann. Entomol. Soc. Am.* 73:665–667.

Augustin, S. 1986. Investigation on the distribution of first instar larvae of *Cryptococcus fagisuga* Lind. (Homoptera, Coccoidea) in a beech forest. *J. Appl. Entomol.* 102:178–194.

Baker, H. G. 1965. Characteristics and modes of origin of weeds. Pages 147–172 in Baker, H. G., and G. L. Stebbins (eds.), *The Genetics of Colonizing Species*. New York: Academic Press.

Beardsley, J. W., Jr., and R. H. Gonzalez. 1975. The biology and ecology of armored scales. *Annu. Rev. Entomol.* 20:47–73.

Bliss, C. I., A. W. Cressman, and B. M. Broadbent. 1935. Productivity of the camphor scale and the biology of its eggs and crawler stages. *J. Agric. Res.* 50:243–266.

Brown, C. E. 1958. Dispersal of the pine needle scale, *Phenacaspis pinifoliae* (Fitch), (Diaspididae: Homoptera). *Can. Entomol.* 90:685–690.

Burns, D. P. and D. E. Donley. 1970. Biology of the tulip tree scale, *Toumeyella liriodendri* (Homoptera: Coccidae). *Ann. Entomol. Soc. Am.* 63:228–235.

Cobb, N. S., and T. G. Whitham. In prep. Deme formation and population cycling in the pinyon pine needle scale.

Cornwall, P. B. 1960. Movements of the vectors of virus diseases on cacao in Ghana. II. Wind movements and aerial dispersal. *Bull. Entomol. Res.* 51:175–201.

Davidson, J. A., D. R. Miller, and S. Nakahara. 1983. The white peach scale, *Pseudaulacaspis pentagona* (Targioni-Tozzetti) (Homoptera: Diaspididae): Evidence that current concepts include two species. *Proc. Entomol. Soc. Wash.* 85:753–761.

DeBach, P., D. Rosen, and C. E. Kennett. 1971. Biological control of coccids by introduced natural enemies. Pages 165–194 in Huffaker, C. B. (ed.), *Biological Control*. New York: Plenum.

Denno, R. F. 1983. Tracking variable host plants in space and time. Pages 291–341 in Denno, R. F., and M. S. McClure (eds.), *Variable Plants and Herbivores in Natural and Managed Systems*. New York: Academic Press.

Doane, C. C. 1966. Evidence for a sex attractant in females of the red pine scale. *J. Econ. Entomol.* 59:1539–1540.

Edmunds, G. F. 1973. Ecology of black pineleaf scale (Homoptera: Diaspididae). *Environ. Entomol.* 2:765–777.

Edmunds, G. F., and D. N. Alstad. 1978. Coevolution in insect herbivores and conifers. *Science* 199:941–945.

———. 1981. Responses of black pineleaf scales to host plant variability. Pages 29–38 in Denno, R. F., and H. Dingle (eds.), *Insect Life History Patterns*. New York: Springer-Verlag.

Felsenstein, J. 1988. Sex and the evolution of recombination. Pages 74–86 in Michod, R. E., and B. R. Levin (eds.), *The Evolution of Sex: An Examination of Current Ideas*. Sunderland, Mass.: Sinauer.

Flanders, S. E. 1970. Observations on host plant induced behavior of scale insects and their endoparasites. *Can. Entomol.* 102:913–926.

Futuyma, D. J. 1986. *Evolutionary Biology,* 2nd ed. Sunderland, Mass.: Sinauer.

Futuyma, D. J., and G. C. Mayer. 1980. Non-allopatric speciation in animals. *Syst. Zool.* 29:254–271.

Futuyma, D. J., and S. C. Peterson. 1985. Genetic variation in the use of resources by insects. *Annu. Rev. Entomol.* 30:217–238.

Futuyma, D. J., and T. E. Philippi, 1987. Genetic variation and covariation in responses to host plants by *Alsophila pometaria* (Lepidoptera: Geometridae). *Evolution* 41:269–279.

Gadgil, M. 1971. Dispersal: Population consequences and evolution. *Ecology* 52:253–261.

Greathead, D. J. 1972. Dispersal of the sugar-cane scale *Aulacaspis tegalensis* (Zhnt.) (Hem., Diaspididae) by air currents. *Bull. Entomol. Res.* 61:547–558.

Gould, F. 1979. Rapid host evolution in a population of the phytophagous mite *Tetranychus urticae* Koch. *Evolution* 33:791–802.

Goux, L. 1944. Contribution to the study of the biology of *Eriococcus* (Hemiptera: Coccidae) diet, alimentary specificity and biological races. *Bull. Mus. Hist. Nat. Mars. T.IV* Nos. 3–4: 135–152.

Habib, A. 1957. The morphology and biometry of the *Eulecanium corni*-group, and its relation to host-plants. *Bull. Soc. Entomol. Egypte* XLI:381–410.

Hamilton, W. D., and R. M. May. 1977. Dispersal in stable habitats. *Nature* 269:578–581.

Hanks, L. M., and R. F. Denno. In prep. The relative importance of demic adaptation and habitat effects in the distribution of an armored scale insect.

Howell, J. O., and M. L. Williams. 1976. An annotated key to the families of scale insects (Homoptera: Coccoidea) of America, north of Mexico, based on characteristics of adult females. *Ann. Entomol. Soc. Am.* 69:181–189.

Hughes, I. W. 1960. Host relationships and natural enemies of the white peach scale, *Pseudaulacaspis pentagona* (Targioni), (Homoptera, Coccoidea). M.S. thesis, Univ. of Florida, Gainesville, Fla., 69 pp.

Kareiva, P. 1983. Influence of vegetation texture on herbivore populations: Resource concentration and herbivore movement. Pages 259–289 in Denno, R. F., and M. S. McClure (eds.), *Variable Plants and Herbivores in Natural and Managed Systems*. New York: Academic Press.

Kuris, A. M., and S. F. Norton. 1985. Evolutionary importance of overspecialization: Insect parasitoids as an example. *Am. Nat.* 126:387–391.

Lellakova-Duskova, F. 1963. Morphology, metamorphosis and the life cycle of the scale insect *Quadraspidiotus gigas* (Thiem et Gerneck) (Homoptera, Coccoidea, Diaspididae). *Acta. Entomol. Mus. Natl. Prague* 35:611–648.

Levins, R. 1968. *Evolution in Changing Environments*. Princeton, N.J. Princeton Univ. Press.

Mattson, W. J., R. K. Lawrence, R. A. Haack, D. A. Herms, and P. Charles. 1988. Defensive strategies of woody plants against different insect-feeding guilds in relation to plant ecological strategies and intimacy of association with insects. Pages 3–38 in Mattson, W. J., J. Levieux, and C. Bernard-Dagan (eds.), *Mechanisms of Woody Plant Defenses Against Insects: Search for a Pattern*. New York: Springer-Verlag.

McCambridge, W. F., and D. A. Pierce. 1964. Observations on the life history of the pinyon needle scale *Matsucoccus acalyptus* (Homoptera, Coccoidea, Margarodidae). *Ann. Entomol. Soc. Am.* 57:197–200.

McClure, M. S. 1977. Dispersal of the scale *Fiorinia externa* (Homoptera: Diaspididae) and effects of edaphic factors on its establishment on hemlock. Environ. Entomol. 6:539–544.

———. 1979. Spatial and seasonal distribution of disseminating stages of *Fiorinia externa* (Homoptera: Diaspididae) and natural enemies in a hemlock forest. *Environ. Entomol.* 8:869–873.

———. 1983. Reproduction and adaptation of exotic hemlock scales (Homoptera: Diaspididae) on their new and native hosts. *Environ. Entomol.* 12:1811–1815.

———. 1986. Population dynamics of Japanese hemlock scales: a comparison of endemic and exotic communities. *Ecology* 67:1411–1421.

Melis, A. 1943. Contributo alla conoscenza dell' *Aspidiotus perniciosus*. *Redia* 29:1–170.

Miller, D. R., and M. Kosztarab. 1979. Recent advances in the study of scale insects. *Annu. Rev. Entomol.* 24:1–27.

Mitter, C., and D. J. Futuyma. 1983. An evolutionary genetic view of host-plant utilization by insects. Pages 413–495 in Denno, R. F., and M. S. McClure (eds.), *Variable Plants and Herbivores in Natural and Managed Systems*. New York: Academic Press.

Moran, V. C., B. H. Gunn, and G. H. Walter. 1982. Wind dispersal and settling of first-instar crawlers of the cochineal insect *Dactylopius austrinus* (Homoptera: Coccoidea: Dactylopiidae). *Ecol. Entomol.* 7:409–419.

Nielsen, D. G., and N. E. Johnson 1973. Contribution to the life history and dynamics of the pine needle scale, *Phenacaspis pinifoliae*, in central New York. *Ann. Entomol. Soc. Am.* 66:32–43.

Quayle, H. J. 1916. Dispersion of scale insects by wind. *J. Econ. Entoml.* 9:486–492.

Rabkin, F. B., and R. R. Lejeune. 1954. Some aspects of the biology of the pine tortoise scale, *Toumeyella numismaticum* (Petit and McDaniel) (Homoptera: Coccidae). *Can. Entomol.* 86:570–575.

Rausher, M. D. 1984. Trade offs in performance on different hosts: Evidence from within- and between-site variation in the beetle *Deloyala guttata*. *Evolution* 38:582–595.

———. 1988. Is coevolution dead? *Ecology* 69:898–901.

Rice, W. R. 1983. Sexual reproduction: An adaptation reducing parent–offspring contagion. *Evolution* 37:1317–1320.

Rice, R. E., and D. S. Moreno. 1970. Flight of male California red scale. *Ann. Entomol. Soc. Am.* 63:91–96.

Rosenthal, G. A., and D. H. Janzen (eds.), 1979. *Herbivores: Their Interaction with Secondary Plant Metabolites*. New York: Academic Press.

Rotundo, G., E. Tremblay, and P. Papa. 1980. Short-range orientation of *Pseudococcus calceolariae* (Mask.) males (Homoptera: Coccoidea) in a wind tunnel. *Boll. Lab. Entomol. Agric. <<F. Silvestri>>*. 37:1–37.

Selander, R. K. 1970. Behavior and genetic variation in natural populations. *Am. Zool.* 10:53–66.

Smith, M. A., and H. V. Cornell. 1979. Hopkins host-selection in *Nasonia vitripennis* and its implication for sympatric speciation. *Anim. Behav.* 27:365–370.

Stephens, G. R., and D. A. Aylor. 1978. Aerial dispersal of red pine scale, *Matsucoccus resinosae* (Homoptera: Margarodidae). *Environ. Entomol.* 7:556–563.

Tashiro, H., and J. B. Beavers. 1968. Growth and development of the California red scale, *Aonidiella aurantii*. *Ann. Entomol. Soc. Am.* 61:1009–1014.

Tashiro, H., and C. Moffitt. 1968. Reproduction in the California red scale, *Aonidiella aurantii*. II. Mating behavior and post insemination female changes. *Ann. Entomol. Soc. Am.* 61:1014–1020.

Templeton, A. R. 1980. Modes of speciation and inferences based on genetic distances. *Evolution* 34:719–729.

———. 1981. Mechanisms of speciation—A population genetic approach. *Annu. Rev. Ecol. Syst.* 12:23–48.

Unruh, T. R., and R. F. Luck. 1987. Deme formation in scale insects: A test with pinyon pine scale. *Ecol. Entomol.* 12:439–450.

Via, S. 1984. The quantitative genetics of polyphagy in an insect herbivore: I. Genotype-environment interaction in larval performance on different host plant species. *Evolution* 38:881–895.

Via, S., and R. Lande. 1985. Genotype-environment interaction and the evolution of phenotypic plasticity. *Evolution* 39:505–523.

Wainhouse, D. 1979. Dispersal of the beech scale (*Cryptococcus fagi* Baer) in relation to the development of beech bark disease. *Bull. Soc. Entomol. Suisse* 52:181–183.

———. 1980. Dispersal of first instar larvae of the felted beech scale, *Cryptococcus fagisuga*. *J. Appl. Ecol.* 17:523–532.

Wainhouse, D., and R. Deeble. 1980. Variation in susceptibility of beech (*Fagus* spp) to beech scale (*Cryptococcus fagisuga*). *Ann. Sci. For.* 37:279–289.

Wainhouse, D., and R. S. Howell. 1982. Distribution of attack by beech scale *Cryptococcus fagisuga* in beech progeny trials. *Proc. I.U.F.R.O. Beech Bark Disease Working Party Conf.* 33–37.

———. 1983. Intraspecific variation in beech scale populations and in susceptibility of their host *Fagus sylvatica*. Ecol. Entomol. 8:351–359.

Wainhouse, D., I. M. Gate, and D. Lonsdale. 1988. Beech resistance to the beech scale: A variety of defenses. Pages 277–293 in Mattson, W. J., J. Levieux, and C. Bernard-Dagan (eds.), *Mechanisms of Woody Plant Defenses Against Insects: Search for a Pattern*. New York: Springer-Verlag.

Ward, S. A. 1987. Optimal habitat selection in time-limited dispersers. *Am. Nat.* 129:568–579.

Washburn, J. O., and G. W. Frankie. 1985. Biological studies of iceplant scales, *Pulvina-*

riella mesembryanthemi and *Pulvinaria delottoi* (Homoptera: Coccidae) in California. *Hilgardia* 53:1–27.

Whitham, T. G. 1983. Host manipulation of parasites: Within-plant variation as a defense against rapidly evolving pests. Pages 15–41 in Denno, R. F., and M. S. McClure (eds.), *Variable Plants and Herbivores in Natural and Managed Systems*, New York: Academic Press.

21

PATTERNS OF GENETIC VARIATION AND THE ECOLOGY OF THE YELLOW FEVER MOSQUITO, *AEDES AEGYPTI**

Walter J. Tabachnick

U.S. Department of Agriculture, Agricultural Research Service
Arthropod-borne Animal Disease Research Laboratory

INTRODUCTION

Ecological genetics concerns the relationship between an organism's ecology and population genetic structure. Progress in this area of research has depended on a variety of molecular and quantitative genetic methods to measure genetic variation within natural populations. Although molecular techniques have been extensively applied to study genetic variation, molecular analyses are usually inappropriate for many ecological traits, due to the lack of appropriate genetic markers and the genetic complexity of the traits. Ecological genetic studies need to employ both molecular and quantitative analyses.

Quantitative and Molecular Analyses of Genetic Variation

A number of quantitative techniques have been used to analyze genetic variation of complex traits in insects. One method uses isofemale lines to characterize population variation. Each isofemale line represents a strain descended from an individual field-collected female. Isofemale lines have been used to characterize genetic variation for traits in species of *Drosophila*, that is, vagility, photoresponse, ethanol tolerance, and so on (de la Rosa et al., 1984; Rockwell, 1980; Parsons, 1977, 1980). Isofemale lines have also been utilized to characterize variation in vector competence in mos-

*This article reports the results of research only. Mention of a proprietary product does not constitute an endorsement or a recommendation for its use by USDA, nor does it imply registration under FIFRA as amended.

quitoes (Collins et al., 1986; Tabachnick et al., 1982). A second approach has been to isolate homozygous chromosomes (isochromosomal) using chromosomal systems that suppress recombination. This technique enables one to test the effects of specific chromosomal regions on quantitative variation. Isochromosomal lines of *D. melanogaster* Meigen were used to determine interchromosomal effects on quantitative variation of several enzyme levels (Laurie-Ahlberg et al., 1980; Maroni et al., 1982). Sakai and Baker (1972) used the technique to detect hidden variation in the mosquito, *Culex tritaeniorhynchus* Giles. Finally, artificial selection and the statistics of quantitative genetics have been used to estimate genetic variation in insects (reviewed by Lewontin, 1974). Among ecological traits examined are insect diapause (Dingle, 1974; Tauber and Tauber, 1986), resource preference (reviewed by Futuyma and Peterson, 1985), and life-history traits (Palmer and Dingle, 1986; Tauber et al., 1986).

The application of molecular biology to studies of genetic variation within and among insect populations continues to provide valuable information concerning genetic differentiation, genetic relatedness, and population structure. Molecular techniques include isozyme electrophoresis (Berlocher, 1984), restriction enzyme analysis of mitochondrial DNA (Lansman et al., 1981; DeSalle et al., 1987), and nucleotide sequencing of individual genes (Kreitman, 1983). At present, isozyme electrophoresis has been used in the overwhelming majority of studies.

Genetic studies of insect populations have demonstrated that genetic variation within insect populations may be structured within small geographic areas. Taylor and Powell (1977) studied isozyme microgeographic variation in populations of *Drosophila persimilis* Dobzhansky and Epling. Coluzzi et al. (1977) demonstrated differences in chromosomal inversion frequencies between indoor and outdoor habitats of *Anopheles gambiae* Giles. Diehl and Bush (1984) reviewed numerous examples of genetic variation within insect populations that may be related to host plant utilization (also reviewed by Futuyma and Peterson, 1985).

It has been extremely difficult to relate allozyme variation to environmental variability. Tabachnick and Wallis (1985) distinguished two approaches to relate genetic variation in natural populations to ecological factors. Molecular genetic variation may be due to adaptation to the environment, or such variation may simply be a secondary consequence of the recent history of the populations sampled. Historical explanations, if available, should be used instead of complex adaptational hypotheses, especially when additional evidence concerning the adaptive significance of isozyme polymorphisms is unavailable (Selander and Whittam, 1983).

In this chapter, I will summarize genetic studies of the yellow-fever mosquito, *Aedes aegypti* (Linnaeus), in which a variety of methods have been used to understand ecological factors and their effect on patterns of genetic variation.

POPULATION GENETICS OF *AEDES AEGYPTI*

Genetic variation in *Aedes aegypti* (Linnaeus) has been well characterized by using isozyme electrophoresis (Tabachnick and Powell, 1979; Wallis et al., 1983). Microgeographic differentiation within selected populations of the species is well documented. Tabachnick et al. (1979) found genetic differences between sympatric populations of the subspecies *A. ae. aegypti* and *A. ae. formosus* (Walker) in coastal Kenya. It is believed that reduced gene flow between the subspecies in Kenya is maintained by different habitat preferences (Tabachnick et al., 1979; Petersen, 1977). Due to the reduced gene flow between indoor village *A. ae. aegypti* and surrounding outdoor *A. ae. formosus*, genetic differentiation was detected between indoor populations in villages that were only a few kilometers apart (Tabachnick and Powell, 1978).

Caribbean populations of *A. aegypti* were found to show extensive genetic heterogeneity when isozyme electrophoresis was used (Wallis et al., 1984). Tabachnick and Wallis (1985) believe that the heterogeneity of Caribbean populations is most easily explained by the recent history of mosquito control programs in the region. Mosquito control activities directly reduce mosquito population size. The history in the Caribbean is one of successful control followed by cessation of control efforts and subsequent increases in mosquito populations. It is believed that the fluctuations of population size affect the genetic heterogeneity of Caribbean mosquito populations (Tabachnick and Wallis, 1985). There is a significant correlation between average expected heterozygosity and estimates of the size of mosquito populations in the region. Islands having large and poorly controlled mosquito populations have higher genetic heterozygosity than smaller populations that presumably result from smaller island size and effective mosquito control programs (Tabachnick and Wallis, 1985).

There are several examples of panmictic *A. aegypti* populations. Studies show little microgeographic genetic variation in *A. aegypti* populations subject to poor mosquito control. Genetic differentiation was not detectable among *A. aegypti* populations sampled throughout the city of New Orleans prior to the start of current control efforts (Tabachnick, 1982). Mosquitoes collected within different houses from two Kenyan villages showed little differentiation from one another, demonstrating that the indoor populations of these villages are panmictic (Tabachnick and Powell, 1978). The results of population genetic studies of *A. aegypti* are thus consistent with the hypothesis that a major factor determining the genetic structure of this species is the human effort aimed at controlling mosquito populations.

Microgeographic Variation on the Island of Anguilla

The Caribbean island of Anguilla provides an opportunity to investigate factors that are important in maintaining genetic heterogeneity among populations of the yellow-fever mosquito. Parker et al. (1983) reported that Anguilla contains a large population of *A. aegypti* breeding in rock holes found throughout the island. Anguilla also has typical domestic *A. aegypti* that breed in and around houses. Rock hole populations (nondomestic) are not common to the New World but have been reported in Puerto Rico (Fox et al., 1960), the Caymans (Nathan and Giglioli, 1982), and originally on Anguilla by Belkin and Heinemann (1976). Anguilla is unique, because the rock hole populations are numerous and some are found 1–2 kilometers from any human habitation.

Electrophoretic Variation

Wallis and Tabachnick (1990) examined genetic variation by using isozyme electrophoresis in seven Anguillan *A. aegypti* populations in order to determine the nature of genetic differentiation on the island. Genetic differentiation was detected at two loci consistent with differences between rock hole and domestic habitats. The amount of genetic divergence on Anguilla, as estimated by the average genetic distance among all populations on the island, was not significantly different when compared with divergence estimates observed among mosquito populations within larger Caribbean islands (Wallis and Tabachnick, 1990). Although Anguilla is a small island and has a vigorous mosquito control program, the *A. aegypti* populations maintain as much genetic variation as much larger islands with little mosquito control, that is, Jamaica and Puerto Rico (Tabachnick and Wallis, 1985). We believe that the availability of two habitats plays a major role in the genetic structure of the mosquito populations on Anguilla (Wallis and Tabachnick, 1990). A detailed study of Anguilla *A. aegypti* will provide valuable information concerning how ecological factors shape genetic variation in mosquito populations.

Variation in Developmental Rate

Experiments on the immature stages have shown that rock hole Anguillan *A. aegypti* develop faster than domestic Anguillan *A. aegypti*. Table 21.1 shows the number of pupae recorded per day from two rock hole collections reared in the laboratory compared with a domestic collection. All three populations were reared at the same time under identical conditions of food and temperature. Both rock hole collections developed significantly faster than the domestic collection ($X^2 = 131.26$, df $= 6$, $p < 0.01$). The experiment was repeated by using the F_1 offspring of adults that had been collected from the field as larvae. Each collection was divided as first-instar larvae into five different rearing pans with 25 larvae each. The pans were placed at 25°C and 75% relative humidity, and were provided with identical amounts of a guinea pig chow mixture for food each day. A two-way analysis of variance (Zar, 1984) showed a significant difference in development between the two populations from day to day ($F_{(6,56)} = 6.59$, $p < 0.01$) (Table 21.2).

Table 21.1. Larval Development Rate of Anguillan *Aedes aegypti* in the Laboratory

Population	Number of Eggs Hatched	Number of Larvae	Number Pupated, Day 6	7	8	9
Rock hole I	101	101	78	23	...	...
Rock hole II	99	99	77	22	...	...
Domestic	180	56	3	33	13	7

Table 21.2. Mean Larval Development Rates Using 5 Replicates of 25 Larvae per Replicate in Anguilla *Aedes aegypti*

	Mean Number of Pupae per Day							
	Day							
Population	7	8	9	10	11	12	13	>13
Rock hole (Crocus Bay)	1.4	9.2	5.4	4.2	3.0	0.6	0.6	0.6
Domestic (Island Harbor)	0	1.0	2.0	3.8	4.4	3.2	3.0	7.6

Ecological Factors

Genetic and developmental studies of Anguillan *A. aegypti* demonstrate differences and restricted gene flow among populations in the two different habitats. In contrast to observations in other populations, that is, *A. aegypti* in New Orleans and villages in Kenya (Tabachnick, 1982; Tabachnick and Powell, 1978), Anguillan populations are not panmictic. Mosquito populations on the island are genetically structured, largely along the available habitats on the island.

In view of the significance of human mosquito control efforts on *A. aegypti* population structure in the Caribbean (Tabachnick and Wallis, 1985), it is likely that mosquito control on Anguilla is involved in maintaining the genetic structure on the island. Mosquito control on Anguilla has been extensive for the past several years (Parker et al., 1983). Control consisted initially of the larvicide abate (1% temephos) in domestic water containers and more recently by stocking many of the large water cisterns with the mosquito larvae-eating guppy, *Poecilia reticulata* (Parker and Richardson, 1990, personal communication). Rock hole mosquito populations have not been subjected to mosquito control due to the enormous number of rock holes on the island. Work by Dr. George Georghiou (Georghiou et al., 1987; 1990, personal

Table 21.3. Insecticide Resistance of Anguilla *Aedes aegypti*[a]

	Control Strain Susceptible		Anguilla Domestic (Southhill)		Anguilla Rock hole Limestone (Bay)	
			Resistance Ratios[b] at			
Insecticide	LC_{50} (ppm)	LC_{95} (ppm)	LC_{50}	LC_{95}	LC_{50}	LC_{95}
Malathion	0.240	0.329	1.55	1.95	0.921	1.32
Temephos	0.012	0.020	4.47	4.65	0.894	1.10
Fenthion	0.019	0.029	2.76	3.61	1.52	2.28
DDT	0.054	0.189	89.1	586.0	1.75	6.78

[a]Data for malathion, temephos, and fenthion from Georghiou et al., 1987. Data for DDT from G. P. Georghiou, University of California—Riverside, personal communication, July 1987.
[b]Resistance ratio = LC of field strain/LC of control (susceptible) strain.

communication) showed differences in insecticide resistance between a rock hole and domestic collection from Anguilla. Georghiou found that the domestic collection had significantly higher resistance to several insecticides (Table 21.3).

Behavioral Genetics of Oviposition Site Preference

Domestic populations of Anguillan *A. aegypti* must be under intense selection to oviposit in available untreated oviposition sites. Detectable differences in the oviposition sites in rock hole and domestic habitats would act to maintain restricted gene flow between the habitats. To explore this possibility, we used artificial selection to determine if any differences existed between habitats and the extent of genetic variation for oviposition site preference in a rock hole and domestic collection. The greater the genetic variation for the trait is within a population, the greater will be the response of that population to artificial selection (Hartl, 1980).

Fifty blood-fed Anguillan domestic *A. aegypti* females (F_3 generation in laboratory colony) were placed in a large insect-rearing cage (about 50 cm × 40 cm × 40 cm), with two oviposition cups (10-ml capacity) containing 0.5% abate and two oviposition cups containing only water. After six days, the numbers of eggs oviposited in each cup were recorded; those in the abate cups were rinsed several times with water to remove the insecticide, and the abate and nonabate eggs were hatched separately and reared to the adult stage. The adults were blood-fed, and a maximum of 25 females of each group were tested for oviposition preference as before. An Anguillan rock hole population (F_3 generation) was treated in the same manner.

The results of the selection scheme are shown in Table 21.4. Typical of artificial selection, inbreeding depression occurred, resulting in the loss of some lines after a few generations. The Anguillan rock hole and domestic strains were not significantly different in oviposition preference under laboratory conditions. Adult females from both Anguillan strains did not initially prefer to oviposit in untreated containers ($P > 0.05$). Selection increased the proportion of eggs oviposited in the untreated containers, demonstrating genetic variation for this preference in both populations. However, preference for untreated containers increased in all selected lines, including the two lines selected for ovipositing in insecticide-treated containers. I was unable to select for oviposition preference for insecticide containers demonstrating little genetic variation for the trait in either population. The absence of variation for insecticide oviposition preference is expected, since such variation would be quickly selected against on Anguilla. I believe that the increase in noninsecticide preference in the abate-selected lines may be due to two possibilities. It is consistent with strong preference to avoid insecticide under our laboratory conditions, which may increase with continued exposure. In addition, abate-selected oviposition preference may be selected against in the laboratory, due to other genetic modifiers. Additional behavioral genetic studies of this nature are necessary to define genetic differentiation among these populations. Laboratory studies must be repeated under field conditions. For example, Moore (1977) found insecticide avoidance for both abate and malathion by ovipositing *A. aegypti* during field studies in Puerto Rico. Oviposition

Table 21.4. Artificial Selection for Oviposition Site Preference in Anguilla *Aedes aegypti*.

	Rock hole Population (Crocus Bay)				*Domestic Population (Island Harbor)*			
Selection for	*Abate Preference*		*Nonabate Preference*		*Abate Preference*		*Nonabate Preference*	
Generation	N	% Non-abate Eggs	N	% Non-abate Eggs	N	% Non-abate Eggs	N	% Non-abate Eggs
P_1	1,584	51.5	1,584	51.5	1,383	47.4	1,383	47.4
F_1	1,453	53.5	1,706	88.3	218	38.1	213	83.6
F_2	2,487	84.4	2,074	77.4	284	43.7	1,000	64.5
F_3	1,637	92.4	1,638	63.9	407	61.4	317	64.4
F_4	203	70.0	386	67.1	54	87.0	751	70.2
F_5	130	80.8	620	78.7	...	...	...	...
F_6	...	...	492	77.2	...	...	...	...
F_7	...	...	616	80.3	...	...	...	...

N = total number of eggs counted.

site preference variation need not be based solely on insecticide but could be dependent on other environmental factors associated with the rock hole and domestic habitats, that is, pH, temperature, light, leaf litter, and so forth.

CONCLUSIONS

Ecological genetic studies are necessary to understand the mechanisms maintaining insect population genetic structure. *A. aegypti* populations on Anguilla provide a situation for exploring the relationship between ecological factors and genetic variation. Population structure on Anguilla is most likely due to the availability of alternative habitats that are not subjected to mosquito control activities. It will be necessary to utilize several molecular and quantitative methods to confirm this. Molecular biological techniques will continue to play an important role in analyzing population structure, in determining genetic relatedness, and in defining evolutionary relationships of mosquito populations. However, studies of the genetic variation of complex traits are necessary. The genetics of life-history traits and ecologically important behaviors, such as oviposition behavior, need to be explored in natural populations. Such work will, by necessity, employ such quantitative techniques as artificial selection, as well as isofemale and isochromosomal analyses to quantify genetic variation within and between populations. The application of molecular biological techniques such as DNA restriction fragment polymorphisms (RFLPs) to these studies will allow genetic analyses of these important traits for the first time.

Acknowledgments

I would like to express my appreciation to R. Delanois, J. Duval, and L. Marshall for their technical assistance. G. P. Georghiou generously provided unpublished data on insecticide resistance. I would like to thank T. H. G. Aitken for providing his data on developmental rate. F. R. Holbrook, B. R. Miller, C. Moore, T. E. Walton, and W. C. Wilson provided useful comments on the manuscript. L. Richardson and A. Parker of the Anguilla Mosquito Control Unit generously provided A. aegypti *for these studies. Portions of this work were accomplished in the Department of Biology, Loyola University of Chicago, and were supported by funds from NIH BRSG Grant #007120.*

REFERENCES

Belkin, J. W., and S. J. Heinemann. 1976. Collection records of the project "Mosquitoes of Middle America" on 4 leeward islands: Anguilla (ANG), Barbuda (BAB), Montserrat (MNT), Nevis (NVS), St. Kitts (KIT). *Mosq. Syst.* 8:23–62.

Berlocher, S. H. 1984. Insect molecular systematics. *Annu. Rev. Entomol.* 29:403–433.

Collins, F. H., R. K. Sakai, K. D. Vernick, S. Paskewitz, D. C. Seeley, L. H. Miller, W. E. Collins, C. C. Campbell, and R. W. Gwadz. 1986. Genetic selection of a *Plasmodium*-refractory strain of the malaria vector *Anopheles gambiae*. *Science* 234:607–610.

Coluzzi, M., A. Sabatini, V. Petrarca, and M. A. diDeco. 1977. Behavioral divergence between mosquitoes of the *Anopheles gambiae* complex. *Nature* 266:832–833.

de La Rosa, M. E., M. I. Gaso, F. Gonzales, J. Guzman, L. Levine, O. Olvera, and R. F. Rockwell. 1984. Chromosomal and behavioral studies of Mexican *Drosophila*. II. Temporal changes of vagility cnaracteristics of three populations of *D. pseudoobscura*. *J. Hered.* 75:281–284.

DeSalle, R., A. Templeton, I. Mori, S. Pletscher, and J. S. Johnston. 1987. Temporal and spatial heterogeneity of mtDNA polymorphism in natural populations of *Drosophila mercatorum*. *Genetics* 116:215–223.

Diehl, S. R., and G. Bush. 1984. An evolutionary and applied perspective of insect biotypes. *Annu. Rev. Entomol.* 29:471–504.

Dingle, H. 1974. The experimental analysis of migration and life-history strategies in insects. Pages 329–342 in Browne, L. B. (ed.), *Experimental Analysis of Insect Behavior*. Berlin: Springer-Verlag.

Fox, I., A. H. Boike, Jr., and I. Garcia-Moll. 1960. Notes on rock-hole breeding and resistance of *Aedes aegypti* on Puerto Rico. *Am. J. Trop. Med. Hyg.* 9:425–429.

Futuyma, D. J., and S. C. Peterson. 1985. Genetic variation in the use of resources by insects. *Annu. Rev. Entomol.* 30:217–238.

Georghiou, G. P. 1990. Personal communication.

Georghiou, G. P., M. Wirth, H. Tran, F. Saume, and A. B. Knudsen. 1987. Potential for organophosphate resistance in *Aedes aegypti* (Diptera: Culicidae) in the Caribbean area and neighboring countries. *J. Med. Entomol.* 24:290–294.

Hartl, D. L. 1980. *Principles of Population Genetics*. Sunderland, Mass.: Sinauer.

Kreitman, M. 1983. Nucleotide polymorphism at the alcohol dehydrogenase locus of *Drosophila melanogaster*. *Nature* 304:412–417.

Lansman, R. A., R. O. Shade, J. F. Shapira, and J. C. Avise. 1981. The use of restriction endonucleases to measure mitochondrial DNA sequence relatedness in natural populations. III. Techniques and potential applications. *J. Mol. Evol.* 17:214–226.

Laurie-Ahlberg, C. C., G. Maroni, G. C. Bewley, J. C. Lucchesi, and B. S. Weir. 1980. Quantitative genetic variation of enzyme activities in natural populations of *Drosophila melanogaster*. *Proc. Natl. Acad. Sci. USA* 77:1073–1077.

Lewontin, R. 1974. *The Genetic Basis of Evolutionary Change*. New York: Columbia Univ. Press.

Maroni, G., C. C. Laurie-Ahlberg, D. A. Adams, and A. N. Wilton. 1982. Genetic variation in the expression of Adh in *Drosophila melanogaster*. *Genetics* 101:431–446.

Moore, C. 1977. Insecticide avoidance by ovipositing *Aedes aegypti*. *Mosq. News* 37:291–293.

Nathan, M. B., and M. E. C. Gigliolo. 1982. Eradication of *Aedes aegypti* on Cayman Brac and Little Cayman, West Indies with abate (temephos) in 1970–1971. *Pan Am. Health Org. Sci. Publ.* 16:28–39.

Palmer, J. O., and H. Dingle. 1986. Direct and correlated responses to selection among life-history traits in milkweed bugs (*Oncopeltus fasciatus*). *Evolution* 40:767–777.

Parker, A. G., M. E. C. Giglioli, S. Mussington, A. B. Knudsen, R. A. Ward, and R. Arons. 1983. Rock hole habitats of a feral population of *Aedes aegypti* on the island of Anguilla, West Indies. *Mosq. News* 43:79–81.

Parker, A., and L. Richardson. 1990. Anguilla Mosquito Control Unit. Personal communication.

Parsons, P. A. 1977. Isofemale lines and quantitative traits in natural populations of *Drosophila*. *Am. Nat.* 111:613–621.

———. 1980. Larval response to environmental ethanol in *Drosophila melanogaster*: Variation within and among populations. *Behav. Genet.* 10:183–190.

Petersen, J. L. 1977. Behavioral differences in two subspecies of *Aedes aegypti* (L.) (Diptera: Culicidae) in East Africa. Ph.D Dissertation, Univ. of Notre Dame.

Rockwell, R. F. 1980. Photobehavioral differentiation in natural populations of *Drosophila*: Changes of photoresponse over time in *D. pseudoobscura* and *D. persimilis*. *Behav. Genet.* 10:521–535.

Sakai, R. K., and R. H. Baker. 1972. A method for detecting and measuring concealed variability in the mosquito, *Culex tritaeniorhynchus*. *Genetics* 71:287–296.

Selander, R. K., and T. S. Whittam. 1983. Protein polymorphism and the genetic structure of populations. Pages 89–114 in Nei, M., and R. K. Koehn (eds.), *Evolution of Genes and Proteins*. Sunderland, Mass.: Sinauer.

Tabachnick, W. J. 1982. Geographic and temporal patterns of genetic variation of *Aedes aegypti* in New Orleans. *Am. J. Trop. Med. Hyg.* 31:849–853.

Tabachnick, W. J., and J. R. Powell. 1978. Genetic structure of the East African domestic populations of *Aedes aegypti*. *Nature* 272:535–537.

———. 1979. A world-wide survey of genetic variation in the yellow fever mosquito, *Aedes aegypti*. *Genet. Res.* 34:215–229.

Tabachnick, W. J., and G. P. Wallis. 1985. Genetic structure of *Aedes aegypti* populations in the Caribbean: Ecological considerations. Pages 371–381 in Lounibos, L. P., J. R. Rey, and J. F. Frank (eds.), *Ecology of Mosquitoes: Proceedings of a Workshop.* Entomol. Soc. Am. Publ., College Park, Md.

Tabachnick, W. W. M., W. J., L. E. Munstermann, and J. R. Powell. 1979. Genetic distinctness of sympatric forms of *Aedes aegypti* in East Africa. *Evolution* 33:287–295.

Tabachnick, W. J., T. H. G. Aitken, B. J. Beaty, B. R. Miller, J. R. Powell, and G. P. Wallis. 1982. Genetic approaches to the study of vector competency in *Aedes aegypti*. Pages 413–432 in Steiner, W. J. Tabachnick, K. S. Rai, and S. Narang (eds.), *Recent Developments in the Genetics of Insect Disease Vectors*, Champaign, Ill.: Stipes Publ.

Tauber, C. A., and M. J. Tauber. 1986. Genetic variation in all-or-none life-history traits of the lacewing *Chrysoperla carnea*. *Can. J. Zool.* 64:1541–1544.

Tauber, M. J., C. A. Tauber, and S. Masaki. 1986. *Seasonal Adaptations of Insects*. New York: Oxford Univ. Press.

Taylor, C. E., and J. R. Powell. 1977. Microgeographic differentiation of chromosomal and enzyme polymorphisms in *Drosophila persimilis*. *Genetics* 85:681–695.

Wallis, G. P., and W. J. Tabachnick. 1990. Genetic analysis of rock hole and domestic *Aedes aegypti* on the Caribbean island of Anguilla. *J. Am. Mosq. Contr. Assoc.* 6:625–630.

Wallis, G. P., W. J. Tabachnick, and J. R. Powell. 1983. Macrogeographic genetic variation in a human commensal: *Aedes aegypti*, the yellow fever mosquito. *Genet. Res.* 41: 241–258.

———. 1984. Genetic heterogeneity among Caribbean populations of the yellow fever mosquito, *Aedes aegypti*. *Am. J. Trop. Med. Hyg.* 33:492–498.

Zar, J. H. 1984. *Biostatistical Analysis*. Englewood Cliffs, N.J.: Prentice-Hall.

SECTION V

FUTURE PERSPECTIVES

22

IMPACT OF INTRASPECIFIC VARIATION ON INSECT PEST MANAGEMENT

George G. Kennedy

North Carolina State University

INTRODUCTION

Whereas intraspecific variation is the stock-in-trade of evolutionary biologists and population geneticists, and an important phenomenon to be studied and accounted for by behaviorists and ecologists, it is an extremely important, yet very interesting, thorn in the side of insect pest managers. Intraspecific, phenotypic variation among, as well as within, populations is the rule rather than the exception. Whether genetically or nongenetically (environmentally) based, such variation is a major factor determining the effectiveness of particular pest management approaches. Recognizing the existence of intraspecific variation and understanding its nature and scope is essential at all stages in the development and implementation of effective pest management. Only when such variation is accounted for is it possible (1) to understand and predict, with acceptable accuracy, the spatial and temporal occurrence of insect pest problems; (2) to devise effective management strategies; (3) to characterize pest population responses to control measures; and (4) to assess the potential for the development of resistance to the management effort. My objective is to provide the reader with a recognition of some of the many important ways in which intraspecific variation affects integrated pest management (IPM).

NONGENETIC VARIATION AND ITS IMPLICATIONS FOR IPM

Within the constraints of existing genotypes, individuals in a population respond physiologically and behaviorally to both the components of their life system and to changes in those components. Although only genetically based variation can serve as the basis of selection for adaptation to new or altered environments, nongenetically based variation, or phenotypic plasticity, is an important element in the dynamics and persistence of insect populations (Barbosa and Baltensweiler, 1987). Both endogenous and exogenous factors influence the expressed phenotype of a given genotype and such factors are the determinants of nongenetic, intraspecific variation that is important to pest management. Several examples will serve to illustrate some of the types of nongenetic variation relevant to pest management.

Environmental Influences

Developmental rates in insects can be expressed as a function of thermal-unit accumulations. Numerous temperature-driven insect development models have been developed to predict the temporal occurrence of pest populations for use in pest management programs (see Tauber et al., 1986, Chap. 10, for a general discussion and references to original literature). To be most useful, these models must take into account the variation in the development rate versus the temperature relationship that exists within a population of insects. This variation exists even when insects are reared under carefully controlled laboratory conditions. It undoubtedly has both genetic and environmental causes and can be used in the model to generate an output that reflects the temporal distribution of a particular developmental event (e.g., adult emergence). Such information is essential if the model predictions are to be used for scheduling pest management inputs, such as scouting for eggs or larvae, applying insecticides, or releasing natural enemies en masse.

In practice, temperature-driven models generally utilize as input temperature data from a central, often regional, weather station. As a result, the temperatures do not reflect accurately the microclimatic temperatures experienced by the insects over the entire area for which the predictions are to be used. Different host plants, or common host plants in different stages of growth, may provide larvae with different amounts of insulation or exposure to warming sunlight based on their location, moisture, and growth habits. Hilly or mountainous areas may show large variations in thermal-unit accumulations and, hence, in the developmental time of the target species, between ridge tops and valley floors or between north- and south-facing slopes over relatively short distances. Thus, in areas characterized by large differences in microclimate, it may be necessary to scale the network of temperature collection stations to reflect microclimatic differences or otherwise to correct model predictions for those differences.

Numerous nongenetic factors other than temperature are known to influence the developmental rate of insects as well. Moisture levels, for example, are an important component of the nutritional ecology of phytophagous insects (Mattson and Scriber, 1987). In the case of the European corn borer, *Ostrinia nubilalis* (Hübner), free water is essential for the completion of postdiapause development (Beck, 1967).

Host Influences

The host plants on which an insect develops also profoundly influence not only developmental rate but also longevity and fecundity (Painter, 1951; Maxwell and Jennings, 1980; Slansky and Rodriguez, 1987). For highly mobile, polyphagous and oligophagous species in diverse agroecosystems containing a large biomass of several host species, differential developmental rates on different hosts can result in extended periods of adult emergence and oviposition on an area-wide basis, as compared to areas lacking such host diversity. One consequence of this may be the extensive overlapping of generations.

Similarly, for predators and parasitoids, life-history parameters of importance in determining reproductive rate (i.e., capacity for increase) and presumably the ability to suppress prey or host populations are affected by the type and availability of prey and, in some cases, supplementary food sources (Doutt, 1964; Hagen, 1987; Vinson and Barbosa, 1987). In the case of phytoseiid mites, oviposition rate (mean eggs/female/day) can vary dramatically, depending upon the species of prey fed upon (McMurtry and Rodriguez, 1987).

Parasitoids and predaceous arthropods are also affected by the host plants of their host or prey species. Such effects, when adverse, may alter the effectiveness of biological control. The mechanisms by which plant chemical factors mediate the suitability of herbivores as hosts (prey) for their natural enemies fall into two broadly overlapping categories. The first involves host plant compounds present in herbivores, which are distasteful or toxic to parasitoids or predators. Presence of plant chemicals in herbivores may result from sequestration of plant compounds (Duffey, 1980; Duffey et al., 1986; Roeske et al., 1976) or the simple presence of the plant compounds in the host's gut (Eisner, 1970). Plant compounds sequestered by phytophagous insects may act on parasitoids and predators as acute toxins or repellants (e.g. Brower, 1969; Roeske et al., 1976; Bernays and Chapman, 1978), or they may exert chronic effects (Campbell and Duffey, 1979, 1981). Host plant influences can be observed even on the fourth trophic level. Orr and Boethel (1986) demonstrated that both a predaceous sting bug [*Podisus maculiventris* (Say)] and its egg parasitoid *Telenomus podisi* Ashmead were adversely affected when *P. maculiventris* fed on larvae of soybean looper [*Pseudoplusia includens* (Walker)], reared on resistant soybean foliage. The effects were more pronounced for *T. podisi*, which manifested reductions in both fecundity and intrinsic rate of increase.

The second category involves the nutritional quality of the host plant mediating herbivore utilization by parasitoids and predators (Price, 1986; Vinson and Iwantsch, 1989; Vinson and Barbosa, 1987). Changes in herbivore size due to host plant effects, for example, have been associated with differences in size and sex ratio of parasitoids (Smith, 1957; Vinson and Iwantsch, 1980), and differences in functional responses of parasitoids (Luck and Podoler, 1985). The involvement of nutrition in these effects is strongly suggested by correlations between levels of plant nutrients, especially plant nitrogen and insect performance (e.g., Fraenkel, 1953; Mattson, 1980; McNeill and Southwood, 1978).

In addition to its potential for influencing biological control of insect pest species (Bergman and Tingey, 1979; Van Emden, 1986; Boethel and Eikenbary, 1986), the host plant on which an insect feeds can also influence its susceptibility to insecticides. Numerous host plant chemicals are known to induce detoxication enzymes in insects (Brattsten, 1979; Terriere, 1984; Yu, 1983; Yu et al., 1979). This induction can result in elevated levels of tolerance in the insect not only to host plant allelochemicals but also to insecticides. For example, *Heliothis zea* (Boddie) larvae held on foliage of the wild tomato species *Lycopersicon hirsutum* f. *glabratum* (accession PI 134417) or hatching from eggs incubated on that foliage had elevated levels (three- to fourfold) of tolerance to carbaryl insecticide when compared with control larvae on the cultivated tomato *L. esculentum*. This effect was attributed to the induction of elevated levels of cytochrome P-450 isozymes following exposure to 2-tridecanone, a volatile chemical found in abundance in the glandular trichomes that occur on the foliage of PI 134417 (Kennedy, 1984; Kennedy et al., 1987a). In PI 134417 2-tridecanone confers a high level of resistance to the tobacco hornworm [*Manduca sexta* (L.)] and the Colorado potato beetle [*Leptinotarsa decemlineata* (Say)] (Kennedy, 1986; Kennedy et al., 1985; Kennedy et al., 1987b). Because of this, PI 134417 is a potentially valuable source of insect resistance in tomato. However, breeding for elevated levels of 2-tridecanone in the foliage of the commercial-type tomato cultivars to achieve resistance to *M. sexta* and *L. decemlineata*, without concurrent selection for resistance to *H. zea*, could exacerbate problems with *H. zea* by making it more tolerant to insecticides.

Patterns of host plant utilization by individual insects in a population may be influenced by a variety of nongenetic factors. Among these are experience; age of the insect; temperature, photoperiod, and other climatic factors; and conspecific insect density and maternal effects. The reader should consult Papaj and Rausher (1983) for a detailed discussion of these effects. Maternal effects apparently play an important role in the population dynamics of gypsy moth *Lymantria dispar* (Linnaeus). The host plant on which gypsy moth females have fed as larvae affects the size of the eggs they produce. Large eggs have greater yolk reserves than small eggs, and larvae from large eggs are larger and have a greater tendency toward repeated dispersal as first instars. This variation, due to the diet of the female parent, appears to contribute to the ability of her progeny to colonize new sites within a habitat well before overexploitation of currently infested sites occurs. These differences in dispersal capacity may also influence numerical change in both established and expanding populations (Barbosa and Baltensweiler, 1987).

Pathogen Influences

Pathogens can also cause significant levels of intraspecific variation, especially when infections are widespread in a population. Microsporidial infections can be widespread in populations of many insect species and typically produce chronic infections in their hosts. Characteristic symptoms of chronic infections are abnormal feeding, irregular growth or retarded development, color changes in localized areas of the body, incomplete metamorphosis, formation of deformed pupae or adults, or lowered adult fecundity and longevity. In many cases, these pathogens are transmitted transovarially, as well as per os, and the symptoms of infection are usually most severe when infected individuals are exposed to additional stress factors, such as competition or poor-quality host plants (Brooks, 1988; Maddox, 1987). Because of this, microsporidian-infected individuals may respond differently to a variety of pest management tactics than healthy individuals. Similarly, due to the frequently chronic nature of microsporidial infections, these pathogens often infect laboratory colonies of insects maintained for research purposes. Experimental results obtained with infected insects may be highly variable, if not distinctly different, from those obtained with healthy insects.

Phenotypic Plasticity

Barbosa and Baltensweiler (1987), in their treatment of the relationship between phenotypic plasticity and herbivore outbreaks, convincingly argue that phenotypic plasticity can provide sufficient variation to permit a population to closely track changing environmental resources and enhance a population's potential for increase. However, because, under certain conditions, differential mortality of specific phenotypes may enhance population declines, they predict that many species that exhibit phenotypic plasticity should also exhibit dramatic shifts in density. Since many of our most serious agricultural and forest pests are characterized by dramatic shifts in population density, understanding the contribution of phenotypic plasticity to population changes may provide new insights into predicting and managing population outbreaks. Similarly, since insect pest management relies heavily on changing the pest's environment to make it less suitable for population increase, an understanding of the phenotypic plasticity that, helps the pest population accommodate changes in its environment may help us to anticipate the pest's responses to various management approaches.

GENETIC VARIATION AND ITS IMPLICATIONS FOR PEST MANAGEMENT

Like the nongenetic variation discussed earlier, genetically based variation both within and among populations provides a means by which a species can deal with both spatial and short-term temporal variation in its environment. Unlike nongenetic variation, genetically based variation provides the basis for an evolutionary response to selective pressures exerted by environmental factors that differentially affect the fitness of individuals making up the population. The relevance to pest management is obvious. In most cases, pest management tactics operate to reduce the fitness of the target insect over at least a portion of its range. In other, more specialized situations, such as the introduction of exotic natural enemy species for the biological control of an insect or weed pest, genetic variation in the natural enemy to be introduced is an important determinant of its ability to adapt to a new environment and to effect control of the target pest.

Intraspecific, Genetically based Variation and Pest Status

Many agriculturally important insects are native to the areas in which they initially become pests following the introduction of particular crops or production practices into an area (e.g., northern and western corn rootworms—Krysan, Chapter 18 in this volume; Colorado potato beetle, *Leptinotarsa decemlineata*—Casgrande,

1987). In many cases, the pest species was undoubtedly preadapted to the agricultural changes that took place in its environment. These changes made a new resource available for exploitation and brought the insect into conflict with humans, thereby creating a pest. In other cases, insects adapted, through natural selection, to exploit crop plants made available by introduction of new crops or cropping practices into an area. Once associated with one or more agricultural crops, production of the crop host has often provided a conduit over which the insect pest could expand its range (e.g., boll weevil, *Anthonomis grandis* Boheman—Burke et al., 1986) until stopped by other (nonhost) determinants of environmental suitability, such as temperature, photoperiod, or moisture (e.g., European corn borer, *Ostrinia nubilalis*—Showers, Chapter 14 in this volume).

Adaptation to Habitat

Numerous species have become pests following their introduction into new areas. Successful colonization of new areas involves not only the initial introduction and survival but also its establishment and range expansion. Genetic structure of the source population can have profound effects not only on the probability of successful early establishment of an organism in a new environment but also on the postintroduction development of the colony and its subsequent spread from the initial introduction site (Remington, 1968). The ability to adapt to the new environment is a function of the additive genetic variance for appropriate fitness traits in the colonizing population. The importance of genetic variability to the success of the colonizing population increases with the degree of differences between the native and exotic homes of the colonizing population (Scorza, 1983). Although DeBach (1965) has stated that entomophagous species successfully introduced for biological control have rarely undergone a period of adaptation following establishment and prior to reaching "outbreak" populations, this is not the case for phytophagous species introduced for biological control of weeds (Myers, 1987). Myers (1987) provides several examples of introduced phytophagous species that show a "refractory" period between introduction and outbreak, which, presumably, reflects a period of adaptation to climatic or other environmental factors. Indeed, distribution of exotic insects used in the biological control of weeds to new sites from populations established at an initial introduction site more frequently results in population outbreaks than do subsequent introductions of the same species from its native lands, presumably because the former is better adapted to the new environment than the latter.

The invasion of a new habitat by a pest is typically followed by intensive research on the bionomics of the pest in its new habitat. It is rare, however, that the source population from which the pest emigrated is studied to determine those aspects of its genetic structure that might be relevant to the initial establishment and subsequent adaptation of the pest in its new land (Scorza, 1983).

A major determinant of the successful establishment and ultimate spread of an introduced pest is its ability to adapt to the seasonal patterns of suitability in the physical environment and the availability of essential resources (e.g., host plants) in its new home. Phenological asynchrony with its environment can result in an organism's failure to become established in, or spread to, a new area because it enters diapause at an inappropriate time, fails to enter or terminate diapause at an appropriate time, or has a developmental rate inappropriate to the seasonal availability of essential resources (e.g., host plants) in the new area. Phenological studies of intraspecific variation in a number of insect species over significant portions of their range have demonstrated clines of variability in phenological adaptation, including latitudinal variation in critical photoperiod and diapause depth, as well as altitudinal variation in critical photoperiod. Such variation is generally polygenic in nature (Tauber and Tauber, 1978) and clearly demonstrates the close adaptation of insects to the seasonality of their environment. The reader should consult Tauber and Tauber (1978) and Tauber et al. (1986) for an excellent in-depth treatment of seasonal adaptations in insects.

Adaptation to environmental seasonality appears to have played a major role in determining the rate of spread of the European corn borer (*Ostrinia nubilalis*) from the sites of initial introduction. Showers (1979; Chapter 14 in this volume) presented circumstantial evidence suggesting that the southward range expansion of *O. nubilalis*, which was characterized by an advance and hold pattern, proceeded at a rate determined by its ability to adapt its diapause response to latitudinal differences in photoperiod and temperature. He hypothesized that, initially, the southward spread of this insect was interrupted by a maladaptive summer diapause, induced by the photoperiod and temperature conditions prevailing at specific "barrier latitudes." These interruptions were followed by adaptation to the prevailing environmental conditions at those latitudes, which enabled the populations to use the warm temperatures of the more southern areas to offset shorter day lengths and thereby avoid summer diapause. This enabled the species to expand its range further southward until another "barrier latitude," characterized by temperature and photoperiod conditions that induced summer diapause, was encountered.

Adaptation to Hosts

Just as adaptation to physical aspects of the environment can determine the pest status of a particular insect species, so, too, can adaptation to particular host plants in an area. Adaptation to new hosts, whether by exotic or native insects, can involve both genetically based behavioral and physiological changes that enable the insect to utilize new hosts, as well as changes in seasonal adaptations that sychronize the insects' periods of activity with availability of the host plant. This broad topic has been reviewed by Diehl and Bush (1984), Futuyma and Peterson (1985), and Gould (1983).

A knowledge of the degree of genetic variation relative to host utilization both within and among populations is important, not only in understanding the pest status of an insect on a particular crop in a particular area, but also in predicting the potential for pest adaptation to manipulated changes in host quality (Gould, 1983). High levels of genetic variation for host adaptations are common, and both Diehl and Bush (1984) and Gould (1983) concluded that in some cases there appears to be as much variation in host preference and survival abilities within sympatric populations of a species as among widely separated allopatric populations (see Gould, Chapter 3 in this volume). Futuyma and Peterson (1985) concluded that the occurrence of sympatric-host-associated genotypes is well documented for cyclically or permanently parthenogenetic species, such as aphids, because of reduced genetic recombination relative to sexually reproducing species. For the latter, they considered the evidence for host-associated genotypes (i.e., races or biotypes) to be generally less robust, except in a few cases for which there is strong, or "strongly suggestive" evidence.

Host plant adaptations can play an important role in the pest status and, consequently, the management of insect species. This is illustrated by the examples that follow.

Geographically isolated populations of Colorado potato beetle (CPB) have been shown to differ in ability to survive on several wild hosts, especially *Solanum eleagnifolium* and *S. carolinense* (Hsiao, 1978, 1982; Hare and Kennedy, 1986). *S. carolinense*, or horse nettle, is an important weedy species common on disturbed sites and in and around agricultural fields in North Carolina. Although horse nettle occurs all along the eastern portion of the United States, populations of CPB from the East Coast differ in their ability to utilize this host. In a comparative study of populations from Connecticut, New Jersey, Virginia, and North Carolina, Hare and Kennedy (1986) demonstrated that CPB from North Carolina showed high survival on horse nettle, while those from the remaining populations showed uniformly low survival. They further demonstrated heritable variation in the ability to survive on horse nettle. The ability to utilize horse nettle as a host is an important component in the seasonal dynamics of the CPB in North Carolina but not in other areas in the East. In North Carolina, the potato crop, on which the CPB is a pest, is planted in March and harvested in June. The potato crop is available as a host to only one and, perhaps, a portion of a second generation of CPB. Following the potato harvest in June, the beetles pass an additional two generations on horse nettle and, where available, volunteer potato plants growing in crops rotated or double cropped with potato. Throughout the summer, horse nettle is far more abundant and widespread than are volunteer potato plants. Larvae and adults of the summer generations of CPB are very commonly observed on horse nettle plants in the potato-growing region of eastern North Carolina and elsewhere in the state. Were it not for the well-developed ability of the North Carolina CPB populations to utilize horse nettle as a host, a concerted effort to prevent the occurrence of volunteer potato plants during the summer months might suppress the oversummering populations to such a low level that damaging populations would rarely develop the following spring. The perennial and weedy nature, abundance, and wide distribution of horse nettle in North Carolina preclude such an approach to CPB management.

In the more northern potato-growing areas of the eastern United States, the CPB is better synchronized with the potato-growing season and there is, likewise, no opportunity for managing the CPB through the destruction of alternate hosts. In New England, overwintering CPB emerge from diapause about the time the potato crop is sprouting. Two generations are passed on potato, and the second enters diapause about the time the potato crop is harvested (September) (Hare, 1983). Along the eastern shore of Virginia, the spring potato crop is planted and harvested at nearly the same time as the North Carolina potato crop; however, a second fall crop is grown and has been for many years (e.g., Soule and Panatter, 1907). Thus, because of double cropping, the CPB seems to be nearly as well synchronized with the seasonal availability of potato in Virginia as in New England (Hare and Kennedy, 1986).

The codling moth [*Laspeyresia pomonella* (Linnaeus)], an important pest of apples, was introduced into North America from Europe around 1750; it reached California around 1870, where it attacked apples and pears. Persian walnuts were grown extensively in southern California for about 20 years before the first reports of codling moth attacking walnut trees (around 1910). Within two years, codling moth infestations had become serious in both coastal and subcoastal walnut acreages, and by 1931 the inland walnut acreage of California was generally infested. During the 1930s, the codling moth also developed into a serious pest of plums in California (Phillips and Barnes, 1975). Phillips and Barnes (1975) considered the apple, walnut, and plum populations in Tulare County, California, to be host races. They hypothesized, on the basis of the infestation history and oviposition preferences, that the walnut race arose from the apple race and that the plum race arose from the walnut race. They provided evidence purporting to show genetic differences among races in oviposition preferences. However, those results are confounded by the effects of host conditioning and, thus, are not conclusive. Phillips and Barnes (1975) also provided much more convincing evidence for the close adaptation of the various races to the seasonal phenologies of their respective hosts. In the field, plum is phenologically two to three weeks ahead of walnut and apple. At the end of the season, plums are harvested before walnuts, and walnuts are harvested before apples. The plum race emerges in the spring one and one-half to two weeks earlier than either the apple or walnut race; it also requires fewer heat units to complete development and enters diapause earlier in the season before the plums are harvested. Similarly, the walnut race enters diapause earlier than the apple race and before the walnut harvest. These differences in the time of diapause induction result from different responses to common photoperiod and temperature conditions. They clearly show that the races are distinct and closely adapted to the phenologies of their respective hosts.

The greenbug [*Schizaphis graminum* (Rondani)], long known as an important pest of many graminaceous crops, was not a pest of grain sorghum in the United States until a new race or biotype appeared in the mid-1960s (Harvey and Hackerott, 1969). The 1968 outbreak was accompanied by a shift in greenbug phenology. Prior to 1968, winged aphids were caught on sticky traps only in the spring and fall; after 1972, they were caught primarily in spring and summer. During the period of peak densities on sorghum (1968–1972), peak numbers were caught during August, a time when sorghum, but not wheat, is highly susceptible to greenbug attack (Daniels, 1977). Adaptation to sorghum appeared to involve both a greater preference for sorghum over wheat (Harvey and Hackerott, 1969) and an ability to tolerate higher summer temperatures (Wood et al., 1972) than the wheat-adapted biotypes. Although a genetic analysis of the differences between these biotypes has not been conducted, the absence of any apparent association between the appearance of sorghum biotype and changes in cultural practices or environmental factors in sorghum, combined with the persistence of differences among biotypes over several generations, strongly suggests genetic differences (Mitter and Schneider, 1987). The greenbug biotype adapted to sorghum (Biotype C—Harvey and Hackerott, 1969) was initially managed on sorghum through the use of insecticides. However, sorghum hybrids resistant to Biotype C became available in 1976 and have since become an important component of IPM programs on grain sorghum in the Midwest (Teetes, 1980). Since that time, however, a new greenbug biotype, Biotype E, able to colonize plants resistant to Biotype C, has appeared (Porter et al., 1982; Dreyer and Campbell, 1984).

The fall armyworm, *Spodoptera frugiperda* (J. E. Smith), is a polyphagous species that attacks over 60 species of plants. It is a pest primarily on graminaceous crops, especially corn and Bermuda grass, although damaging populations often develop on a number of other crops. This species typically overwinters in southern Florida, southern Texas, and throughout the Caribbean region. During the late spring and summer months, populations spread northward, often as far as Canada (Mitchell, 1979). Although among population differences in insecticide resistance have been recognized (Young, 1979), the entire fall armyworm population has generally been considered to be uniform with respect to host range. Recently, however, Pashley and coworkers (Pashley, 1986; Pashley et al., 1987) demonstrated the existence of two host-associated races or strains—one associated with rice and Bermuda grass, and another associated with corn. Both races occur sympatrically in Puerto Rico and portions of the U.S. mainland, and perhaps other areas as well. The reader should consult Pashley (Chapter 17 in this volume) for a more complete discussion of the differences between these two strains.

The host-associated genetic structuring of fall armyworm populations has important implications for both insecticide resistance management (NAS, 1986; Roush and McKenzie, 1987) and for the use of host plant resistance. In the latter case, the two races responded very differently to different Bermuda grass varieties (Pashley et al., 1987). The Bermuda grass variety Tifton 292 was resistant to the corn but not the rice fall armyworm race, which is the predominant race on Bermuda grass. Clearly, in the selection of crops for fall armyworm resistance, it is important to use the appropriate race. More generally, this finding serves as a caution that insects used in host plant resistance work should be collected from the target crop, preferably shortly before the research is conducted, and should be generally representative of the population on that crop (Pashley et al., 1987).

As described earlier, intraspecific genetic variation and genetic change can contribute significantly to the pest status of insect species. The possibility that temporal changes in the genetic makeup of local populations can play an important role in population dynamics of both vertebrates and invertebrates has long been discussed (e.g., reviews by Wilson, 1967; Birch, 1971; Krebs, 1978; Berry, 1979). In a recent treatment of this subject, Mitter and Schneider (1987) found very little evidence for an association of genetic changes within populations with outbreak episodes. They found no conclusive evidence that rapid temporal genetic changes influenced population fluctuations. They concluded that a genetic causation for rapid population fluctuations was not plausible because of the drastic selective differences required to produce major changes in gene frequency over at most a few generations. They did not, however, discount the long-term contribution of genetic change over many generations (i.e., microevolution) to population outbreaks.

CONSEQUENCES OF INTRASPECIFIC VARIATION FOR PEST MANAGEMENT

Resistance

Both the type and amount of intraspecific genetic variation within, as well as among, populations of a target pest species have important implications for its management. Within-population variation is an important determinant of the uniformity of the response of a pest population to a pest management tactic, as well as a major element in the longer term evolutionary response of the pest population to that management tactic. Variation among populations can play a major role in determining the degree to which pest management techniques are uniformly effective against all populations. Thus, for example, northern corn rootworm populations with a one-year diapause are effectively managed by a two-year crop rotation, whereas populations with a two-year diapause are not (Krysan, Chapter 18 in this volume). Similarly, large differences in insecticide resistance occur among populations of a number of important pest species (Georghiou, 1986), and this determines which insecticides will provide effective control. With some species that do not show extensive wide-area movement, such as the Colorado potato beetle, large differences in insecticide resistance can occur over relatively short distances. Populations of potato beetle in Carteret County, North Carolina, for example, are highly resistant to carbofuran (sixty- to a hundredfold) and fenvalerate (twentyfold) and cannot be controlled with these insecticides, while those from Pamlico County (12 miles north of Carteret) are resistant to fenvalerate but not to carbofuran. Beetles from Tyrrell County (40 miles North of Pamlico County) are resistant to neither compound (Heim et al., 1990).

Monitoring Insect Populations

The effective use of insecticides or other pest population supression techniques to prevent economic damage requires the ability to predict or to monitor the occurrence of pest populations and to determine when they reach threatening levels. The prediction of seasonal events (e.g., adult emergence, oviposition, pupation, diapause) in the life history of insect pest species has involved modeling diapause induction and termination, as well as postdiapause development and development in nondiapausing stages and generations (Tauber et al., 1986). Genetically based variation within and among populations in response to development- and diapause-related factors can greatly influence the general predictability of seasonal events. Two examples are presented here to illustrate the potential significance of such variability. First, in the cabbage rootfly, [*Delia radicum* (L.)], genetically based variation in the rate of postdiapause development causes considerable difficulty in predicting adult emergence in some populations (Finch and Collier, 1983). After storage at 4°C, some geographic populations of *D. radicum* require 14 days at 20°C to complete postdiapause development, while other populations have large numbers of flies requiring greater than 100 days at 20°C.

A second example involves the European corn borer (*O. nubilalis*) in New York, where both univoltine and bivoltine populations exist (Eckenrode et al., 1983). Flights of the univoltine populations occur at day degree (DD) accumulations of 600 to 700, whereas the first and second flights of the bivoltine populations occur at DD accumulations of about 400 and 1,100, respectively. Thus, in New York, accurate prediction of European corn borer moth flights based on thermal-unit accumulations requires a knowledge of whether the target population is uni- or bivoltine. It is made more difficult by the fact that uni- and bivoltine populations coexist at some locations but not at others and their relative spatial distributions can shift over time (Eckenrode et al., 1983).

The ability to monitor flight activity periods of the European corn borer using sex-pheromone-baited traps is also confounded by the occurrence of genetically based differences in the isomeric composition of 11-tetradecenyl acetate (tda) used as a sex pheromone in premating communication between the sexes. The sex pheromone of most corn borer populations in the United States is an isomeric blend of about 97Z:3E 11-tda (Z-race), while populations in portions of New York, New Jersey, Pennsylvania, North Carolina, South Carolina, and perhaps other areas along the eastern seaboard use an isomeric blend of about 97E:3Z 11-tda (E-race) (Carde et al., 1975; Klun et al., 1975; Kennedy and Anderson, 1980; Roelofs et al., 1985; DuRant and Manley, 1987; Glover et al., 1987). In eastern North Carolina, where the E-race predominates and the Z-race is relatively rare, moth activity periods can be monitored using only traps baited with an isomeric blend of 97E:3Z 11-tda. In western North Carolina, where the Z-race predominates and the E-race is relatively rare, moth activity periods can be monitored using only traps baited with an isomeric blend of 97Z:3E 11-tda. The central portion of North Carolina represents a transition zone in which both pheromone races are common. Since the flight periods of the two races are not coincident, both traps baited with 97Z:3E and traps baited with 97E:3Z are required to obtain an accurate picture of corn borer flight periods (Sorenson, Kenendy, and Van Duyn, unpublished). In New York, monitoring corn borer moth activity periods using sex-pheromone-baited traps is even more complex because of differences in voltinism as well as sex pheromone blend. Roelofs et al. (1985) have described populations from New York that are bivoltine Z, bivoltine E, and univoltine Z, as well as populations that are mixed.

Biological Control

Naturally occurring biological control is an important contributor to the management of arthropod pests. Many species, which are not pests and are not particularly abundant where they are native, develop extremely high population densities and become severe pests in other areas following their introduction. An important factor contributing to such population outbreaks is the lack of adapted natural enemies of the introduced pest species in its new home. Although not well documented experimentally, over time at least some native species of parasitoids and predators likely adapt to include introduced species within their host range and thus provide some degree of population suppression. There is circumstantial evidence that this has occurred in the case of the oriental fruit moth, *Grapholitha molesta* (Busck) (OFM), which was introduced into North America from Asia. Between 1916 and 1933, a race of the native parasitoid, *Macrocentris ancylivoris* Roh., which parasitized OFM larvae, appeared in the mid-Atlantic region of the United States. Over time, this race increased its ability to parasitize OFM and, assisted by a release program, has become the dominant parasitoid providing a substantial level of control over much of the OFM-infested area of the United States and Canada (Wilson, 1965). The rate and extent to which native parasitoids and predators adapt to include exotic species within their host range depends in part on the amount of genetic variation that exists in their populations.

Control of exotic pest species has often been achieved through classical biological control that involves the introduction of parasitoids or predators (DeBach, 1964; Huffaker and Messenger, 1976). A key element in the success of classical biological control is the adaptation of the introduced natural enemy population to both the target pest and the habitat in which it is introduced. In most cases of successful classical biological control using entomophagous species, the introduced natural enemy has been preadapted to its new habitat and host. Few cases of apparent postcolonization adaptation have been recorded (DeBach, 1965), probably because the very limited numbers of parasitoids released did not possess sufficient relevant genetic variation to permit adaptation. The need to select, to the extent possible, preadapted natural enemies and to maximize the possibility for postcolonization adaptation places a premium on recognizing and exploiting relevant intraspecific genetic diversity of the natural enemy species.

Because of intraspecific variation in any of a variety of characters, different populations of a natural enemy species may exhibit markedly different abilities to exert control of the target pest following introduction. Indeed, biological control research has revealed numerous examples of morphs, races, semispecies, and sibling species that differ in biological (e.g., sex ratios, development period, emergence patterns, number of progeny and females, host range) but not morphological characteristics and, as a result, differ in their ability to exert biological control (Schlinger and Doutt, 1964; Delucchi et al., 1976). The significance of such differences is illustrated by *Comperiella bifasciata* Howard, an endoparasitoid of citrus red scale, *Aonidiella aurantii* (Maskell), and citrus yellow scale, *A. citrina* (Coquillett). *C. bifasciata*, introduced into California from Japan, prefers yellow scale and does not complete development in red scale, whereas *C. bifasciata*, introduced into California from China, prefers red scale but can develop in yellow scale. The race introduced from China is an important factor in the control of red scale, whereas the "Japanese race" became established on yellow scale and provided effective control in southern California (Bennett et al., 1976; Diehl and Bush, 1984).

The occurrence of intraspecific variation for host range is particularly important in the use of phytophagous species for the biological control of weeds. A key feature of biological weed control involves the careful selection and testing of candidate species to avoid the introduction of species or populations that will damage nontarget plants or that might possess sufficient genetic variation to permit subsequent host shifts in released populations (Huffaker, 1964). Intraspecific variation in host range also creates the need to match biotypes of the phytophagous species to the species, variety, or form of the weed to be controlled. There are numerous examples in which inappropriate insect biotype-weed host pairings have resulted in failures of colonized insects to establish or to effect acceptable control of the target weed (Goeden and Kok, 1986). Unruh and Goeden (1987) thoroughly documented one such case involving *Rhinocyllus conicus* Froelich, a weevil introduced from Europe into North America in order to control several introduced European thistles. Attempts to introduce *R. conicus* from musk thistle in Europe for control of milk thistle in northern California failed, despite the fact that milk thistle is a widespread host of the weevil in southern Europe. In contrast, when *R. conicus* was introduced into California from milk thistle in Italy, it readily became established. Subsequent studies of host plant associations in Europe have shown the existence of distinct host biotypes or races and that the weevil biotypes established in California differ markedly in their abilities to utilize each other's host plants. The genetic basis for these differences has been substantiated by electrophoretic analyses of allozyme patterns that also provide evidence that one of the two races introduced into California has shifted its host range to include native thistles (Unruh and Goeden, 1987).

Adaptation of the introduced natural enemy to the climatic condition of its new home is extremely important (Messenger, 1970). For example, an introduction of the parasitoid *Trioxys pallidus* (Haliday) from Nice, France, into California in order to control the walnut aphid, *Chromaphis juglandicola* (Kaltenbach), resulted in effective biological control in the coastal areas of southern California. However, the level of control was less in the intermediate valleys, where summer temperatures were higher and the air was drier; no control occurred in the inland walnut-growing areas, where summer heat and, perhaps, dryness were apparently limiting. A second introduction of *T. pallidus* from an area in Iran climatically similar to the inland walnut-growing areas of California was successful in providing effective biological control of the walnut aphid in those areas (Messenger, 1970).

For those natural enemy species that will not become established in the release areas but that exert control through repeated inoculative or innundative releases, preadaptation to conditions in the agroecosystem during the period in which they are to be effective is particularly important. Such species, of course, have no opportunity to respond to selection operating in the agroecosystem. *Edovum puttleri* Grissell is an egg parasitoid of the Colorado potato beetle that shows promise for use in biological control programs involving inoculative or innundative releases. Two different biotypes of this parasitoid are available for release. They differ in susceptibility to various pesticides used in potato production (Obrycki et al., 1986) and in ability to tolerate low temperatures (Obrycki et al., 1987). Based primarily on its ability to better tolerate low temperatures, Obrycki et al. (1987) considered the "Mexican biotype" a better candidate for use in the potato-growing areas of the northeastern United States.

The importation of insects for biological control is a complex process that involves foreign exploration and the collection, shipment, quarantine, and release of potential biological control agents (DeBach, 1964). To maximize the likelihood that the biological control agent is preadapted to the area of release, foreign exploration is usually concentrated in areas within the pest's range that are bioclimatically similar to the release area (Messenger, 1970). To maximize the potential for postintroduction adaptation to the habitat and host population (or race) in the release site, it is desirable to maximize the genetic diversity of the natural enemy population to be released. With this objective in mind, Bartlett and van den Bosch (1964) recommended that the shipments of a natural enemy be as large as practicable and include collections from areas as widely different as possible. Based on theory, Remington (1968) argued for sampling from central (genetically diverse) and marginal (genetically homogenous) populations. Mackauer (1976), however, pointed out a number of practical difficulties in identifying these populations and suggested that a founding colony consisting of about 500 individuals, collected over a conveniently large area, should include a level of genetic diversity that characterizes the parental population to a degree that would be adequate for most biological control projects.

Following importation, a colony of the natural enemy must be established and maintained in quarantine and then increased prior to release. For a variety of reasons that will be discussed later, the genetic composition of the insects in culture can change during colony maintenance.

The Need to Preserve Genetic Variation in Laboratory Colonies

Insects produced in small (1,000 or less) laboratory cultures are commonly used in ecological, as well as pest-management-related, research, in which population parameters, host relationships, behavior, and responsiveness to environmental variables are measured. Similarly, insects produced in mass production facilities (hundreds of thousands or millions of individuals) are used in various insect control programs [e.g., innundative releases of parasitoids (Stinner, 1977), autocidal control programs (Scruggs, 1978; Novy, 1978; Whitten, 1970)]. Regardless of the size of the insect colony, the rearing facility represents a vastly altered habitat that can have profound phenotypic and genotypic consequences for the insect population in culture (Mason et al., 1987). The genetic composition of an insect colony is influenced not only by founder effects but also by inbreeding, random genetic drift, and inadvertent selection for adaptation to the rearing environment (Mackauer, 1976; Chambers, 1977; Mason et al., 1987). Some of the numerous types of changes observed in insect colonies are illustrated later. All are relevant to both ecological and pest-management-related research.

The ability of insects to respond normally to their host plants is frequently lost during colony maintenance as a result of both environmental effects (e.g.,

Schoonhoven, 1967) and genetic changes in the colony. For example, boll weevils (*A. grandis*) from a colony mass reared for over 10 years were unable to discriminate among cotton lines that were readily distinguished by weevils from a second colony not in mass production and to which wild individuals were added yearly; the former weevils were unsuitable for use in host plant resistance research (McCarty and McGovern, 1987). Similarly, the European corn borer maintained on an artificial diet for 54 generations were no longer able to survive on corn. Survival on corn equal to that of the wild borers could be recovered by one generation of backcrossing to wild borers (Guthrie and Carter, 1972).

In other insects, changes in the reproductive parameters of the colony have occurred over time. A culture of the Mediterranean fruit fly, *Ceratitis capitata* (Wiedemann), maintained for 12 years, had a lower net reproductive rate (Ro) and a longer generation time than a field-collected population kept in culture for 12 generations (8 months) (Rossler, 1975). Selection for adaptation to the rearing environment can be rapid, however. Raulston (1975), for example, reported quite drastic changes in mating and ovipostion behavior of *Heliothis virescens* during the first six or seven generations in culture.

The success of autocidal control programs is dependent upon the competitiveness with which mass-reared insects mate with their wild counterparts following release. Competitiveness depends on a number of difficult-to-define and difficult-to-measure attributes, such as predator avoidance, habitat selection, dispersal, mating success, activity rhythms, longevity, and so on (Chambers, 1977). An important factor contributing to the breakdown of the sterile insect release program for control of the screwworm, *Cochliomyia hominivorax* (Coquerell), in 1972 (Richardson, 1978) was a loss in competitiveness of mass-reared flies due to fixation of an allele that was rare in the wild flies and that controlled a temperature sensitive enzyme needed for strong flight (Bush, 1978). As a result, mass-reared flies were active in the field only in the late afternoon, whereas wild flies were active throughout the day; mating between wild flies may have been completed before the released flies became active. This shift apparently was the result of selection for adaptation to the high temperatures maintained in the mass-rearing facility to speed development. Other factors were also involved in the breakdown of the screwworm control program (Richardson et al., 1982).

Periodic renewal of an insect colony with a sufficient number of healthy, wild individuals is one way to minimize divergence of the colony from the wild populations. However, changes in the colony can occur fairly quickly (Raulston, 1975), and, depending upon the use to which the insects are to be put, it may be important to monitor changes in the genetic composition of the colony. Starch gel electrophoresis has been proposed as a relatively easy way of detecting genetic shifts early (Mason et al., 1987). It is easier, less variable than bioassays of phenotypes, and is capable of detecting genetic changes directly. A limitation lies in the fact that, although electrophoretic shifts indicate shifts in the genetic composition of the colony, it is difficult to interpret those shifts in terms of relevant phenotypes.

Amount and Kind of Genetic Variation in Target Pest Population Determines Its Potential Response to Pest Management Measures

One of the basic tenets of modern ecology and evolutionary biology is that any factor that reduces fitness imposes selection pressure for the ability to resist or to avoid that factor, provided sufficient relevant genetic variation exists in the populations. By their very nature, most pest management tactics reduce the fitness of their target pests. Thus, the evolutionary process itself is a major constraint on the potentiality of pest management.

The intensity of selection pressure imposed by a pest management regime is determined to a large degree by the size of the management unit relative to the pest population. In practice, for pest management purposes, the size of the pest population is defined by the size of the management unit, typically a field or a farm, which is set on the basis of economic and crop-related factors; it may bear no meaningful relationship to the actual pest population (or populations) through which gene flow occurs. Yet, the amount of gene flow between selected and unselected portions of a population (or populations) is an important determinant of the response to selection.

The amount of movement among selected (i.e., managed) and nonselected populations, and portions of populations, greatly influences the patterns of gene flow; yet, movement patterns of most pest species—indeed most insect species—are poorly known (Rabb and Kennedy, 1979; MacKenzie et al., 1985). Pest species vary greatly in their mobility relative to the size of potential pest management units. These differences influence not only the types of management practices that are effective against a particular pest but also the likelihood of an evolutionary response (i.e., resistance) to the management practices.

Years of experience in insect control have demonstrated, unequivocally, the ability of pest species to develop resistance to a broad spectrum of insect control tactics, often with severe economic and public health consequences. Krysan (Chapter 18 in this volume) has described the development in northern corn rootworm of resistance to a cultural control, crop rotation, based on altered diapause. He discusses basic differences in the nature of diapause in the northern and western corn rootworm, which explain why a similar response to crop rotation was not selected in the western corn rootworm. More generally, cultural control techniques, which involve alteration of the cropping system or specific production practices to reduce pest populations or avoid crop injury (Herzog and Funderburk, 1986), do not always result in high levels of pest mortality. For this reason, in many, if not most, cases, cultural controls will not exert the same high level of selection pressure for "resistance" as some of the more lethal insect control techniques, such as the use of insecticides or crop cultivars highly resistant to pest attack.

Although parasitoids and predators introduced in classical biological control programs have very effectively suppressed populations of a large number of pest species to subeconomic levels (DeBach, 1964; Huffaker and Messenger, 1976), there have been few documented cases in which established biological control has broken down because the pest species or population has developed resistance to the natural enemy. It is not likely that in all cases this can be attributed to the lack of appropriate genetic variation in the pest species, since there are a number of examples of resistance in insects to one or more of their natural enemies. The parasitoid *Bathyplectes curculionis* (Thomson) readily parasitizes and is a major mortality factor for the alfalfa weevil, *Hypera postica* (Gyll), in the western United States. The populations of *H. postica* in the eastern United States have a well-developed ability to encapsulate *B. curculionis* eggs, greatly reducing its effectiveness as a control agent. Even among individuals of eastern *H. postica*, there is considerable variation in the ability of their progeny to encapsulate *B. curculionis* eggs (Puttler, 1967). In contrast, another parasitoid of the alfalfa weevil, *Tetrasticus incertus* (Ratzeburg), became readily established in the eastern United States but not in the western United States. Volker and Simpson (1975) attributed this differential establishment to an evasive, defensive behavior of western *H. postica* larvae in response to *T. incertus* adults. This behavior was of much lower frequency and intensity in the eastern *H. postica* population that they studied. Their contention remains unproven, however, since they failed to investigate within-population variation in the defensive behavior in the eastern and western weevils, and did not consider the consequences of interspecific competition with *B. curculionis*. Two other examples described by Turnock et al. (1976) illustrate that biological control is not immune to being negated by genetic changes in the pest populations. The first involves the negation of partial biological control of *Rhyacionia bushnelli* (Bsk.) in Nebraska by the parasitoid *Campoplex frustranae* Cushman when the host was replaced by a subspecies not parasitized by *C. frustranae*. The second involves loss of biological control of *Pristophora erichsonii* (Hartig) following the appearance and spread of a strain that was able to encapsulate eggs of the parasitoid *Mesoleius tenthredinis* Morley, which previously had not been effectively encapsulated.

The specific reasons why resistance to natural enemies used in biological control has not occurred more frequently probably vary from case to case. Along with many other factors, the positive density-dependent nature of mortality caused by the natural enemies used in biological control programs (Huffaker and Messenger, 1964) and the capability of the natural enemies to respond through selection to changes in their hosts' "resistance" to them may play a role (Pimentel, 1961, 1968).

Genetic variation in the response to insect pathogens has also been demonstrated (Watanabe, 1987). There are many reports of differences between two or more isolated populations of an insect species in their response to pathogens, of increased pathogen resistance among progeny of survivors of epizootics, and of increased resistance to pathogens following selection for resistance (Watanabe, 1987). This has important implications for pest management, as it clearly indicates the potential for inadvertently selecting pest populations resistant to microbial controls (Harper, 1987).

Resistance to other important insect control tactics has been well documented (e.g., host plant resistance—Maxwell and Jennings, 1980; autocidal control—Richardson et al., 1982; chemical control—NAS, 1986; Roush and McKenzie,

1987). The frequency with which such resistance is encountered, especially to insecticides and to pest resistant crop varieties, has prompted considerable research into ways to avoid or to delay such resistance. A detailed discussion of this topic is beyond the scope of this chapter, and the reader is referred to NAS (1986), Gould (1978, 1986a, 1986b), and Kennedy et al. (1987).

Exploiting Genetic Variation

Although intraspecific genetic variation poses many difficulties for pest management, it also provides some opportunities. Analysis of patterns of genetic variation through the techniques of quantitative genetics and the use of genetic markers can provide valuable information regarding the structure and dynamics of pest populations, including migration patterns (Pashley and Bush, 1979; Pashley, 1988; Pashley et al., 1985; Stock, 1979). Genetic variation also provides the raw material for selecting or engineering improved strains of parasitoids, predators, and pathogens for use in biological and microbial control (Hoy, 1985; Kirschbaum, 1985), as well as for autocidal control programs (Whitten, 1970).

CONCLUDING REMARKS

Natural insect populations are not uniform. Indeed, intraspecific variability is one of the most important characteristics of insects, and those engaged in the study or management of insects can ill afford to ignore that variability. Important intraspecific variation can be nongenetic as well as genetic. Recognizing and understanding such variation can help to elucidate the genesis of pest problems. It can also provide a basis for predicting both the near-term and long-term response of a target pest population to a particular pest management approach and thereby aid in the development of appropriate, lasting pest management strategies.

Acknowledgments

I would like to thank Fred Gould, Ron Stinner, Jim Walgenbach, and Jim Barbour for their constructive criticism of earlier drafts of this manuscript. This is Paper No. 11765 of the Journal Series of the North Carolina Agricultural Research Service, Raleigh, N.C. 27695-7643.

REFERENCES

Barbosa, P., and W. Baltensweiler. 1987. Phenotypic plasticity and herbivore outbreaks. Pages 469–503 in Barbosa, P., and J. C. Schultz (eds.), *Insect Outbreaks*. New York: Academic Press.

Bartlett, B. R., and R. van den Bosch. 1964. Foreign exploration for beneficial organisms.

Pages 283–304 in DeBach, P. (ed.), *Biological Control of Insect Pests and Weeds*. New York: Reinhold Publishing Co.

Beck, S. D. 1967. Water intake and termination of diapause in the European corn borer, *Ostrinia nubilalis*. *J. Insect Physiol.* 13:739–750.

Bennett, F. D., D. Rosen, P. Cochereau, and B. J. Wood. 1976. Biological control of pests of tropical fruits and nuts. Pages 359–395 in Huffaker, C. B., and P. S. Messenger (eds.), *Theory and Practice of Biological Control*. New York: Academic Press.

Bergman, J. M., and W. M. Tingey. 1979. Aspects of interaction between plant genotypes and biological Control. *Bull. Entomol. Soc. Am.* 25:275–279.

Bernays, E. A., and R. F. Chapman. 1978. Plant chemistry and acridoid feeding behavior. Pages 99–142 in Harbourne, J. B. (ed.), *Biochemical Aspects of Plant and Animal Coevolution*. London: Academic Press.

Berry, R. J. 1979. Genetical factors in animal population dynamics. *Symp. Br. Ecol. Soc.* 20:53–80.

Birch, L. C. 1971. The role of environmental heterogeneity and genetical heterogeneity in determining distribution and abundance. *Proc. Adv. Study Inst. Dym. Numbers Popul.*, 1970, pages 109–128.

Boethel, D. J., and R. D. Eikenbary (eds.). 1986. *Interactions of Plant Resistance and Parasitoids and Predators of Insects*. Chichester, England: Ellis Horwood Ltd.

Brattsten, L. B. 1979. Biochemical defense mechanisms in herbivores against plant allelochemicals. Pages 199–270 in Rosenthal, G. A., and D. H. Janzen (eds.), *Herbivores Their Interaction with Secondary Plant Metabolites*. New York: Academic Press.

Brooks, W. M. 1988. Entomogenous protozoa. Pages 1–149 in Ignoffo, C. M. (ed.), *Microbial Pesticides*. Boca Raton, Fla.: CRC Press.

Brower, J. S. 1969. Ecological chemistry. *Sci. Am.* 220(2):22–29.

Burke, R. R., W. E. Clark, J. R. Cate, and P. A. Fryxell. 1986. Origin and dispersal of the boll weevil. *Bull. Entomol. Soc. Am.* 32:228–238.

Bush, G. L. 1978. Planning a rational quality control program for the screwworm fly. Pages 37–47 in Richardson, R. H. (ed.), *The Screwworm Problem*. Austin: Univ. of Texas Press.

Campbell, B. C., and S. S. Duffey. 1979. Tomatine and parasitic wasps: Potential incompatibility of plant antibiosis with biological control. *Science* 205:700–702.

———. 1981. Alleviation of α-tomatine-induced toxicity to the parasitoid, *Hyposotor exiguae*, by phytosterols in the diet of the host, *Heliothis zea*. *J. Chem. Ecol.* 7:927–946.

Cardé, R. T., J. Kochansky, J. F. Stimmel, A. G. Wheeler, and W. L. Roelofs. 1975. Sex pheromone of the European corn borer: cis and trans responding males in Pennsylvania. *Environ. Entomol.* 4:413–419.

Casagrande, R. A. 1987. The Colorado potato beetle: 125 years of mismanagement. *Bull. Entomol. Soc. Am.* 33:142–150.

Chambers, D. L. 1977. Quality control in mass rearing. *Annu. Rev. Entomol.* 22:289–308.

Daniels, N. E. 1977. Seasonal populations and migration of the greenbug in the Texas Panhandle. *Southwest. Entomol.* 2:20–26.

DeBach, P. (ed.). 1964. *Biological Control of Insect Pests and Weeds*. New York: Reinhold Publishing, Co.

———. 1965. Some biological and ecological phenomena associated with colonizing entomophagous insects. Pages 287–300 in Baker, H. G., and G. L. Stebbins (eds.), *The Genetics of Colonizing Species*. New York: Academic Press.

Delucchi, V., D. Rosen, and E. I. Schlinger. 1976. Relationship of systematics to biological control. Pages 81–91 in Huffaker, C. B,. and P. S. Messenger (eds.), *Theory and Practice of Biological Control*. New York: Academic Press.

Diehl, S. R., and G. L. Bush. 1984. An evolutionary and applied perspective of insect biotypes. *Annu. Rev. Entomol.* 29:471–504.

Doutt, R. L. 1964. Biological characteristics of entomophagous adults. Pages 145–167 in DeBach, P. (ed.), *Biological Control of Insect Pests and Weeds*. New York: Reinhold Publishing Co.

Dreyer, D. L., and B. C. Campbell. 1984. Association of the degree of methylation of intercellular pectin with plant resistance to aphids and with the induction of aphid biotypes. *Experientia* 40:224–226.

Duffey, S. S. 1980. Sequestration of plant natural products by insects. *Annu. Rev. Entomol.* 25:447–477.

Duffey, S. S., K. A. Bloem, and B. C. Campbell. 1986. Consequences of sequestration of plant natural products in plant-insect-parasitoid interactions. Pages 31–60 in Boethel, D. J., and R. D. Eikenbary (eds.), *Interactions of Plant Resistance and Parasitoids and Predators of Insects*. Chichester, England: Ellis Horwood Ltd.

DuRant, J. A., and D. G. Manley. 1987. Relative efficiencies of two European corn borer sex pheromone blends in South Carolina. *J. Agric. Entomol.* 4:82–86.

Eckenrode, C. J., P. S. Robbins, and J. T. Andaloro. 1983. Varation in flight patterns of European corn borers (Lepidoptera: Pyralidae) in New York. *Environ. Entomol.* 12:393–396.

Eisner, T. 1970. Chemical defense against predation in arthropods. Pages 157–217 in Sondheiner, E., and J. B. Simeone (eds.), *Chemical Ecology*. New York: Academic Press.

Finch, S., and R. H. Collier. 1983. Emergence of flies from overwintering populations of cabbage rootfly pupae. *Ecol. Entomol.* 8:29–36.

Fraenkel, G. 1953. The nutritional value of green plants. *Proc. 9th Int. Cong. Entomol.* 2:90–100.

Futuyma, D. J., and S. C. Peterson. 1985. Genetic variation in the use of resources by insects. *Annu. Rev. Entomol.* 30:217–238.

Georghiou, G. P. 1986. The magnitude of the resistance problem. Pages 14–43 in National Academy of Sciences, *Pesticide Resistance Strategies and Tactics for Management*. Washington, D.C.: National Academy Press.

Glover, T. J., X. H. Tang, and W. L. Roelofs. 1987. Sex pheromone blend discrimination by male moths from E and Z strains of European corn borer. *J. Chem. Ecol.* 13:143–151.

Goeden, R. D., and L. T. Kok. 1986. Comments on a proposed new approach for selecting agents for biological control of weeds. *Can. Entomol.* 118:51–58.

Gould, F. 1978. Predicting the future resistance of crop varieties to pest populations: A case study of mites and cucumbers. *Environ. Entomol.* 7:622–626.

———. 1983. Genetics of plant-herbivore systems: Interactions between applied and basic study. Pages 599–653 in Denno, R. F., and M. McClure (eds.), *Variable Plants and Herbivores in Natural and Managed Systems*. New York: Academic Press.

———. 1986a. Simulation models for pedicting durability of insect resistant germplasm: A deterministic diploid, two-locus model. *Environ. Entomol.* 15:1–10.

———. 1986b. Simulation models for predicting durability of insect resistant germplasm: Hessian fly-resistant winter wheat. *Environ. Entomol.* 15:11–23.

Guthrie, W. D., and S. W. Carter. 1972. Backcrossing to increase survival of larvae of a laboratory culture of the European corn borer on field corn. *Annu. Entomol. Soc. Am.* 65:108–109.

Hagen, K. S. 1987. Nutritional ecology of terrestrial insect predators. Pages 533–577 in Slansky, F., Jr., and J. G. Rodriguez (eds.), *Nutritional Ecology of Insects, Mites, Spiders and Related Invertebrates.* New York: John Wiley & Sons.

Hare, J. D. 1983. Seasonal variation in plant-insect associations: Utilization of *Solanum dulcamara* by *Leptinotarsa decemlineata. Ecology* 64:345–361.

Hare, J. D., and G. G. Kennedy. 1986. Genetic variation in plant-insect associations: Survival of *Leptinotarsa decemlineata* populations on *Solanum carolinense. Evolution* 40:1031–1043.

Harper, J. D. 1987. Applied epizootiology: Microbial control of insects. Pages 473–496 in Fuxa, J. R., and Y. Tanada (eds.), *Epizootiology of Insect Diseases.* New York: John Wiley & Sons.

Harvey, T. L., and H. L. Hackerott. 1969. Recognition of greenbug biotype injurious to sorghum. *J. Econ. Entomol.* 62:776–779.

Heim, D. C., G. G. Kennedy, and J. W. van Duyn. 1990. Survey of insecticide resistance among North Carolina Colorado potato beetle (Coleoptera: Chrysomelidae) populations. *J. Econ. Entomol.* 83:1229–1235.

Herzog, D. C., and J. E. Funderburk. 1986. Ecological basis for habitat management and pest cultural control. Pages 217–250 in Kogan, M. (ed.), *Ecological Theory and Integrated Pest Management Practice.* New York: John Wiley & Sons.

Hoy, M. A. 1985. Recent advances in genetics and genetic improvement of the phytoseiidae. *Annu. Rev. Entomol.* 30:345–370.

Hsiao, T. H. 1978. Host plant adaptations among geographic populations of the Colorado potato beetle. *Entomol. Exp. Appl.* 24:237–247.

———. 1982. Geographic variation and host plant adaptation of the Colorado potato beetle. Pages 315–324 in Visser, J. H., and A. K. Minks (eds.), *Proceedings Fifth International Symposium on Insect Plant Relationships.* Wageningen, The Netherlands: Pudoc.

Huffaker, C. B. 1964. Fundamentals of biological weed control. Pages 631–679 in DeBach, P. (ed.), *Biological Control of Insect Pests and Weeds.* New York: Reinhold Publishing Co.

Hufffaker, C. B., and P. S. Messenger. 1964. The concept and significance of natural control. Pages 74–117 in DeBach, P. (ed.), *Biological Control of Insect Pests and Weeds.* New York: Reinhold Publishing.

———. (eds.). 1976. Theory and Practice of Biological Control. New York: Academic Press.

Kennedy, G. G. 1984. 2-Tridecanone, tomatoes and *Heliothis zea*: Potential incompatibility of plant antibiosis with insecticidal control. *Entomol. Exp. Appl.* 35:305–311.

———. 1986. Consequences of modifying biochemically mediated insect resistance in *Lycopersicon* species. Pages 130–141 in Green, M. B., and P. A. Hedin (eds.), *Natural Resistance in Plants to Pests: Roles of Allelochemicals.* ACS Symposium Series 296.

Kennedy, G. G., and T. E. Anderson. 1980. European corn borer trapping in North Carolina with various sex pheromone component blends. *J. Econ. Entomol.* 73:642–646.

Kennedy, G. G., C. E. Sorenson, and R. L. Fery. 1985. Mechanisms of resistance to Colorado potato beetle in tomato. Pages 107–116 in Ferro, D. N., and R. H. Voss (eds.), *Proceedings of the Symposium on the Colorado Potato Beetle XVII International Congress of Entomology*. Mass Agr. Exp. Stn. Res. Bull. 704.

Kennedy, G. G., R. R. Farrar, Jr., and M. R. Riskallah. 1987a. Induced tolerance in *Heliothis zea* neonates to host plant allelochemicals and carbaryl following incubation of eggs on foliage of *Lycopersicon hirsutum* f. *glabratum*. *Oecologia* 73:615–620.

Kennedy, G. G., J. Nienhuis, and T. Helentjaris. 1987b. Mechanisms of arthropod resistance in tomatoes. Pages 145–154 in Nevins, D. J., and R. A. Jones (eds.), *Tomato Biotechnology*. Plant Biology, Vol. 4. New York: Alan R. Liss.

Kirschbaum, J. B. 1985. Potential implication of genetic engineering and other biotechnologies to insect control. *Annu. Rev. Entomol.* 30:51–70.

Klun, J. A., and cooperators. 1975. Insect sex pheromones: Intraspecific pheromonal variability of *Ostrinia nubilalis* in North America and Europe. *Environ. Entomol.* 4:891–894.

Krebs, C. J. 1978. A review of the Chitty hypothesis of population regulation. *Can. J. Zool.* 56:2463–2480.

Luck, R. F., and H. Podoler. 1985. Competitive displacement of *Aphytis lignanensis* by *A. melinus*: The role of host size and female progeny production. *Ecology* 66:904–913.

Mackauer, M. 1976. Genetic problems in the production of biological control agents. *Annu. Rev. Entomol.* 21:369–385.

MacKenzie, D. R., C. S. Barfield, G. G. Kennedy, and R. D. Berger (eds.). 1985. *The Movement and Dispersal of Agriculturally Important Biotic Agents*. Baton Rouge, La.: Claitor's Publishing.

Maddox, J. V. 1987. Protozoan diseases. Pages 417–452 in Fuxa, J. R., and Y. Tanada (eds.), *Epizootiology of Insect Diseases*. New York: John Wiley & Sons.

Mason, L. J., D. P. Pashley, and S. J. Johnson. 1987. The laboratory as an altered habitat: Phenotypic and genotypic consequences of colonization. *Fla. Entomol.* 70:49–58.

Mattson, W. J. 1980. Herbivory in relation to plant nitrogen content. *Annu. Rev. Ecol. Syst.* 11:119–161.

Mattson, W. J., and J. M. Scriber. 1987. Nutritional ecology of insect folivores of woody plants: Nitrogen, water, fiber and mineral considerations. Pages 105–146 in Slansky, F., Jr., and J. G. Rodriguez (eds.), *Nutritional Ecology of Insects, Mites, Spiders and Related Invertebrates*. New York: Academic Press.

Maxwell, F. G., and P. R. Jennings. 1980. *Breeding Plants Resistant to Insects*. New York: John Wiley & Sons.

McCarty, J. C., Jr., and W. L. McGovern. 1987. Oviposition preferences in two laboratory colonies of boll weevils (Coleoptera: Curculionidae). *J. Econ. Entomol.* 80:366–368.

McMurtry, J. A., and J. A. Rodriguez. 1987. Nutritional ecology of phytoseiid mites. Pages 609–644 in Slansky, F., Jr., and J. G. Rodriguez (eds.), *Nutritional Ecology of Insects, Mites, Spiders and Related Invertebrates*. New York: Academic Press.

McNeill, S., and T. R. E. Southwood. 1978. The role of nitrogen in the development of insect/plant relationships. Pages 77–98 in Harborne, J. (ed.), *Biochemical Aspects of Plant and Animal Coevolution*. New York: Academic Press.

Messenger, P. S. 1970. Bioclimatic inputs to biological control and pest management programs. Pages 84–102 in Rabb, R. L., and F. E. Guthrie (eds.), *Concepts of Pest Management*. Raleigh: University Graphics, North Carolina State Univ.

Mitchell, E. R. 1979. Migration by *Spodoptera exigua* and *S. frugiperda*, North American style. Pages 386–393 in Rabb, R. L., and G. G. Kennedy (eds.), *Movement of Highly Mobile Insects: Concepts and Methodology in Research*. Raleigh: University Graphics, North Carolina State Univ.

Mitter, C., and J. C. Schneider. 1987. Genetic change and insect outbreaks. Pages 505–532 in Barbosa, P., and J. C. Schultz (eds.), *Insect Outbreaks*. New York: Academic Press.

Myers, J. H. 1987. Population outbreaks of introduced insects: Lessons from the biological control of weeds. Pages 173–193 in Barbosa, P., and J. C. Schultz (eds.), *Insect Outbreaks*. New York: Academic Press.

National Academy of Sciences (NAS). 1986. *Pesticide Resistance Strategies and Tactics for Management*. Washington, D.C.: National Academy Press.

Novy, J. E. 1978. Operation of a screwworm eradication program. Pages 19–36 in Richardson, R. H. (ed.), *The Screwworm Problem*. Austin: Univ. of Texas Press.

Obrycki, J. J., M. J. Tauber, and W. M. Tingey. 1986. Comparative toxicity of pesticides to *Edovum puttleri* (Hymenoptera: Eulophidae), an egg parasitoid of the Colorado potato beetle (Coleoptera: Chrysomelidae). *J. Econ. Entomol.* 79:948–951.

Obrycki, J. J., M. J. Tauber, C. A. Tauber, and B. Gollands. 1987. Developmental responses of the Mexican biotype of *Edovum puttleri* (Hymenoptera: Eulophidae) to temperature and photoperiod. *Environ. Entomol.* 16:1319–1323.

Orr, D. B., and D. J. Boethel. 1986. Influence of plant antibiosis through four trophic levels. *Oecologia* 70:242–249.

Painter, R. H. 1951. *Insect Resistance in Crop Plants*. Lawrence: Univ. Press of Kansas.

Papaj, D. R., and M. D. Rausher. 1983. Individual variation in host location by phytophagous insects. Pages 77–124 in Ahmad, S. (ed.), *Herbivorous Insect Host Seeking, Behavior and Mechanisms*. New York: Academic Press.

Pashley, D. P. 1986. Host associated genetic differentiation in fall armyworm (Lepidoptera: Noctuidae): A sibling species complex? *Annu. Entomol. Soc. Am.* 79:898–904.

———. 1988. Quantitative genetics, development and physiological adaptation in sympatric host strains of fall armyworm. *Evolution* 42:93–102.

Pashley, D., and G. L. Bush. 1979. The use of allozymes in studying insect movement with special reference to the codling moth, *Laspeyresia pomonella* (L.). Pages 333–341 in Rabb, R. L., and G. G. Kennedy (eds.), *Movement of Highly Mobile Insects: Concepts and Methodology in Research*. Raleigh: University Graphics, North Carolina State Univ.

Pashley, D. P., S. J. Johnson, and A. N. Sparks. 1985. Genetic population structure of migratory moths: The fall armyworm (Lepidoptera: Noctuidae). *Annu. Entomol. Soc. Am.* 76:756–761.

Pashley, D. P., S. S. Quisenberry, and T. Jamjanya. 1987. Impact of fall armyworm (Lepidoptera: Noctuidae) host strains on the evaluation of Bermuda grass resistance. *J. Econ. Entomol.* 80:1126–1130.

Phillips, P. A., and M. M. Barnes. 1975. Host race formation among sympatric apple, walnut and plum populations of the codling moth, *Laspeyresia pomonella*. *Annu. Entomol. Soc. Am.* 68:1053–1060.

Pimentel, D. 1961. Animal population regulation by the genetic feed-back mechanism. *Am. Nat.* 95:65–79.

———. 1968. Population regulation and genetic feedback. *Science* 159:1432–1437.

Porter, K. B., G. L. Peterson, and O. Vise. 1982. A new greenbug biotype. *Crop Sci.* 22:847–850.

Price, P. W. 1986. Ecological aspects of host plant resistance and biological control: Interactions among three trophic levels. Pages 11–30 in Boethel, D. J., and R. D. Eikenbary (eds.), *Interactions of Plant Resistance and Parasitoids and Predators of Insects*. Chichester, England: Ellis Horwood Ltd.

Puttler, B. 1967. Interrelationship of *Hypera postica* (Coleoptera.: Curculionidae) and *Bathyplectes curculionis* (Hymenoptera: Ichneumonidae) in the eastern United States with particular reference to encapsulation of parasite eggs by the weevil larvae. *Annu. Entomol. Soc. Am.* 60:1031–1038.

Rabb, R. L., and G. G. Kennedy. 1979. *Movement of Highly Mobile Insects: Concepts and Methodology in Research*. Raleigh: University Graphics, North Carolina State Univ.

Raulston, J. R. 1975. Tobacco budworm: Observation on the laboratory adaptation of a wild strain. *Annu. Entomol. Soc. Am.* 68:139–142.

Remington, C. L. 1968. The population genetics of insect introductions. *Annu. Rev. Entomol.* 13:415–426.

Richardson, R. H. (ed.). 1978. *The Screwworm Problem*. Austin: Univ. of Texas Press.

Richardson, R. H., J. R. Ellison, and W. W. Averhoff. 1982. Autocidal control of screwworms in North America. *Science* 215:361–370.

Roelofs, W. L., J. W. Du, X. H. Tang, P. S. Robbins, and C. J. Eckenrode. 1985. Three European corn borer populations in New York based on sex pheromones and voltinism. *J. Chem. Ecol.* 11:829–836.

Roeske, C. N., J. N. Seiber, L. P. Brower, and C. N. Moffit. 1976. Milkweed cardenolides and their comparative processing by monarch butterflies. Pages 93–167 in Wallace, J. W., and R. L. Mansell (eds.), *Biochemical Interactions Between Plants and Insects*. Rec. Adv. Phytochem. New York: Plenum.

Rossler, Y. 1975. Reproductive differences between laboratory-reared and field collected populations of the Mediteranean fruit fly, *Ceratitis capitata*. *Annu. Entomol. Soc. Am.* 68:987–991.

Roush, R. T., and J. A. McKenzie. 1987. Ecological genetics of insecticide and acaricide resistance. *Annu. Rev. Entomol.* 32:361–380.

Schlinger, E. I., and R. L. Doutt. 1964. Systematics in relation to biological control. Pages 247–280 in DeBach, P. (ed.), *Biological Control of Insect Pests and Weeds*. New York: Reinhold Publishing.

Schoonhoven, L. M. 1967. Loss of host plant specificity by *Manduca sexta* after rearing on artificial diet. *Entomol. Exp. Appl.* 10:270–272.

Scorza, R. 1983. Ecology and genetics of exotics. Pages 219–238 in Wilson, C. L., and C. L. Graham (eds.), *Exotic Plant Pests and North American Agriculture*. New York: Academic Press.

Scruggs, C. G. 1978. The origin of the screwworm control program. Pages 11–18 in Richardson, R. H. (ed.), *The Screwworm Problem*. Austin: Univ. of Texas Press.

Showers, W. B. 1979. Effect of diapause on the migration of the European corn borer into the southeastern United States. Pages 420–429 in Rabb, R. L., and G. G. Kennedy (eds.), *Movement of Highly Mobile Insects: Concepts and Methodology in Research*. Raleigh: University Graphics, North Carolina State Univ.

Slansky, F., Jr., and J. G. Rodriguez (eds.). 1987. *Nutritional Ecology of Insects, Mites, Spiders and Related Invertebrates*. New York: John Wiley & Sons.

Smith, J. M. 1957. Effects of the food plant of the California red scale, *Aonidiella aurantii* (Mask.), on reproduction of its hymenopterous parasites. *Can. Entomol.* 89:219–230.

Soule, A. M., and P. O. Panatter. 1907. The cultivation of potatoes. *Va. Agric. Exp. Stn. Bull.* No. 174. 16:125–129.

Stinner, R. E. 1977. Efficacy of innundative releases. *Annu. Rev. Entomol.* 22:516–531.

Stock, M. W. 1979. Genetic markers in applied research on forest Lepidoptera. Pages 328–332 in Rabb, R. L., and G. G. Kennedy (eds.), *Movement of Highly Mobile Insects: Concepts and Methodology in Research*. Raleigh: University Graphics, North Carolina State Univ.

Tauber, M. J., and C. A. Tauber. 1978. Evolution of phenological strategies in insects: A comparative approach with ecophysiological and genetic considerations. Pages 53–71 in Dingle, H. (ed.), *Evolution of Insect Migration and Diapause*. New York: Springer-Verlag.

Tauber, M. J., C. A. Tauber, and S. Masaki. 1986. *Seasonal Adaptations of Insects*. New York: Oxford Univ. Press.

Teetes, G. L. 1980. Breeding sorghums resistant to insects. Pages 457–485 in Maxwell, F. G., and P. R. Jennings (eds.), *Breeding Plants Resistant to Insects*. New York: John Wiley & Sons.

Terriere, L. C. 1984. Induction of detoxication enzymes in insects. *Annu. Rev. Entomol.* 29:71–88.

Turnock, W. J., K. L. Taylor, D. Schroder, and D. L. Dahlsten. 1976. Biological control of pests of coniferous forests. Pages 289–311 in Huffaker, C. B., and P. S. Messenger (eds.), *Theory and Practice of Biological Control*. New York: Academic Press.

Unruh, T. R., and R. D. Goeden. 1987. Electrophoresis helps to identify which race of the introduced weevil, *Rhinocyllus cornicus* (Coleoptera: Curculonidae), has transferred to two native southern California thistles. *Environ. Entomol.* 16:979–983.

Van Emden, H. F. 1986. The interaction of plant resistance and natural enemies: Effects on populations of sucking insects. Pages 138–150 in Boethel, D. J., and R. D. Eikenbary (eds.), *Interactions of Plant Resistance and Parasitoids and Predators of Insects*. Chichester, England: Ellis Norword, Ltd.

Vinson, S. B., and P. Barbosa. 1987. Interrelationships of nutritional ecology of parasitoids. Pages 673–695 in Slanksy, F., Jr., and J. G. Rodriguez (eds.), *Nutritional Ecology of Insects, Mites, Spiders and Related Invertebrates*. New York: John Wiley & Sons.

Vinson, S. B., and G. F. Iwantsch. 1980. Host suitability for insect parasitoids. *Annu. Rev. Entomol.* 25:397–419.

Volker, K. C., and R. G. Simpson. 1975. Behavior of alfalfa weevil larvae affecting the establishment of *Tetrastichus incertus* in Colorado. *Environ. Entomol.* 4:742–744.

Watanabe, H. 1987. The host population. Pages 71–112 in Fuxa, J. R., and Y. Tanada (ed.), *Epizootiology of Insect Diseases*. New York: John Wiley & Sons.

Whitten, M. J. 1970. Genetics of pests in their management. Pages 119–137 in Rabb, R. L., and F. E. Guthries (eds.), *Concepts of Pest Management*. Proceedings of a Conference Held at North Carolina State University, Raleigh, North Carolina, March 25–27, 1970. University Graphics.

Wilson, F. 1965. Biological control and the genetics of colonizing species. Pages 307–329 in

Baker, H. G., and G. L. Stebbins (eds.), *The Genetics of Colonizing Species*. New York: Academic Press.

———. 1967. Insect abundance: Prospect. *Symp. R. Entomol. Soc. Lond.* 4:143–155.

Wood, J. R., E. A. Starks, and K. J. Starks. 1972. Effect of temperature and host plant interaction on the biology of three biotypes of the greenbug. *Environ. Entomol.* 1:230–234.

Young, J. R. 1979. Assessing the movement of the fall armyworm (*Spodoptera frugiperda*) using insecticide resistance and wind pattern. Pages 344–351 in Rabb, R. L., and G. G. Kennedy (eds.), *Movement of Highly Mobile Insects: Concepts and Methodology in Research*. Raleigh: University Graphics, North Carolina State Univ.

Yu, S. J. 1983. Induction of detoxifying enzymes by allelochemicals and host plants in the fall armyworm. *Pest. Biochem. Physiol.* 19:330–336.

Yu, S. J., R. E. Berry, and L. C. Terriere. 1979. Host induced stimulation of detoxifying enzymes in phytohpagous insects. *Pest. Biochem. Physiol.* 12:280–284.

23

BIOLOGY OF VARIATION: EPILOGUE

Ke Chung Kim
Bruce A. McPheron

The Pennsylvania State University

INTRODUCTION

Infraspecific variation has been the subject of genetic research since the dawn of modern genetics (Dunn, 1951), which has resulted in a large volume of works on genetic diversity (e. g., Lewontin, 1974; Nevo et al., 1984). For the last 15 years, the subject of variation within species has become a target of intensive entomological and ecological research using molecular and morphometric techniques with regard to arthropod pests (Kim, Chapter 1; Steiner, Chapter 2, in this volume).

Infraspecific variation includes (1) nongenetic or ecological variation, such as homeostasis and phenocopy, which are phenotypic adaptation to environmental conditions within the limit of the norm of reactions, and (2) genetic variation in many different biological traits. Genetic traits range from single genes through polygenic variation, expressed in traits such as behavior, morphology, and virulence. However, the ultimate issue is how this variation affects fitness under natural selection (e.g., Futuyma and Peterson, 1985) and how genes and phenotypes, genes and fitness, and phenotype and fitness interact (Scharloo, 1984). In many instances, genetic variation is assessed without the knowledge of the corresponding fitness variation, which may affect the interpretation of molecular data.

Insects are highly polymorphic (Selander, 1976), and this variation is found in diverse biological traits, both genetic and phenotypic characters. A wide range of genetic diversity has been documented for many insect pests, and the pattern of genetic variation varies greatly with different species (Kim, Chapter 1 in this volume). Geographic and temporal variation is found in almost all species of insect pests. Local environments differ greatly in levels of anthropogenic stresses. It is increasingly evident that insect pests have evolved rapidly in their stressed environments (Diehl and Bush, 1984; NRC, 1986; Parsons, 1988; Hudon et al., 1989; Green et al., 1990; Bethke et al., 1991; Chapters 1, 2, 14–22 in this volume).

The differentiation of populations within species may be initiated by changes in genetic or environmental systems, or both. A change in environmental conditions destabilizes the genotype environment equilibrium, and deterministic selection may lead to adaptation to the new environment (Templeton 1980, 1981; Barker and Thomas, 1987). Variation in allele frequencies involves both genic and chromosomal mechanisms (Beaverstock and Adams, 1987; Nevo, 1988). Differentiation by colonization involves niche shift, invasion of disturbed or newly created habitats by anthropogenic causes, or immigration and founder effect (Parsons, 1983; Carson and Templeton, 1984). Subdivision of the populations of a widely distributed species by founder effects is often relatively rapid (Wilson and Graham, 1983).

Rapid differentiation within speices has not only raised interesting research avenues but also has posed serious problems for conventional strategies in the management of arthropod pests and vectors. This differentiation also threatens the value of information on insects pest species in the scientific literature, because a species name (binomen) by itself does not predict accurate biological information about a genetically substructured species. In other words, the citation of a scientific work on a species no longer provides the kind of biological information conventionally expected by the scientific community.

In this concluding chapter, we will summarize the subject of variation within species and propose a taxonomy of infraspecific taxa.

SPECIES AND GENETIC VARIATION

A central theme of this book has been the demonstration of significant variation below the level of the nominal species. In many cases, this variation, whether phenotypic, genotypic, or both, was identified because of the pest status of the organism. Detailed research on individual agricultural pests led to the realization that heterogeneity in one or more biologically significant characters was inconsistent with the notion that the species of interest represented a single, uniform gene pool. If we now recognize the sometimes substantial genetic variation within a single species, can we develop a consistent method for referring to this variation?

The desire to name homogeneous entities is certainly not novel. The evolution of the species concept itself is a case in point. The typological species concept worked very well when the naturalist's philosophical framework did not require the recognition of variation. However, as naturalists systematists noted extensive diversity within their typological species, there was a growing recognition of the role of biological factors in defining the species boundary; Mayr (1982) traces the roots of the reproductive isolation criterion in species thinking back to Ray and Linnaeus.

Acceptance of the biological species concept did not solve the problem of unambiguously defining all variant taxa. A huge variety of terms have been coined to describe infraspecific (beneath the category of species) variation. However, efforts to use nearly all of these terms consistently among or even within taxa have failed (e.g., Diehl and Bush, 1984; Berlocher, 1989; Pashley, Chapter 17 in this volume). This should not be surprising, given that even application of the species concept varies substantially among taxa and among biologists (e.g., Mayr, 1982, Templeton, 1989).

Several problems must be addressed from the perspective of infraspecific variation. First, how do we identify important variants? We must distinguish between polymorphism across the range of a species and variation that is structured in some manner below the species level. For instance, host-associated variation, clines, and geographically disjunct variation are all patterns that may be important in devising a successful pest management scheme. Second, given that many techniques are available to study infraspecific variation, how are we to interpret the results of such studies? Cataloguing patterns of phenotypic variation provides substantially different information than a quantitative genetic study of a particular trait that, in turn, addresses different features of genetic structure than does a population survey using some genetic marker. In order to progress in the study of infraspecific variation, we must understand the strengths and limitations of these varying approaches. Third, if we can identify important variants and develop an understanding of genotypic and phenotypic aspects of the traits, is there some consistent manner in which to refer to these variants? The species, in addition to serving as a philosophical and biological concept, provides a retrieval mechanism. However, if an insect species actually represents two or more distinct groups of organisms that, while not reproductively isolated, differ in their use of host plants or their resistance to insecticides, everyone from evolutionary biologists to applied entomologists would benefit from some unique identifier.

TAXONOMY OF INFRASPECIFIC TAXA

Systematists have been perplexed with the definition and designation of infraspecific taxa for the last 50 years (e.g., Dobzhansky, 1937, 1970; Mayr, 1942, 1969, 1970; Amadon and Short, 1976; Davis and Heywood, 1973; Van Valen, 1976; Diehl and Bush, 1984; Walters, 1989; Hewitt, 1989; Raffa, 1989). The problem is complex, because it involves the delimitation of the degree of infraspecific population differentiation, on the one hand, and the matter of nomenclature, on the other hand.

A major difficulty lies in the fact that the process of differentiation of populations is complex and the rate of evolutionary change is different by populations and habitats. Mosaic differentiation results in infraspecific taxa with different degrees of genetic divergence that are difficult to measure precisely. The delimitation of such infraspecific taxa is at best arbitrary. A species name simply does not predict the complex biological information on infraspecific variation, which is very important for applied ecology and entomology. Thus, it is necessary to define infraspecific taxa, because such designation becomes an information marker in the vast scientific literature.

Formation of Infraspecific Taxa

The pattern of geographic variation varies greatly with different species and different populations within species. Genetic divergence is found in many biological traits

(Kim, Chapter 1 in this volume), such as behavior, morphology, physiology, developmental rate, fecundity, dispersal ability, seasonal adaptations, host and feeding preference, and resistance to pesticides and toxins (Denno and Dingle, 1980; Daly, 1985; Tauber et al., 1986; Taylor and Karban, 1986; Hudon et al., 1989), as well as chromosomal and genic traits, such as inversion rates, allozyme polymorphism, and DNA sequences (Lewontin, 1974; Nevo et al., 1984; Parsons, 1988; Steiner, Chapter 2 in this volume).

Genetic divergence occurs in response to changes in the environment and genetic system. In any locality, the change in environmental conditions (Templeton, 1980, 1981; Barker and Thomas, 1987) or variation in gene frequency and/or chromosomal structure (Beaverstock and Adams, 1987; Nevo 1988) destabilizes the ecological equilibrium between genetic and environmental systems, and then deterministic selection pressure will lead to adaptation to the new environment (Kirkpatrick, 1982).

The genetic differentiation of populations largely depends upon the interacting balance between gene flow as a homogenizing force and drift and natural selection as diversifying mechanisms (Nevo et al., 1984; Futuyma and Peterson, 1985; Parsons, 1991). Individual local populations experience differential interaction of genetic and environmental systems. These processes therefore produce diverse infraspecific taxa with variable levels of genetic divergence, which have been classified as biotype, ecotype, Formenkreis, Rassenkreis, race, or subspecies.

Infraspecific Categories

Populations of a species that occur in different localities and in different environmental conditions may be described in terms of morphometric and behavioral differences, relative allele frequences, or relative proportions of chromosomal types in connection with specific adaptive changes (Eastop, 1973; Russell, 1978; Kim, Chapter 1 in this volume). Different terms have been applied to infraspecific taxa with varying degrees of biological differences (Davis and Heywood, 1973; Endler, 1977; Gonzalez et al., 1979; Diehl and Bush, 1984; Pashley, Chapter 17 in this volume).

The terminology that has been applied to infraspecific taxa is summarized as follows:

General

1. **Population.** Any group of individuals considered together at any one time because of common features (Davis and Heywood, 1973).
2. **Mendelian (breeding) population.** A reproductive community of sexual and cross-fertilizing individuals that shares a common gen pool (Dobzhansky, 1970).
3. **Deme.** A local population (Endler, 1977).

Ecotypes

1. **Ecomorph.** A population that shows phenotypic differences from other conspecific populations with the same genotype under different environmental conditions (Diehl and Bush, 1984).
2. **Ecotype.** A population distinguished by morphological and physiological characters, most frequently of a quantitative nature (Turesson, 1922; Davis and Heywood, 1973).

Biotypes

1. **Genemorph.** A population that shows genetic differences from other conspecific populations (Davis and Heywood, 1973).
2. **Strain.** A sympatric and synchronic population with genetic variation adapted to specific host plants that differs from other conspecific populations (Pashley, 1989; Pashley, Chapter 17 in this volume); a population that shows biological differences from other conspecific populations (authors).
3. **Biotype.** An assemblage of individuals with the same or essentially the same genotypes (authors); a population of individuals of similar genetic composition for some defined biological attribute (Gonzalez et al., 1979).
4. **Host race.** A population that is partially reproductively isolated from other conspecific populations as a direct consequence of adaptation to a specific host (Diehl and Bush, 1984).
5. **Biological race.** A noninterbreeding sympatric population that differs in biological characteristics but not, or scarcely, in morphology (Mayr, 1970).
6. **Race.** A group of individuals or populations characterized by a more or less distinctive combination of biological traits that are transmitted in descent (Endler, 1977); a cluster of local populations that differs from other clusters in the frequency of some alleles or chromosomal structure (Dobzhansky, 1970).
7. **Semispecies.** Conspecific populations that are partially reproductively isolated (DeBach, 1969).

Subspecies

1. **Subspecies.** A geographically defined aggregate of local populations that differ taxonomically from other such subdivisions of the species (Mayr, 1969).
2. **Geographic race.** A subspecies (Mayr, 1969).
3. **Rassenkreis.** A group of subspecies connected by clines (Rensch, 1929).

These terms, related to infraspecific taxa as they are generally applied, are not very useful, because they simply define an infraspecific entity without providing any practical criteria. Therefore, the following definitions and use of terms are proposed for practical application:

Ecotype. A population or an aggregate of populations distinguished by morphological and/or physiological characters from other such conspecific populations where the genetic basis of this difference is unknown.

Biotype. A population or an aggregate of populations with similar genetic composition distinguished by morphological or other phenotypic characters that are genetically different from other such conspecific populations.

Host Race. A population that is partially reproductively isolated from other conspecific populations as a direct consequence of adaptation to a specific host.

Subspecies. A geographically defined aggregate of local populations that differ by distinct morphological characters from other such conspecific subdivisions.

Nomenclature of Infraspecific Taxa

The subspecies (trinomen) is the only infraspecific taxon so recognized by the International Code of Zoological Nomenclature (ITZN, 1985). However, the category of subspecies has not been widely used in entomology, because, in most cases, the level of population differentiation within a species is not clearly determined and subspecies populations are difficult to identify.

The recognition of infraspecific taxa is obviously important for pest management and biological control, and a formal designation of such taxa is useful in scientific communication. In scientific literature, the species name alone does not convey any information about infraspecific taxa. A literature search usually yields a large volume of references for a pest species, many of which may not carry relevant biological information about the species because of infraspecific differentiation. At present, there is no way to avoid this "unproductive" literature search. For this reason, we propose a formal designation of infraspecific taxa without infringing upon the Zoological Code.

The European corn borer, *Ostrinia nubilalis* (Hübner), which was introduced into North America approximately 75 years ago, has become a major economic pest of corn in the northcentral United States for the last 50 years. It is widely distributed from eastern North America to Montana, and from Canada to Texas, Louisiana, and Florida (Hudon et al., 1989). The colonization began in two areas—one in the Boston area and the other in the area surrounding Lake Erie. European corn borer populations in different localities, perhaps originally two strains with voltinism and host range differences, rapidly adapted to local environments through mixing of the strains and genetic changes (Hudon et al., 1989; Showers, Chapter 14 in this volume). As a result, they differ in diapause, host plants, morphology (Mutuura and Munroe, 1970; Kim et al., 1967; Chiang et al., 1970; Showers, 1981) and other characteristics (Hudon et al., 1989). Morphometric distinctness is well correlated with geographic population groupings (Kim et al., 1967; Chiang et al., 1970). Differences in voltinism and diapause among biotypes are persistent in identifical environments and are genetically controlled by sex-linked, nondominant polygenes (Showers et al., 1975; Showers, 1979, 1981). Presently, European corn borer populations in North America are classified into three biotypes: (1) the northern univoltine biotype (Northern Minnesota, Quebec, parts of New York, Ontario), (2) the central bivoltine biotype (Central Corn Belt, southwestern Ontario), and (3) the southern multivoltine biotype (Alabama, Georgia, Missouri, and North Carolina with characteristics (Showers et al., 1975; Showers, Chapter 14 in this volume). They are characterized by morphological variation and differences in biological characteristics, such as responses to photoperiods and thermoperiods (Beck, 1989), rates of growth, feeding habits, diapause response (Chiang, 1975; Showers et al., 1975; Showers, 1979), and pheromone response (Liebherr and Roelofs, 1975; Roelofs et al., 1985).

Recently, the sweetpotato whitefly, *Bemisia tabaci* (Gennadius), has become a serious pest of field-grown crops in California. This nonindigenous species has been in California since the 1920s (Natwick and Zalom, 1984) but did not cause any major injuries until 1981. Although it is polyphagous, attacking as many as 350 plant hosts (Mound and Halsey, 1978), the population growth and developmental rates of *B. tabaci* differ by host plants (Coudriet et al., 1985). *Bemisia tabaci* populations on cotton have increased and have evolved resistance to pesticides (Bellows and Arakawa, 1988; Nuessly et al., 1989). More recently, *B. tabaci* has become a serious pest of some ornamental plants, including poinsettia (Byrne et al., 1990). This recent increase in economic injuries is attributed to pesticide resistance (Bellows and Arakawa, 1988; Nuessly et al., 1989) and genetic alteration of the poinsettia population from the cotton population (Bethke et al., 1991).

What is the status of populations and so-called ecotypes in the European corn borer? At what stage of differentiation are the cotton and poinsettia populations of sweetpotato whitefly? These examples illustrate the problems for identification and designation of infraspecific taxa. According to the information available for infraspecific taxa of these species, the geographic populations of the European corn borer and the poinsettia and cotton populations of the sweetpotato whitefly are biotypes. It is proposed that they are designated by the scientific name (binomen) with superscript designation of infraspecific category:

Generic name + specific epithet$^{\textbf{Infraspecific category}}$

Thus, the Minnesota biotype of the European corn borer is designated

Ostrinia nubilalis$^{\text{B/MN}}$

and the poinsettia biotype of the sweetpotato whitefly is designated as

Bemisia tabaci$^{\text{B/POINSETTIA}}$

This system is convenient and flexible without complicating zoological nomenclature. Superscripts may indicate those infraspecific taxa recognized here: ecotype, biotype, and host race; for example, B is used for biotype, E is for ecotype, and HR for host race. Specific infraspecific type may be so designated; for example MN for

Minnesota. This nomenclature is an information vehicle for infraspecific variation, and this system can also be applied to cytospecies or cytotypes for sibling species that are not described with recognizable morphological characteristics. The utility of this system is dependent upon the consistent application of these infraspecific terms, which, in turn, depends on the ability to delimit this infraspecific variation.

The most basic requirement for successfully applying this system is the voucher designation and formal deposition of specimens for all infraspecific taxa formally recognized (Pashley, Chapter 17 in this volume). The preparation and preservation of voucher samples must follow the standards established for diverse techniques; for example, Steyskal et al. (1986) for pinned and alcohol preserved specimens and Dessauer et al. (1990) for specimens for molecular analysis. All voucher specimens must be deposited in the recognized museums and collections that are equipped with conventional collection facilities and deep freezers.

IDENTIFICATION OF VARIATION AND VARIANT POPULATIONS

The issue of recognition of genetic variation plagues applied entomologists and evolutionary biologists. Often, the first hint of infraspecific variation comes from pest managers, ecologists, or behaviorists. Two different classes of genetic variability below the species level are important for understanding evolutionary responses of insects. The first is simply genetic polymorphism across a species' range. Polymorphism has been variously described as an adaptive response to environmental heterogeneity (Nevo et al., 1984) or as the molecular outcome of neutral evolutionary processes (Kimura, 1983). Much of the early isozyme literature in insects and other taxa was directed at the exposition of levels of polymorphism.

A second, potentially more interesting, class of genetic variability is the nonrandom distribution of genetic polymorphism across a species' range. Into this class of variation we can lump such phenomena as clines in allelic variation, host-associated variation, and geographic substructuring of gene pools. In studies of this type of variation, many authors have identified actual or potential selective forces operating to influence the patterns of genetic variation identified. Structured polymorphism, in contrast to widespread polymorphism, is often indicative of some reduction in panmixis within a species. This observation has strong implications for the strategies employed to manage a pest species with such a genetic architecture.

Technology springing from advances in genetics has made possible the discrimination of population structure below the level of the species. A variety of genetic markers are now employed to investigate specific cases that have proved intractable by other means. Some knowledge regarding the nature of these markers is necessary. We frequently assume that genetic markers, such as isozymes, restriction fragment length polymorphisms (RFLPs), or DNA sequence variants, are behaving as if they are neutral to the effects of natural selection. However, it is to our advantage to confirm this wherever possible, although many of the pest species of interest are difficult subjects for genetic study. We can determine statistically if these markers segregate in a manner consistent with Hardy-Weinberg expectations, but this null hypothesis is fairly resilient, and conformance with Hardy-Weinberg expectations may disguise the fact that significant departures from such assumptions as no migration, negligible mutation, or no effects of natural selection could be occurring (Hartl and Clark, 1989). The fact remains that we are, in general, not looking directly at the genes responsible for whatever ecological, behavioral, or physiological traits vary within the species. Rather, we are looking at markers that differ diagnostically or in frequency of occurrence between subpopulations because a restriction in gene flow has led to some independence in evolution in the subpopulations. This disruption in gene flow is caused, we assume, by the variant traits we have identified. A brief review of some of the major genetic techniques used to discriminate among infraspecific variants will highlight the utility of the different approaches and also point out some of the limitations.

Isozymes

Perhaps the most widely used genetic technology in insect studies has been isozymes (reviewed in Wagner and Selander, 1974; Berlocher, 1984; Menken and Ulenberg, 1987; Loxdale and den Hollander, 1989). Isozymes still serve a useful function in studies of population level variation. It is a relatively inexpensive technique that has displayed excellent sensitivity in many case studies. The genetic basis of isozyme variation has been established for many insects (reviewed in Menken and Ulenberg, 1987), and enzyme genes have been mapped to chromosomes or linkage groups in a number of species (e.g., Saul, 1986; Feder et al., 1989; Hagen and Scriber, 1989).

Isozyme analysis has been successfully used to test specific hypotheses regarding population structure within a species. Steiner et al. (1985) demonstrated that isozyme patterns in corn leaf aphid populations from the central United States were inconsistent with the widely held assumption that these populations are founded annually by migration from the southern United States. Isozyme analysis of the apple maggot fly, a species often cited as an example of the process of sympatric speciation via host race formation, shows that this species does, in fact, consist of populations exhibiting at least partial reproductive isolation on sympatric host plants (Feder et al., 1988; McPheron et al., 1988). Pashley (Chapter 17 in this volume, and references therein), initially studying population structure of a migratory, polyphagous moth, has found that different host-associated populations appear to represent sibling species.

Isozyme variation can range from diagnostic differences (often characteristic of sibling species) to consistent frequency differences (e.g., the apple maggot host races discussed earlier) to a lack of differentiation, even though significant phenotypic variation exists in other characters (e.g., Wellso et al., 1988).

Mitochondrial DNA Sequence Variation

A variety of other genetic approaches are gaining popularity as the techniques become more accessible. Among these are studies of sequence variation in mitochondrial DNA (mtDNA). The advantages of using mtDNA for analysis of infraspecific variation lie primarily in two features. First, mtDNA is maternally inherited, so it provides an accurate picture of evolutionary patterns from a different perspective than do genes of nuclear origin. Second, mtDNA is relatively amenable to genetic analysis, particularly via restriction enzyme analysis. Digestion of mtDNA from different populations with restriction enzymes may yield RFLPs. RFLP studies of honey bee mtDNA have proven useful for identifying interactions among different geographic races or subspecies brought together in a new habitat. Of particular note are recent studies of African- and European-origin honey bees in South and Central America (Hall and Muralidharan, 1989; Smith et al., 1989; Sheppard et al., 1991). Pashley (1989) examined mtDNA RFLPs in fall armyworm host-associated strains and showed that mtDNA analyses increased the resolution of genetic differences between these taxa.

Nuclear DNA Sequence Variation

As with analysis of mtDNA sequence variation, much of the work done using nuclear DNA comparisons at infraspecific levels to date has relied on indirect estimates of sequence divergence. RFLP analyses of regions of rapidly evolving DNA have proven useful in some cases for separating taxa below the species. Beach et al. (1989) demonstrated substantial interpopulation length heterogeneity in a study of ribosomal DNA variation in the malaria vector *Anopheles albimanus* Wiedemann. This species shows substantial polymorphism in a variety of ecological characters, including incidence of *Plasmodium* infection (Collins et al., 1976), so clear understanding of infraspecific genetic structure is vital to proper design of epidemiological and management research.

Interpretation of Molecular Data

Resolving infraspecific variation in any character into a meaningful description of the evolutionary status of the taxa involved is a difficult matter. A number of suggestions for partitioning variation into different classes and thereby identifying a population by a certain name have been proposed (e.g., Jaenike, 1981; Diehl and Bush, 1984; Berlocher, 1989; Pashley, Chapter 17 in this volume). Information on population structure is reported by a relatively few summary statistics (reviewed by Weir, 1990). One common approach is the calculation of genetic distance (or identity), which gives an idea of the total evolutionary change between the taxa under consideration (at least, change at the loci studied; the issue of how representative a few loci are of the entire genome is beyond the scope of this chapter). Attempts have been made to generalize about the amount of genetic divergence that characterizes taxa at different levels (Ayala, 1975), but such a generalization makes substantial assumptions about the constancy of evolutionary change across taxa with entirely different phylogenetic histories. Genetic distance is, in general, more reliable when levels of interbreeding among the study units are restricted, so, for our purposes, it applies more to analyses of sibling species complexes or geographically isolated subspecies. A contrasting approach, more applicable to the study of populations that experience some level of gene flow, is the use of Wright's F-statistics to characterize how genetic polymorphism is organized among and within populations. F_{ST} values, the particular F-statistic that describes patterns of genetic variation among populations, have been summarized for a variety of insect species (Pashley et al., 1985; McCauley and Eanes, 1987). These F_{ST} values can be used to estimate levels of gene flow (Slatkin, 1985), but, as with genetic distance measures, it is impossible to generalize to an expected level of gene flow that would define a given infraspecific taxon.

The best use of molecular markers is as a relative, rather than an absolute, indicator of evolutionary status. Most molecular studies are conducted because other observations have indicated heterogeneity among populations of the species under consideration. This is the true strength of molecular techniques for identifying infraspecific variants: in cases in which traditional approaches have provided suggestive but inconclusive evidence for structure within a species, it is often possible to address the issue with genetic techniques. A combination of classical, quantitative, and molecular genetics methods may permit the actual basis of infraspecific variation to be identified and studied, greatly expanding our understanding of underlying evolutionary mechanisms, as well as our ability to devise appropriate management tools.

FUTURE PERSPECTIVES AND STRATEGIES

The importance of infraspecific variation goes far beyond the realm of basic genetics. The study of variation below the level of the nominal species contributes greatly to the understanding of dispersal and distribution patterns, habitat characteristics, population dynamics, species packing and resource utilization, and other ecological processes. This provides a more precise definition of pest problems and demands a dynamic systems approach to IPM based on the genetics and evolution of infraspecific variation of the pest species.

This book has demonstrated the tremendous range of infraspecific variation and the rapid rate of evolutionary changes among insect and arachnid pest species. The extent of infraspecific variation ranges from polyphagy to geographic populations of polytypic species. Many populations are host specific, forming host races. Yet, for many other species, practically nothing is known about variation below the level of the nominal species or the biology of these populations. Although the interaction between genetic structure and environmental conditions is known to produce infraspecific variation, the relationship between this variation and fitness is not known.

Much biological information for major economic pests now in the literature does not aid in the study of infraspecifc taxa. Such general biological data about the species are of historical value but are not relevant and applicable to the geographically and temporally-isolated populations that differ in genetic and ecological characteristics from their recent ancestors. If IPM is to succeed, consistent biological definition and delimitation of infraspecific variation for major economic pests and vectors must be accelerated and new biological information must be obtained. With this information, new strategies can be developed for effectively containing major economic pests.

REFERENCES

Amadon, D., and L. L. Short. 1976. Treatment of subspecies approaching species status. *Syst. Zool.* 25:161–167.

Ayala, F. J. 1975. Genetic differentiation during the speciation process. *Evol. Biol.* 8:1–78.

Beach, R. F., D. Mills, and F. H. Collins. 1989. Structure of ribosomal DNA in *Anopheles albimanus* (Diptera: Culicidae). *Ann. Entomol. Soc. Am.* 82:641–648.

Barker, J. S. F., and R. H. Thomas. 1987. A quantitative genetic perspective on adaptive evolution. Pages 3–23 in Loeschcke, V. (ed.), *Genetic Constraints on Adaptive Evolution.* Berlin: Springer-Verlag.

Beaverstock, P. R., and M. Adams. 1987. Comparative rates of molecular, chromosomal and morphological evolution in some Australian vertebrates. Pages 175–188 in Campbell, K. S. W., and M. F. Day (eds.), *Rates of Evolution.* London: Allen & Unwin.

Beck, S. D. 1989. Developmental and seasonal biology of *Ostrinia nubilalis*. Pages 45–82 in Russell, G. E. (ed.), *Biology and Population Dynamics of Invertebrate Crop Pests.* Andover, Haphshire: Intercept, Ltd.

Bellows, T. S., Jr., and K. Arakawa. 1988. Dynamics of preimaginal populations of *Bemisia tabaci* (Homoptera: Aleyrodidae) and *Eretmocerus* spp. (Hymenoptera: Aphelinidae) in southern California cotton. *Environ. Entomol.* 17:483–487.

Berlocher, S. H. 1984. Insect molecular systematics. *Annu. Rev. Entomol.* 29:403–433.

———. 1989. The complexities of host races and some suggestions for their identification by enzyme electrophoresis. Pages 51–68 in Loxdale, H.D., and J. den Hollander (eds.), *Electrophoretic Studies on Agricultural Pests.* Syst. Assoc. Special Vol. 39. New York: Clarendon Press.

Bethke, J. A., T. D. Paine, and G. S. Nuessly. 1991. Comparative biology, morphometrics, and development of two populations of *Bemisia tabaci* (Homoptera: Aleyrodidae) on cotton and poinsettia. *Ann. Entomol. Soc. Am.* 84:407–411.

Byrne, D. N., T. S. Bellows, Jr., and M. P. Parrella. 1990. Whiteflies in agricultural systems. Pages 227–261 in Gerling, D. (ed.), *Whiteflies: Their Bionomics, Pest Status and Management.* Andover, Hampshire, England: Intercept.

Carson, H. L., and A. R. Templeton. 1984. Genetic revolutions in relation to speciation phenomena; the founding of new populations. *Annu. Rev. Ecol. Syst.* 15:97–131.

Chiang, H. C. 1975. Applied ecological investigations on the European corn borer in Minnesota. *Report of the International Project on Ostrinia nubilalis, Agricultural Research Institute of the Hungarian Academy of Sciences, Martonvasar, Phase II Results,* pp 14–17.

Chiang, H. C., K. C. Kim, and B. W. Brown. 1970. Morphometric variability related to ecological conditions of three biotypes of *Ostrinia nubilalis* in the North Central United States. *Ann. Entomol. Soc. Am.* 63:1013–1016.

Collins, W. E., J. C. Skinner, M. Warren, and B. Richardson. 1976. Studies on human malaria in *Aotus* monkeys. VII. Comparative infectivity of two strains of *Plasmodium vivax* to *Anopheles freeborni, A. maculata* and four strains of *A. albimanus. J. Parasitol.* 64:190–194.

Coudriet, D. L., N. Prabhaker, A. N. Kishaba, and D. E. Meyerdirk. 1985. Variation in developmental rate on different hosts and overwintering of the sweetpotato whitefly *Bemisia tabaci* (Homoptera: Aleyrodidae). *Environ. Entomol.* 14:516–519.

Daly, H. V. 1985. Insect morphometrics. *Annu. Rev. Entomol.* 30:415–438.

Davis, P. H. and V. H. Heywood. 1973. *Principles of Angiosperm Taxonomy.* Huntington, N.Y.: Krieger.

DeBach, P. 1969. Uniparental, sibling and semi-species in relation to taxonomy and biological controls. *Israel J. Entomol.* 4:11–28.

Denno, R. F., and H. Dingle. 1980. *Insect Life History Patterns.* New York: Springer-Verlag.

Dessauer, H. C., C. J. Cole, and M. S. Hafner. 1990. Collection and storage of tissues. Pages 25–41 in Hillis, D. M., and C. Moritz. *Molecular Systematics.* Sunderland, Mass.: Sinauer Assoc.

Diehl, S. R., and G. L. Bush. 1984. An evolutionary and applied perspective of insect biotypes. *Annu. Rev. Entomol.* 29:471–504.

Dobzhansky, T. 1937. *Genetics and the Origin of Species.* New York: Columbia Univ. Press.

———. 1970. *Genetics of the Evolutionary Process.* New York: Columbia Univ. Press.

Dunn, L. C. (ed.). 1951. *Genetics in the Twentieth Century.* New York: Macmillan.

Eastop, V. F. 1973. Biotypes of aphids. *Bull. Entomol. Soc. N. Z.* 2:40–41.

Endler, J. A. 1977. *Geographic Variation, Speciation, and Clines.* Princeton, N.J.: Princeton Univ. Press.

Feder, J. L., C. A. Chilcote, and G. L. Bush. 1988. Genetic differentiation between sympatric host races of *Rhagoletis pomonella. Nature* 336:61–64.

———. 1989. Inheritance and linkage relationships of allozymes in the apple maggot fly. *J. Hered.* 80:277–283.

Futuyma, D. J., and S. C. Peterson. 1985. Genetic variation in the use of resources by insects. *Annu. Rev. Entomol.* 30:217–238.

Gonzalez, D., G. Gordh, S. N. Thompson, and J. Adler. 1979. Biotype discrimination and its importance to biological control. Pages 129–136 in Hoy, M. A., and J. J. McKelvey, Jr. (eds.), *Genetics in Relation to Insect Management.* New York: Rockfeller Foundation.

Green, N. G., H. M. LeBaron, and L. W. K. Moberg. 1990. *Managing Resistance to Agrochemicals: From Fundamental Research to Practical Strategies.* Washington, D. C.: American Chemical Society.

Hagen, R. H., and J. M. Scriber. 1989. Sex-linked diapause, color, and allozyme loci in *Papilio glaucus:* Linkage analysis and significance in a hybrid zone. *J. Hered.* 80:179–185.

Hall, H. G., and K. Muralidharan. 1989. Evidence from mitochondrial DNA that African honey bees spread as continuous maternal lineages. *Nature* 339:211–213.

Hartl, D., and A. G. Clark. 1989. *Principles of Population Genetics,* 2nd ed. Sunderland, Mass.: Sinauer Assoc.

Hewitt, G. M. 1989. The subdivision of species by hybrid zones. Pages 85–133 in Otte, D., and J. A. Endler (eds.), *Speciation and Its Consequences.* Sunderland, Mass.: Sinauer Assoc.

Hudon, M., E. J. LeRoux, and D. G. Harcourt. 1989. Seventy years of European corn borer *(Ostrinia nubilalis)* research in North America. Pages 1–44 in Russell, G. E. (ed.), *Biology and Populations Dynamics of Invertebrate Crop Pests.* Andover, Hampshire, England: Intercept.

International Trust for Zoological Nomenclature (ITZN). 1985. *International Code of Zoological Nomenclature.* 3rd ed. London: International Trust for Zoological Nomenclature.

Jaenike, J. 1981. Criteria for ascertaining the existence of host races. *Am. Nat.* 117:830–834.

Kim, K. C., H. C. Chiang, and B. W. Brown, Jr. 1967. Morphometric differences among four biotypes of *Ostrinia nubilalis*. *Ann. Entomol. Soc. Am.* 60:796–801.

Kimura, M. 1983. *The Neutral Theory of Molecular Evolution*. New York: Cambridge Univ. Press.

Kirkpatrick, M. 1982. Quantum evolution and punctuated equilibria in continuous genetic characters. *Am. Nat.* 119:833–848.

Lewontin, R. C. 1974. *The Genetic Basis of Evolutionary Changes*. New York: Columbia Univ. Press.

Liebherr, J., and W. Roelofs. 1975. Laboratory hybridization and mating period studies using two pheromone strains of *Ostrinia nubilalis*. *Ann. Entomol. Soc. Am.* 68:305–309.

Loxdale, H. D., and J. den Hollander (eds.). 1989. *Electrophoretic Studies on Agricultural Pests*. Syst. Assoc. Special Vol. 39. New York: Clarendon Press.

Mayr, E. 1942. *Systematics and the Origin of Species*. New York: Columbia Univ. Press.

———. 1969. *Principles of Systematic Zoology*. New York: McGraw-Hill.

———. 1970. *Populations, Species, and Evolution*. Cambridge, Mass.: Harvard Univ. Press.

———. 1982. *The Growth of Biological Thought: Diversity, Evolution, and Inheritance*. Cambridge, Mass.: Belknap Press.

McCauley, D. E., and W. F. Eanes. 1987. Hierarchical population structure analysis of the milkweed beetle, *Tetraopes tetraophthlamus* (Forster). *Heredity* 58:193–201.

McPheron, B. A., D. C. Smith, and S. H. Berlocher. 1988. Genetic differences between host races of the apple maggot fly. *Nature* 336:64–66.

Menken, S. B. J., and S. A. Ulenberg. 1987. Biochemical characters in agricultural entomology. *Agric. Zool. Rev.* 2:305–360.

Mound, L. A., and S. H. Halsey. 1978. *Whiteflies of the World: A Systematic Catalogue of Aleyrodidae (Homoptera) with Host Plant and Natural Enemy Data*. British Museum (Nat. Hist.). New York: John Wiley & Sons.

Mutuura, A., and E. Munroe. 1970. Taxonomy and distribution of the European corn borer and allied species: Genus *Ostrinia* (Lepidoptern: Pyralidae). *Mem. Entomol. Soc. Canada* 71:1–8, 37–40.

National Research Council (NRC). 1986. *Pesticide Resistance: Strategies and Tactics for Management*. Committee on Strategies for the Management of Pesticide Resistant Pest Populations. Washington, D. C.: National Academy Press.

Natwick, E. T., and F. G. Zalom. 1984. Surveying sweetpotato whitefly in the Imperial Valley. *Calif. Agr.* 38 (3 and 4):11.

Nevo, E. 1988. Genetic diversity in nature: Patterns and theory. *Evol. Biol.* 23:217–246.

Nevo, E., A. Beiles, and R. Ben-Shlomo. 1984. The evolutionary significance of genetic diversity: Ecological, demographic and life-history correlates. Pages 13–213 in Mani, G. S. (ed.), *Evolutionary Dynamics of Genetic Diversity*. Berlin: Springer-Verlag.

Nuessly, G. S., T. J. Henneberry, and H. H. Perkins, Jr. 1989. Effects of sweetpotato whitefly population densities on cotton fibre stickiness and reducing sugars. Page 281 in *Proceedings, Beltwide Cotton Production and Research Conference*, National Cotton Council, Nashville, Tenn.

Parsons, P. A. 1983. *The Evolutionary Biology of Colonizing Species*. Cambridge: Cambridge Univ. Press.

———. P. A. 1988. Evolutionary rates: Effects of stress upon recombination. *Biol. J. Linn. Soc.* 35:49–68.

———. 1991. Evolutionary rates: Stress and species boundaries. *Annu. Rev. Ecol. Syst.* 22:1–18.

Pashley, D. P. 1989. Host-associated differentiation in armyworms (Lepidoptera: Noctuidae): An allozymic and mitochondrial DNA perspective. Pages 103–114 in Loxdale, H. D., and J. den Hollander (eds.), *Electrophoretic Studies on Agricultural Pests.* Syst. Assoc. Special Vol. 39. New York: Clarendon Press.

Pashley, D. P., S. J. Johnson, and A. N. Sparks. 1985. Genetic population structure of migratory moths: The fall armyworm (Lepidoptera: Noctuidae). *Ann. Entomol. Soc. Am.* 78:756–762.

Raffa, K. F. 1989. Genetic engineering of trees to enhance resistance to insects. *BioScience* 39(8):524–534.

Rensch, B. 1929. *Das Prinzip geographischer Rassenkreis und das Problem der Artbildung.* Berlin: Brontraeger.

Roelofs, W. L., J.-W. Du, X.-H. Tang, P. S. Robbins, and C. J. Eckenrode. 1985. Three European corn borer populations in New York based on sex pheromones and voltinism. *J. Chem. Ecol.* 11:829–836.

Russell, G. E. 1978. *Plant Breeding for Pest and Disease Resistance.* Boston: Butterworth.

Saul, S. H. 1986. Genetics of the Mediterranean fruit fly *(Ceratitis capitata)* (Wiedeman). *Agric. Zool. Rev.* 1:73–108.

Scharloo, W. 1984. Genetics of adaptive reactions. Pages 5–15 in Woehmann, K., and V. Loeschcke (eds.), *Population Biology and Evolution.* Berlin: Springer-Verlag.

Selander, R. K. 1976. Genic variation in natural populations. Pages 21–44 in Ayala, F. J. (ed.), *Molecular Evolution.* Sunderland, Mass.: Sinauer Associates.

Sheppard, W. S., T. E. Rinderer, J. A. Mazzoli, J. A. Stelzer, and H. Shimanuki. 1991. Gene flow between African- and European-derived honey bee populations in Argentina. *Nature* 349:782–784.

Showers, W. B. 1979. Effect of diapause on the migration of the European corn borer in the southwestern United States. Pages 420–430 in Rabb, R. L., and G. G. Kennedy (eds.), *Movement of Highly Mobile Insects: Concepts and Methodology in Research.* Raleigh, N.C.: University Graphics.

———. 1981. Geographic variation of the diapause response in the European corn borer. Pages 97–111 in Denno, R. F., and H. Dingle (eds.), *Insect Life History Patterns: Habitat and Geographic Variation.* New York: Springer-Verlag.

Showers, W. B., H. C. Chiang, A. J. Keaster, R. E. Hill, G. L. Reed, A. N. Sparks, and G. J. Musick. 1975. Ecotypes of the European corn borer in North America. *Environ. Entomol.* 4:753–760.

Slatkin, M. 1985. Gene flow in natural populations. *Annu. Rev. Ecol. Syst.* 16:393–430.

Smith, D. R., O. R. Taylor, and W. M. Brown. 1989. Neotropical Africanized honey bees have African mitochondrial DNA. *Nature* 339:213–215.

Steiner, W. W. M., D. J. Voegtlin, and M. E. Irwin. 1985. Genetic differentiation and its bearing on migration in North American populations of the corn leaf aphid *Rhopalosiphum maidis* (Fitch) (Homoptera: Aphididae). *Ann. Entomol. Soc. Am.* 78:518–525.

Steyskal, G. C., W. L. Murphy, and E. M. Hoover (eds.). 1986. *Insects and Mites: Techniques for Collection and Preservation.* U. S. Dept. of Agric., Misc. Publ. No. 14423. 103 pp.

Tayler, H., and R. Karban (eds.). 1986. *The Evolution of Insect Life Cycles.* New York: Springer-Verlag.

Tauber, M. J., C. A. Tauber, and S. Masaki. 1986. *Seasonal Adaptation of Insects*. New York: Oxford Univ. Press.

Templeton, A. R. 1980. The theory of speciation via the founder principle. *Genetics* 94:1011–1038.

———. 1981. Mechanisms of speciation—A population genetic approach. *Annu. Rev. Ecol. Syst.* 12:23–48.

———. 1989. The meaning of species and speciation: A genetic perspective. Pages 3–27 in Otte, D., and J. A. Endler (eds.), *Speciation and Its Consequences*. Sunderland, Mass.: Sinauer Assoc.

Turesson, G. 1922. The species and variety as ecological units. *Hereditas* 3:100–113.

Van Valen, L. 1976. Ecological species, multispecies, and oaks. *Taxon* 25:233–239.

Wagner, R. P., and R. K. Selander. 1974. Isozymes in insects and their significance. *Annu. Rev. Entomol.* 19:117–138.

Walters, S. M. 1989. Experimental and orthodox taxonomic categories and the deme terminology. *Pl. Syst. Evol.* 167:35–41.

Weir, B. S. 1990. *Genetic Data Analysis*. Sunderland, Mass.: Sinauer Assoc.

Wellso, S. G., D. J. Howard, J. L. Adams, and J. Arnold. 1988. Electrophoretic monomorphism in six biotypes and two populations of the Hessian fly (Diptera: Cecidomyiidae). *Ann. Entomol. Soc. Am.* 81:50–53.

Wilson, C. L., and C. L. Graham. 1983. *Exotic Plant Pests and North American Agriculture*. New York: Academic Press.

INDEX